CONDENSED MATTER
DISORDERED SOLIDS

CONDENSED MATTER
DISORDERED SOLIDS

Editors

S K Srivastava
D. A. Univ., Indore, India

N H March
Univ. Oxford, UK

World Scientific
Singapore • New Jersey • London • Hong Kong

Published by

World Scientific Publishing Co. Pte. Ltd.
P O Box 128, Farrer Road, Singapore 9128
USA office: Suite 1B, 1060 Main Street, River Edge, NJ 07661
UK office: 57 Shelton Street, Covent Garden, London WC2H 9HE

CONDENSED MATTER – DISORDERED SOLIDS

Copyright © 1995 by World Scientific Publishing Co. Pte. Ltd.

All rights reserved. This book, or parts thereof, may not be reproduced in any form or by any means, electronic or mechanical, including photocopying, recording or any information storage and retrieval system now known or to be invented, without written permission from the Publisher.

For photocopying of material in this volume, please pay a copying fee through the Copyright Clearance Center, Inc., 222 Rosewood Drive, Danvers, Massachusetts 01923, USA.

ISBN: 981-02-1924-5

Printed in Singapore.

PREFACE

This volume by prominent contributors is the second publication of the International Disordered-Systems Associates Society (INDIAS). Its first publication is the proceedings of the International Conference INDIAS−91 on DISORDERED MATERIALS held at D A University, Indore, India in February 1991 which contains the research and invited papers related to different disordered materials such as liquid crystals, glassy materials and disordered alloys-complex ions, *etc.*

The present context of condensed matter physics deals with the developments of the structure and properties of few disordered materials.

The ordering and disordering in materials-systems depends on the arrangements and distributions of atoms and molecules. There have been generally in crystalline materials, the periodicity of the atoms which develops the long-range ordered structure. As the periodicity is disturbed, there appears the short-range disordered structure which follows random behaviour. The disordered states are stable only to a corresponding fixed temperature.

In the case of liquids and amorphous materials, the stability is generally observed in the temperature region $T > \theta_D$ and $T \leq \theta_D$, respectively. Magnetically ordered states and superconducting states have been observed in the case of some metallic glasses. The vibrational modes also exist in amorphous materials and the structural change of materials is a microscopic behaviour.

The structure and properties of the materials of solid state are directional dependent. Few operations such as phase transition varies with the structural behaviour. The three controlling factors — pressure, volume and temperature affect the environmental setup of the atomic positions. Whenever a solid is heated up, the atomic distances become large and the solid turns into liquid form. It creates random distribution of molecules and isotropicity. At melting point, a metal is called liquid metal which possesses the disordered structure. Nevertheless, near melting temperature region, there is a similarity in the electronic properties of a solid metal and liquid metal because it is called disordered solid later.

The first chapter of this book deals with the experimental observations of the liquid and amorphous materials. The electron and phonon distribution and their dynamics for short-range ordered state have been described in Chapter 2. A very little deviation in the electronic and ionic characteristics occurs in the the case of the materials at melting point. The differences in between non-transition and transition disordered solids (metals and alloys) have been discussed on the basis of their structure and behaviours in Chapters 3 and 4. The electronic structure plays a dominant role in this direction. The transport behaviours in materials like liquid metals, amorphous solids and binary complexes have been discussed in Chapters 5 and 6 which differentiate very well the understanding of transport behaviour in conductors and semiconductors in one component and two component systems of disordered states on the basis of electrical and thermoelectric behaviours. The electron correlation effects play an important role in ordered and disordered states of matter as visualized in Chapter 7. The final chapter of this volume highlights very well the super structure of the tunnelling behaviour of a new disordered material-fullerene lattices.

CONTENTS

Preface		v
Chapter 1	Structural Investigation of Disordered Materials Liquids and Amorphous Solids *Yoshio Waseda*	1
Chapter 2	Electron Distribution, Phonon States and Short-Range Order *N. H. March*	95
Chapter 3	Condensed Matter — Liquid Transition Metals and Alloys *Toshio Itami*	123
Chapter 4	Condensed Matter — Non Transition Liquid Metals and Alloys *S. K. Srivastava*	251
Chapter 5	Transport Properties: Mainly in Liquid Metals and Amorphous Silicon *N. H. March*	317
Chapter 6	Electrical and Thermoelectric Properties of Disordered Metallic Binary Continuous Solid Solutions *A. T. Burkov and M. V. Vedernikov*	361
Chapter 7	Interplay Between Electron Correlation and Disorder *Fabio Siringo*	425
Chapter 8	High Resolution Scanning Tunnelling Microscopy of Defect Structures and Distortion of the Carbon Cage C_{60} Forming Fullerene Lattice *A. V. Narlikar et al.*	441

CHAPTER 1

STRUCTURAL INVESTIGATION OF DISORDERED MATERIALS LIQUIDS AND AMORPHOUS SOLIDS

Yoshio Waseda

Institute for Advanced Materials Processing (SOZAIKEN)

Tohoku University, Aoba–ku, Sendai 980, Japan.

CONTENTS

1.1. Introduction	2
1.2. Brief Background of the Present Requirement for Structural Study of Disordered Systems	3
1.3. Structure Factors of Liquid Metals in Low Wavevector Region	11
1.4. Distribution of Valence Electrons Around an Ion in Liquid Metals	18
1.5. Compound Forming and Structural Inhomogeneity in Liquid Alloys	33
1.6. Structure of Oxide Melts and Glasses	47
1.7. Relatively New Techniques for Determining the Local Chemical Environment Structure of Disordered Systems	67
Acknowledgement	84
References	86

1.1 Introduction

The method of structural analysis for crystalline substances have been well established for a long time. owever, the understanding of liquids and amorphous solids has lagged behind that of the crystalline state, even though much of the data on thermodynamic and electronic properties are available [see for example, Hultgren et al[1], Richardson[2], Kubaschewski and Alcock[3]]. On the other hand, there is a vast amount of research on new materials going on today and the utmost importance of the structure of such advanced materials at a microscopic level has been well recognized as one of the most important research subjects from both basic and applied science points of view. The evolution of our understanding of physics and chemistry of new materials is known to rely heavily upon their structural characterization. The development of disordered materials such as amorphous alloys and ferromagnetic glasses stimulates current interest in this rapidly growing field. There is also an increasing need for understanding the structure-property relationships of disordered systems, because of the novelty of the physics related to the particular non-periodicity in their atomic arrangements.

In the 1980's, the systematic structural investigation of disordered systems, liquids and amorphous solids, by x-ray and neutron diffraction was carried out and then several advances have been made only recently, although this research field itself has long been studied in the past.

The main purpose of this chapter is to describe the current experimental information on the structure of liquids and amorphous solids which is not covered in the monographs and review chapter previously published. The subject matter in this chapter is treated selectively rather than comprehensively and the examples are taken mainly from the results obtained by our group in the Institute for Advanced Materials Processing, Tohoku University; structure factors of liquid metals in the low wave vector region, the distribution of valence electrons around an ion in liquid metals, compound forming and structural inhomogeneity in liquid alloys, and structure of oxide melts and glasses.

The description of the atomic scale structure of disordered systems usually employs the concept of the radial distribution function (RDF) which can be obtained by diffraction experiments of x-rays and neutrons [see for example, Warren[4], Klug and Alexander[5]]. Since the usual RDF data is known to be only one dimensional and average information, a large amount of experimental and theoretical effort have been devoted to the development of techniques for obtaining the partial structural functions of the individual chemical constituents or the local chemical environment structure around a specific element in multicomponent disordered systems. For example, neutron diffraction using various isotopes, a combination of x-ray and neutron diffractions, the extended x-ray absorption fine structure (EXAFS) measurement and multiwavelength diffraction making use of anomalous (resonance) x-ray scattering. The anomalous x-ray scattering (AXS) enables us to extract information about local chemical environment of the desired elements [see for example, Waseda[6], Bienenstock[7]] which is of course quite important for quantitative discussion of particular properties of liquids and amorphous solids at a microscopic level. Such environment structural information obtained by this technique is very similar to the results from the EXAFS measurement. However, we remain convinced that the anomalous x-ray scattering technique is much more straightforward, at least theoretically, and the environmental structural information including the so-called intermediate range ordering may be evaluated as a function of radial distance with much higher reliability than the EXAFS method. For this reason, fundamentals together with some results of the anomalous x-ray scattering will also be presented.

1.2 Brief Background of the Present Requirement for Structural Study of Disordered Systems

All atomic positions in crystalline systems are known to be well-described with a few parameters of distances and angles. On the other hand, disordered systems such as liquids and amorphous solids lack the long range structural periodicities like crystalline systems. The concept of the radial distribution functions is useful, because these functions cannot be used to specify completely the positions and chemical identities of the constituent elements in a desired system. For this reason, the structure of liquids and amorphous solids can be quantitatively described in terms of the **radial distribution function (RDF)** indicating the average probability of finding another atom in a specified volume from an origin atom as a function of radial distance [see for example, Cole[8]]. The RDF provides spherically averaged information on the atomic correlation as one-dimensional data. However, the RDF gives unique quantitative information for describing the atomic scale structure of disordered systems. A description of the principles, and their utility of the RDF has already been given in detail [see for example, Warren[4], Wagner[9], Waseda[10]], so that we need give here only the essential points of the RDF analysis of disordered materials for convenience of discussion in this chapter.

In contrast to a simple system containing only one kind of element, the RDF analysis and its interpretation are more complicated for multicomponent disordered systems. However, when the compositionally averaged functions are introduced, a similar approach to that for the simple one-component case can be applied to the case of disordered systems containing more than two kinds of elements. The essential equations are as follows.

$$\rho(r) = \sum_{i=1}^{n} \sum_{j=1}^{n} c_i f_i f_j \rho_{ij}(r) / <f>^2 \qquad (1.2.1)$$

$$<f>^2 = (\sum_{i=1}^{n} c_i f_i)^2 \qquad (1.2.2)$$

$$<f^2> = \sum_{i=1}^{n} c_i f_i^2 \qquad (1.2.3)$$

where c_i is the atomic fraction of i-type atom and $\rho_{ij}(r)$, called the partial radial density function, corresponds to the number of i-type atoms found at a radial distance of r from a j- type atom at the origin. f_i is the atomic scattering factor for atom i, and here its Q-dependence is excluded for simplification. Equation (1.2.1) implies that the average radial density function $\rho(r)$ of multicomponent disordered systems can be given by the summation of the partial radial density function of $\rho_{ij}(r)$ with a weighting factor expressed by the atomic scattering factor and concentration. By using the approximate equations of eqs.(1.2.1)–(1.2.3), the coherent x-ray scattering intensity per atom $I_a^{coh}(Q)$ and the structure factor $a(Q)$, generally called total structure factor or total interference function, for a multicomponent disordered system are given by the following;

$$I_a^{coh}(Q) = <f^2> + <f>^2 \int_0^\infty 4\pi r^2 [\rho(r) - \rho_0] \frac{\sin(Q \cdot r)}{Q \cdot r} dr \qquad (1.2.4)$$

$$a(Q) = [\, I_a^{coh}(Q) - (<f^2> - <f>^2)\,] / <f>^2 \qquad (1.2.5)$$

Thus, the following common relation is readily obtained.

$$G(r) = 4\pi r [\rho(r) - \rho_0] = \frac{2}{\pi} \int_0^\infty Q\,[a(Q) - 1] \sin(Q \cdot r)\, dQ \qquad (1.2.6)$$

This equation provides the relation between the atomic scale structure and measured intensity data for disordered systems including more than two kinds of atoms, although the information of $G(r)$ cannot be used to describe completely the positions and chemical identities of the constituents. The knowledge of structural functions of individual pairs like $\rho_{ij}(r)$ is an essential requirement for discussing various characteristic properties of multicomponent disordered systems of interest at a microscopic level.

The conventional diffraction experiments for multicomponent dis-ordered systems only allow one to obtain the total structure factor which represents a weighted sum of the partial structure factors. Therefore, the determination of partial structural functions is undoubtedly one of the most important research subjects for liquids and amorphous solids. Some essential points are given below with respect to the separation of the partial structure factors from measured scattering intensity data. For simplification, let us consider a binary disordered system using the definition proposed by Faber and Ziman[11].

The total structure factor $a(Q)$ corresponding to the structurally sensitive part in eq.(1.2.4) is expressed by the three partial structure factors $a_{ij}(Q)$ as follows:

$$a(Q) = w_{11}a_{11}(Q) + w_{22}a_{22}(Q) + 2w_{12}a_{12}(Q) \qquad (1.2.7)$$

$$w_{ij} = c_i c_j f_i f_j \ / <f>^2 \qquad (1.2.8)$$

The partial structure factors $a_{ij}(Q)$ are connected with the partial RDFs by the following equations.

$$a_{ij}(Q) = 1 + \int_0^\infty 4\pi r^2 \rho_o \left[g_{ij}(r) - 1 \right] \frac{\sin (Q \cdot r)}{Q \cdot r} dr \qquad (1.2.9)$$

$$4\pi r^2 \rho_o \left[g_{ij}(r) - 1 \right] = r\, G_{ij}(r) = \frac{2r}{\pi} \int_0^\infty Q\, [a_{ij}(Q) - 1]\, \sin (Q \cdot r)\, dQ \qquad (1.2.10)$$

where $g_{ij}(r) = \rho_{ij}(r)/(c_j \rho_o)$. Regarding the partial RDFs, it is also customary to use the following relation.

$$G(r) = w_{11}G_{11}(r) + w_{22}G_{22}(r) + 2\, w_{12}G_{12}(r) \qquad (1.2.11)$$

The separation of three individual structural functions [$a_{ij}(Q)$ or $G_{ij}(r)$] is the present objective. We will discuss here only the partial structure factors $a_{ij}(Q)$, because $G_{ij}(r)$ is straightforward as calculated by the simple Fourier transformation in the manner of eq.(1.2.10).

As easily seen from eq.(1.2.7), the total structure factor $a(Q)$ of a binary disordered system is given by the summation of two like atom pairs (1–1 and 2–2) and one unlike atom pair (1– 2), and the coefficient w_{ij}, frequently called weighting factor, depends on the concentrations and the atomic scattering factors. Then, the individual partial structure factors may be estimated by making available at least three independent scattering experiments for which the weighting factors are varied without any change in the RDF. For example, when the scattering ability is changed in the component of 1, the following matrix form can be obtained.

$$\begin{bmatrix} a_{11}(Q) \\ a_{22}(Q) \\ a_{12}(Q) \end{bmatrix} = \begin{bmatrix} c_1^2 f_1^2 & c_2^2 f_2^2 & 2c_1c_2 f_1 f_2 \\ c_1^2 (f_1')^2 & c_2^2 f_2^2 & 2c_1c_2 f_1^* f_2 \\ c_1^2 (f_1'')^2 & c_2^2 f_2^2 & 2c_1c_2 f_1^{**} f_2 \end{bmatrix}^{-1} \begin{bmatrix} a(Q) \\ a^*(Q) \\ a^{**}(Q) \end{bmatrix} \quad (1.2.12)$$

This can be done by several methods. Although they are not trivial tasks in practice, the following classification may be suggested in principle [see for example, Waseda[6]].

(A) Three radiation technique where the weighting factors are varied using x-rays, neutrons and electrons.

(B) Isotope substitution technique for neutron diffraction in which the scattering powers of the components are varied by using different isotopes.

(C) Polarized neutron diffraction technique which is applicable only to ferromagnetic substances.

(D) Anomalous X-ray scattering technique using the anomalous dispersion (resonance) effect at the close vicinity of the absorption edge for a constituent element. Anomalous neutron scattering technique is also possible for several isotopes.

Of course, an assortment of the above techniques such as the combination of x-ray and neutron diffraction experiments with the polarized neutron technique [Sadoc and Dixmier[12]] has also been used in the literature. The isomorphous substitution technique [see for example, Chipman et al[13], Cargill III and Tsuei[14]] in which one constituent is partially or totally replaced by physically and chemically similar elements, has also been used to estimate the partial structure factors in a binary or pseudo-binary disordered systems. This technique is possible way with which to obtain some useful information on the partial structure factors of liquids and amorphous solids in cases where other techniques are found to be technically difficult. However, note that its applicability strongly depends on the assumption of structural identity of samples in which at least one component is substituted by physically and a chemically similar elements.

The four techniques above-mentioned have, of course, own advantage and disadvantage and they have previously been discussed [Enderby[15], Waseda[6,10]] and are not duplicated here. However, the following intrinsic point is considered to be worthy of note. The technique (A) requires different samples in size; bulk ($\approx$ mm) for neutron, foil ($\approx$ μm) for x-rays and thin film ($\approx$ nm) for electrons. In technique (B), the structure is automatically assumed to remain identical upon substitution by the isotopes. Thus, the fundamental idea of these two techniques (A) and (B) should be attributed to not only chemical but also structural identity. However, we should keep in mind that this structural identity may not remain in a thermodynamically metastable state such as glasses, because some properties of the as-quenched glasses particularly metallic glasses are known to vary from one run of production to the other and from one portion of the ribbon to another, clearly suggesting the question of the structural identity at the microscopic level. With respect to this point, the techniques (C) and (D) are free from this ambiguity, because it is possible to vary the weighting factors without the use of different samples. Nevertheless, the neutron scattering amplitudes of different isotopes of one element are very different, so that the isotope substitution technique for neutron diffraction enables us to provide the useful partial structural functions at the present time by obtaining the sufficiently sizable variation in weighting factors when we tacitly accept the assumption that the structure of samples of interest is unchanged by the isotope substitution. Following the pioneering work of liquid Cu-Sn alloys by this method of Enderby, Egelstaff and North[16], the partial structural functions have been determined for various liquid alloys, metallic glasses [see for example, Steeb and Lamparter[17]] and molten salts [see for example, Enderby and Neilson[18,19]]. For example, this isotope substitution technique of neutron diffraction has been successfully applied by Lamparter et al[20] to determine the partial structural functions of the $Ni_{81}B_{19}$ glass by using the combination of the ^{62}Ni isotope with negative scattering amplitude (-0.87×10^{14} m) and ^{60}Ni isotope with small but positive scattering amplitude (0.28×10^{-14} m). In this case, the preparation of nickel isotope mixture with zero scattering amplitude (so-called zero alloy) is allowed and then the neutron diffraction experiment of such zero alloy provides directly the partial function of B-B pairs. Their elegant results are given in **Fig.1.2.1.**

On the other hand, anomalous x-ray scattering (AXS) method has recently received attention when coupled with a high intensity white x-ray source such as the synchrotron radiation by making the use of an energy range in which the anomalous dispersion is sufficiently large and thus the quantitative accuracy of the AXS results is markedly improved compared to the previous results obtained by the characteristic radiations only. For example, De Lima et al[21] determined the partial structural functions of the Ni_2Zr glass as exemplified by the results of **Fig.1.2.2**. Similar results were reported for the $CuZr_2$ glass [Laridjani and Sadoc[22]]. These results are very encouraging, but the use of this direct anomalous x-ray scattering technique is still limited, at the time of writers, for binary cases mainly due to the experimental difficulties. With respect to such inconvenience, the energy dependence of the anomalous x-ray scattering intensity is recently found to give the environmental structure around a specific element in multicomponent disordered systems as a function of radial distance, without carrying out

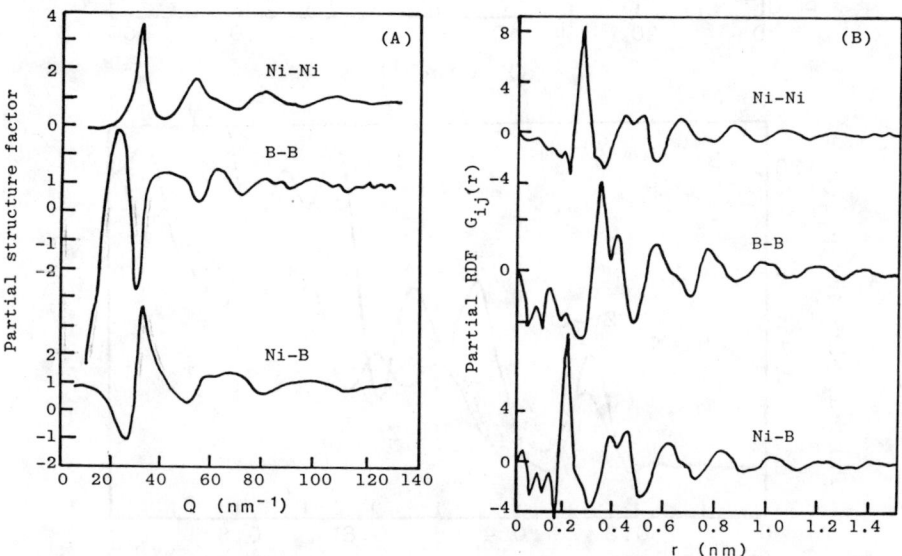

Fig.1.2.1　　Partial structural functions of amorphous $Ni_{81}B_{19}$ obtained by the isotope substitution method [Lamparter et al[20]].

the complete separation of all partial structural functions. This energy derivative technique provides almost an order of magnitude better stability of the solutions than the direct AXS method [Munro[23]]. Such relatively new information will be given in later sections with some selected examples.

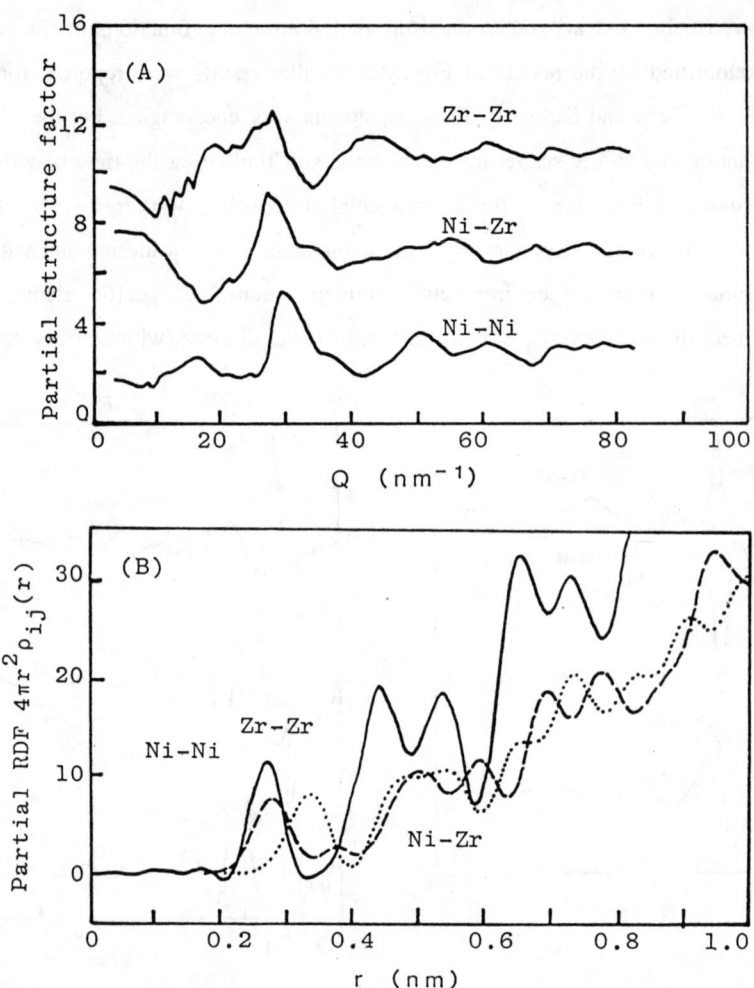

Fig.1.2.2 Partial structural functions of amorphous Ni_2Zr obtained by the anomalous x-ray scattering method [De Lima et al[21]].

1.3 Structure Factors of Liquid Metals in Low Wavevector Region

A large amount of valuable information about the atomic scale structure of liquid metals have been obtained from diffraction experiments of X-rays and neutrons [Wagner[9], Waseda[10]] by determining the structure factor, S(Q), in wide wave vector region, $Q=4\pi\sin\theta/\lambda$, from which the RDF data is calculated. On the other hand, the need for structural data concerning the low Q region has been frequently emphasized as a rigorous test of the electron transport theory and the approximate statistical theory for liquid metals [Greenfield[24], Ballentine and Jones[25]]. The low Q structure factor of liquid metals is again recognized to be of considerable interest, in parallel with recent theoretical progress in liquid metals [Evans and Slucklin[26], Ohkoshi et al[27], MacLaughlin and Young[28], Matthai and March[29]]. However, such low Q structure factors are available for only a few metallic elements [Egelstaff et al[30], North and Wagner[31], Greenfield et al[32], Huijben and van der Lugt[33]] and the long wavelength limit value of S(Q) derived from these experimental structural data appears to agree well with that calculated from the isothermal compressibility. However, the systematic information of the low Q structure factors for liquid metals are still sparse.

For this reason, the purpose of this section is to describe the low Q structure factors of various liquid metals including transition metals at temperatures above their respective melting points carefully and systematically measured by x-ray diffraction.

The low Q structure factors are usually determined in the following procedure [Waseda[34]]. In the first step, the conventional experiment of transmission mode (hereafter to be referred to as large angle measurements) is carried out to obtain the structure factor S(Q) as a function of Q with emphasis on the region larger than 5 nm^{-1} for obtaining the usual RDF data. The results are given in **Fig.1.3.1** for liquid Na at 378K as an example. The increase in the value of S(Q) below Q=5 nm^{-1} is the so-called spurious effect related to the sensing of the primary beam by the detector system of a scintillation counter in the present case. In order to overcome this spurious effect, a more accurate low angle measurement, in the second step, is made using a very narrow incident beam. The low angle results are superimposed on the large angle results obtained in the first step as exemplified in Fig.1.3.1. The uncertainty in the normalization procedure for determining the structure factor from the low angle measurements can be reduced when overlapping results are obtained in the region including the first peak using the two steps of low angle and large angle measurements. The following point may be noticed. Compared with the previous works made in the limited Q region, it is possible to obtain more accurate information down to a value of 0.8 nm^{-1}, by the present technique of two measurements including the first peak region in the structure factor.

Since the experimental uncertainty in the low angle results themselves is difficult to estimate, only the following points are suggested. The accumulated counts varying from 5×10^4 in the low angle region (Q≤50 nm^{-1}) to 1×10^5 (Q≥50 nm^{-1}) in the high angle region were chosen so that the counting statistics were approximately uniform. The normalization for the present large angle measurements is in error by less than 2.2

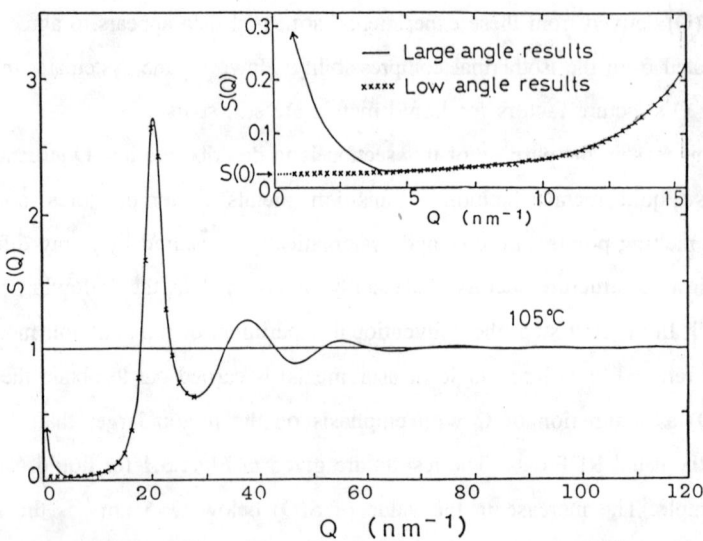

Fig.1.3.1 Structure factor of liquid Na at 378k. The solid line denotes the results of large angle measurements and the crosses are the results of low angle meaurements.[34] The insert shows the values in the low wave vector region.

% by Rahman's method [Rahman[35)]] and the resultant structure factors appear to agree well with the values determined by reflection mode [Waseda[10)]]. According to the detailed discussion given by Greenfield et al[32)] and Malet et al[36)], the maximum error in the atomic scattering factor, of the Compton scattering and of the multiple scattering. these quantities seems not to exceed 1 % for metals presently investigated. Therefore, the total uncertainty of the structure factors obtained from the large angle measurements in this work is estimated to be 3.2 %. On the other hand, the mismatch in the superimposition of two x-ray scattering intensities (low angle and high angle measurements) was found of the order of 0.7 % over the range of Q presently investigated. Thus an uncertainty of 3.9 % may be suggested for the low Q structure factors of liquid metals given here.

The experimental results are known to indicate a smooth variation of the structure factor in low wavevector region as a function of Q for all liquid metals at temperatures near melting point. Such structural feature is easily seen from **Fig.1.3.2** using the results of liquid Na as an example. The low Q structure factor shows a slowly increasing function of Q, particularly in the range of $Q \leq Q_1/2$ where Q_1 is the first peak position in the structure factor, while it rises rapidly to its peak value for $Q \geq Q_1/2$. It may also be noticed that the low Q structure factor at temperature close to the melting point gives a nearly constant value in the Q range less than $Q_1/4$, as first suggested by Egelstaff et al[30)].

As shown in Fig.1.3.1, the present low Q structure factor of liquid metals for $Q \leq Q_1/2$ can be expressed by the following polynomial form;

$$S(Q) = a_0 + a_1 Q + a_2 Q^2 \tag{1.3.1}$$

where a_0, a_1 and a_2 are the coefficients. The importance of the linear term of $a_1 Q$ in eq.(1.3.1) has been stressed for liquid metals by Matthai and March[29)]. In addition, the higher terms such as $a_3 Q^3$ and $a_4 Q^4$ are, of course, required for fitting the measured structure factor in a wider Q region. As easily seen from the results of Fig.1.3.1, the experimental structure factors can be extrapolated smoothly to the long wavelength value of S(0) calculated from the well-known relation;

$$S(0) = \rho_0 k_B T \chi_T \qquad (1.3.2)$$

where ρ_0 is the number density of atoms, k_B is the Boltzmann constant, T is the absolute

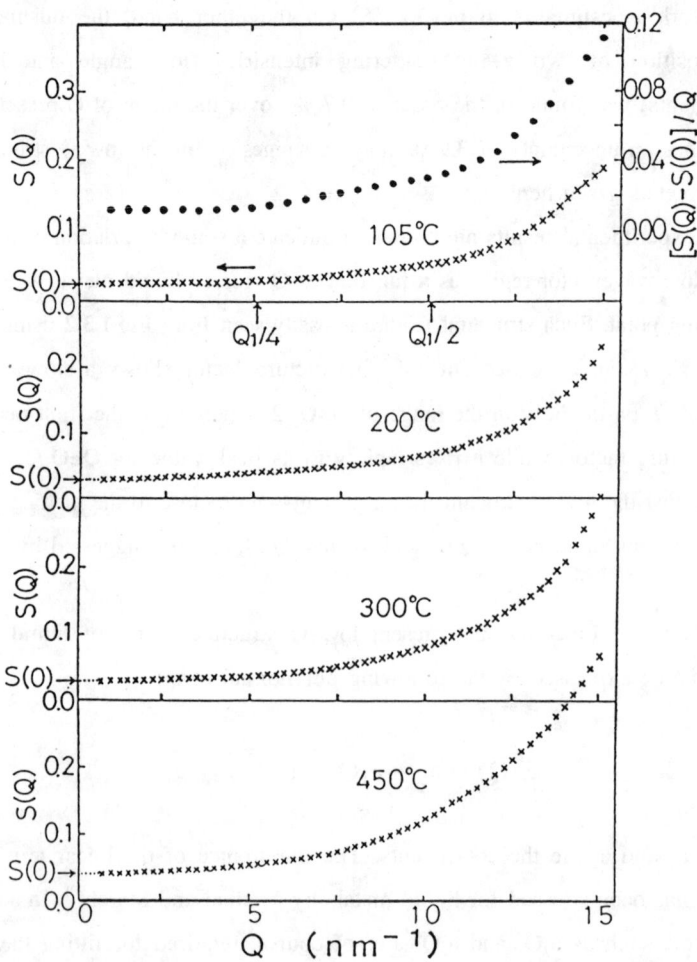

Fig.1.3.2   Temperature dependence of the low wave vector structure factor of liquid Na. Solid dotts denote the plot of [S(Q)−S(0)]/Q as a function of Q.

temperature and χ_T is the isothermal compressibility [see for example, Landau and Lifschitz[37]]. Thus, the coefficient a_0 in eq.(1.3.1) is often replaced by the S(0) value, although the S(0) value calculated from eq.(1.3.2) is subject to some ambiguity (about 10 %), part arising from the uncertainty in the isothermal compressibility value. Conversely, the S(0) value can be estimated directly from the measured low Q structure factor by the following manner.

In the Q range less than $Q_1/4$, the following simplification may be made on the basis of experimental data;

$$S(Q) = S(0) + a_1 Q \quad (Q \leq Q_1 /4) \quad (1.3.3)$$

Equation (3.3) could also be rewritten in the form;

$$S(Q) / Q = S(0) / Q + a_1 \quad (Q \leq Q_1 /4) \quad (1.3.4)$$

Then, one can obtain, with reasonably good accuracy, the value of S(0) and possibly a_1 from the plot of S(Q)/Q as a function of 1/Q in the low Q region less than $Q_1/4$. The coefficient a_1 should be determined by using the structure factor in a wider Q region, so that the present subject is restricted only within the determination of the S(0) value directly from the measured low Q structure factor.

A plot of S(Q)/Q against 1/Q is given in **Fig.1.3.3** using the results of Na at 105 °C as example. The plotted points are the experimental data and the straight lines are the fit to the form of eq.(1.3.4). The results of Fig.1.3.3 clearly show that it is possible to determine the S(0) value from measured structural data alone using eq.(1.3.4). A similar feature of the plot of S(Q)/Q against 1/Q are also observed in other liquid metals. The S(0) values estimated in this procedure for 19 liquid non-transition metals are summarized in **Table 1.3.1** together with the corresponding values derived from the isothermal compressibility data using eq.(1.3.2). The overall agreement between S(0) and $\rho_o k_B T \chi_T$ is rather surprisingly good for 19 non-transition metals, although the difference exceeds 20 % in three cases of liquid Ga, Si and Sn [Waseda and Ueno[38]]. The origin of the difference between S(0) and $\rho_o k_B T \chi_T$ of these three liquid metals cannot be certainly identified at the present time. However, Egami and Srolovitz[39], on the basis of the atomic level stresses in disordered system, suggest that the microscopic isothermal compressibility related to the local fluctuation in liquids is not necessarily equal to the macroscopic isothermal compressibility and then such situation may account for part of the difference, as well as the experimental uncertainties in both S(0) and χ_T.

On the other hand, the isothermal compressibility of liquid transition metals are

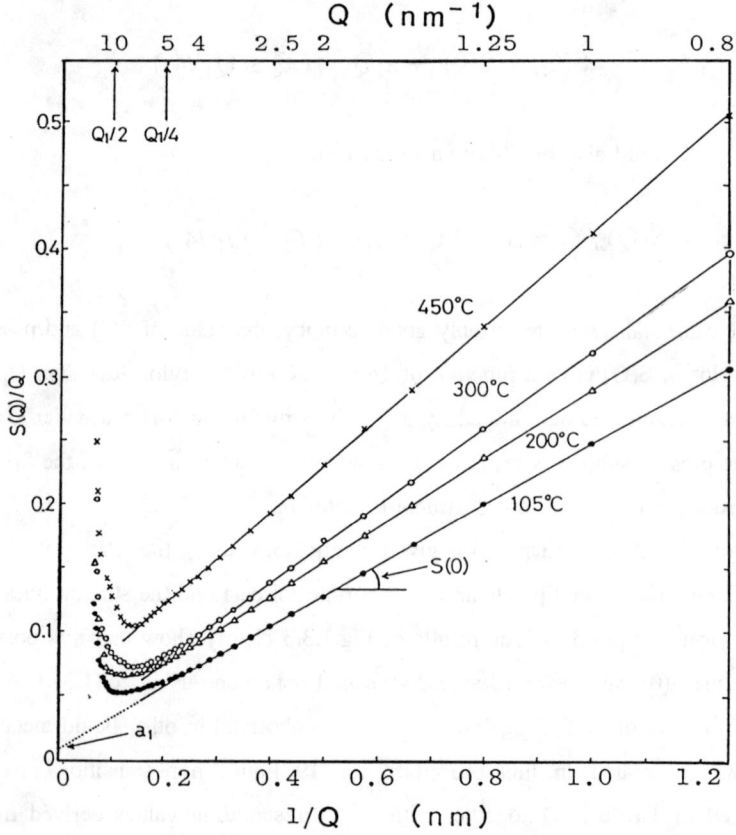

Fig.1.3.3   Relationship between S(Q)/Q and 1/Q for liquid Na at several temperatures. The plotted points are the experimental data and the slight lines are the fit to the form of eq.(1.3.4).

Table 1.3.1 Comparison of the long wavelength value of S(0) obtained from diffraction data and those calculated from the isothermal compressibility.[38]

Element	Temp. (°C)	Density (g/cm^3)	χ_T(cm^2/dyn) × 10^{-12}	$\rho_o k_B T \chi_T$	S(0)
Li	190	0.512	10.2	0.0290	0.031
Na	105	0.928	18.6	0.0234	0.023
K	70	0.826	38.2	0.0231	0.024
Rb	40	1.476	49.3	0.0221	0.022
Cs	30	1.838	68.8	0.0239	0.027
Mg	680	1.545	5.06	0.0255	0.025
Ca	850	1.37	11.0	0.0349	0.031
Sr	780	2.38	13.1	0.0310	0.032
Ba	730	3.32	17.9	0.0362	0.035
Zn	450	6.91	2.50	0.0159	0.014
Cd	350	7.95	3.24	0.0119	0.013
Al	670	2.34	2.42	0.0166	0.019
Ga	50	6.082	2.19	0.0051	0.0098
In	160	7.03	2.96	0.0065	0.0068
Si	1460	2.59	3.96	0.0526	0.014
Ge	980	5.56	3.22	0.0257	0.021
Sn	250	6.93	2.71	0.0069	0.010
Cu	1150	7.97	1.45	0.0215	0.018
Ag	1000	9.27	1.86	0.0169	0.016

very limited. Thus, the direct comparison between S(0) obtained in this procedure and $\rho_o k_B T \chi_T$ is rather difficult in liquid transition metals, although the recent experimental results of the isothermal compressibility for liquid Fe, Co and Ni [Tsu et al[40]] provide the $\rho_o k_B T \chi_T$ values, of 0.0199 for Fe, 0.0191 for Co and 0.0190 for Ni, respectively, in substantial agreement with the S(0) values. For this reason, the values of the isothermal compressibility estimated from the structure factor in the long wavelength limit S(0) in this work using eq.(1.3.2) are listed in **Table 1.3.2**. The detailed discussion may not be given here about these χ_T values for liquid transition metals. However, this information would be valuable in cases where the conventional measurements for the isothermal compressibility are found to be technically difficult.

Table 1.3.2 The long wavelength value of S(0) obtained from diffraction data and the isothermal compressibility χ_T calculated from S(0) values.[38] *taken from the work of Tsu et al.[40]

Element	Temp. (°C)	Density (g/cm³)	S(0)	χ_T(cm²/dyn) × 10⁻¹²	
Sc	1560	2.92	0.036	3.64	
Ti	1700	4.15	0.020	1.40	
V	1900	5.36	0.025	1.31	
Cr	1900	6.27	0.021	1.10	
Mn	1260	5.97	0.024	1.73	
Fe	1560	7.01	0.020	1.05	(1.04)*
Co	1550	7.70	0.019	0.96	(0.97)*
Ni	1500	7.72	0.020	1.03	(0.98)*
Pd	1580	10.5	0.020	1.32	
La	970	5.95	0.019	4.29	
Ce	870	6.67	0.021	4.64	
Pr	950	6.61	0.025	5.23	
Nd	1050	6.92	0.018	3.41	
Eu	830	4.61	0.018	6.46	
Gd	1330	6.91	0.018	3.07	
Tb	1360	7.24	0.021	3.40	
Dy	1430	8.14	0.021	2.97	
Ho	1480	8.25	0.014	1.92	
Er	1520	8.37	0.017	2.28	
Yb	850	6.20	0.015	4.48	
Lu	1680	9.18	0.016	1.88	

1.4 Distribution of Valence Electrons Around an Ion in Liquid Metals

Liquid metals in their pure states are known to be binary mixtures of ions and electrons. On the other hand, neutrons scatter only on nuclei of the center of ions. X-rays scattering experiment also provides the distribution of ions, but it is known to be obtained through the scattering on electrons surrounding ions. Such difference in the scattering mechanism in x-ray and neutrons suggests that small but a certain difference should be observed in the structural data determined by two scattering methods. This is first indicated by Egelstaff, March and McGill[41], but their suggestion was explicitly cited very recently, mainly due to the limited experimental accuracy of the neutron results for liquid metals. The structure factors for several liquid metals were determined by neutron diffraction with sufficient accuracy and then a significant difference in the structure factor has been well recognized between the two curves, as exemplified by **Fig.1.4.1** using the results of liquid Zn where the qualitative coincidence appears with respect to the structural profile, Such differences, larger than the experimental uncertainties, have been quantitatively confirmed for several liquid metals, as well as a slight variation of the first peak position. With these facts in mind, a method has recently been proposed for estimating the ion-electron correlation function in liquid metals from measured structural data of x-rays and neutrons. A few efforts have been

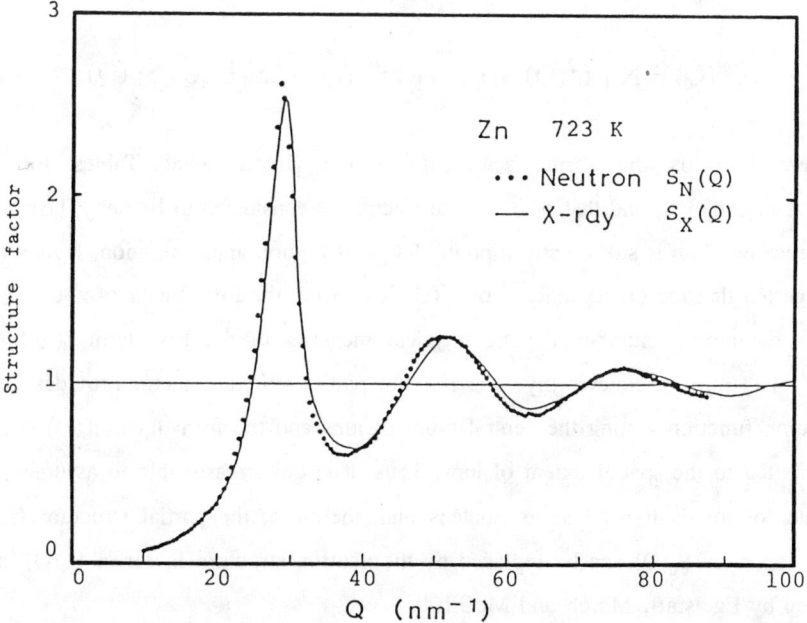

Fig.1.4.1   Structure factors of liquid Zn determined by x-ray and neutron diffraction.

devoted to discuss the difference and to estimate the ion–electron or electron–electron correlation functions for a system composed of ions and electrons [Watabe and Hasegawa[42], Chihara[43], Cusack et al[44]]. However, these previous works are not necessary for convenience of direct estimation of the ion–electron correlation function in liquid metals from the experimental structural data. Therefore, essential equations required for the present purpose are given below.

Let us consider a liquid metal containing N ions and zN valence electrons, where z is the number of valence electrons per atom. In other words, the core electrons are excluded here and a liquid metal is considered to be a mixture of ions and conduction electrons. Thus, there are three correlation functions, ion–ion, ion–electron and electron–electron correlation functions. When we use the Ashcroft–Langreth form [see for example, Ashcroft and Langreth[45], Waseda[10]] of the partial structure factors for these three correlation functions, the coherent x-ray scattering intensity $I_X^{coh}(Q)$ is given as follows.

$$I_X^{coh}(Q) = N [f_i^2(Q) S_{ii}(Q) + zS_{ee}(Q) + 2z^{1/2}f_i(Q) S_{ie}(Q)] \quad (1.4.1)$$

where $f_i(Q)$ is the form factor of an ion [International Tables for X–ray Crystallography[46]] and that of a valence electron is considered to be unity, because each valence electron is sufficiently a point charge as a good approximation. It may also be suggested that the Q–dependence of $f_i(Q)$ is close to the form factor of z–th valent ion with the atomic number Z. The physical meaning of the first term $f_i(Q)S_{ii}(Q)$ in eq.(1.4.1) may be noteworthy. The ion–ion partial structure factor provides the correlation function among the central point of ions and the form factor $f_i(Q)$ should be attributed to the spatial extent of ions. Thus, it is quite reasonable to assume that the center of ion is located at its nucleus and, therefore, the partial structure factor of ion–ion pairs $S_{ii}(Q)$ can be replaced by the neutron structure factor of $S_N(Q)$, as first given by Egelstaff, March and McGill[41].

A conventional expression for the structure factor obtained by x–ray diffraction, $S_X(Q)$, is given by the following equation.

$$I_X^{coh}(Q) = Nf_a^2(Q)S_X(Q) \qquad (1.4.2)$$

where $f_a(Q)$ is the form factor of a free atom and the Hartree–Fock method is generally used for its calculation. Using the above two equations, one can readily obtain the following equation, with respect to the ion–electron partial structure factor in liquid metals.

$$S_{ie}(Q) = \frac{1}{2z^{1/2}f_i(Q)} [\, f_a^2(Q)S_X(Q) - f_i^2(Q)S_N(Q) - zS_{ee}(Q)\,] \qquad (1.4.3)$$

This equation corresponds to the starting equation to describe the pseudo–binary system of ions and valence electrons in monatomic liquid metals. It is also worth mentioning that the first term of eq.(1.4.3), $f^2(Q)S_X(Q)$, is just the measured coherent x–ray scattering intensity per atom. On the other hand, the electron–electron correlations in metals have long been studied [see for example, Ichimaru[47]], and the partial structure factor of electron–electron pairs, $S_{ee}(Q)$, is well established for various densities of conduction electrons. In the present case, the distribution of valence electrons is considered to be the electron jellium in metallic density as the first approximation and it is relatively easy to calculate $S_{ee}(Q)$, as shown in **Fig.1.4.2** using the results of Ustumi and Ichimaru[43]. Therefore, eq.(1.4.3) enables us to estimate the ion–electron partial structure factor, $S_{ie}(Q)$ directly from measured structural data of $S_X(Q)$ and $S_N(Q)$, when coupled with the theoretical values of $S_{ee}(Q)$.

In the strict sense, the contribution of ionic distribution in liquid metal to the electron–electron correlation is excluded in the $S_{ee}(Q)$ values calculated by the Utsumi–Ichimaru scheme, although such contribution is considered not to be so significant except for the long wavelength limit (Q=0), as suggested by Chihara[43]. As also seen in eq.(2.4.3), the value of z, corresponding to the coefficient of $S_{ee}(Q)$, is considerably smaller than those of $f_a(Q)$ and $f_i(Q)$. Numerical examples are Z=2, $f_a(Q) \approx f_i(Q) \approx 200$, where Q=60 nm^{-1} in the case of liquid Zn. Thus, the basic profile of $S_{ie}(Q)$ obtained in this procedure would be considered not to be critically affected by the values of $S_{ee}(Q)$ employed.

The corresponding radial distribution function $g_{ie}(r)$ is straightforwardly obtained by the following Fourier transformation.

$$g_{ie}(r) = 1 + \frac{1}{2\pi^2 \rho_o r} \int_0^\infty z^{-1/2} \, S_{ie}(Q) \, Q \sin(Qr) \, dQ \qquad (1.4.4)$$

where $\rho_o = zN/V_M$, N and V_M are the molar volume and the Avogadro number, respectively.

According to Chihara[49], the static structure factor of valence electrons is expressed as follows.

$$S_{ee}(Q) = |\rho(Q)|^2 / z \, S_{ii}(Q) + S_{ee}^0(Q) , \qquad (1.4.5)$$

where $\rho(Q)$ is the Fourier transform of the charge density of valence electrons around a given ion and $S_{ee}(Q)$ is the structure factor of uniformly distributed electron system

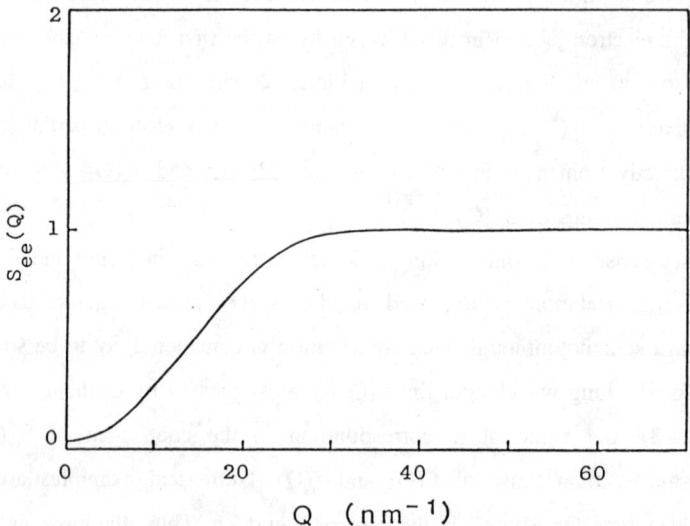

Fig.1.4.2 Partial structure factor of $S_{ee}(Q)$ calculated by Utsumi and Ichimaru[48] using the Wigner–Seitz radius of 2.25 a.u. as an example.

and usually expressed by,

$$S_{ee}^{0}(Q) \equiv \frac{1}{\pi} \int_{-\infty}^{\infty} \beta \hbar / (1 - e^{-\beta \hbar}) \times \text{Im} \frac{1}{\tilde{\epsilon}(Q, \omega)} d\omega \qquad (1.4.6)$$

where $h = h/2$, $\beta = 1/k_B T$ and $\tilde{\epsilon}$ is the dynamical dielectric function. The total scattering intensity by x-ray diffraction at $Q=0$ is therefore, written in the following form,

$$I_M(Q)/N = |f_i(Q)|^2 S_{ii}(Q) + 2f_i(Q)z^{1/2}S_{ie}(Q) \\ + zS_{ee}^{0}(Q) + (Z - z)S_{inc}^{i}(Q) \qquad (1.4.7)$$

where $S_{inc}^{i}(Q)$ is the incoherent scattering factor of ion and Z is the atomic number. Using the following relation,

$$ZS_{inc}^{M}(Q) \approx zS_{ee}^{0}(Q) + (Z - z) S_{inc}^{i}(Q) \qquad (1.4.8)$$

where Siinc being the total incoherent scattering factor in liquid metal. Chihara[49] indicates the coherent x-ray scattering intensity $I_X^{coh}(Q)$, instead of eq.(1.4.1), in the following form.

$$I_X^{coh}(Q) = N[f_i^2(Q)S_{ii}(Q) + 2z^{1/2} f_i(Q)S_{ie}(Q) + |\rho(Q)|^2 S_{ii}(Q)] \qquad (1.4.9)$$

$$= N[f_i^2(Q)S_{ii}(Q) + 2z^{1/2} f_i(Q)S_{ie}(Q) + z^{1/2}\rho(Q)S_{ie}(Q)] \qquad (1.4.10)$$

here, the following relation is also used.

$$|\rho(Q)|^2 / z\, S_{ii}(Q) = \rho(Q) / z^{1/2} S_{ie}(Q) \qquad (1.4.11)$$

Therefore, eq.(1.4.10) is rewritten as follows.

$$I_X^{coh}(Q) = N \, [\, |f_i(Q)|^2 S_{ii}(Q) + 2 f_i(Q) z^{1/2} \\ \times [(1 + \rho(Q))/2 f_i(Q)]\, S_{ie}(Q)\,] \qquad (1.4.12)$$

In the usual experimental wavevector region ($Q \approx 100$ nm^{-1}), the order of magnitude of $\rho(r)/2f_i(Q)$ may be close to $1/2(z/Z)$ and less than 0.10 in liquid metals such as Zn. Thus, it is not unreasonable that eq.(1.4.12) is approximated to the following form.

$$I_X^{coh}(Q) = N[\,|f_i(Q)|^2 S_{ii}(Q) + 2f_i(Q)\,z^{1/2}\,S_{ie}(Q)\,] \qquad (1.4.13)$$

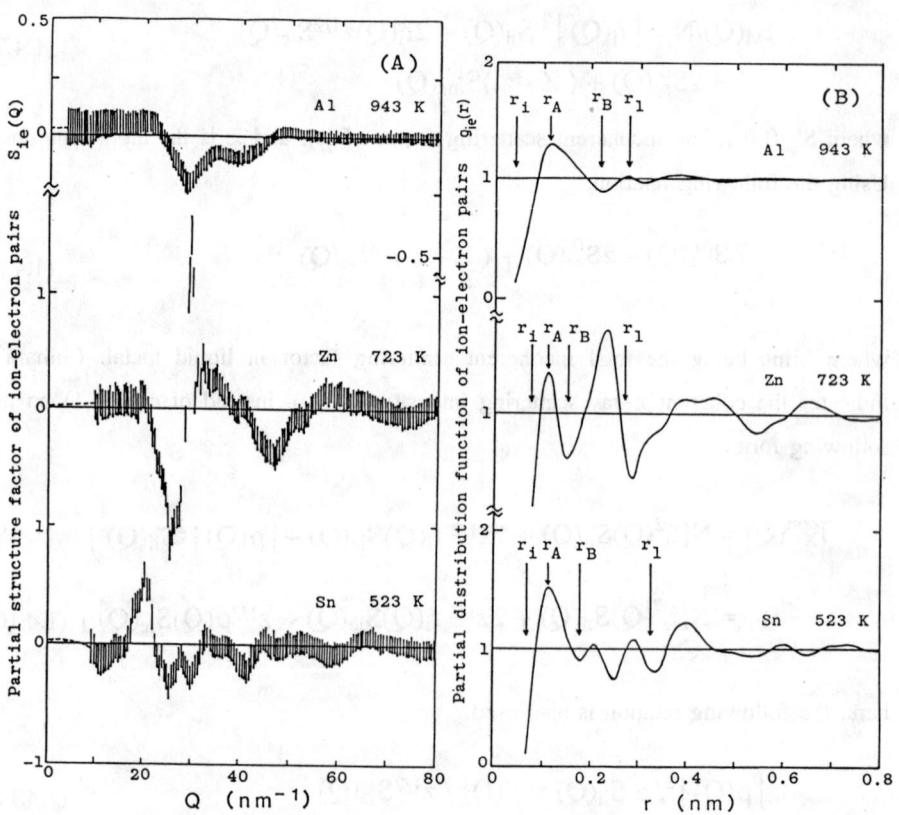

Fig.1.4.3 Partial structural functions of ion–electron pairs for liquid Al, Zn and Sn estimated directly from measured structural data by x–rays and neutrons. (A) $S_{ie}(Q)$, (B) $g_{ie}(r)$. r_i: ionic radius, r_A: first peak position of $g_{ie}(r)$, r_B: midpoint of the nearest neighbor ions, r_1: distance of the nearest neighbor ions.

Therefore, there is no difference between eqs.(1.4.1) and (1.4.13), for deriving $S_{ie}(Q)$ numerically.

Along the line of this relatively new procedure with eqs. (1.4.3) and (1.4.4), the distribution of valence electrons around an ion has been obtained for seven liquid metals. The ion–electron correlation functions of $S_{ie}(Q)$ and $g_{ie}(r)$ are exemplified by **Fig.1.4.3** using the results of liquid Al, Zn and Sn [Takeda et al[50,51]]. The vertical lines in Fig.1.4.3(A) corresponds the experimental uncertainty which is of the order of ±0.2, mainly arising from the fact that the difference between $S_X(Q)$ and $S_N(Q)$ is relatively small, compared with the uncertainties of $S_X(Q)$ and $S_N(Q)$ themselves. A series test using the computer technique of numerical calculation was carried out and we confirmed the significance of these results and of principal structural profile of $S_{ie}(Q)$.

Some interesting and important observations can be made from these data about the ion–electron correlation functions in liquid metals. First, a sharp drop of distribution

Table 1.4.1 Ionic radius r_i and its relevant information obtained from the experimental ion–electron correlation function in unit of nm.

	r_i	ionic radius (Pauling)	r_A	r_B	r_1	n_{ve}
Na^+	0.105	0.095	0.165	0.225	0.381	0.7
Zn^{2+}	0.074	0.074	0.110	0.141	0.268	1.7
Al^{3+}	0.045	0.050	0.115	0.225	0.282	2.8
Ga^{3+}	0.070	0.062	0.118	0.156	0.282	2.2
Tl^{3+}	0.080	0.095	0.110	0.143	0.328	0.9
Sn^{4+}	0.072	0.071	0.096	0.167	0.323	2.7
Pb^{4+}	0.080	0.084	0.108	0.160	0.333	1.9
Bi^{5+}	0.075	0.074	0.094	0.172	0.338	1.5

n_{ve}: Number of valence electrons localized in an ion.

function occurs at a certain small value of r, as easily seen in Fig.1.4.3(B). For example, the curve of $g_{ie}(r)$ for Zn drops at the distance of 0.074 nm, which exactly agrees with the well-known ionic radius (0.074 nm) of Zn^{2+} ion proposed by Pauling[52]. Similar agreement is found in other metals, as summarized in **Table 1.4.1** and this is, from the present author's view, rather surprisingly good. However, the number of valence electrons localized at the close vicinity of an ion, n_{ve}, can also be calculated from the

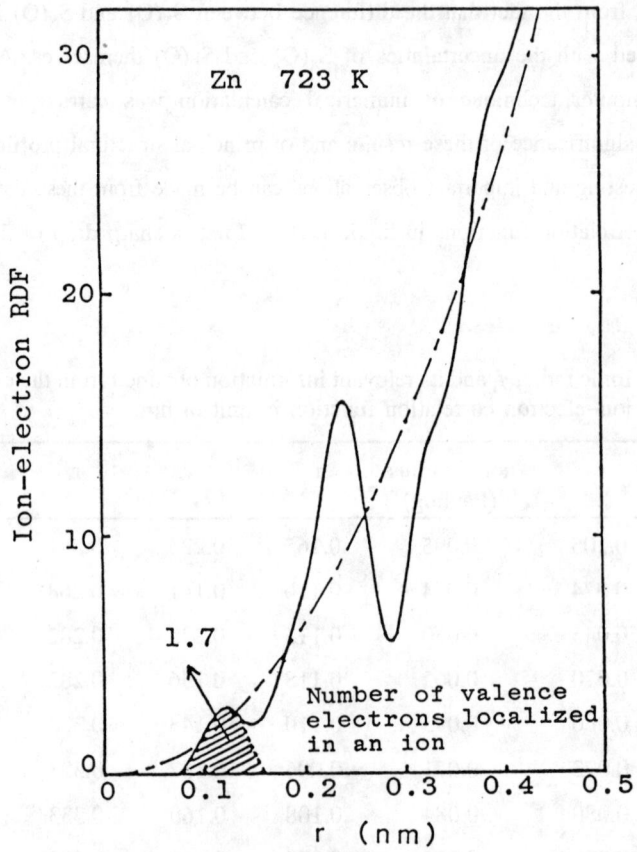

Fig.1.4.4   Ion-electron radial distribution function (RDF) of liquid Zn. Hatched area corresponds to the number of valence electrons closely located around an ion.

ion-electron RDF given in **Fig.1.4.4** using the result of Zn as an example. The hatched area around the first peak of Fig.1.4.4 corresponds to the number of valence electrons which surround an ion closely and this number is estimated to be 1.7 in the case of liquid Zn. As is well-known, the number of valence electrons is of course, two for Zn. Thus, most of the valence electrons (1.7) associate with ions, whereas only small number of them (0.3) exist in the intermediate region among ions in liquid Zn. The information of n_{ve} is also summarized in Table 1.4.1 and the different electron localization is suggested among Al, Ga and Tl or between Sn and Pb for elements having the same valence number.

The first peak of $g_{ie}(r)$ means the distribution of the electron cloud around an ion located at the origin and the second peak indicates the electron cloud around the neigh

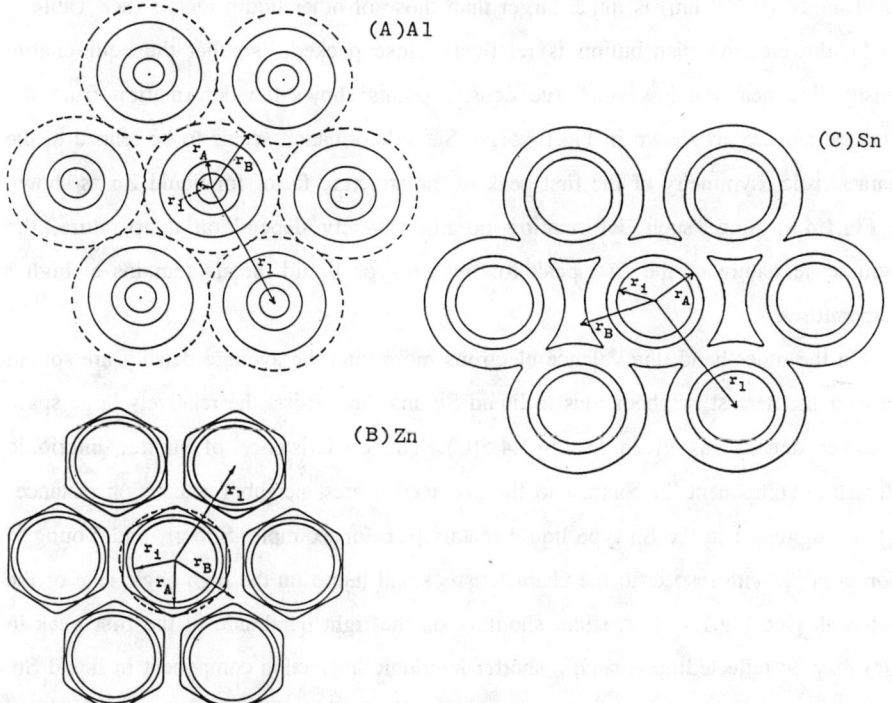

Fig.1.4.5 Schematic diagram for the equi-charge density map of valence electron around an ion in liquid Al, Zn and Sn.

boring ions. Thus, the spatial distribution of the valence electrons around ions in liquid metals may be schematically obtained by drawing the equi-charge density map. The equi-charge density map in liquid metals are illustrated in **Fig.1.4.5**. Here, the fcc type structural coordination is considered for Al and Zn, whereas the bcc type structural coordination is employed for Sn, based on the experimental data of the nearest neighbor coordination number determined from the ordinary RDF data of ion-ion pairs. It may also be suggested that liquid Tl and Pb [Al type, see Waseda[10]] shows the electron distribution map similar to liquid Al and a quite similar nature of liquid Sn is observed in liquid Ga and Bi [Sn type]. Some notable points which have emerged from these equi-charge density maps of the valence electrons around an ion are as follows.

Since the ratio between the ionic diameter of Zn^{2+} (0.148 nm) and the nearest neighbor distance (0.268 nm) is much larger than those of other liquid metals (see Table 1.4.1), the electron distribution is relatively close packed, so that the equi-charge density line near the lowest charge density points shows the deformation from the spherical surface, as shown in Fig.1.4.5(B). Such deformation seems to be related to the characteristic asymmetry of the first peak of the structure factor for liquid Zn as shown in **Fig.1.4.6**. Since such deformation do not severely depend on temperature, the asymmetric nature of the first peak for the Zn type liquid metals remains at higher temperatures.

On the other hand, the valence electrons more than the average density are spread out over the nearest neighbor ions in liquid Sn and there exists the relatively large space of lower density, as given in Fig.1.4.5(C). The co-existence of shorter interionic interaction component, in contrast to the averaged nearest neighbor interaction distance (r_1) is suggested in the Sn type liquid metals [see for example, Silbert and Young[53], Mon et al[54]], with respect to the characteristic small hump on the high angle side of the first peak [see Fig.1.4.6]. A small shoulder on the right hand side of the first peak in $g_{ie}(r)$ may be reflected upon such a shorter interionic interaction component in liquid Sn, because the electron cloud relevant to the small shoulder is able to screen a strong Coulomb repulsion between tin ions located in the near neighbor region. The inhomogeneous distribution of valence electrons including the relatively large space of lower density in liquid Sn at about the midpoint between the nearest neighbor ions seems to be one of the origins attributed to the characteristic small hump on the high angle side of the first peak in the structure factor of the Sn type liquid metals. The ionic vibrations increase at higher temperatures and then the relatively large space of lower density is diminished. This may be related to the disappearance of the small hump of the Sn type liquid metals.

In contrast to these notable points found in the inhomogeneous distribution of the valence electrons around an ion for liquid Zn and Sn, the rather symmetrical nature is appreciated in the valence electron distribution map for liquid Al, Pb and Tl, although the valence electrons are distributed with the relatively low density around ions and also distributed near the tetrahedral region among ions, as shown in Fig.1.4.5(A). This

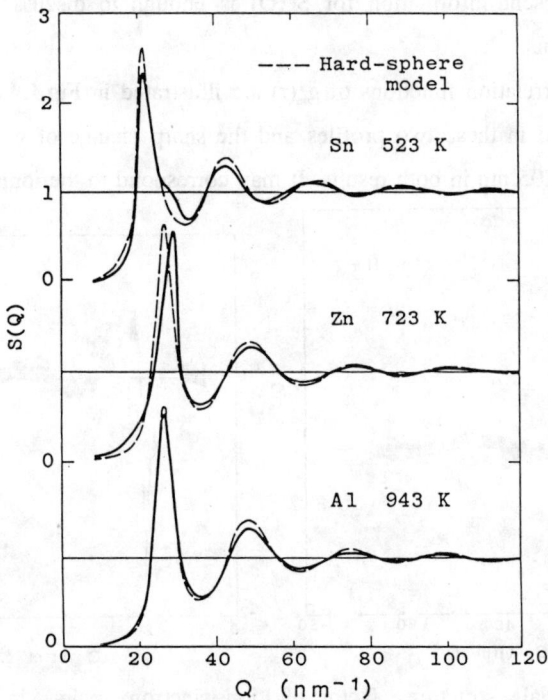

Fig.1.4.6 Structure factors of liquid Al, Zn and Sn near the melting point with hard sphere sturcture factor for comparison[10].

symmetrical nature in the ion–electron correlation can most likely attributed to the simple structural features for the Al type liquid metals.

The partial structure factor of ion–electron pairs, $S_{ie}(Q)$, for liquid Na has recently been obtained by the similar procedure [Takeda et al[55]]. The results are given in **Fig.1.4.7**. The vertical lines in this figure indicate the experimental uncertainty which is of the order of ±0.15 estimated from the counting statistical errors of intensity measurements of x-rays and neutrons. In these two sets of data, we found again good coincidence except for the difference of the positive amplitude of the oscillations in the region of Q=20 nm^{-1}. The difference in the structure factors between $S_X(Q)$ and $S_N(Q)$ disappear beyond about Q=55 nm^{-1} as shown in Fig.1.4.7 and the atomic form factor $f_a(Q)$ is in accord with the ionic form factor $f_i(Q)$ beyond Q=56 nm^{-1} for sodium. Therefore, the present information for $S_{ie}(Q)$ is enough to discuss the partial pair correlation functions.

The partial correlation functions of $g_{ie}(r)$ are illustrated in **Fig.1.4.8**. The essential agreement is found in these two profiles and the sharp change of $g_{ie}(r)$ in liquid Na occurs at about 0.105 nm in both results. It may correspond to the ionic radius of Na$^+$,

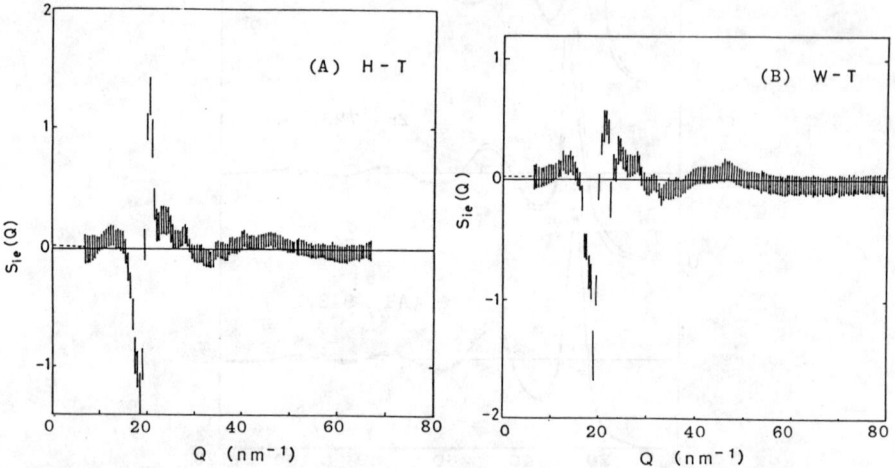

Fig.1.4.7   Partial structure factor of ion–electron pairs in liquid Na. (A) estimated from x-ray data of Huijben and van der Lugt[33], (B) estimated from x-ray data of Waseda[10]

although this observation slightly larger than the value proposed by Pauling[52] as listed in Table 1.4.1. However, it is noticed that $g_{ie}(r)$ increases sharply with decreasing the distance inside of r_i, which is different behavior from those of liquid polyvalent metals. The present author maintain the view that this is caused by the penetration of electron charge from outside to the inside of the ion core. Thus, it may be, more or less, that eq.(1.4.3) is not sufficient enough to completely describe the electron and ion distributions in liquid metals, particularly for metals, where the electron charge distributes both outside and inside of the ion core. It may be added that such behavior found in liquid Na is consistent with the recent theoretical calculation of liquid Li by Chihara[56].

The first peak at 0.165 nm in $g_{ie}(r)$ means the electron charge distribution around a sodium ion located at the origin. The second peak at about 0.310 nm suggests the electron cloud around the nearest neighbor ions. Since the nearest neighbor ions

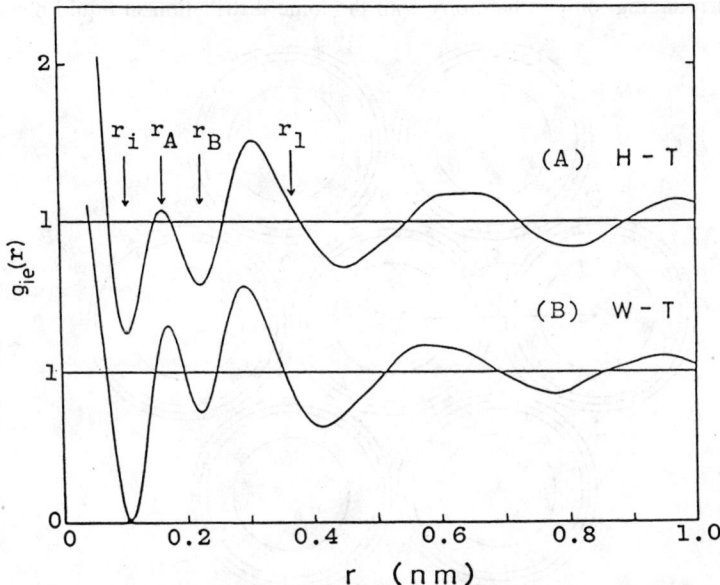

Fig.1.4.8 Partial distribution function of ion−electron pairs in liquid Na calculated from the data of Fig.1.4.7. r_i: the position of $g_{ie}(r)$ shows the minimum point, r_A: first peak position of $g_{ie}(r)$, r_B: first minimum position between the neighbor ions, r_1: distance of the nearest neighbor ions.

coordinate around the distance of r=0.370 nm, on the basis of the ordinary ion–ion correlation function, the distance of the first peak in $g_{ie}(r)$ at r=0.165 nm, corresponds to the middle region among the nearest neighbors. Thus, the electrons more than the average density are considered to be distributed over the intermediate region among the nearest neighbor ions. This implies that the valence electron charges in liquid Na are not bounded at a certain sodium ion, but rather spread out over the intermediate region among the nearest neighbors. Such behavior is consistent with a nearly free electron picture and the electron charges are loosely distributed around ions. For convenience, the schematic diagram for the valence electron distribution around ions in liquid Na drawn from the present ion–electron correlation function data is given in **Fig.1.4.9**. It may be added that about two third of valence electrons (0.7) associate with ions and reminders exist in the intermediate region among ions in liquid Na.

These selected examples clearly indicate the inhomogeneous distribution of valence electrons surrounding ions, in harmony with the ionic distribution in liquid metals and

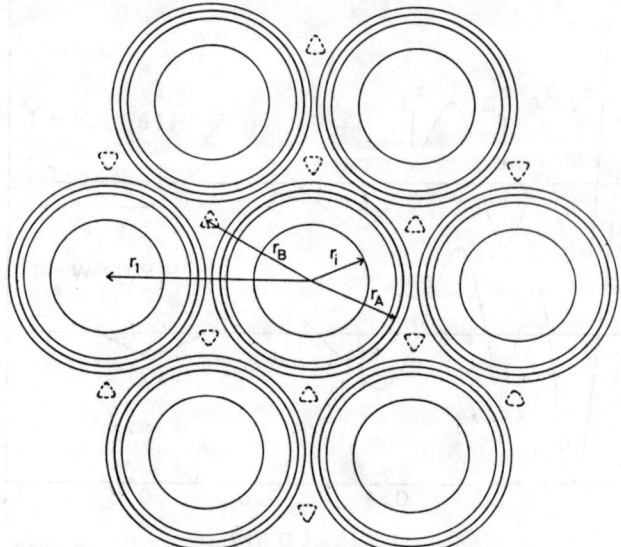

Fig.1.4.9　　Schematic diagram for the equi-charge density map of valence electron around an ion in liquid Na. Dotted area are the minimum region of electron distribution among the ions. The symbols denote the same quantities of Fig.1.4.8.

they are found to be consistent with the characteristic structural features, such as the asymmetry of the first peak in liquid Zn or the small hump on the high angle side of the first peak in liquid Sn which have long been suggested. Therefore, it may be interesting to extend the present approach to estimate the ion-electron correlation functions in other metals, particularly liquid transition metals and rare earth metals from measured x-ray and neutron diffraction data.

1.5 Compound Forming and Structural Inhomogeneity in Liquid Alloys

The structural aspects of disordered systems such as liquids and glasses should be characterized by the atomic short-range order directly related to the RDF. However, anyone who studies disordered systems, also knows the following points. The characteristic features of long-range atomic periodicity found in the crystalline state disappear in the disordered state and the increase in the freedom of atomic configurations due to the increase of the vacant space contributes to the construction of universal short-range ordering and then the structural functions of various disordered systems, metallic, semiconducting and insulating liquids, look essentially similar [Waseda[10]]. This results in the fact that their basic structural features could be explained more or less by the model of **dense random packing of hard spheres** [see for example, Bernal[57], Cargill III[58]] and the structural homogeneity is well recognized in disordered systems, particularly in the liquid state. In other words, disordered systems are characterized by this structural homogeneity. However, we also frequently find the experimental results suggesting that a completely random mixing is not obtained, particularly in the disordered state of condensed matter, so that the distribution of constituent atoms around the atoms of each component likely differs from the average value. Such deviation from simple mixtures probably approximated by the hard sphere mixtures of different sizes [see for example, Ashcroft and Langreth[45]] is usually conceivable, when disordered systems consists of more than two kinds of atoms having different physico-chemical properties such as size, charge number and electronegativity.

This deviation from the simple average indeed produces the structural inhomogeneity in multicomponent disordered systems.

No unique definition is available for describing completely the atomic short range order in disordered systems and then there is frequently some confusing about the concept of structural inhomogeneity in liquids. With respect to this point, the concentration–concentration structure factor $S_{CC}(0)$ in the long wavelength limit ($Q \rightarrow 0$) in the Bhatia–Thornton form [Waseda[10], Wagner[59]] is almost undoubtedly one of the best methods. $S_{CC}(0)$ give a useful physical significance indicating the mean square fluctuation in the concentration in liquids of interest. According to Bhatia–Thornton[60], the concentration–concentration structure factor in the long wavelength limit, $S_{CC}(0)$, is given by the following equation providing a direct link with the thermodynamic activity data of the desired disordered system.

$$S_{CC}(0) = (1 - c_j) [(\partial \ln a_j / \partial c_j)_{T,P}]^{-1} \tag{1.5.1}$$

where a_j is the thermodynamic activity of j–component. It can be easily found that $S_{CC}(0)$ for ideal mixing (the concentration fluctuation is non–exist) is expressed by the following simple form.

$$S_{CC}(0) = c_j (1 - c_j) \tag{1.5.2}$$

Figure 1.5.1 shows the function of $S_{CC}(0)$ for some liquid alloys as a function of concentration estimated from measured thermodynamic activity data. For example, a negative deviation from ideal mixing is pronounced at a certain composition such as A_4B or AB_2 in liquid Na–Sn [Tamaki et al[61]], Na–Pb [Matsunaga et al[62]] and Mg–Bi [McAlister et al[63]] alloys. This qualitatively indicates the formation of chemical complexes in the melt near this composition. On the other hand, a large positive deviation from ideal mixing suggesting phase separation in the melt is also detected at the composition near AB_3 in liquid Ga–Na [Tamaki and Cusack[64]] and Bi–Zn [Tamaki et al[65]] alloys. These appreciable deviations imply the structural inhomogeneity in these alloy liquids. More detailed structural information about these particular liquids obtained recently is given below with some selected examples.

Thermodynamic quantities of mixing, electrical resistivity and magnetic susceptibility of several alloy liquids of alkali–polyvalent metal, alakaline earth–polyvalent metal or metal–chalcogens strongly suggest the formation of chemical complexes in the close vicinity of a certain composition. As for the theoretical studies of such liquids with strong chemical interacting systems, Copestake et al[66,67] first proposed an ionic model with hard-sphere Yukawa interactions for the ordering potential and applied it to the liquid Li_4Pb alloy. Hafner et al[68] has extended this approach to liquid Na–Pb alloys. With these facts in mind, some structural investigation of these compound forming liquids have been made. Such structural information is exemplified by the results of liquid Na–Pb alloys including the possible interpretation for their characteristic features.

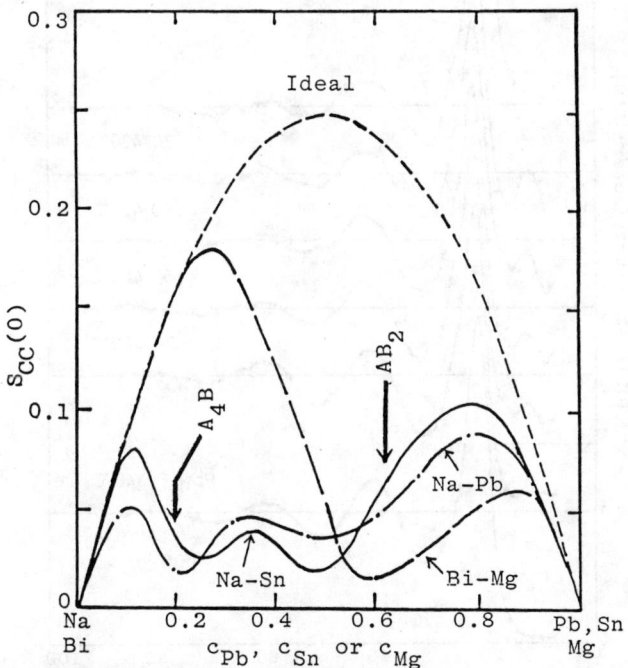

Fig.1.5.1　Concentration–Concentration fluctuation function of some liquid binary alloys suggesting compound forming.

The total structure factors of liquid Na–Pb alloys determined by neutron diffraction are illustrated in **Fig.1.5.2** [Takeda et al[69]]. The first peak positions of the structure factors of alloys appear to differ from those of pure Na and Pb and they shift to higher wave vector in all alloy compositions presently investigated. The interesting structural aspects are as follows. There exists an additional small peak at about $Q=12.5$ nm^{-1} and

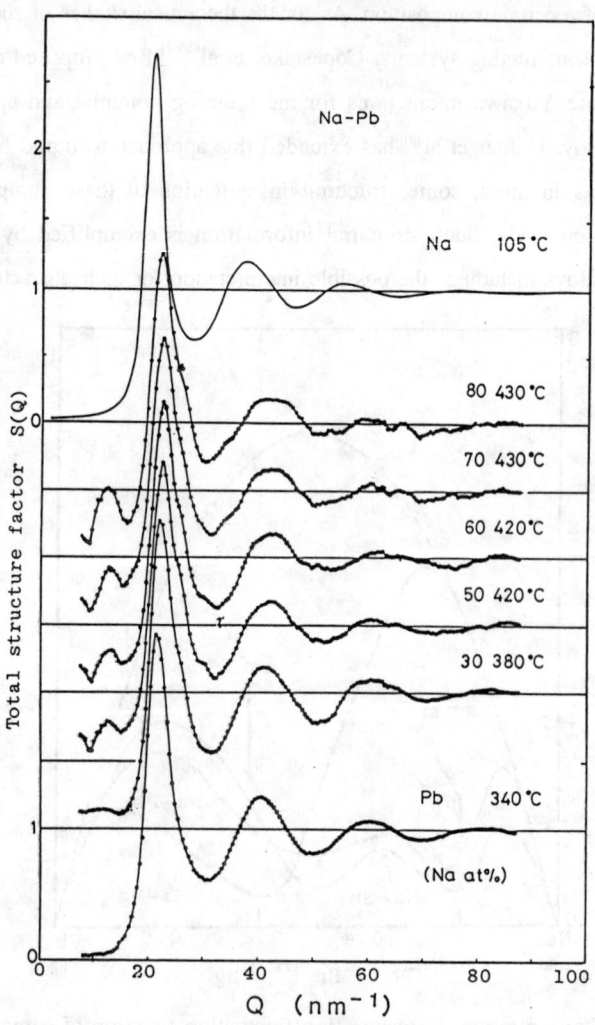

Fig.1.5.2   Total structure factors of liquid Na–Pb alloys by neutron diffraction.

its height increases with Na concentration. The third peak of the structure factor located around 60 nm^{-1} decreases and the damping behavior of its oscillating profile becomes rather complicate beyond the third peak with the increase in Na concentration. A small shoulder at about Q=15 nm^{-1} can also be seen behind the first peak at about Q=25 nm^{-1} in alloys and this shoulder becomes distinct in the Na–rich region. It may also be added that a small angle scattering effect can be observed in the Na–rich region as Q approaches to Q=0.

The existence of a certain chemical complex is quite feasible in liquid Na–Pb alloys, because the additional small peak in the low Q side of the first peak in the structure factor is clearly detected and the similar additional small peak has been observed in other alkali–polyvalent metal alloys such as Na–Sn [Alblas et al[70]] and K–Pb [Reijers et al[71]] alloys. Based on measured thermodynamic and electronic properties of liquid Na–Pb alloys, the Na$_4$Pb type chemical short range order has been anticipated and then it is quite realistic that this alloys are considered as a ternary mixture composed of the atomic association of Na$_4$Pb and its dissociated atoms of Na and Pb. In this model, a chemical interaction between the unlike atom pairs is somewhat squeezed in the formation of "molecule", and the measured structure factors can be calculated by

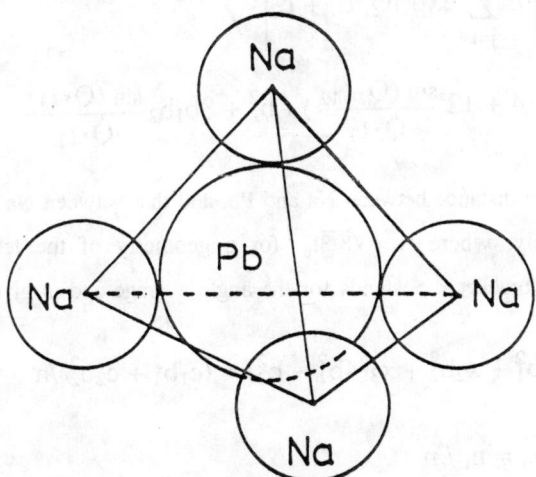

Fig.1.5.3   Schematic diagram of local ordering unit of Na$_4$Pb.

applying the ternary hard sphere mixing model developed by Hoshino[72]. The most probable configuration of the molecular unit may be a tetrahedron with Pb at the center as schematically shown in **Fig.1.5.3** and the distance between Pb and Na atoms is estimated to be 0.33 nm from the distribution function data of 80 at% Na alloy. According to Hoshino[72], the total neutron diffraction structure factor of ternary hard sphere mixture can be expressed in the following.

$$S(Q) = [\, x_1 b_1^2 S_{11}(Q) + x_2 b_2^2 S_{22}(Q) + x_3 [\, <b_3(Q)>^2 (S_{33}(Q) - 1)$$
$$+ <b_3^2(Q)> \,] + 2\sqrt{x_1 x_2}\, b_1 b_2 S_{12}(Q) + 2\sqrt{x_1 x_3}\, b_1 <b_3(Q)> S_{13}(Q)$$
$$+ 2\sqrt{x_2 x_3}\, b_2 <b_3(Q)> S_{23}(Q) \,] / <b^2> \qquad (1.5.3)$$

where $S_{ij}(Q)$ (i, j=1, 2 and 3) in eq.(1.5.3) are the partial structure factors [1:Na, 2:Pb and 3:Na$_4$Pb] and the following approximate relations are also employed.

$$<b_3(Q)> = <b_1 \sum_{j=1}^{4} \exp(iQ, l_j) + b_2> = 4 b_1 \frac{\sin(Q \cdot l_0)}{Q \cdot l_0} + b_2, \qquad (1.5.4)$$

$$<b_3^2(Q)> = \left\langle \left| b_1 \sum_{j=1}^{4} \exp(iQ, l_j) + b_2 \right|^2 \right\rangle$$

$$= b_1^2 \left(4 + 12 \frac{\sin(Q \cdot l_1)}{Q \cdot l_1} \right) + b_2^2 + 8 b_1 b_2 \frac{\sin(Q \cdot l_1)}{Q \cdot l_1} \qquad (1.5.5)$$

where l_0 and l_1 are the distance between Na and Pb, and that between Na and Na in the "molecule" respectively, where $l_1 = \sqrt{8/3}\, l_0$ for a geometry of the tetrahedral unit structure. The angular bracket $< >$ stands for the angle average and is given as follows.

$$<b^2> \equiv x_1 b_1^2 + x_2 b_2^2 + x_3 (4 b_1^2 + b_2^2) = (c_1 b_1^2 + c_2 b_2^2)/n \qquad (1.5.6)$$

$$x_i = n_i / n \qquad (1.5.7)$$

Here, the relation of $n_1 = c_1 - 4 n_3$, $n_2 = c_2 - n_3$, and $n = 1 - 4 n_3$ are used, where c_1, c_2 and the

Table 1.5.1   Several parameters of model calculation for the $Na_{80}Pb_{20}$ alloy.[69]

σ_{Na} (nm)	σ_{Pb} (nm)	σ_{Na_4Pb} (nm)	η	x_{Na}	x_{Pb}	x_{Na_4Pb}
0.296	0.326	0.87	0.51	0.762	0.191	0.048

nominal concentration of Na and Pb, respectively. The orientational correlation among Na_4Pb, Na and Pb are neglected in this model, because the Na_4Pb molecule is assumed to be a hard sphere of with an effective size of Na_4Pb. The hard sphere diameters were chosen so as to reproduce the positions of both the first peak and the small prepeak in the measured structure factors and each fraction of Na, Pb and Na_4Pb were estimated from thermodynamic data [Takeda et al[69]]. They are listed in **Table 1.5.1**. A slight large value of the packing fraction, compared with the usual value of 0.45, was employed in this case and it is rather realistic since a considerable volume contraction

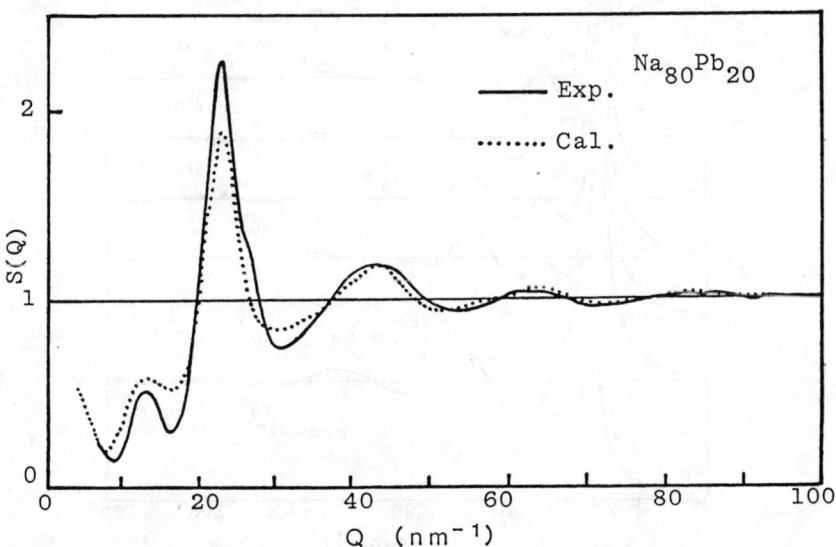

Fig.1.5.4   Comparison of the calculated structure factor of liquid $Na_{80}Pb_{20}$ with the experimental data.

is observed in Na–Pb alloys [Hesson et al[73], McAlister[74]].

As shown in **Fig.1.5.4**, the overall agreement between calculation and experiment in the structure factors can be recognized. The oscillating profiles in the larger Q region beyond 60 nm^{-1} are also quite in phase with measured structure factors. Although there are some differences in detail between calculation and experiment such as the mismatch in the heights of the first peak and the characteristic prepeak, the simple hard sphere mixture model basically work well by reproducing the characteristic structural features

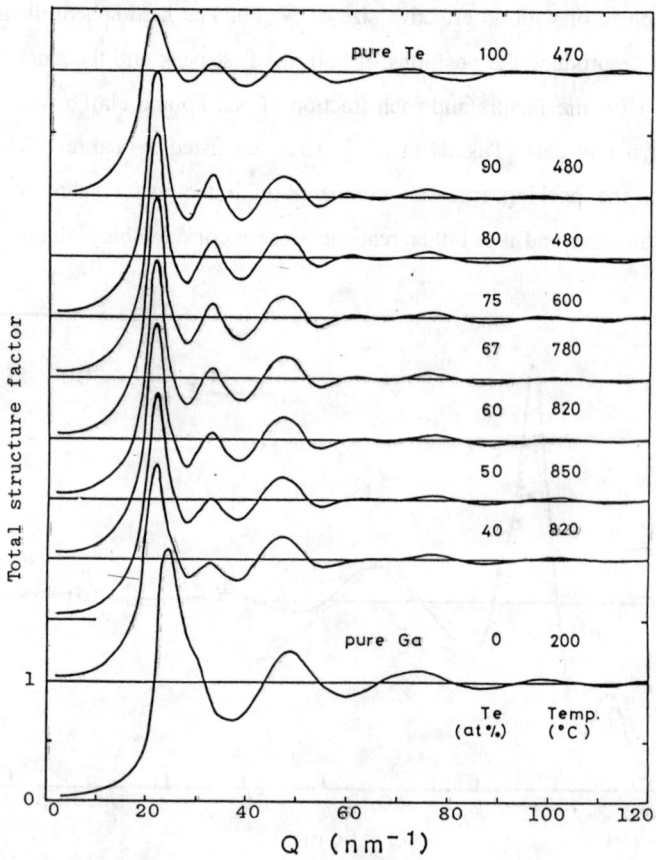

Fig.1.5.5   Total structure factors of liquid Ga–Te alloys by x-ray diffraction.

of liquid Na–Pb alloys. In other words, the compound forming behavior detected in several alloy liquids of alkali–polyvalent metal and alakaline earth– polyvalent metal may be interpreted by the relatively simple picture of a particular atomic association such as A_4B and its dissociated atoms of the constituents, A and B.

The compound forming character is also suggested in liquid metal–chalcogen alloys such as Te–based or Se–based alloys by measuring electronic and thermodynamic

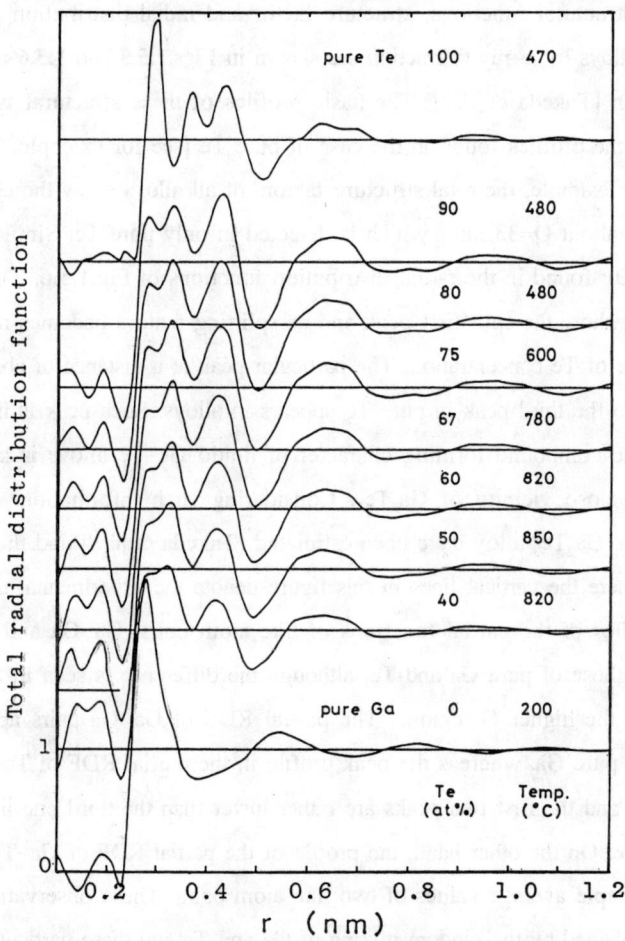

Fig.1.5.6　　Total radial distribution functions of liquid Ga–Te alloys.

properties [see for example, Glazov et al[75], Cutler[76]] and a feasible chemical complex explaining their particular structural features has been proposed using the defected-zinc-blend type ordering units observed in the crystalline state. These concluding remarks slightly differs from those observed in liquid alkali-polyvalent metal alloys above-mentioned. Therefore, some essential points are given using the results of Ga-Te alloys as an example.

The total structural functions, structure factor and radial distribution function, of liquid Ga-Te alloys by x-ray diffraction are shown in **Figs.1.5.5** and **1.5.6** as a function of concentration [Takeda et al[77]]. The basic profiles of these structural functions are rather close to the profiles found in the case of pure Te [see for example, Waseda and Tamaki[78]]. For example, the total structure factors of all alloys show the characteristic second peak at about $Q=33$ nm^{-1} which is detected in only pure Te. Similar structural aspects are easily found in the radial distribution functions of Fig.1.5.6. The measured RDFs of alloys show the split first peak and its splitting feature becomes more intense with an increase of Te concentration. The particular peak at a distance of about 0.43 nm corresponding to the third peak in pure Te appears in alloys. Such peak is not detected in pure Ga. The compound forming character of liquid Ge-Te alloys is known to be distinct in the close vicinity of Ga_2Te_3. Considering such information, the partial structures of the Ga_2Te_3 alloy have been estimated [Takeda et al[77]] and they are given in **Fig.1.5.7**, where the vertical lines in this figure denote the experimental uncertainties. The basic profiles of the partial functions of like atom pairs, Ga-Ga and Te-Te, are very similar to those of pure Ga and Te, although the difference is seen in the phase of oscillations in the higher Q region. The partial RDF of Ga-Ga pairs appears to be close to that of pure Ga, whereas the peak profile in the partial RDF of Te-Te pairs is much the same and the first two peaks are rather lower than the third one in contrast to those of pure Te. On the other hand, the profile of the partial RDF of Ga-Te pairs does not lie in the simple average values of two like atom pairs. These observations can not certainly be explained by the random mixing of Ga and Te and these particular structural features should be attributed, more or less, by the crystal-like chemical ordering formed by the constituents, here Ga_2Te_3.

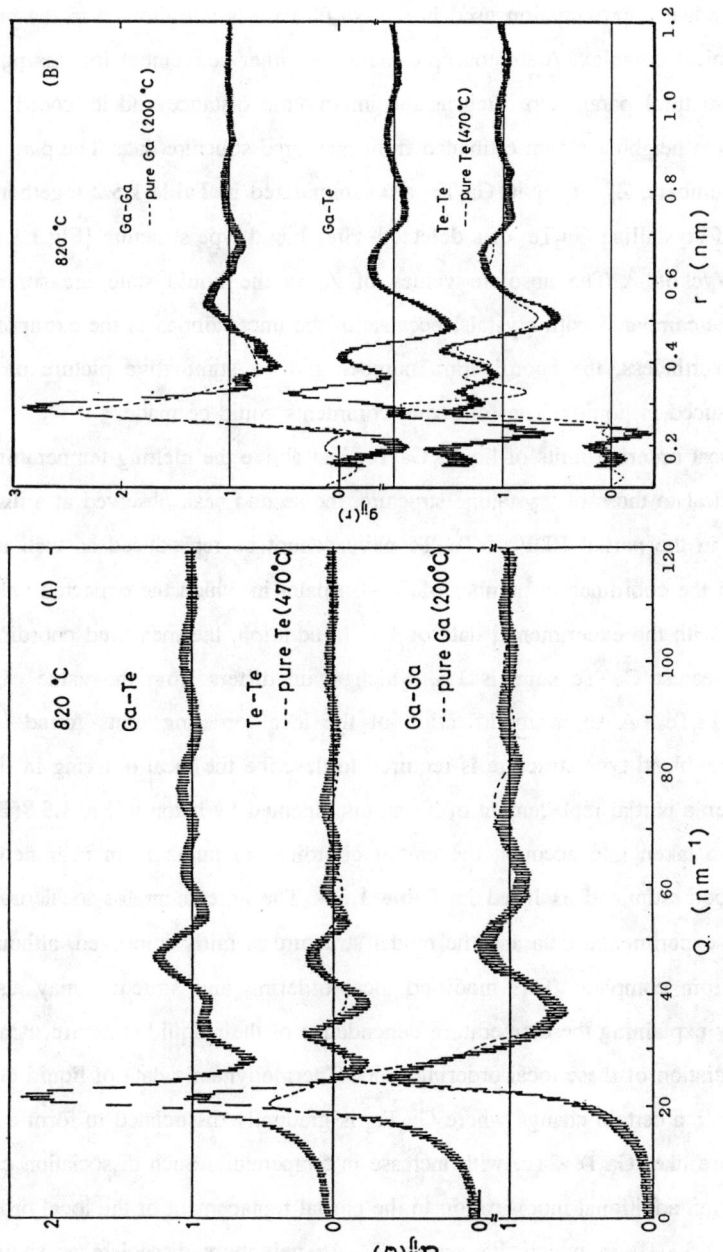

Fig.1.5.7 Partial structural functions of liquid Ga_2Te_3.

The spherical approximation used in the Na$_4$Pb case is not allowed in the present Ga$_2$Te$_3$ chemical complex. A stereoscopic model is rather convenient for this purpose using the structural parameters such as the interatomic distance and its coordination number in near neighbor region estimated from measured structure data. The partial coordination numbers, Z_{ij}, of liquid Ga$_2$Te$_3$ are summarized in **Table 1.5.2** together with the values of crystalline Ga$_2$Te$_3$ of a defected-zinc-blend type structure [**Fig.1.5.8(A)**] given by Wyckoff[79]. The absolute values of Z_{ij} in the liquid state are somewhat ambiguous as compared with crystals, because of the uncertainties in the extrapolation process. Nevertheless, the coordination number gives a stimulative picture of local ordering deduced in liquids. The following comments could be made.

If the local ordering units of liquid Ga$_2$Te$_3$ just above the melting temperature are exactly identical to those of crystalline structure, the second peak observed at a distance of 0.34 nm in the partial RDF of Te-Te pairs cannot be reproduced as well as the difference in the coordination number of Ga-Ga pairs in which the expected value of 1 compared with the experimental data of 1.9. In addition, the measured coordination number of nearest Ga-Te pairs is 1.6, which again differs from the value of 3 in crystalline Ga$_2$Te$_3$. A slight modification of the local ordering units found in the defected-zinc-blend type structure is required to describe the local ordering in liquid Ga$_2$Te$_3$. When a partial replacement of Te atoms, denoted by hatch in **Fig.1.5.8(B)**, by Ga atoms are taken into account, the partial coordination numbers in near neighbor region can be estimated as listed in **Table 1.5.3**. The agreement for local ordering between the experimental data and the model structure is fairly improved, although it is still far from complete. This modified local ordering unit structure may also be supported by explaining the temperature dependence of their liquid structure using the partial dissociation of these local ordering units. Thermodynamic data of liquid Ga$_2$Te$_3$ clearly indicate a certain change where Ga$_2$Te$_3$ is gradually dissociated to form a more simple mixture like (Ga$_2$Te+2Te) with increase in temperature. Such dissociation can be explained by an additional modification to the partial replacement of the local ordering units of Fig.1.5.8(B) in which the nearest Te-Te neighbors dissociate as shown in **Fig.1.5.8(C)**, with an increase of temperature. This additional modification yields the

Table 1.5.2 Comparison of the experimental partial coordination numbers of liquid Ga_2Te_3 with those of a defected-zinc-blend type local ordering crystal structure suggested by Wyckoff.[79] The distance of r is given in unit of nm.[77]

Ga–Ga					Ga–Te					Te–Te			
Observed		Estimated			Observed		Estimated			Observed		Estimated	
r	N_{GaGa}	r	Ga–Ga		r	N_{GaTe}	r	Ga–Te	Te–Ga	r	N_{TeTe}	r	Te–Te
0.29	1.9	0.28	1		0.28	1.6	0.281	3	2 or 3	0.28	1.4	0.281	2
0.36	0.7				0.33	3.2	0.344	1	1 or 2	0.34	1.7		
							0.397		2 or 1				
0.43	1.1	0.444	2		0.41	8.1	0.444	2	2 or 1	0.43	4.8	0.444	4

Table 1.5.3 Comparison of the experimental partial coordination numbers of liquid Ga_2Te_3 with those of a model structure employed in this work. When considering the partial dissociation, the first coordination numbers are given in the bottom line in this table. The distance of r is given in unit of nm.[77]

Ga–Ga					Ga–Te					Te–Te			
Observed		Estimated			Observed		Estimated			Observed		Estimated	
r	N_{GaGa}	r	Ga–Ga		r	N_{GaTe}	r	Ga–Te	Te–Ga	r	N_{TeTe}	r	Te–Te
0.29	1.9	0.281	2		0.28	1.6	0.281	2 or 3	2 or 1	0.28	1.4	0.281	2
0.36	0.7	0.397	1 or 0		0.33	3.2	0.344	1 or 0	1 or 0	0.34	1.7	0.344	1 or 0
0.43	1.1	0.444	1				0.397	2 or 1	1 or 2			0.397	1 or 0
					0.41	7.8	0.444	5 or 4	4 or 3	0.43	4.8	0.444	3 or 2
0.29	1.9	0.281	2.0		0.28	1.6	0.281	1.72		0.28	1.4	0.281	1.55

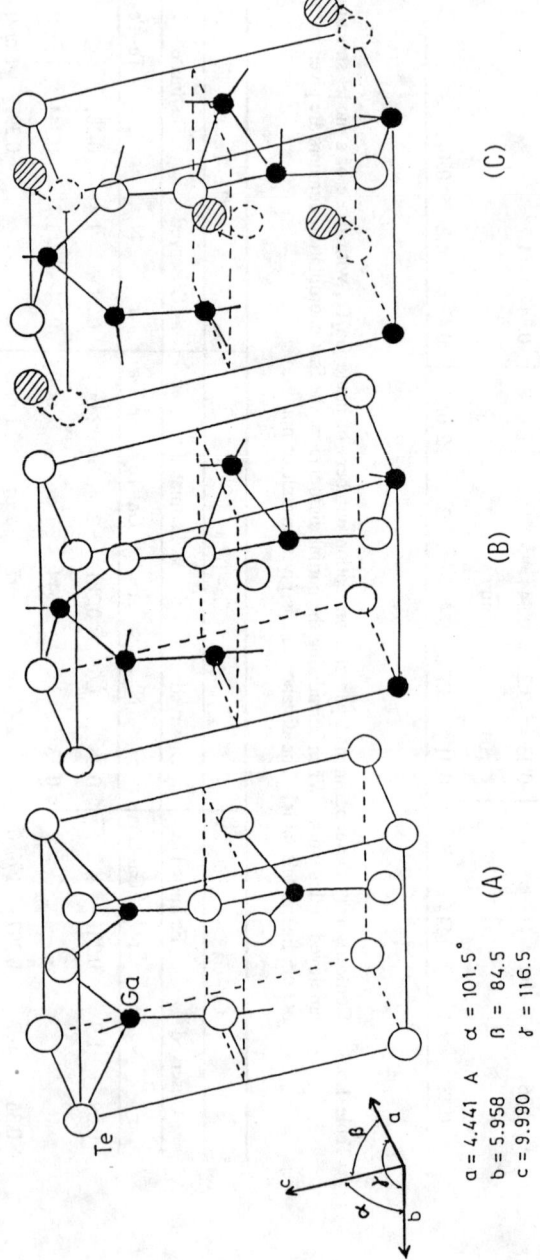

Fig.1.5.8 Schematic diagrams of local ordering units for liquid Ga_2Te_3, (A) crystal [Wyckoff[4]], (B) partially substituted model, (C) dissociated model.

first coordination numbers of three atomic pairs listed in the bottom of Table 1.5.3. The agreement with the experimental data at higher temperature of 1333K is well recognized.[77] Thus, it may be concluded that a feasible structural model for explaining the particular features of liquid Ga_2Te_3 is close to the defected-zinc-blend type local ordering units observed in crystalline Ga_2Te_3 with some modification. Similar approach has been well approved in other liquid metal-chalcogen alloys such as In-Te, Sn-Te [Takeda[80]] and Pb- Te [Takeda et al[81]].

1.6 Structure of Oxide Melts and Glasses

Structural studies of oxide melts, particularly silicates, are important from a petrological perspective of magma formation and from a metallurgical view of slag-metal reactions. The recent growth in the technology of advanced ceramics and glasses has necessitated a better understanding of the various properties of oxide melts at the microscopic level. For this purpose, structural information about the melt is essential.

The apparent profiles of both structure factor (or interference function) and RDF of disordered systems are known to be quite similar. This is consistent with the fact that the characteristic structural features of their respective crystal structures become obscure on melting. The resulting increase in freedom of the atomic configuration contributes to the formation of universal short-range order. However, in contrast to the structure of metallic melts and glasses, the following point may be noted with respect to the structural features of oxide melts and glasses. There exists a distinct local ordering within a narrow region and a complete loss of positional correlation at a few nearest neighbor distance away from any origin. For this reason, the determination of such distinct local ordering unit structure and its distribution is one of the important components in the structural study of oxide melts and glasses. The oxygen coordination number for a metallic ion is also of interest in discussing their structural aspects. These characteristic structural features of oxide melts and glasses request the slightly different data processing for measured intensity data, although the principle of the RDF analysis is unchanged. The essential points are given below.

When a compositionally averaged structural function is introduced, an approach similar to the RDF analysis for a simple one-component case can be extended to oxide melts and glasses containing more than two kinds of elements. The definition proposed by Mozzi and Warren[82] is employed here.

A reduced interference function in electron units, i(Q), as a function of the wave vector Q, is related to the structurally sensitive part of the total scattering intensity of X-rays directly determined by experiments. It is defined by the following equation;

$$i(Q) = [\ I_{eu}(Q) / N - \sum_{uc} f_i^2\ ] / f_e^2 \tag{1.6.1}$$

where I_{eu}/N is the intensity of unmodified scattering in electron units per unit of composition (uc), f_j and f_e are the usual atomic scattering factor and the average scattering factor per electron, respectively. The electron radial distribution function (RDF) can readily be estimated from the interference function, i(Q), by Fourier transformation. The use of pair functions is convenient for interpretation of the RDF data as suggested by Mozzi and Warren[82]. When the pair functions are employed, the following useful relation can be obtained with respect to the theoretical and experimental RDFs.

$$\sum_{uc} \sum_{i} \frac{N_{ij}}{r_{ij}} \int_0^{Q_{max}} \frac{f_i f_j}{f_e f_e} e^{-\alpha^2 Q^2} \sin(Q \cdot r_{ij}) \sin(Q \cdot r)\, dQ$$

$$= 2\pi^2 r \rho_e \sum_{uc} Z_j + \int_0^{Q_{max}} Q\, i(Q)\, e^{-\alpha^2 Q^2} \sin(Q \cdot r)\, dQ \tag{1.6.2}$$

where r_{ij} and N_{ij} are the distance and its coordination number of i-j pairs, respectively, ρ_e the average number density of electrons and Z_j the atomic number of the j-species. Equation (1.6.2) has been employed by Mozzi and Warren[82] under the name of Pair

Function Distribution (PDF). The left-hand side of eq.(1.6.2) provides the theoretical RDF, whereas the right-hand side of eq.(1.6.2) corresponds to the experimental RDF data. The term $\exp[-\alpha^2 Q^2]$ in eq.(1.6.2) is a convergence factor, introduced to minimize the truncation error and weigh down the uncertainties in the higher wave vector region. This artificial parameter α does not have to be critically selected. However, the value is zero in the calculation of the experimental RDF using eq.(1.6.2). The theoretical RDF is generally calculated using a value of $\alpha = 0.03–0.05$, based on the previous studies on various oxide glasses [Wright and Leadbetter[83]].

The interatomic distances are easily determined from the positions of the peaks in the experimental RDF data obtained from eq.(1.6.2). The number of neighbors can be determined using a least-squares analysis, by finding the quantity of N_{ij} which must be used to bring the calculated RDF to the best fit with the experimental RDF. In this process, the variation of r_{ij} and N_{ij} in the theoretical RDF are ±0.001 nm and ±0.2 atom, respectively. Although the present pair function method is effective for only a few near neighbor correlations, the structural parameters in the near neighbor region can be quantified with a much higher reliability compared with those obtained in the previous investigation. Thus, the present data processing is quite useful for the structural study of oxide melts and glasses by reducing the ambiguity in the procedure for estimating the area under each peak in the experimental RDF [see Waseda[10]]. It may be noticed that the root mean square displacements $(\Delta r_{ij}^2)^{1/2}$ corresponding to the magnitude of the peak broadening of the distribution of i–j pairs are also frequently estimated for discussing the structural aspects of disordered systems.

Figure 1.6.1 indicates a comparison between the RDF_{exp} and the RDF_{cal} in the near neighbor region of a quartz glass [Sugiyama et al[84]]. The numerical values in Fig.1.6.1 denote the coordination numbers of respective pairs estimated by use of the pair function method. The numerical values for Si–O pairs clearly suggest that each silicon is surrounded by four oxygens. The coordination number for O–O pairs is consistent with the value expected from the geometry of the tetrahedra formed by oxygens whose center is occupied by silicon. The coordination number of Si–Si pairs which corresponds to the correlation of SiO_4 tetrahedral units is estimated to be about four. This shows no significant inconsistencies with the atomic arrangements observed in the beta-quartz type crystal structure [Galasso[85]]. The RDFs calculated from the values of N_{O-O} equal to 5.0 and 6.0 are also given in a small figure inserted into Fig.1.6.1. As easily seen from these results, the coordination number can be determined by this processing method to within a variation of less than $N_{ij} \pm 0.5$ atom.

The following point is also worthy of note. In the present case, three partial structural functions are superimposed, so that the area under each peak in the RDF might be affected, more or less, by a few kinds of atomic pairs. The peak observed at about 0.21nm in the RDF of a quartz glass [see in Fig.1.6.1(B)] has been considered as a spurious ripple arising mainly from the truncation effect in the Fourier transformation. However, it is now interpreted by the pair function analysis as a summation of three

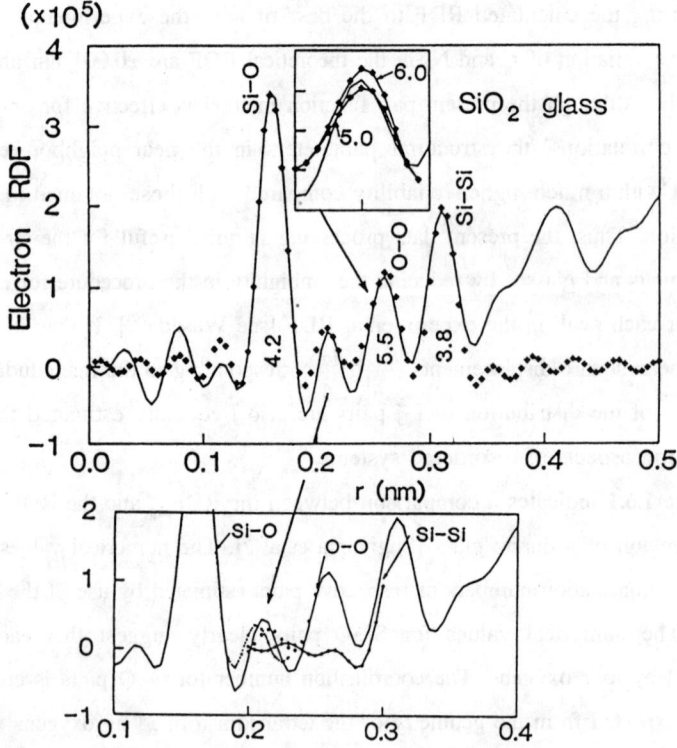

Fig.1.6.1   Electron RDF of quartz glass [Sugiyama et al[84]].

pair correlation tails and their enhancement.

The interference function refining technique [Busing and Levy[86)]] is also employed in the structural analysis of oxide melts and glasses. This technique is based on the characteristic structural features of silicate melts and glasses; namely the contrast between the narrow distribution of local ordering and a complete loss of positional correlation in the longer distance. In other words, the average number of j elements around i- elements, N_{ij}, is separated by an average distance, r_{ij}. The distribution can be approximated by a discrete Gaussian like distribution with a mean-square variation $2\sigma_{ij}$. The distribution for higher neighbor correlations is approximately expressed by a continuous distribution with an average number density of a system. These features can provide the following expression with respect to the reduced interference function, i(Q);

$$[f_e]^2 i(Q) = \sum_{i=1}^{m} \sum_{k} N_{ik} \exp(-\sigma_{ik} Q^2) f_i f_k \frac{\sin(Q \cdot r_{ik})}{Q \cdot r_{ik}}$$

$$+ \sum_{\alpha=k}^{m} \sum_{\beta=1}^{m} [\exp(-\sigma'_{\alpha\beta} Q^2) f_\alpha f_\beta 4\pi\rho_o (Qr'_{\alpha\beta} \cos Qr'_{\alpha\beta} - \sin Qr'_{\alpha\beta})] / Q^3 \quad (1.6.3)$$

The quantities $r'_{\alpha\beta}$ and $\sigma'_{\alpha\beta}$ are the parameters of the boundary region which need not be sharp [Busing and Levy[86)], Narten[87)]]. The structural parameters for near neighbor correlations are determined by a least-squares analysis so as to fit the experimental interference function by iteration. This interference function refining technique has frequently been used for structural investigation of inorganic liquid and glasses. However, it should be kept in mind that this method is not a unique mathematical procedure, but a semiempirical one for the resolution of the peaks in the RDF of disordered systems. In the case of neutron diffraction, the atomic scattering factor should be changed to the neutron scattering amplitude which is constant, because the dimensions of the scattering nuclei are much smaller than the wavelength of neutrons.

Figure 1.6.2 show the interference function data Qi(Q) and the resultant RDFs, respectively[Waseda and Toguri[88)]], for silica(SiO_2), and metasilicates of $MgSiO_3$

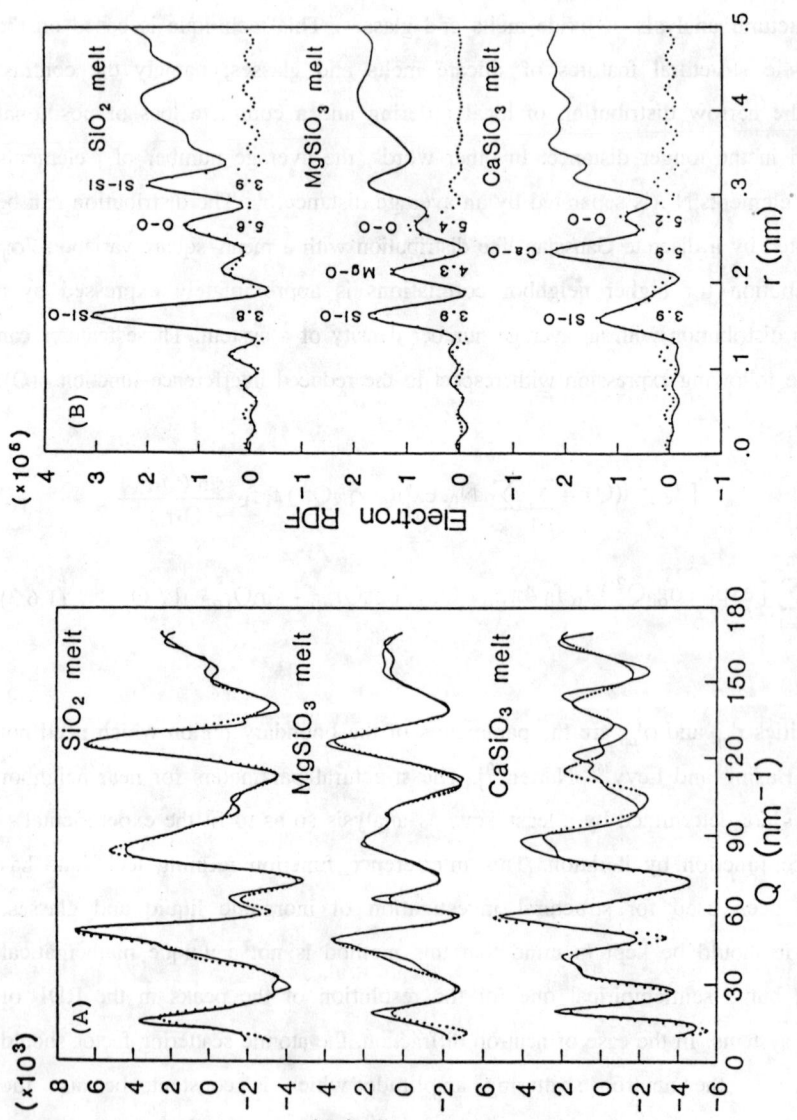

Fig.1.6.2 Interference functions (A) and electron RDFs (B) of SiO_2, $MgSiO_3$ and $CaSiO_3$ melts. (A) Solid line: experimental data, dotted line: calculation by the interference function refining technique. (B) Solid line: experimental data, dotted line: calculation by the pair function method.

Table 1.6.1 Near neighbor correlations in silicate melts determined by X-ray diffraction.[88]

	pair	r_{ij} (nm)	N_{ij} (atom)	$(\Delta r_{ij}^2)^{1/2}$ (nm)
SiO$_2$ Quartz	Si–O	0.162	3.8	0.0098
	O–O	0.265	5.6	0.0126
	Si–Si	0.312	3.9	0.0205
CaSiO$_3$ Wollastonite	Si–O	0.161	3.9	0.0127
	Ca–O	0.235	5.9	0.0171
	O–O	0.267	5.2	0.0206
	Si–Si	0.321	3.1	0.0264
MgSiO$_3$ Enstatite	Si–O	0.162	3.9	0.0109
	Mg–O	0.212	4.3	0.0151
	O–O	0.265	5.4	0.0215
	Si–Si	0.316	3.3	0.0282

(enstatite) and CaSiO$_3$ (wollastonite). These are typical rock forming and metallurgical slag components. The solid lines in Fig.1.6.2(B) are the experimental data and the dotted lines correspond to the theoretical predictions using the pair function method. The structural parameters concerning the near neighbor correlations in these silicate melts are summarized in **Table 1.6.1**. The area under each peak corresponds to the coordination number which gives rise to a local ordering in the melt structure. The numerical values in Fig.1.6.2(B) for the Si–O coordination numbers indicate that each silicon is surrounded by four oxygens at a distance of 0.162 nm in silicate melts. These results clearly suggest that the SiO$_4$ tetrahedra is a realistic fundamental local ordering unit not only in silica but also enstatite and wollastonite melt structures. Thus, these melts can be viewed as mixtures of small ordering unit structures such as SiO$_4$ tetrahedra, although their correlations decay rapidly at larger distances.

Some systematic structural studies of binary and ternary silicate melts has been

Table 1.6.2 Variation of the coordination number in silicate melts determined by X-ray diffraction(± 0.2 atoms).[88]

CaO(mol%)	Temp.(°C)	Si–O	Ca–O	O–O	Si–Si
0	1750	3.8	0	5.6	3.9
34	1700	4.0	5.2	5.6	3.6
41	1600	3.8	5.4	5.4	3.4
45	1600	3.9	5.8	5.3	3.3
50	1600	3.9	5.9	5.2	3.1
57	1750	3.7	6.2	5.1	3.2
MgO(mol%)	Temp.(°C)	Si–O	Mg–O	O–O	Si–Si
0	1750	3.8	0	5.6	3.9
44	1700	4.1	4.1	5.7	3.5
51	1700	3.9	4.3	5.4	3.3
56	1790	3.8	4.4	5.3	2.9

reported [see for example, Waseda and Toguri[88]]. The essential points of these results are summarized below. As shown in Fig.1.6.2(B), the SiO_4 tetrahedron is an acceptable form of the fundamental local ordering unit in quartz and metasilicate melts of MgO and CaO. This has also been confirmed from the systematic results of the $MgO-SiO_2$ and $CaO-SiO_2$ binary silicate melts over a wide composition range. The concentration dependence of the coordination number of near neighbor correlations is listed in **Table 1.6.2**. An almost constant value of the oxygen coordination number of four around a silicon quantitatively confirms the formation of SiO_4 tetrahedra in binary silicate melts with CaO and MgO over a wide concentration range. Consequently, the change in the fundamental local ordering units is less sensitive to the change in the concentration of the melt due to the addition of CaO or MgO. However, beyond the equimolar composition, the coordination number of Si–Si pairs corresponding to the inter-SiO_4 tetrahedral units decreases from 4 to 3 in both binary silicate melts, as shown in Table

1.6.2. This change implies that the SiO$_4$ tetrahedral units become more loosely packed with the break-down of the so-called network structure of pure SiO$_2$ due to the addition of CaO or MgO. The slight decrease in the coordination number of O-O pairs could also be attributed to such break-down of the network structure. However, the depolymerized anionic species [see for example, Bockris et al[89]] can not be identified from these experimental data alone.

The following comments may also be worthy of note. The dotted curves in Fig.1.6.2(A) are the resultant interference function calculated from eq.(1.6.3) with the

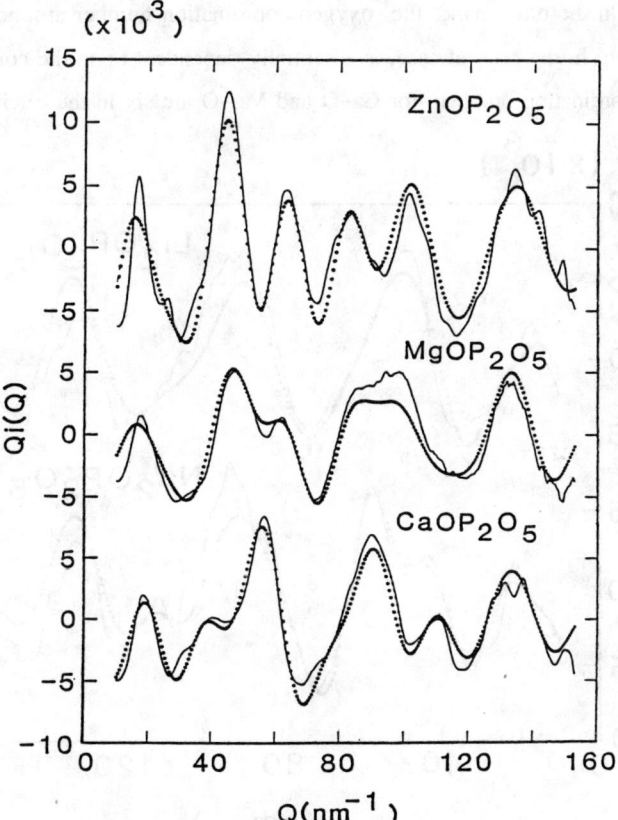

Fig.1.6.3   Reduced Interference function of Qi(Q) of alkali metaphosphate glasses by x-ray diffraction. Solid line: experimental data, dotted line: calculation by the interference function refining technique.

initial structural parameters determined by the pair function method. The agreement appears to be satisfactory. Thus, the structural parameters determined by the pair function method can also be considered quite realistic. The effect of temperature on the structure of silicate melts is also one of the interesting subjects. However, as temperature increases, only a broadening of the peak profile in the RDF is detected in some oxide melts and the essential structural features of silicate melts are found to be rather insensitive to temperature at least within the temperature range (1500–1800°C) presently investigated [Waseda and Toguri[88]]. In this respect, they clearly differ from the case of metallic melts. On the other hand, the oxygen coordination number around a metallic element, Ca or Mg in the present case, is essentially dependent upon the concentration. However, the coordination numbers for Ca–O and Mg–O are six in the calcium silicate

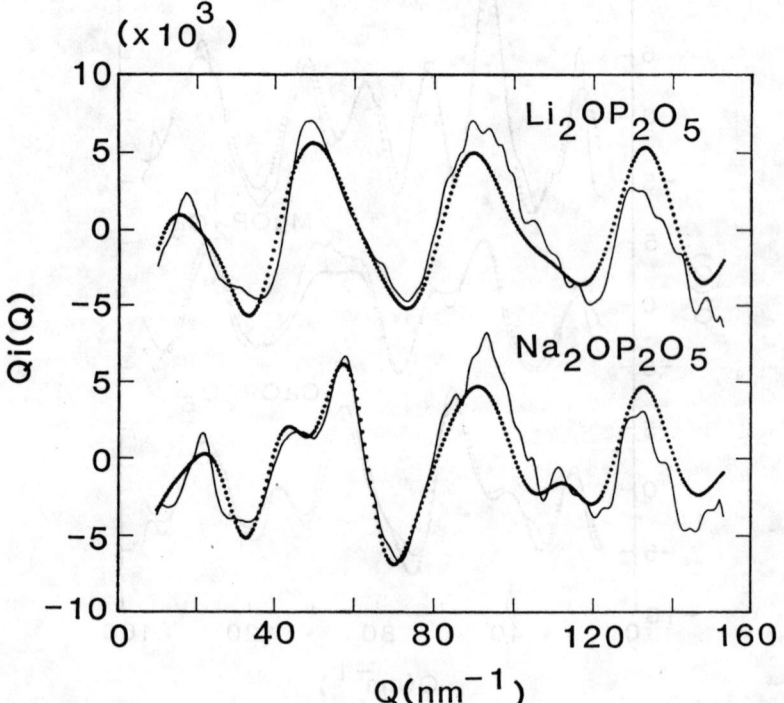

Fig.1.6.4 Reduced Interference function of Qi(Q) of alkaline-earth metaphosphate glasses by x-ray diffraction. Solid line: experimental data, Dotted line: calculation by the interference function refining technique.

Table 1.6.3 Structural parameters of local ordering and densities of metaphosphate glasses.[95]

pair		Pair function method		Refining technique		Density
		r_{ij}(nm)	N_{ij}(atom)	r_{ij}(nm)	N_{ij}(atom)	(Mg/m^3)
Li$_2$OP$_2$O$_5$	P–O	0.154	4.1	0.154	3.9	
	Li–O	0.203	4.0	0.203	4.0	2.34
	O–O	0.250	3.8	0.249	4.3	
NaOP$_2$O$_5$	P–O	0.154	4.0	0.154	3.9	
	Na–O	0.239	3.9	0.238	4.4	2.50
	O–O	0.255	3.9	0.253	4.0	
ZnOP$_2$O$_5$	P–O	0.153	4.0	0.152	4.1	
	Zn–O	0.196	4.5	0.196	4.2	2.90
	O–O	0.249	4.2	0.250	4.0	
MgOP$_2$O$_5$	P–O	0.154	3.8	0.154	4.0	
	Mg–O	0.203	4.5	0.203	4.3	2.49
	O–O	0.250	4.0	0.249	4.4	
CaOP$_2$O$_5$	P–O	0.154	4.0	0.154	4.1	
	Ca–O	0.237	4.9	0.237	5.2	2.66
	O–O	0.254	4.3	0.255	4.5	

and four in magnesium silicate at the equimolar composition, respectively.

Recently, there is an increasing need for the understanding of various properties of phosphate glasses, because their low transformation temperature and relatively large thermal expansion coefficient indicate a good applicability of glass to metal seals involving aluminum alloys and stainless steels [Minami and Mackenzie[90], Wilder Jr[91]]. The interest in phosphate glasses has also been intensified in parallel with recent progress in biological materials. For example, one of the main components of bone and teeth is calcium phosphate and then glasses including calcium phosphate are quite promising for application to artificial bond and tooth materials [Akao et al[92], De With et al[93], Sakka and Kokubo[94]]. Based on these requirements, a systematic structural investigation of binary phosphate glasses of M$_2$O–P$_2$O$_5$ and MO–P$_2$O$_5$, where M= Li, Na, Zn, Mg, and Ca has been done and the results are summarized below.

Figures 1.6.3 and **1.6.4** show the reduced interference function Q i(Q) of five phosphate glasses determined by x-ray diffraction [Waseda et al[95]]. The general feature of the reduced interference function Q i(Q) for all glasses is similar, although there are some differences in detail. Namely, the profile of Q i(Q) is composed of the first peak at about $Q=20$ nm^{-1} followed by a number of peaks, which contrasts to the cases of the metallic glasses where the damping behavior of the function Q i(Q) is rapid and monotonous. The structural parameters obtained from the RDF data by the pair function method are listed in **Table 1.6.3** together with the measured density data. The number of oxygens around phosphorus in these five phosphate glasses is found to be four at a distance of 0.152–0.154 nm and then the PO$_4$ tetrahedra is quantitatively confirmed as

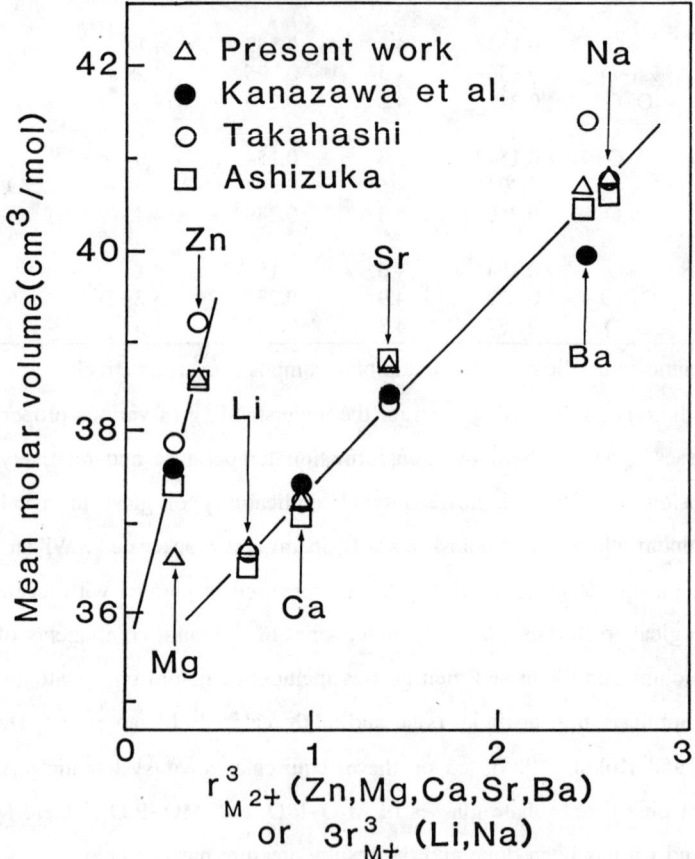

Fig.1.6.5　Mean molar volumes estimated from measured density data as a function of the cube of cation radius for divalent cations or as a function of three times the cube of cation radius for monovalent cations.

a local ordering unit structure in these glasses. The dotted lines in Figs. 1.6.3 and 1.6.4 are the resultant interference function calculated by the interference function refining technique, which are also listed in Table 1.6.3. The consistency between the experimental and calculated interference functions appears to be satisfactory.

It will be of interest to recall the description about the oxygen coordination numbers around M element in phosphate glasses systematically obtained in this work. **Figure 1.6.5** shows the mean molar volumes of various phosphate glasses estimated from measured density data as a function of the cube of cation radius for divalent cations or as a function of three times the cube of cation radius for monovalent cations. Here, Pauling's ionic radius[52] was employed. The factor of three is taken from the reason that one mole alkaline earth metal phosphate produce one gram-ion cations, whereas one mole alkali metal phosphates produce two gram-ion cations. In the latter case, two alkali metal cations most likely occupy more than two times their own volume in phosphate glasses and it is known to be approximated by about three times the cube of cation radius, as first suggested by Tomlinson et al[96].

The linear relationship is easily seen in Fig.1.6.5, which implies that the accommodation of the cations is approximately constant and the volumes of the anions are comparable in these phosphate glasses. However, the linear relationship is classified into two groups, for Li, Na, Ca, Sr and Ba and for Mg and Zn. This is consistent with the previous discussion for phosphate glasses in terms of the concepts of the normal and abnormal types as proposed by Kordes and his colleagues[98,99]. It is generally recognized that the oxygen coordination number around cation in oxides is proportional to the cation radius and phosphate glasses containing Li, Na, Ca ,Sr and Ba are , of course, included in this category as being of the normal type. On the other hand, magnesium phosphate and zinc phosphate glasses are classified as being of the abnormal type attributed to four oxygens around Mg^{+2} and Zn^{+2} cations. The present structural information summarized in Table 1.6.3 supports the discussion of Kordes et al[98,99] about the role of Zn or Mg cation and rather contrasts with the suggestion of Isozaki et al[99] who proposed the six coordinated magnesium in phosphate glass, based on the measurements of x-ray emission spectra. However, the origin of the abnormal behavior for phosphate glasses is not decesively identified at the present time, although it is plausible that the differences in the oxygen coordination number around cations play an important role in the normal or abnormal classification of phosphate glasses.

The scattering power of x-rays for light elements such as oxygen is small relative to that of metallic elements, because x-rays are scattered by the outer electrons of atomsand then the scattering power is known to increase in magnitude with increasing the atomic number. On the other hand, scattering of neutrons arises from the magnetic scattering process as well as that of nucleus and thus neutron diffraction experiments provide useful information, since the scattering power of oxygen is larger than that of other elements, Na, Mg, Ca etc. For this reason, it is no doubt that neutron diffraction is a powerful method for obtaining structural information of oxide melts and glasses. In addition, one of the most important requirements in the structural study of disordered systems is to determine accurately the local ordering structure and its distribution. This is particularly true of oxide systems. For this purpose, the higher wave vector (momentum transfer) measurements of the structural function are quite useful for providing the high resolution RDF using the method of Time of Flight (TOF) pulsed neutron diffraction [see for example, Sinclair et al[100], Suzuki[101]] or Energy Dispersive

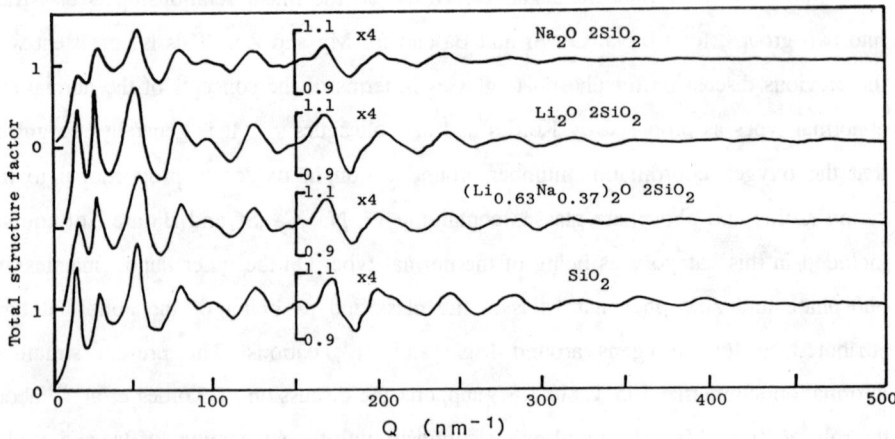

Fig.1.6.6 Structure factors of silicate glasses obtained by TOF pulsed neutron diffraction [Misawa et al[104]].

X-ray Diffraction (EDXD) [see for example, Egami[102,103]]. These two methods could obtain structural function in high wave vector regions, up to 300 nm^{-1} which is beyond the limit (usually 150 nm^{-1}) for conventional X-ray and neutron diffraction methods. The structural study of oxide melts by these devised techniques is not available yet. However, for example, Misawa et al[104] reported the high resolution RDFs of some silicate glasses by TOF pulsed neutron diffraction method and the results are given in **Figs. 1.6.6** and **1.6.7**. The information about local ordering in near neighbor region obtained in this TOF work is found to be consistent with the previous conclusion which was deduced from the results using conventional method. Nevertheless, these high

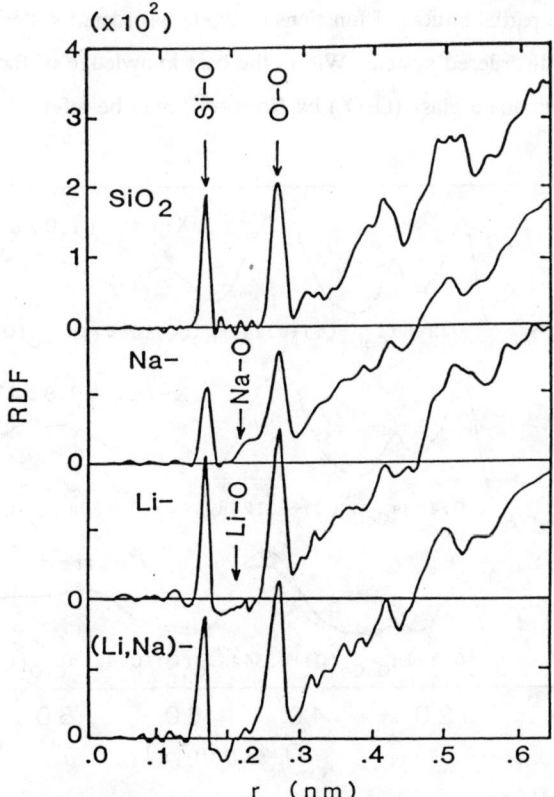

Fig.1.6.7 Radial distribution function of silicate glasses calculated from the data of Fig.1.6.6 [Misawa et al[104]].

resolution RDF results are believed to be quantitatively accurate. Thus, TOF pulsed neutron diffraction technique is one way to determine the local structure of oxide melts and glasses systems, although there are some problems related to the corrections for static approximation, Placzek effect and others for neutron diffraction [Enderby[15], North et al[105]].

As discussed previously, the ordinary structural functions obtained directly from conventional diffraction experiments provide only a compositionally averaged structural data of disordered systems. Although we have now obtained quantitatively accurate partial structural functions of various metallic melts and glasses or molten salts. However, available partial structural functions of oxide disordered systems is still limited even for a binary disordered system. Within the best knowledge of the present author, the results of a germanate glass (GeO_2) by Bondot[106] may be referred to as the unique

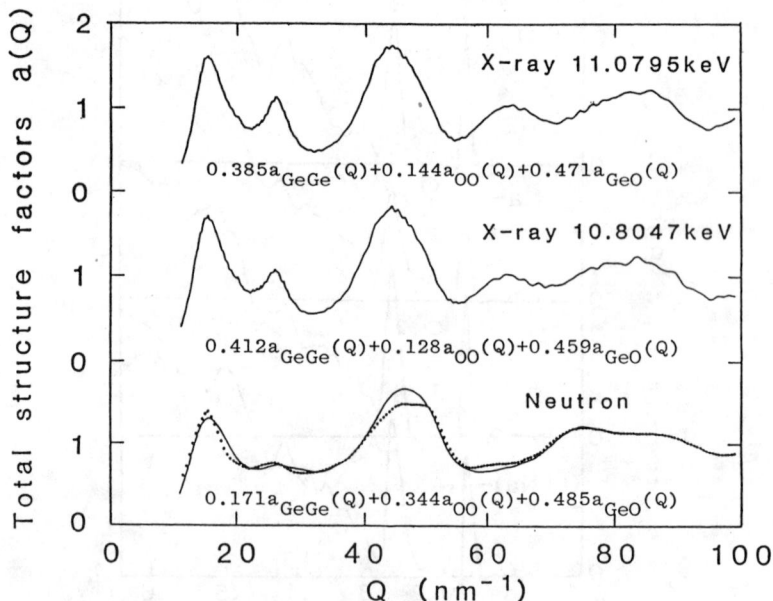

Fig.1.6.8 Three sets of the experimental data of GeO_2 glass obtained by the anomalous X-ray scattering and neutron diffraction measurements[107]. Solid dotts indicate the results of neutron diffraction by Ueno et al[110].

example by determining the partial structural functions of oxide disordered system, although they appear not to be completed for allowing quantitative discussion, mainly due to the experimental difficulties in that stage. For these reasons, the following information is of interest with respect to the partial structural functions newly determined from the anomalous x-ray scattering data at the Ge K absorption edge and conventional neutron diffraction result.

Figure 1.6.8 shows the structure factor currently obtained by neutron diffraction

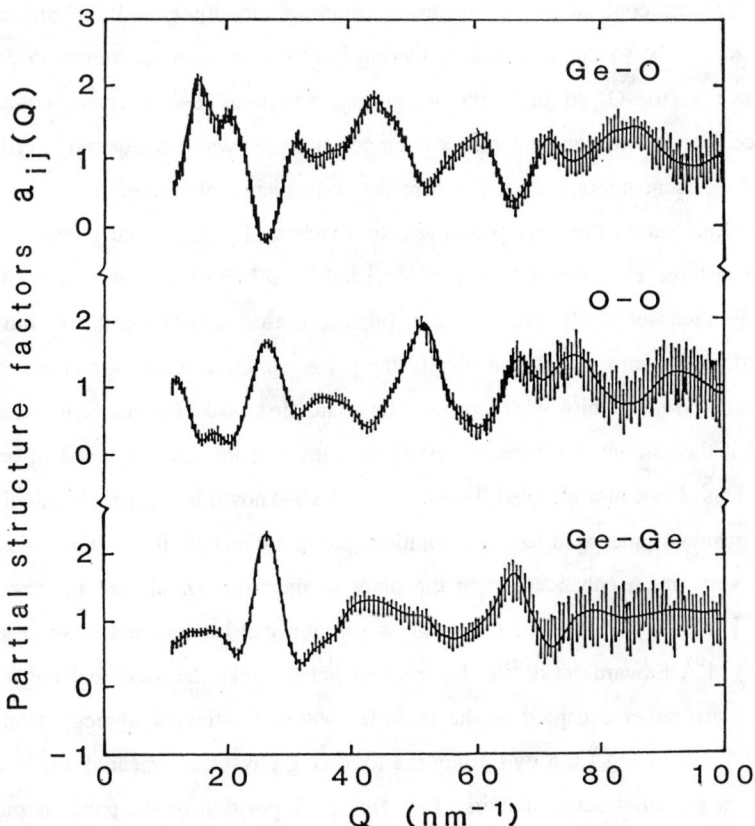

Fig.1.6.9 Three partial structure factors of GeO_2 glass estimated directly from the data of Fig.1.6.8. The vertica line denotes the experimental uncertainties. solid lines denote the results obtained by applying the refining sequence with eqs.(1.6.4)–(1.6.6).

experiment [Waseda et al[107]] together with the structure factors by the anomalous X-ray scattering measurements using two energies [Matusbara et al[108]]. The recent neutron diffraction data are found to agree well with the previous results by neutron diffraction [Lorch[109], Ueno et al[110], Erwin Desa et al[111]]. Such agreement is exemplified in Fig.1.6.8 using the result of Ueno et al[110]. The different peak profiles observed in these three structure factors should be attributed to the differences in weighting factors arising from the X-ray anomalous dispersion effect and neutron scattering amplitude as inserted in this figure. In contrast to the neutron scattering amplitude and the anomalous dispersion terms, the so-called atomic scattering factors of X-rays are known to depend on the wave vector Q, so that the weighting factors of x-rays show a weak Q dependence. For this reason, the values averaged over the wave vector range (10~100 nm^{-1}) of the present interest are given here for convenience of discussion.

Figure 1.6.9 shows the three partial structure factors of a germanate glass estimated directly from three experimental data of Fig.1.6.9 by solving the simultaneous linear equations [Waseda et al[107]]. Their general profiles are essentially similar to those of other disordered systems, i.e. the profiles of the partial structure factors are characterized by oscillations around unity which follow the usual first peak. However, it is readily detected that the estimated values of partial structure factors are dispersed in certain positions. This is not disappointed, because, it is well-known in numerical calculation that the residual uncertainties in solution in simultaneous linear equations are frequently very much enhanced when the pivot of matrix is small [see for example, Stantson[112]] and only a physical meaningless solution can be obtained in some cases [Enderby et al[16], Edwards et al[113]]. The present authors take the view that the results of Fig.1.6.9 are rather accepted as the feasible solution estimated directly from the experimental data of Fig.1.6.8 by finding the following physically meaningful features in the present partial structure factors. The first peak position of the partial structure factors of like atom pairs, O-O and Ge-Ge, is situated at the close vicinity of the first minimum of the unlike atom pair partial structure factor of Ge-O pairs. In addition, the partial structure factor of unlike atom pairs of Ge-O oscillates nearly out of phase with like atom pairs although there are differences in detail. These features are quite similar to the case reported for molten salts [see for example, Edwards et al[113], Mitchell et al[116]]. The iterative method frequently used [Edwards et al[113], Waseda[6]] was also applied. This refining sequence is made so as to satisfy the following physically meaningful conditions that the measured intensity data must always be positive and any atoms do no mutually approach inside the atomic core diameter.

$$c_{Ge} + c_{Ge}^2 (a_{GeGe}(Q) - 1) > 0 \tag{1.6.4}$$

$$c_O + c_O^2 (a_{OO}(Q) - 1) - \frac{[c_{Ge}^2 c_O^2 (a_{GeO}(Q) - 1)^2]}{[c_{Ge} + c_{Ge}^2 (a_{GeGe}(Q) - 1)]} > 0 \tag{1.6.5}$$

$$\int_0^{Q_{max}} Q^2 (a_{ij}(Q) - 1) dQ = - 2\pi^2 \rho_0 \tag{1.6.6}$$

where ρ_0 is the average number density of atoms in a system. This iterative method confirms the results of Fig.1.6.9 clearly explain the detected variations of Fig.1.6.8, corresponding to the differences in weighting factors due to the anomalous X-ray dispersion effect and neutron scattering amplitude, as well as the exclusion of physically meaningless values suggested by eqs.(1.6.4) and (1.6.6). In this process, the partial structure factor values were made by smoothing with 5 points by five repetition was applied, since the general profiles of the structure factors, which are smooth functions without sharp changes of slope near the first peak region, appear to be feasible based on the structural analysis for various disordered systems reported in the past. The resultant partial structure factors of a germanate glass are illustrated in Fig.1.6.9 by solid lines.

The partial radial distribution functions can be calculated by Fourier transformation of the partial structure factors with the density data of 3.60 Mg/m³ and the results are given in **Fig.1.6.10**. The coordination numbers for near neighbor correlations were estimated from the area under the corresponding peaks along the way similar to those employed in various non-crystalline systems and the results are summarized in **Table 1.6.4** together with the values observed in the alpha-quartz type crystal structure. The present partial structural information clearly indicates that each germanium is surrounded by four oxygens with a Ge–O distance of 0.175 nm and the distance of O–O pairs is nearly equal to the edge length of the GeO_4 tetrahedron. It may be added that each oxygen located at the corner of a GeO_4 tetrahedron is likely to be bonded with two germanium and then a network structure formed by these tetrahedra joined at their corners

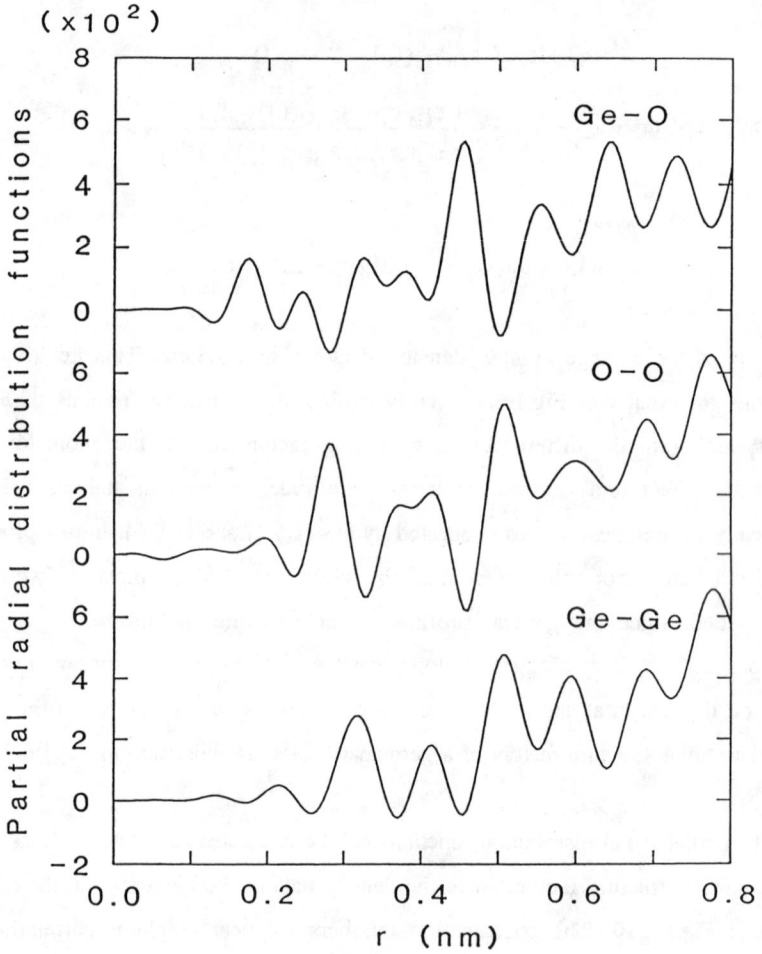

Fig.1.6.10 Three partial radial distribution functions of GeO_2 glass. Density=3.60 Mg/m^3.

Table 1.6.4 Near neighbor correlations of GeO$_2$ glass together with those found in the alpha-quartz crystal sturcture.[107]

	Ge–O		O–O		Ge–Ge	
	r(nm)	N(atom)	r(nm)	N(atom)	r(nm)	N(atom)
glass	0.175	4.0	0.282	6.8	0.318	4.5
crystal	0.1737	2	0.2783	2	0.3153	4
	0.1741	2	0.2805	1		
			0.2860	1		
			0.2902	2		

is quite realistic, because the coordination numbers of Ge–Ge and O–O pairs obtained in the present partial structural functions are about 4 and 6, respectively. Over all agreement between the present results and the values observed in alpha–quartz type structure is also suggested for all distances and coordination numbers of three pairs, as shown in Table 1.6.4, although this comment is, of course, restricted only to the near neighbor correlation. A detailed model structure of a germanate glass has not been established yet. Nevertheless, the partial structural functions newly estimated in this work represent the first effort to make such further quantitative discussion for the structural aspects of a germanate glass.

1.7 Relatively New Techniques for Determining the Local Chemical Environment Structure of Disordered Systems

1.7.1 Fundamentals for the AXS method

The near neighbor atomic correlations of the individual chemical constituents or the local chemical environments around a specific element is required for describing the quantitative structure in multicomponent disordered systems of interest. For this purpose, the Anomalous X-ray Scattering (hereafter to be referred to as AXS) method by applying the so-called anomalous dispersion effect near the absorption edge of the constituent element has recently received much attention [see for example, Ramaseshan and Abraham [115]]. Since the availability of synchrotron radiation has greatly improved both acquisition and quality of the AXS data over those obtainable using conventional x-ray source [see for example, Hosoya et al[116], Bienenstock[7]]. The AXS method will, in the near future, become one of the most powerful tools for structural characterization of disordered systems by obtaining the accurate local ordering data, as well as the isotope substitution technique for neutron diffraction. For this reason, an extended introductory treatise is described in this section for future convenience with respect to the novel application of anomalous x-ray scattering (AXS) to structural characterization of disordered systems.

The absorption edge relevant to the K-shell or L-shell electrons for x-rays is found in respective element at the characteristic energy above which an inner electron can be ejected into the continuum state. Therefore, when the energy of the incident x-ray is close to such an absorption edge of one of the constituent elements, the scattering intensity shows the distinct energy dependence, due to the so-called anomalous dispersion effect and the atomic scattering factor $f(Q,E)$, in practice, should be used in the following form [James[117]].

$$f(Q,E) = f^{\circ}(Q) + f'(E) + i\, f''(E) \tag{1.7.1}$$

where Q is the wave vector and E is the incident x-ray energy, respectively. The $f^{\circ}(Q)$ term is the normal atomic scattering factor for radiation at energy far from any absorption edge and f' and f'' are the real and imaginary components of the anomalous dispersion. These dispersion terms strictly depend upon both the wave vector Q and the energy E. However, the Q-dependence of the anomalous dispersion terms f' and f'', in practice, appears insignificant, because the spatial distribution of inner electrons is considerably smaller than the magnitude of the x-ray wavelength and then the dipole approximation is well accepted.

The energy dependence of f' and f" is illustrated in **Fig.1.7.1** using the iron atom as an example. The real part of f' indicates a sharp negative peak at the absorption edge and its width is typically 50 eV at half maximum. The component of f" exists on either side of the absorption edge, but only the monotonic energy dependence is detected in the lower energy side, although it approaches an approximately constant level for energies a few hundred eV away from the edge.

Since the characteristic absorption edge of various elements are separated by, at least, several hundred eV, the change in scattering intensity can be made with sufficient

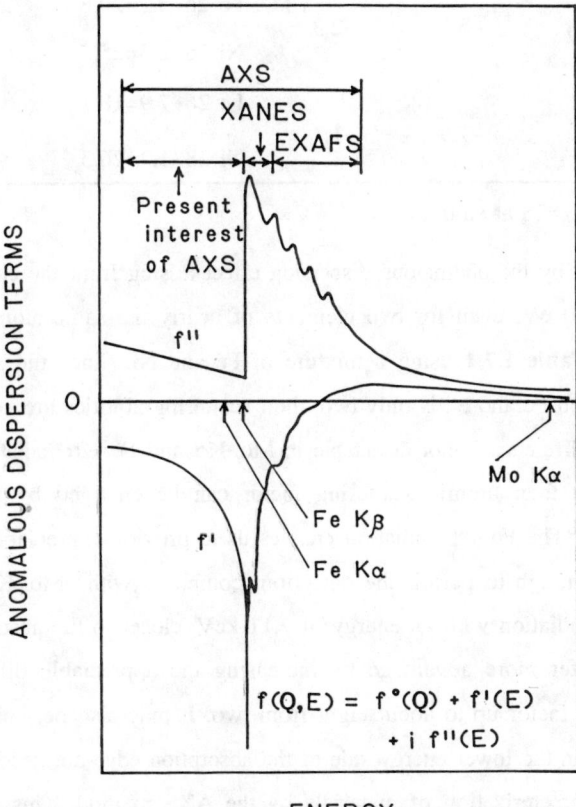

Fig.1.7.1   Schematic diagram for anomalous dispersion factors of Fe near K absorption edge.

Table 1.7.1 Real part of anomalous dispersion factors of Fe and Ni, and total scattering factors for Mo-Kα, Fe-Kα, and Fe-Kβ and energy of 7.11 keV.

Energy(keV)	f'_{Fe}	f'_{Ni}	f° + f'	difference
17.480(Mo-Kα)	0.3	0.3	Fe 26+0.3=26.3 Ni 28+0.3=28.3	≈ 2
6.404(Fe-Kα)	−2.1	−1.3	Fe 26−2.1=23.9 Ni 28−1.3=26.7	≈ 3
7.057(Fe-Kβ)	−5.2	−2.3	Fe 26−5.2=20.8 Ni 28−2.3=25.7	≈ 5
7.110	−7.9	−1.7	Fe 26−7.9=18.1 Ni 28−1.7=26.3	≈ 8

$f°_{Fe}$=26 and $f°_{Ni}$=28 at sinθ/λ=0

atomic sensitivity by the anomalous dispersion effect arising from the energy variation of about 200~300 eV, even for two elements of nearly the same atomic number, as exemplified in **Table 1.7.1** using a mixture of Fe and Ni. Since the atomic number difference between Fe and Ni is only two, their scattering abilities are very similar and thus sufficient difference is not detectable in Mo-Kα and Fe-Kα radiations. However, the difference in their atomic scattering factor can be enlarged by the anomalous dispersion effect. The Fe-Kβ radiation enables us to provide appreciable difference which is large enough to permit the detection, compared with Mo-Kα and Fe-Kα radiations. The radiation with an energy of 7.11 keV, closer to the absorption edge of Fe atom, can offer more advantage by increasing the appreciable difference in the atomic scattering factor up to about eight from two. It may also be worth mentioning that the energies in the lower energy side of the absorption edge are generally employed for structural characterization of materials by the AXS method. This is because the particular near-edge phenomena such as white line, the edge shift and the so-called EXAFS, as well as the significant fluorescent radiation are known to frequently prevent us from obtaining unique structural information from measurements in the higher energy side alone.

The use of AXS for separating the partial functions of disordered systems is described in section 1.2 with the results of a Ni_2Zr glass reported recently by De Lima et al[21]. We describe here another use of AXS for determining the local chemical environment around a specific element in disordered systems.

When the energy of incident x-rays is tuned to the vicinity of the absorption edge of a specific constituent, for example the A element, the detected variation in intensity might be attributed only to the change of the anomalous dispersion terms of f' and f" of the A element. The variation of the imaginary term of f" is also known to be quite small and almost constant in the lower energy side of the absorption edge. Thus the following useful relation is readily obtained for multicomponent disordered systems [Waseda[118]]:

$$\Delta I_A(Q) = I(Q,E_2) - I(Q,E_1) = c_A (f'_A(E_2) - f'_A(E_1))$$

$$\times \sum_{k=1}^{Elements} Re[f_k(Q,E_1) + f_k(Q,E_2)] \int_0^\infty 4\pi r^2 (\rho_{Ak} - \rho_{ok}) \frac{\sin(Q \cdot r)}{Q \cdot r} dr \quad (1.7.2)$$

where $E_{edge} > E_1 > E_2$. $I(Q, E_i)$ is the x-ray scattering intensity after subtracting the average of the square of the atomic scattering factor from the coherent scattering intensity in eu/atom by the usual way. The c_k is the atomic fraction of the k-element and "Re" indicates the real part of the scattering factors, $f_A(Q, E_i)$ and $f_k(Q, E_i)$, $\rho_{Ak}(r)$ the radial density function of a k-element around an A element at a radial distance of r and ρ_{ok} the average number density of k-element in the system. Then, the local chemical environmental structure around an A element, $\rho_A(r)$, can be estimated by Fourier transformation of the quantity of $\Delta I_A(q)$ in eq.(1.7.2).

$$4\pi r^2 \rho_A(r) = 4\pi r^2 \rho_0 + \frac{2r}{\pi} \int_0^\infty \frac{Q \Delta I_A(Q) \sin(Qr)}{c_A [f'_A(E_2) - f'_A(E_1)] W(Q)} dQ \quad (1.7.3)$$

$$W(Q) = \sum_{k=1}^{\text{Elements}} c_j \left[f_k(Q,E_1) + f_k(Q,E_2) \right] \quad (1.7.4)$$

where ρ_o is the overall average number density in the system.

From these equations, an environmental structure around a specific atom can be obtained by measuring the energy dependence of the x-ray scattering intensity with more than two energies in the close vicinity of the absorption edge of a desired element without carrying out the complete separation procedures for partial functions. This energy derivative technique of AXS was first used by Fuoss et al[119] under the name of differential anomalous scattering (DAS) with the results of Ge-Se glasses. Although the basic concept of the energy derivative (or frequency modulated) technique of the AXS method [Hosoya[120], Shevchik[121]] slightly differs from that of the direct AXS method described in section 1.2, the principle of the anomalous scattering is unchanged. The local chemical environmental structure around a specific element in multicomponent systems can be estimated not only by the AXS method but also by the EXAFS measurement. However, we remain convinced that the AXS method is much more straightforward, at least theoretically, because of difficulties of the EXAFS results, such as the chemical and structural dependence of the phase shifts and mean free paths of photoelectrons, and the choice of the final state energy associated with the data analysis [Lee et al[122], Cargill III[123], Teo[124]]. More detailed discussion of the AXS method and its merits and demerits has already been given [see for example, Waseda[6]].

1.7.2 Selected examples obtained by applying the AXS method

The local chemical environmental structures around a specific element in several binary disordered systems have been determined by the energy derivative technique of AXS; for example, Ge-Se [Fuoss et al[119]], Mo-Ni [Aur et al[125]], Mo-Ge [Kortright and Bienenstock[126,127]], Ge-Br [Ludwig Jr et al[128]], Zr-Fe [Krebs et al[129]], Ge-O [Matsubara et al[108]] and Al-Y [Matsubara et al[130]]. Successful application of this relatively new technique to more complex disordered systems are given below.

Studies on numerous aluminum-base amorphous alloys are vigorously carrying out in order to search for new aluminum–base materials whose physical or chemical properties are superior to known aluminum–base crystalline alloys [see for example, Suryanarayana[131]]. On the other hand, the recent experimental evidence of an icosahedral symmetry first reported in rapidly- quenched binary Al-Mn alloys [Shechtman et al[132]] has also received a special interest in last few years, because of their unusual quasi-periodicity implying a five-fold symmetry. This quasi-periodicity has been found in various aluminum–base ternary alloys, such as Al-Mn-Ge and Al-Cu-V ternary alloys. The previous works suggest that there exists a certain chemical ordering, similar to that inferred in their glassy state, in these quasicrystalline alloys [Steinhardt and Ostlund[133], Janot and Dubois[134]]. However, the direct structural evidence is still limited, particularly concerning the structural relation between the icosahedral and amorphous phases. For these reasons, the AXS measurements at the Ge K absorption edge, as well as the ordinary X-ray diffraction measurement, has been made for characterizing the atomic structures of rapidly-quenched $Al_{60}Ge_{30}Mn_{10}$ and

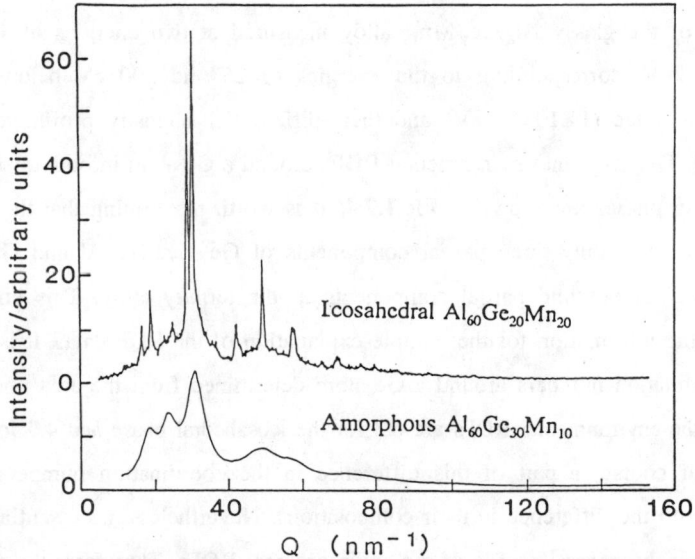

Fig.1.7.2 X-ray diffraction profiles of icosahedral $Al_{60}Ge_{20}Mn_{20}$ and amorphous $Al_{60}Ge_{30}Mn_{10}$ alloys by Mo Kα radiation.

$Al_{60}Ge_{20}Mn_{20}$ alloys which form amorphous and icosahedral phases respectively.

Figure 1.7.2 shows the ordinary x-ray intensity intensity profiles of the two samples [Matsubara et al[135]]. According to the indexing scheme by Bancel et al[136] the sharp peaks observed in the $Al_{60}Ge_{20}Mn_{20}$ alloy were attributed to the icosahedral phase. Although the intensity profile of the $Al_{60}Ge_{30}Mn_{10}$ alloy indicates a diffuse hallow, which implies the non-periodicity in the atomic arrangements, some distinct features are observed in this glassy phase. The main peak is split into three diffuse peaks located at Q=18, 24 and 31 nm^{-1} and the peak with a shoulder at about Q= 50 nm^{-1}, followed by subsequent small and diffuse peaks. Also it was found that the diffuse peaks of the glassy phase and the sharp peaks of the icosahedral phase appear at similar wavevectors. It is imagined from these coincidence that the local structure in a near-neighbor region, similar to the icosahedral chemical ordering, is present even in the glassy phase.

In order to obtain insights into the structural features of these rapidly-quenched ternary Al-Ge-Mn alloys, the AXS technique at the Ge K absorption edge was applied to estimate the environmental RDF around a Ge atom. As an example, the intensity profiles of the glassy $Al_{60}Ge_{30}Mn_{10}$ alloy measured at two energies of 11.0795 and 10.8047 keV, corresponding to the energies of 25 and 300 eV below the Ge K absorption edge (11.1036 keV) and their differential intensity profile are given in **Fig.1.7.3**. The resultant environmental RDFs around a Ge atom in both quasicrystalline and glassy phases are shown in **Fig.1.7.4**. It is worth mentioning that the differential profile contains only three partial components of Ge-Ge, Ge-Al and Ge-Mn pairs instead of six possible partial components in the ternary alloy. This enables us to provide the information for the simple explanation of the RDF data. It is found that the coordination numbers around a Ge atom determined from the area under the first peak in the environmental RDFs are 6.3 for the icosahedral phase and 4.3 for the glassy phase. Of course, a part of this difference in the coordination numbers should be attributed to the difference in their compositions. Nevertheless, no essential difference is found in the general profile of the environmental RDFs. Therefore, the similarity of the local environmental structure around a Ge atom in both icosahedral and glassy phases is first quantitatively confirmed by the present AXS measurements, although

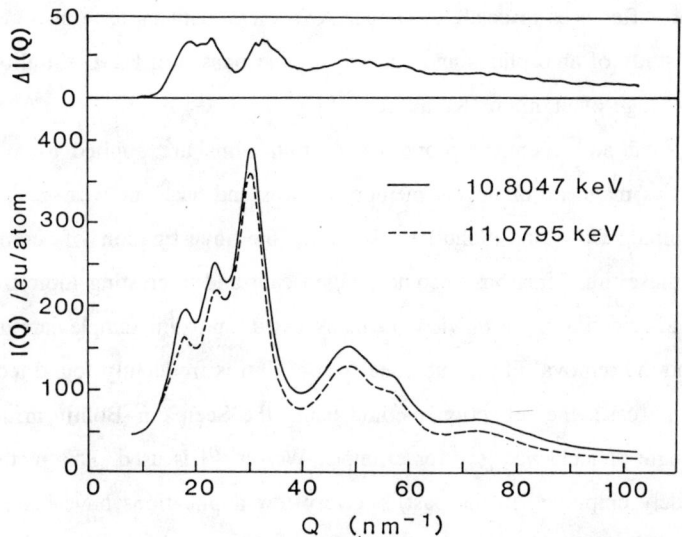

Fig.1.7.3   A differential intensity profile of amorphous $Al_{60}Ge_{30}Mn_{10}$ alloy at the top was obtained from intensities measured at energies of 300 and 25 eV in the bottom below the Ge K absorption edge.

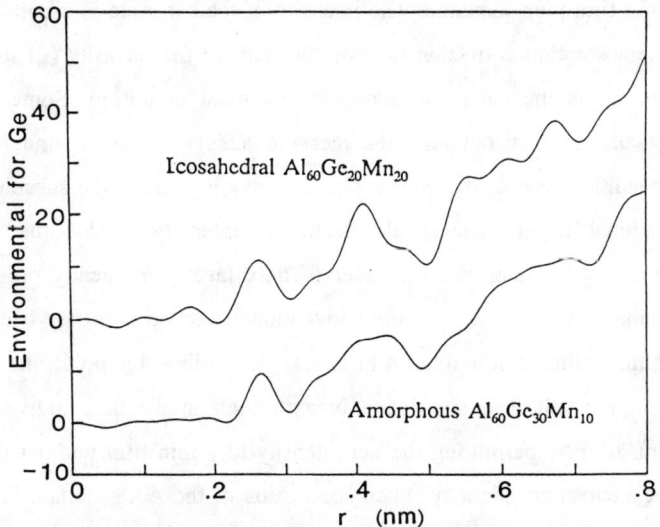

Fig.1.7.4   Environmental RDFs around a Ge atom of icosahedral $Al_{60}Ge_{20}Mn_{20}$ and amorphous $Al_{60}Ge_{30}Mn_{10}$ alloys.

there are some differences in detail. A similar AXS measurements have also been made for structural study of amorphous and quasicrystalline phases of Pd–U–Si alloys using the absorption edge of uranium [Kofalt et al[137]].

Both physical and chemical properties of thin films are applied to many new materials such as magnetic or optical memory devices and heat- or wear-resist coating [see for example, Dow and Schuller[138]]. Thus, the investigation of atomic scale structures of these thin films has become a significant and interesting topic from both engineering and scientific point of view. In many cases, thin film samples are grown on a substrate and the removal of the substrate from a film is frequently found technically difficult. In this case, the reflection method using the Seemann–Bohlin arrangement witha small angle of incidence [see for example, Weiner[139]] is used. This method itself has been widely employed in the past, but very few applications have been actually performed on disordered film mainly due to the following reason.

In the quantitative structural analysis of a thin film grown on a substrate, the scattering intensity from the substrate should be subtracted from the total scattering intensity (I_t) of a film plus substrate. The intensity of the substrate is usually estimated by applying the absorption correction (A_o) by the film to the intensity (I_s) measured in the same material as the substrate under the identical condition. Sometimes it is extremely difficult to carry out both the measurements of I_t and I_s under the exact experimental conditions especially when a single crystal is used as the substrate because of the orientational dependence of the scattering intensities. Also the absorption correction term must be determined precisely. These factors frequently produce some errors for the analysis of the film in the conventional method and they become more serious in a thinner film of sub–micron thickness. Regarding this point, the use of the AXS method may result in a significant breakthrough in the quantitative structural analysis of thin film by permitting the net intensity of a thin film without the subtraction process of a substrate intensity. Such application of the AXS method is described using the results of an oxide film of the Bi–Fe–Ca–O system of 0.3 μm thickness deposited on a Si (100) wafer [Matsubara et al[140]].

It will be useful at this stage to recall the principle of this new method. Let us introduce the anomalous dispersion effect of one of the thin film constituents (Bi in the present case). In the AXS measurements at two energies (E_1 and E_2) in close vicinity of the L_{III} absorption edge (13.418 keV) of Bi, where f of Bi shows a distinct variation, the detected change in intensity should be attributed only to the change in the anomalous dispersion effect of f' of Bi. In this energy region, the atomic scattering factor of the substrate (Si in the present case) is unchanged, because its absorption edge (1.839 keV) is far away from the absorption edge of Bi. This implies that the scattering from a substrate is automatically eliminated by taking the difference ΔI in intensity at two energies (E_1 and E_2) and then the resultant experimental data provides directly the environmental structure around Bi in a thin film without the difficulty in the subtraction process of a substrate intensity which is frequently unsolved by the conventional method.

The two scattering intensities measured at the energies just below the Bi L_{III} absorption edge are shown in **Fig.1.7.5** [Matsubara et al[140]]. The difference between

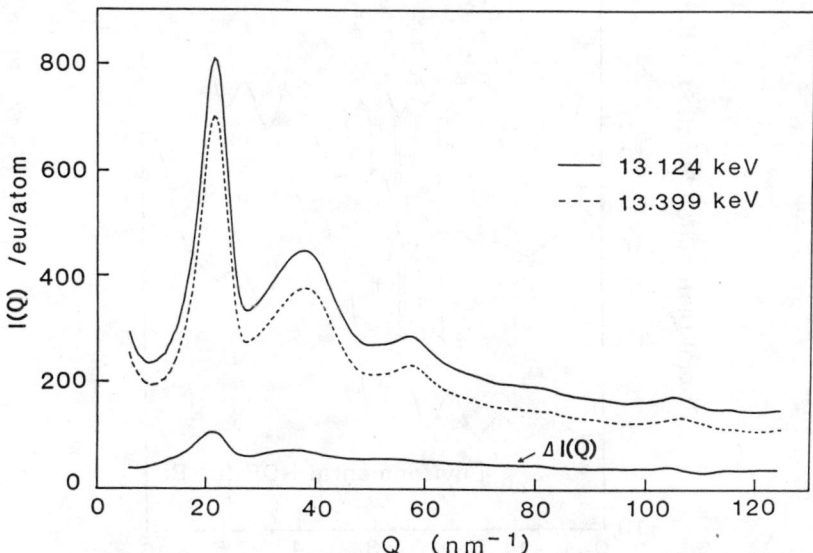

Fig.1.7.5　Intensity profiles of $0.6Fe_2O_3$–$0.25Bi_2O_3$–$0.15CaO$ glass film at 13.124 and 13.399 keV which correspond to the energies of 300 and 25 eV below the Bi L_{III} absorption edge, and a profile of the difference between them.

these two profiles is surprisingly large although the thickness of the film is only 0.3 μm. This unexpectedly large difference is understood by the large anomalous dispersion effect at the L_{III} absorption edge, about four times larger than the values expected at the K-absorption edge. The detected variation corresponds to the local chemical environment around a Bi atom in an amorphous Bi- Fe-Ca-O film. The environmental RDF around Bi estimated from I by Fourier transformation is shown in **Fig.1.7.6** together with the ordinary RDF obtained by the conventional method including the subtraction process of a substrate intensity for comparison. It is again worth mentioning the following points. Since the sample of an amorphous Bi-Fe-Ca-O film contains four elements, there are ten possible atomic pairs; Bi-Bi, Bi-Fe, Bi-Ca, Bi-O, Fe-Fe, Fe-Ca, Fe-O, Ca-Ca, Ca-O, and O-O. The ordinary RDF is reflected on these ten

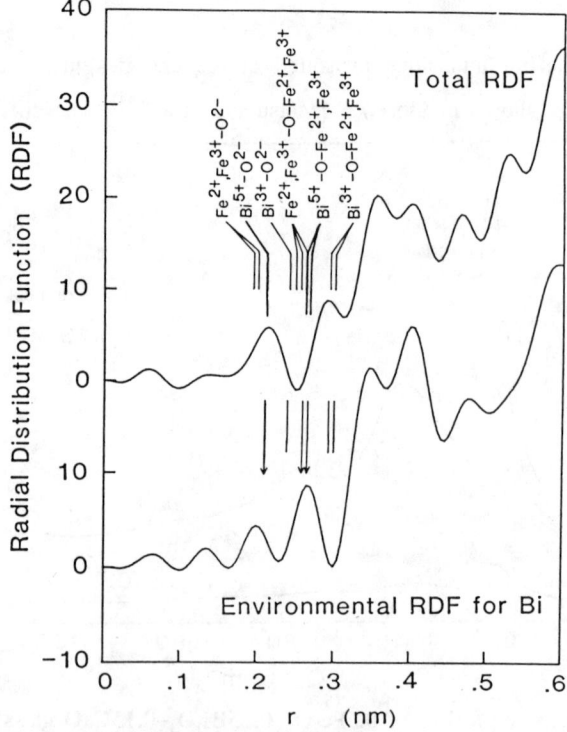

Fig.1.7.6 Environmental RDF around Bi and ordinary RDF of a $0.6Fe_2O_3$–$0.25Bi_2O_3$–$0.15CaO$ glass film.

partial structural information, while the environmental RDF around Bi obtained by the present AXS contains only four partials of Bi–Bi, Bi–Fe, Bi–Ca and Bi–O pairs. Thus, a relatively easy interpretation of the result is allowed in the environmental RDF data of multicomponent systems. This is one of the merits of the energy derivative technique of AXS. A discussion of the detailed structure for an amorphous thin film of the Bi–Fe–Ca–O system is outside the scope of the present article. However, the following essential point is suggested. Referring to the ionic radii values proposed by Shannon and Prewitt[141], the distances of some possible pairs are indicated in Fig.1.7.6. Although (Bi^{3+} or Bi^{5+})–O^{2-}, (Bi^{3+} or Bi^{5+})–O–(Fe^{2+} or Fe^{3+}), or (Bi^{3+} or Bi^{5+})–O–(Bi^{3+} or Bi^{5+}) are possibly included, it is easily seen from the environmental RDF data that the first and second peaks are explained by pairs with Bi^{5+} but not by pairs with Bi^{3+}. Consequently, in this system, Bi atoms become Bi^{5+} instead of Bi^{3+} and it is plausible that a charge transfer could happen in other cations.

The structural investigation of solutions has also been recognized as one of the most important research subjects in both aqueous solution chemistry and hydrometallurgy. Such importance is again emphasized in parallel with recent progress in modern biochemistry. One of the important findings for these studies is the average distribution of water molecules around specific ions which are described by the partial RDF in solutions. This is frequently characterized by the concept of **hydration numbers** in solutions. The energy derivative technique of AXS will again bring about a significant breakthrough in this subject by providing the environmental structure around a specific element as a function of radial distance. For example, the structural sensitive part $F(Q)$ of measured x-ray scattering intensity for the MX_n solution (MX_n – H_2O system) may be expressed by the following ten partial structure factors, $a_{OO}(Q)$, $a_{HH}(Q)$, $a_{OH}(Q)$, $a_{MM}(Q)$, $a_{XX}(Q)$, $a_{MX}(Q)$, $a_{XO}(Q)$, $a_{MO}(Q)$, $a_{MH}(Q)$ and $a_{XH}(Q)$, as already suggested by Soper et al[142]. However, when the energy of the incident x-rays is tuned to close vicinity of the absorption edge of the component M, the observed variation in intensity $F_M(Q)$ contains information of only four partial structure factors, $a_{MM}(Q)$, $a_{MX}(Q)$, $a_{MO}(Q)$ and $a_{MH}(Q)$ and may be given by the environmental structural analysis described here. This idea corresponds to the first-order difference scattering of neutrons with isotope [see for example, Soper et al[142], Enderby and Neilson[18,19]]. For future convenience, the recent results of $ZnCl_2$ solutions [Matsubara and Waseda[143]] are given below.

Intensity profiles of the 0.98 and 2.95 kmol/m³ $ZnCl_2$ aqueous solutions are shown in **Fig.1.7.7**. Solid and dotted lines of each solution describe the scattering intensities at 9.361 and 9.636 keV, respectively, which correspond to energies of 299 and 24 keV below the Zn K-absorption edge (9.660 keV). These profiles essentially show a typical profile of aqueous solutions containing metallic ions which has a broad first peak at about 20 nm^{-1} with a shoulder at about 30 nm^{-1}. In addition to these ordinary features, there is a prepeak. Similar prepeak has been reported in some concentrated aqueous solutions such as $NiCl_2$. The increase in the prepeak intensity and increase in its position with Zn concentrations in Fig.1.7.7 are considered to be closely related to the ionic

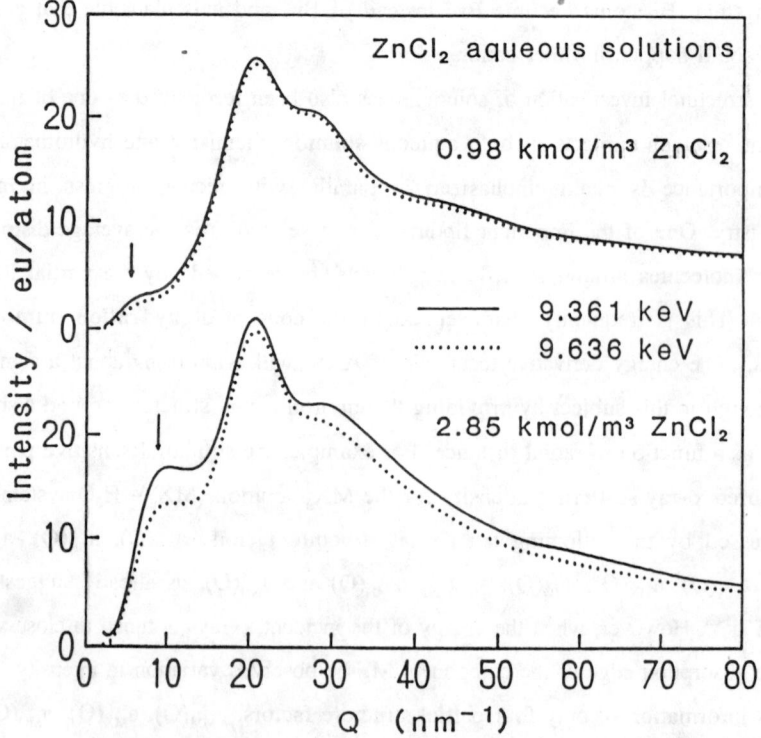

Fig.1.7.7 Intensity profiles measured at 9.361 and 9.636 keV for the aqueous solutions of 0.98 and 2.85 kmol/m³ $ZnCl_2$.

correlations in this solution. However, it has not been known from only these two results whether the zinc ions form any lattice type structure as Neilson et al[144] proposed for the NiCl$_2$ solutions.

The differential intensity functions is shown in **Fig.1.7.8**. For comparison, the ordinary intensity functions are also shown with the dotted lines in the same figure. The profile of Q i(Q) also shows a broad peak at the prepeak position which is indicated with an arrow. This is another evidence to support the assumption on the origin of the prepeak discussed above. Incidentally, it is interesting to notice that the differential

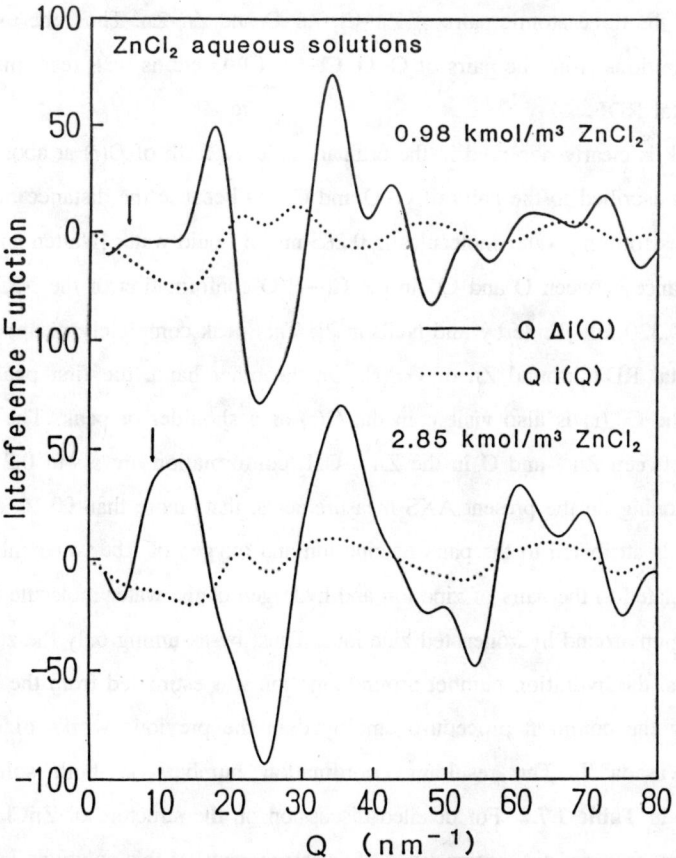

Fig.1.7.8 Intensity functions Qi(Q) (solid lines) and Qi(Q) (dotted lines) for the aqueous solutions of 0.98 and 2.85 kmol/m^3 ZnCl$_2$.

intensity profile of the 2.85 kmol/m^3 ZnCl$_2$ aqueous solution in the bottom of this figure shows the similar profile with the partial structure factor of Ni–Ni determined by Neilson et al[144]. Fourier transformation of the differential intensity function gives the environmental reduced RDF around Zn in this solution. Similarly, the ordinary reduced RDF was computed and the results are given in **Fig.1.7.9**. As it was explained previously, the distribution function around Zn is a sum of the partial distribution functions for the pairs of Zn–O, Zn–H, Zn–Cl and Zn–Zn. Taking accounts of much smaller x–ray scattering factors of H, the environmental RDF around Zn is regarded as a sum of the three atomic pairs of Zn–Cl, Zn–O and Zn–Zn. Thus, the exclusion of other contributions from the pairs of O–O, Cl–Cl, Cl–O etc. is well recognized in this environmental RDF.

The peak is clearly observed in the ordinary reduced RDF of G(r) at about 0.30 nm. This peak is ascribed to the pairs of O–O and Cl$^-$–O because the distance between the oxygen atoms forming water molecules is 0.285 nm in liquid water [Narten and Levy[145]] and the distance between O and Cl$^-$ in the Cl$^-$–H$_2$O conformation of the NiCl$_2$ aqueous solution is 0.320 nm [Enderby and Neilson[18]]. This peak completely disappears in the environmental RDF around Zn of $G_{Zn}(r)$. On the other hand, the first peak at about 0.2 nm in the $G_{Zn}(r)$ is also visible in the G(r) as a shoulder or peak. The suggested distances between Zn^{2+} and O in the Zn^{2+}–OH$_2$ conformation are about 0.2 nm. It is worth mentioning, in the present AXS measurements, that more than 90 % of the total contribution is attributed to the pairs of zinc ion and oxygen of the water molecule and the rest is related to the pairs of zinc ion and hydrogen of the water molecule in the near neighbor region around hydrogenated zinc ions. Thus, by assuming only the zinc ion and oxygen pairs, the hydration number around zinc ion was estimated from the area under the peak by the common procedure employed in the previous works of disordered systems [Waseda[10]]. The resultant coordination numbers in both solutions are summarized in **Table 1.7.2**. For detailed discussion on the structure of ZnCl$_2$ solutions, it is necessary to carry out systematic AXS measurements of this solutions in the wider concentration range from the low concentration to saturation. However, it is out of the scope in the present case. Only some concluding remarks are given here.

Table 1.7.2 Summary of coordination numbers (N) and distances (r) in the concentrated $ZnCl_2$ aqueous solutions.[143]

Molarity kmol/m^3	r (nm)	N (oxygen)
0.98	0.210	5.7
2.85	0.215	6.2

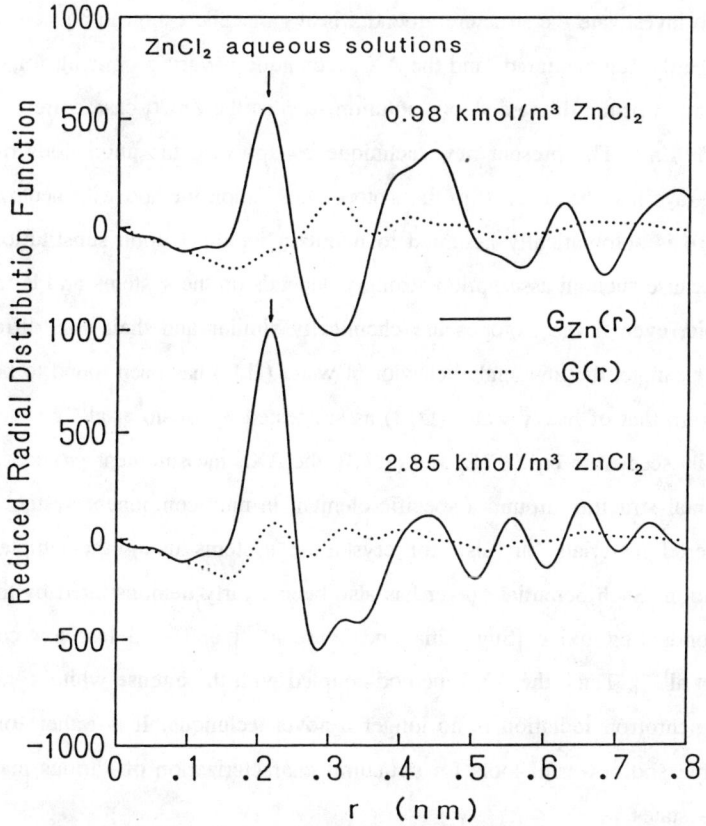

Fig.1.7.9 Environmental reduced radial distribution functions (RDFs) around the Zn^{2+}, $G_{Zn}(r)$, for the aqueous solutions of 0.98 and 2.85 kmol/m^3 $ZnCl_2$. The ordinary reduced RDFs, $G(r)$, are plotted with dotted lines.

The Zn^{2+} is coordinated by 5.8 OH_2 molecules in the solution for concentrations up to at least 1 kmol/m^3 $ZnCl_2$, and the essential configuration around Zn^{2+} does not show the significant difference even in the solution of about 3 kmol/m^3 $ZnCl_2$. It should also be mentioned that the possible interpretation by the formation of certain chemical complexes containing Cl^- around Zn^{2+} is not excluded.

The available experimental results of solutions by the AXS method are still limited for only a small number of compositions [Ludwing Jr et al[146], Matsubara and Waseda[143]]. However, these results suggest that the potential capability of the AXS technique to investigate the structure around a heavy metallic cation in aqueous solutions has been clearly demonstrated and the AXS technique is very promising for structural characterization of the dilute aqueous solution such as the $ZnCl_2$ case corresponding to the 0.6 at% Zn^{2+}. The present new technique by applying the anomalous dispersion effect of x-rays may be superior to the isotope substitution method with neutrons where the structure is automatically assumed to maintain identical upon substitution by the isotope, because such an assumption strongly depends on the systems and is frequently questionable, even if the isotopes are chemically similar and their size difference is small. For example, the low angle behavior of water (H_2O) has been found to be slightly different from that of heavy water (D_2O) as suggested by Bosio et al[147].

As easily seen from eqs.(1.7.2) and (1.7.3), the AXS measurement provides the local environmental structure around a specific element in multicomponent system not only for disordered materials but also for crystalline systems using the simple Fourier transformation. Such potential power has also been clearly demonstrated by the results on superconducting oxide [Sugiyama and Waseda[148]] and sold fast ion conductors [Sakuma et al[149]]. Thus, the AXS method coupled with the intense white x-ray source such as synchrotron radiation is no longer a novel technique. It is rather one of the most reliable and powerful tools for structural characterization of various materials in the various states.

Acknowledgment

A significant part of this article is based on the recent results of a collaboration by

Professors S.Tamaki (Niigata University), S.Takeda (Kyushu University), J.M.Toguri (University of Toronto) and Drs. E.Matsubra and K.Sugiyama (Institute for Advenced Materials Processing, Tohoku University). Their valuable contribution is greatly appreciated. The author wishes to thank the staff of the Photon Factory, National Laboratory for High Energy Physics, Tsukuba, particularly Professors M.Nomura and T.Matsushita, H.Iwasaki, M.Ando, T.Ishikawa (University of Tokyo) and Dr.K.Koyama who made the AXS measurements possible by their services and advice.

REFERENCES

1. Hultgren, R., Orr, R.L., Anderson, P.D. and Kelley, K.K., Selected Values of Thermodynamic Properties of Metals and Alloys, John Wiley, New York,(1963).
2. Richardson, F.D., Physical Chemistry of Melts in Metallurgy, Academic Press, London,(1974).
3. Kubaschewski, O. and Alcock, C.B., Metallurgical Thermochemistry (5th edition), Pergamon Press, Oxford,(1979).
4. Warren, B.E., X-ray Diifraction, Addison-Wesley, Leading, MA, USA,(1969).
5. Klug, H.P., and Alexander, L.E., X-ray Diffraction Procedures for Polycrystalline and Amorphous Materials (2nd edition), John Wiley, New York,(1974).
6. Waseda, Y., Novel Application of Anomalous X-ray Scattering for Structural Characterization of Disordered Materials, Springer-Verlag, Heidelberg,(1984).
7. Bienenstock, A., J. Non-Cryst. Solids, 106, 17(1988).
8. Cole, G.H.A, An Introduction to the Statistical Theory of Simple Dense Fluids, Pergamon Press, Oxford,(1967).
9. Wagner, C.N.J., Liquid Metals, Chemistry and Physics edited by Beer, S.Z., Marcel-Dekker, New York, (1972), p.258.
10. Waseda, Y., The Structure of Non-Crystalline Materials, McGraw-Hill, New York,(1980).
11. Faber, T.E. and Ziman, J.M., Phil. Mag. **11**,153(1965).
12. Sadoc, J.F. and Dixmier, J., Proc. 2nd Inter. Conf. on Rapidly Quenched Metals, Cambridge, Mass. Mater. Sci. Eng. **23**,187(1976).
13. Chipman, D.R., Jennings, L.D., and Giessen, B.C. Bull. Amer. Phys. Soc. **23**,467(1978).
14. Cargill III, G.S. and Tsuei, C.C., Proc. 3rd Inter. Conf. on Rapidly Quenched Metals, The Metals Society (London), No.198, Vol.2, (1978),p.337.
15. Enderby, J.E., Physics of Simple Liquids edited by Temperley, N.H., Rowlinson, J.S. and Rushbrooke, G.S., North-Holland, Amsterdam, (1968),p.612.
16. Enderby, J.E., Egelstaff, P.A. and North, D.M. Phil. Mag. **14**,961(1966).

17. Steeb, S. and Lamparter, P., J. Non-Cryst. Solids, **61/62**, 237(1984).
18. Enderby, J.E. and Neilson, G.W., Advanced in Physics, **29**, 323(1980).
19. Enderby, J.E. and Neilson, G.W., Rep. Progr. Phys. **44**, 593(1981).
20. Lamparter, P., Sperl, W. and Steeb, S., Zeit. für Naturforsch., **37a**, 1223(1982).
21. De Lima, J.C., Tonnerre, J.M. and Raoux, D., J. Non-Cryst. Solids, **106**,38(1988).
22. Laridjani, M. and Sadoc, J.F., J. Non-Cryst. Solids, **106**, 42(1988).
23. Munro, R.G., Phys. Rev. **B25**,5037(1982).
24. Greenfield, A.J., Phys. Rev. Lett., **16**, 6(1966).
25. Ballentine, L.E. and Jones, J.C., J. Phys. **5**, 1831(1973).
26. Evans, R. and Sluckin, T.J., J. Phys. C. **14**, 3137(1981).
27. Ohkoshi, I., Yokoyama, I., Waseda, Y. and Young, W.H. J. Phys. F. **11**,531(1981).
28. MacLaughlin, I.L. and Young, W.H., J. Phys. F. **12**, 245(1982).
29. Matthai, C.C. and March, N.H., Phys. Chem. Liquids, **11**, 207(1982).
30. Egelstaff, P.A., Duffill, C., Rainey, V. and Enderby J.E. Phys. Lett. **21**,286(1966).
31. North, D.M., Enderby, J.E. and Egelstaff, P.A., J. Phys. C. **1**, 784, and 1075(1968).
32. Greenfield, A.J., Wellendorf, J. and Wiser, N., Phys. Rev. **A4**, 1607(1971).
33. Huijben M.J. and van der Lugt, W., Acta Cryst., **A35**, 431(1979).
34. Waseda, Y., Zeit. für Naturforsch., **38a**, 509(1983).
35. Rahman, A., J. Chem. Phys. **42**,3540(1965).
36. Malet, G., Cobos, C., Escande A. and Delord, P., J. Appl. Cryst. **6,** 139(1973).
37. Landau, L.D. and Lifschitz, E.M., Statistical Physics, Pergamon Press, Oxford,(1958).
38. Waseda, Y. and Ueno, S., Sci. Rep. Res. Inst. Tohoku Univ. **34A**, 15(1988).
39. Egami, T. and Srolovitz, D., J. Phys. F. **12,** 2141(1982).
40. Tsu, H. Takano, K. and Shiraishi, Y., Bull. Res. Inst. Min. Met. SENKEN, Tohoku Univ. **41**, 1(1985).

41. Egelstaff, P.A., March, N.H. and McGill, N.C., Can. J. Phys. **52**, 1651(1974).
42. Watabe, M. and Hasegawa, M., Proc. 2nd Inter. Conf. on the Properties of Liquid Metals ed. by Takeuchi, S., Taylor and Francis, London, (1973),p.133.
43. Chihara, J., Proc. 2nd Inter. Conf. on Liquid Metals ed. by Takeuchi, S., Taylor and Francis, London, (1973),p.137.
44. Cusack, S., March, N.H., Parrinello, M. and Tosi, M.P., J. Phys. F. **6**, 749(1976).
45. Ashcroft, N.W. and Langreth, D.C., Phys. Rev. **159**, 500(1967).
46. International Tables for X-ray Crystallography, Vol.IV ed. by Ibers, J.A. and Hamilton, W.C., The Kynoch Press, Birmingham,(1974).
47. Ichimaru, S., Rev. Mod. Phys., **54**, 1017(1982).
48. Utsumi, K. and Ichimaru, S., Phys. Rev. **B22**, 5203(1982).
49. Chihara, J., J. Phys. F. **17**, 295(1987).
50. Takeda, S., Tamaki, S. and Waseda, Y., J. Phys. Soc. Japan, **54**, 2552(1985).
51. Takeda, S., Tamaki, S., Waseda, Y. and Harada, S., J. Phys. Soc. Japan, **55**,184(1986).
52. Pauling, L., The Nature of Chemical Bonds (3rd edition), Cornell Univ. Press, Ithaca, New York,(1960).
53. Silbert, M. and Young, W.H., Phys. Lett., **58A**, 46(1976).
54. Mon, K.K., Ashcroft, N.W. and Chester, G.V., Phys. Rev. **B19**, 5103(1979).
55. Takeda, S. Harada, S., Tamaki, S. and Waseda, Y., J. Phys. Soc. Japan, **58**,3999(1989).
56. Chihara, J., Phys. Rev. A. **40**, 4507(1989).
57. Bernal, J.D., Nature, **183**,141(1959).
58. Cargill III, G.S., Solid State Phys. edited by Ehrenreich, H., Seitz, F. and Turnbull, D. Academic Press, New York, **30**,227(1975).
59. Wagner, C.N.J., Amorphous Metallic Alloys edited by Luborsky, F.E., Butterworths, New York, chapter 5,(1983).
60. Bhatia, A.B. and Thornton, D.E., Phys. Rev. **B2**,3004(1970).
61. Tamaki, S., Ishiguro, T. and Takeda, S., J. Phys. F. **12**, 1613(1982).
62. Matsunaga, S., Ishiguro, T. and Tamaki, S., J. Phys. F. **13**, 587(1983).

63. McAlister, S.P., Crozier, E.D. and Cochran, J.F., J. Phys. C. **6**, 2269(1973).
64. Tamaki, S. and Cusack, N.E., J. Phys. F. **9**, 403(1979).
65. Tamaki, S., Takeda, S., Harada, S., Matsubara, E. and Waseda, Y., J. Phys. Soc. Japan, **55**, 4296(1986).
66. Copestake, A.P., Evans, R. and Telo da Gamma, M.M., J. de Phys. Coll. **41**,145(1980).
67. Copestake, A.P., Evans, R., Ruppersberg, R. and Schirmacher, W. J. Phys. F. **13**,1993(1983).
68. Hafner, J., Pasturel, A. and Hicter, P., J. Phys. F. **14**, 2279(1984).
69. Takeda, S., Harada, S., Tamaki, S., Matsubara, E. and Waseda, Y., J. Phys. Soc. Japan, **56**, 3934(1987).
70. Alblas, B.P., van der Lugt, W, Dijkstra, J., Greertsma, W, and van Dijk, C., J. Phys. F., **13**, 2465(1983).
71. Reijers, H.T.J., van der Lugt, W., van Dijk, C. and Sanboungi, M−L., J. Phys. Cond. Matter, **1**, 5229(1989).
72. Hoshino, K. (1983), J. Phys. F. **13**, 1981.
73. Hesson, J.C., Shimotake, H. and Tralmer, J.M., J. of Metals, **20**, 6(1968).
74. McAlister, S.P., Phil. Mag. **26**, 853(1972).
75. Glazov, V.M., Chizchevskaya, S.N. and Glagoleva, N.N., Liquid Semiconductors, Plenum Press, New York,(1969).
76. Cutler, M., Liquid Semiconductors, Academic Press, New York, (1977).
77. Takeda, S., Tamaki, S. and Waseda, Y., J. Phys. Soc. Japan, **52**, 2062(1983).
78. Waseda, Y. and Tamaki, S., Zeit. für Naturforsch. **30a**, 1665(1975).
79. Wyckoff, R.W.G., Crystal Structure 2 (Inorganic Compounds), Interscience, London,(1964).
80. Takeda, S., Ph. D. Thesis, Hokkaido University,(1983).
81. Takeda, S., Iida, K., Tamaki, S. and Waseda, Y., Bull. Coll. Biomed. Tech. Niigata Univ. **1**, 30(1983).
82. Mozzi, R.L. and Warren, B.E., J. Appl. Cryst. **2**, 164(1969).
83. Wright, A.C. and Leadbetter, A.J., Phys. Chem. Glasses, **17**, 122(1976).

84. Sugiyama, K., Matsubara, E., Suh, I.K., Waseda, Y. and Toguri, J.M., Sci. Rep. Res. Inst. Tohoku Univ., **34a**, 143(1989).
85. Galasso, F.S., Structure and Properties of Inorganic Solids, Pergamon, Oxford,(1970).
86. Busing, W.R. and Levy, H.A., Oak Ridge National Laboratory Report, ORNL-TM-271(1962).
87. Narten, H.A., J. Chem. Phys. **56**, 1905(1972).
88. Waseda, Y. and Toguri, J.M., Materials Science of the Earth's Interior edited by Marumo, F., Terra Sci. Pub. Company, Tokyo, (1990),p.37.
89. Bockris, J.O'M., Mackenzie, J.D. and Kitchner, J.A., Trans. Faraday Soc. **51**,1734(1955).
90. Minami, T. and Mackenzie, J.D., J. Amer. Ceram. Soc. **60**, 232(1977).
91. Wilder Jr, J.A., J. Non-Cryst. Solids, **38/39**, 879(1980).
92. Akao, M., Aoki, H. and Kato, K., J. Mater. Sci. **16**,809(1981).
93. De With, G., Van Dijk, H.J.A., Hattu, H. and Prijs, K., J. Mater. Sci. **16**,1592(1981).
94. Sakka, S. and Kokubo, T., Ceramics Japan, **17**,342(1982).
95. Waseda, Y., Matsubara, E., Sugiyama, K., Suh, I.K., Kawazoe, T., Kasu, O., Ashizuka, M. and Ishida, E., Sci. Rep. Res. Inst. Tohoku Univ. **35A**,19(1990).
96. Tomlinson, J.W., Heynes, M.S.R. and Bockris, J.O'.M., Trans. Faraday Soc., **54**,1822(1958).
97. Kordes, E., Vogel, W. and Feterowsky, R., Zeit. Elektrochem., **57**,282(1953).
98. Kordes, E. and Navarrete, J., Glastechn. Ber. **46**,113(1973).
99. Isozaki, K., Hosono, H., Kokumai, H., Kawazoe, H., Kanazawa, T. and Goshi, Y., J. Mater. Sci. **16**,2318(1981).
100. Sinclair, R.N., Johnson, D.A.G., Dore, J.C., Clark, J.H. and Wright, A.C., Nucl. Instrum. Meth. **117**, 445(1974).
101. Suzuki, K., Bunsen-Gesll. Phys. Chem. **80**, 689(1976).
102. Egami, T., J. Mater. Sci. **13**, 2587(1978).
103. Egami, T., Glassy Metals I edited by Güntherodt, H-J. and Beck, H., Springer-Verlag, Berlin, (1981),p.25.

104. Misawa, M., Price, D.L. and Suzuki, K., J. Non-Cryst. Solids, **37**, 85(1980).
105. North, D.M. and Wagner, C.N.J., J. Appl. Cryst., **2**, 149(1969).
106. Bondot, P., Acta Cryst. A30,470; phys. Stat Sol.(a), **22**,511(1974).
107. Waseda, Y., Sugiyama, K., Matsubara, E. and Harada, K., Mater. Trans. JIM, **31**,421(1990).
108. Matsubara, E., Harada, K., Waseda, Y. and Iwase, M., Zeit. für Naturforsch. **43a**,181(1988).
109. Lorch, E., J. Phys. C. **2**, 229(1969).
110. Ueno, M., Misawa, M. and Suzuki, K., Physica, **12B**, 347(1983).
111. Erwin Desa, J.A., Wright, A.C. and Sinclair, R.N., J. Non-Cryst. Solids, **99**,276(1988).
112. Stantson, R.G., Numerical Methods for Science and Engineering, Pretice-Hall, New York,(1961).
113. Edwards, F.G., Enderby, J.E., Howe, R.A., and Page, D.I., J. Phys. C. **8**,3483(1975).
114. Mitchell, E.W.J., Ponet, P.F.J. and Stewart, R.J., Phil. Mag. **34**,721(1976).
115. Ramaseshan, S. and Abraham, S.C. (Editors), Anomalous Scattering, (Inter. Union of Crystallography), Munksgaard, Copenhagen,(1975).
116. Hosoya, S., Iitaka Y. and Hashizume, H., X-ray Instrumentation for the Photon Factory; Dynamic Analysis of Microstructure in Matter, KTK Sci. Pub. Tokyo(1986).
117. James, R.W., The Optical Principles of the Diffraction of X-rays, G.Bell & Sons, London,(1954).
118. Waseda, Y., ISIJ Inter. **29**, 198(1989).
119. Fuoss, P.H., Eisenberger, P., Warburton, W.K. and Bienenstock, A. Phys. Rev. Lett. **46**,1537(1981).
120. Hosoya, S., Bull. Phys. Soc. Japan, **25**,110 and 288(1970).
121. Shevchik, N.J., Phil. Mag. **35**, 805 and 1289(1977).
122. Lee, P.A. Citrin, P.H., Eisenberger, P. amd Kincaid, B.M. Rev. Mod. Phys. **53**,761(1981).

123. Cargill III, G.S., J. Non-Cryst. Solids, **61/62**, 261(1984).
124. Teo, B.K., EXAFS: Basic Principles and Data Analysis, Springer-Verlag, New York, (1986).
125. Aur, S., Kofalt, D., Waseda, Y., Egami,T., Wang, R., Chen, H.S. and Teo, B.K., Solid State Commun. **48**,111(1983).
126. Kortright, J.B. and Bienenstock, A., J. Non-Cryst. Solids, **61/62**, 273(1984).
127. Kortright, J.B. and Bienenstock, A., Phys. Rev. **B37**, 2909(1988).
128. Ludwing Jr. K.F., Warburton, W.K., Wilson, L. and Bienenstock, A., J. Chem. Phys. **87**, 604 and 613(1987).
129. Krebs, H.V., Biegel, W., Bienenstock, A., Webb, D.J. and Geballe, T.H., Proc. 6th Inter. Conf. on Rapidly Quenched Metals, Elsevier, Amsterdom, (1988),p.163.
130. Matsubara, E., Waseda,Y., Inoue, A., Ohtera, H. and Masumoto, T., Zeit. für Naturforsch. **44a**, 814(1989).
131. Suryanarayana, C., Rapidly Quenched Metals: A Bibliographys, IFI/Plenum Data Company, New York,(1980).
132. Shechtman, D., Blech, I., Graias, D. and Cahn, J.W., Phys. Rev. **53**, 1951(1984).
133. Steinhardt, P.J. and Ostlund, S., The Physics of Quasicrystals, World Scientific, Singapore,(1987).
134. Janot, Ch. and Dubois, J.M., Quasicrystalline Materials, World Scientific, Singapore,(1988).
135. Matsubara, E., Takeda, S. and Waseda, Y., Proc. of the MRS Inter. Meeting on Advanced Materials, Vol.3, (1989),p.291.
136. Bancel, P.A., Heiney, P.A., Stephens, P.W., Goldman, A.I. and Horn, P.M., Phys. Rev. Lett., **54**, 2422(1985).
137. Kofalt, D., Nanao, S., Egami, T., Wong, K.M. and Poon, S.J., Phys. Rev. Lett. **57**, 114(1986).
138. Dow, J.D. and Schuller, I.K., Interfaces, Superlattices and Thin Films, MRS Symp. Proc. Vol.77,(1987).

139. Weiner, K.L., Zeit. für Krist. **123**,315(1966).
140. Matsubara, E., Waseda, Y., Mitera, M. and Masumoto, T., Trans. Japan Inst. Metals, **28**, 697(1988).
141. Shannon, R.D. and Prewitt, C.T., Acta Cryst. **B25**,925(1969).
142. Soper, A.K., Neilson, G.W., Enderby, J.E. and Howe, R.A., J. Phys. C. **10**,1793(1977).
143. Matsubara, Y. and Waseda, Y., J. Phys. Cond. Matter, **1**, 8575(1989).
144. Neilson, G.W., Howe, R.A. and Enderby, J.E., Chem.Phys.Lett. **33**, 245(1975).
145. Narten, A.H. and Levy, H.A., Science **165**, 447(1969).
146. Ludwing Jr, K.F., Warburton, W.K. and Fontaine, A., J. Chem. Phys. **87**,620(1987).
147. Bosio, L., Texeira, J. and Stanley, H.E., Phys. Rev. Letters, **46**,597(1981).
148. Sugiyama, K. and Waseda, Y., Mater. Trans. JIM, **30**,235(1989).
149. Sakuma, T., Sugiyama, K., Matsubara, E. and Waseda, Y., Mater. Trans. JIM, **30**, 365(1989).

Chapter 2

ELECTRON DISTRIBUTION, PHONON STATES AND SHORT-RANGE ORDER

by N.H. March,
Theoretical Chemistry Department,
University of Oxford,
5 South Parks Road,
Oxford OX1 3UB,
U.K.

2.1 Introduction

The purpose of this article is to discuss the nature of both electron and phonon states in relation to atomic structure. In monatomic disordered systems, say liquid Na or amorphous Si, the nuclear-nuclear structure factor $S_{nn}(k) = S(k)$ is accessible via diffraction experiments, as discussed fully by Waseda[1]. Its Fourier transform $g(r)$ is the pair correlation function which is the most basic characterization of the atomic structure of a disordered system. Of course, higher order correlation functions are of considerable importance also, but at the time of writing only limited information about them is accessible via experiment.

2.2 Structure and Forces in Liquid Metals

If we return to the example of liquid metal Na, it has been clear for a long time (see, for example, Johnson and March[2]; Worster and March[3]) that the force law between Na$^+$ ions is very directly dependent on the electronic screening of the ions by the itinerant (3s) electrons which belong to the liquid metal as a whole. This is exemplified very directly in the work of Perrot and March[4]. These authors calculated the pair potential $\phi(r)$ for liquid metal Na near freezing from electron theory, the basic input information being the atomic number $Z = 11$ and the density at the melting temperature. The pair potential $\phi(r)$ obtained in this way is shown in the upper curve at large r of Fig. 2.1.

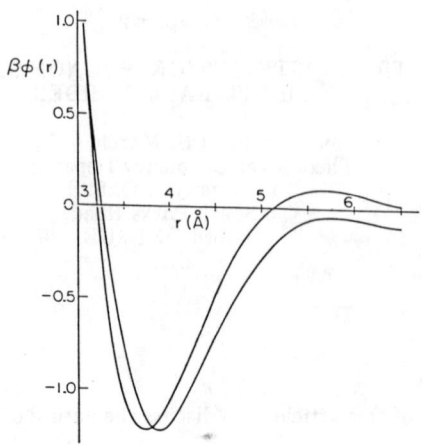

Fig.2.1. Pair potential $\phi(r)$ for liquid Na at melting point T_m in units of $\beta^{-1} = k_B T_m$. Lower curve for larger r is obtained by inversion of structure $S(k)$. Upper curve for large r is electron theory calculation of Perrot and March[4].

But now, the inversion of the structure factor $S(k)$ to yield a pair potential $\phi(r)$ can be carried out using the so-called force equation[5]

$$-\frac{\partial U(r_{12})}{\partial r_1} = -\frac{\partial \phi(r_{12})}{\partial r_1} - \rho \int \frac{\partial \phi(r_{13})}{\partial r_1} \frac{g^{(3)}(r_1,r_2,r_3) dr_3}{g(r_{12})} \qquad (2.1)$$

as emphasized by Johnson and March[2]. However, to do this via liquid structure theories still presents the difficulty that while the potential of mean force $U(r)$ in eqn (2.1) can be extracted from the measured pair function $g(r)$ via

$$g(r) = \exp(-\beta U(r)) : \beta = (k_B T)^{-1}, \qquad (2.2)$$

one must approximate the three-particle correlation function $g^{(3)}$, for reasons already touched on above. For Na, as will be seen below, there is extreme cancellation in the long-range tail of $\phi(r)$ between the potential of mean force term in eqn (2.1) and the three-body piece. Thus, approximations to $g^{(3)}$ can lead to serious errors in the tail of $\phi(r)$. This is the reason why Reatto and coworkers (see Reatto[6] and earlier references there) have bypassed the need to approximate $g^{(3)}$

by invoking computer simulation (see also the review by the present writer, March[7]). This then leads, by inputting the experimental g(r) at melting, to the 'diffraction' pair potential shown in the lower curve at large r of Fig. 2.1. The agreement between these two potentials is quite remarkable, the salient features of the diffraction $\phi(r)$ being predicted accurately by electron theory, as shown in Table 2.1 below.

Table 2.1
Characteristics of diffraction potential compared with electron theory form (after Perrot and March[4])

Positions of turning points and nodes (Å)	Principal minimum	First maximum	Second minimum	First node
Diffraction $\phi(r)$	3.9	5.76	7.44	3.3
Electron theory $\phi(r)$	3.73	5.67	7.37	3.20

This agreement between theory and diffraction $\phi(r)$'s establishes beyond reasonable doubt that, in liquid metal Na, there is an inextricable relation between atomic and electronic structure. This will be the first major focus of the present article.

2.2.1 Dynamical structure factor

So far, one has been concerned with the static structure factor S(k). But in a classical liquid such as argon near the triple point, the atomic dynamics is intimately reflected in the frequency (ω) dependent generalization of S(k), namely the van Hove function or dynamical structure factor S(k, ω), related to S(k) by (see also Appendix.2.3)

$$\int_{-\infty}^{\infty} S(k, \omega) d\omega = S(k). \qquad (2.3)$$

Again, S(k, ω) is accessible to experiment, via inelastic neutron scattering. Indeed the physical interpretation of S(k, ω) is that, in essence, it represents the probability that a neutron incident on the liquid transfers momentum of magnitude $\hbar k$ and energy $\hbar \omega$ to the liquid.

As an example, one can refer to a heavier liquid alkali metal Rb. Here there is a well defined collective mode, as measured by neutron scattering (Copley and Rowe[8]) and simulated on the computer by Rahman[9] with a pair potential having the general features shown in Fig. 2.1. If one assumes that the collective mode, with dispersion relation ω_k, exhausts the sum rule (compare eqn (2.3))

$$\int_{-\infty}^{\infty} \omega^2 S(k,\omega) d\omega = \frac{k_B T k^2}{M} \qquad (2.4)$$

with M the atomic mass, following Feynman's pioneering work on liquid helium four, then the classical analogue of his ansatz (Feynman[10]) for $S(k,\omega)$ is

$$S(k,\omega) = \frac{1}{2} S(k) \left[\delta(\omega - \omega_k) + \delta(\omega + \omega_k) \right]. \qquad (2.5)$$

This is immediately seen to satisfy eqn (2.3). Utilizing eqn (2.4) then connects the dispersion relation ω_k of the collective mode directly with the liquid structure factor through

$$\omega_k^2 = \frac{k^2 k_B T}{MS(k)}. \qquad (2.6)$$

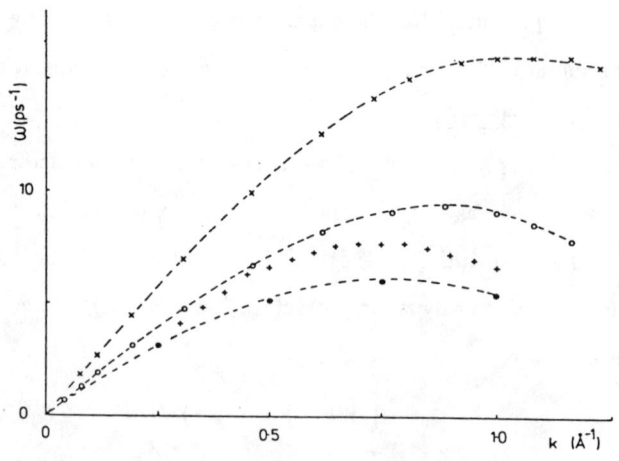

Fig.2.2. Measured collective mode dispersion relation (+) compared with the model prediction (2.6) using experimentally determined liquid Rb structure factor $S(k)$(•). x is same theory but for Na; o for K.

This model prediction (2.6) is compared with the measured ω_k in Fig 2.2. It has the main features of the experiment though, not surprisingly, it cannot match the quantitative accuracy of Rahman's computer simulation.

To summarize, the purpose of the above account has been to emphasize:

(i) That atomic and electronic structure of liquid Na are intimately linked through the fact that electronic screening of the Na$^+$ ions is a crucial element in determining the pair potential $\phi(r)$ shown in Fig 2.1. Naturally, this force law, in turn, generates both S(k) and S(k, ω) for liquid Na near freezing, when employed in a molecular dynamics simulation.

(ii) While S(k) can be used to extract $\phi(r)$ via, essentially, the force equation (2.1), the dynamics embodied in S(k, ω) determines the collective mode, which is so well established by the neutron experiment of Copley and Rowe[8], in the heavier liquid alkali Rb. That is, the 'phonons' (compare Appendix 2.3), are also intimately linked to S(k), via $\phi(r)$, which as seen above, determines S(k, ω).

These two matters (i) and (ii) above provide an important focus of this chapter. However, for what follows, especially on amorphous Si, it will be useful to introduce here, very briefly, the structural description of a two-component liquid. This might be molten NaCl, or BaCl$_2$, or, in the present context, a pure liquid metal like Na.

2.2.2 <u>Pure liquid metal as a two-component system</u>. Liquid metal Na just above its freezing point is a two-component system consisting of Na$^+$ ions (i) and electrons (e). Thus, for the generalization of the ionic pair correlation function $g_{ii}(r) \equiv g(r)$, one needs to add the electron-electron correlation function $g_{ee}(r)$ and the cross-correlation contribution $g_{ie}(r)$. This description was considered within a semiclassical (Thomas-Fermi) framework by Cowan and Kirkwood[11] and in a fully quantum-mechanical theory by March and Tosi[12]. As emphasized by Egelstaff, March and McGill[13], these correlation functions can, at least in principle, be extracted by combining neutron, X-ray and electron diffraction measurements on the same liquid metal. A field of work is now building up in this area (Dobson[14], Johnson et al[15], Tamaki[16]). The corresponding partial structure factors can be set up from the pair correlation functions introduced above; namely $S_{ii}(k) \equiv S(k)$, $S_{ee}(k)$ and $S_{ie}(k)$. These are the quantities accessible through the diffraction experiments.

So far we have dwelt mainly on liquid alkali metals. Let us turn next to a different type of disordered system, dominated by the covalent bond, and typified by amorphous Si. Again, a focal point will be diffraction experiments. For their interpretation, we shall appeal to the above two-component liquid description, for reasons to be explained in section 2.3 below.

2.3 Structure and Chemical Bonding in Amorphous Si

Having introduced the partial structure factors of two-component liquids, we shall now consider diffraction from disordered solids. While neutrons interact with the nuclei, X-rays and electrons, already referred to above in the context though of liquid metals, interact also with the electrons in the amorphous or disordered system. Thus one is exploring then the electron density distribution, as well as nuclear-nuclear pair correlations. For a long time, it proved very difficult to see bonding electrons by diffraction experiments. However, in amorphous Si, with pronounced covalent bonds, it turns out that bonding charge can be clearly seen by studying X-ray and electron diffraction data (Stenhouse et al[17]).

2.3.1 LCAO picture of electron density in amorphous Si.

Stenhouse et al[17] constructed contours of equal electron density for a Si-Si bond as in amorphous Si. They did this by starting from the known, local, tetrahedral bonding, akin to that in crystalline Si. This then prompted the construction of atomic sp³ hybrids (compare Coulson[18]) built from the 3s and 3p wave functions. The linear combination of atomic orbitals (LCAO) method was employed to construct the upper part of Fig. 2.3. In the lower part of this Figure, a superposition of isolated atom densities is

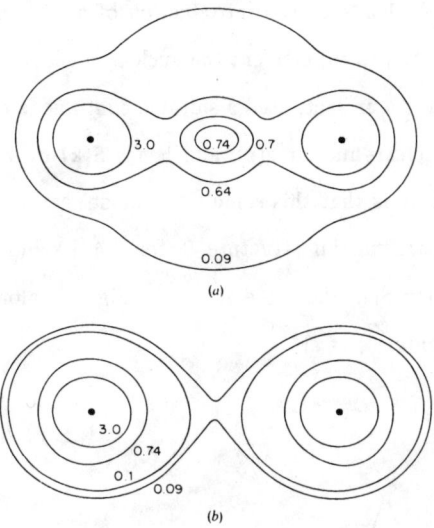

Fig 2.3. Upper part: contours of equal electron density for a Si–Si bond as in amorphous Si, starting from sp³ hybrids built from the 3s and 3p Si atomic wave functions. Linear combination of atomic orbitals (LCAO) method is then employed to construct figure (after Stenhouse et al[17]). Lower part: superposition of isolated atom densities for comparison with bond in upper part of picture.

shown for comparison with the bond density above. The marked difference between the superposition density and that of the LCAO bond density is the absence in the lower part of Fig 2.3 of the closed, almost spherical, contours round the bond centre.

2.3.2 <u>Effects of chemical bonding on diffraction intensities</u>. The LCAO contours indicate that to model the covalent bond one should include, in addition to the superposition of spherical charge clouds on each Si nucleus, a spherical charge distribution at the bond centre, an idea exploited by Phillips[19] and other workers. Whereas Phillips was concerned with dielectric properties and lattice dynamics and was able to model the bond charge as a point charge, the concern here is with modelling diffraction intensities which will evidently require an extended spatial bond charge distribution. Before discussing the form of this, let us emphasize next that this immediately leads to the introduction of a further structural characterization, the bond centre, to supplement the nuclear–nuclear structure factor $S(k)$.

Stenhouse et al[17], in fact, used a suitable continuous random network (CRN) model to get the nuclear–nuclear $g(r)$, and hence $S(k)$ by Fourier transform. The merit of this procedure is that this same CRN model can then be used to calculate the nuclei–bond centre partial structure factor $S_{nb}(k)$ and the bond centre–bond centre structure factor S_{bb}; these are shown in Fig 2.4, along with $S(k)$, all taken from the work of Stenhouse et al[17].

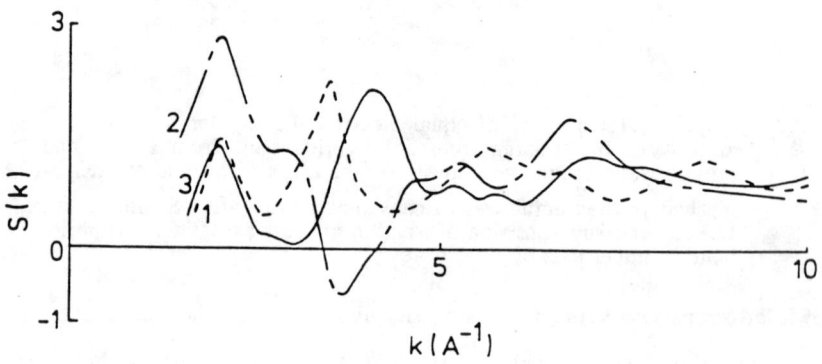

Fig.2.4. Partial structure factors for amorphous Si. (1) $S(k)$ is the nuclear–nuclear structure factor for amorphous Si (2) $S_{nb}(k)$ is nuclei–bond centre structure factor while (3) $S_{bb}(k)$ is that for bond–centre structure.

The essential step in utilizing these three partial structure factors to calculate, say, the X–ray intensity $I_x(k)$, is to replace the bond density in the upper part of Fig 2.3 by a superposition of three spherical 'blobs'; two atomic–like pieces centred on the Si nuclei and a bond charge 'blob'. Stenhouse et al[17] show that little accuracy is thereby sacrificed provided the scattering factors $f_a(k)$ and $f_b(k)$ of the 'atomic' blobs (f_a) and the bond charge (f_b) are chosen as in Fig 2.5.

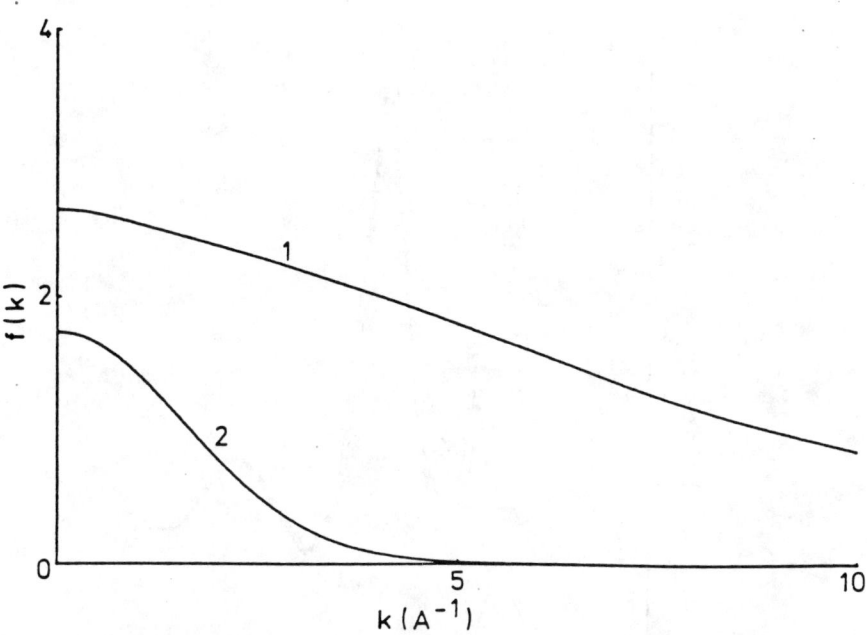

Fig.2.5. Form factors f_a and f_b appearing in intensity eqn (2.7). $f_a(k)$ is upper curve.

Then the scattering is as from a two-component liquid and we can write for the X-ray intensity with suitable normalization[17]

$$I_x(k) = f_c^2(k)S(k) + 2f_b^2(k)S_{bb}(k) + 2(S_{nb}(k) - 1)f_c(k)f_b(k) \tag{2.7}$$

with $f_c = 4f_a$. Similarly, the electron scattering intensity is given by

$$I_e(k) = \text{const } k^{-4}\{[Z - f_c(k)]^2 S(k) + 2f_b(k)^2 S_{bb}(k) - 2[S_{nb}(k) - 1][Z - f_c(k)]f_b(k)\}. \tag{2.8}$$

Equations (2.7) and (2.8) are the basic equations for modelling the diffraction intensities, to allow for the effects of chemical bonding. Reference to Figures 2.6 and 2.7 show how the inclusion of the bond charge brings theory and experiment into substantial agreement for both X-ray and electron diffraction.

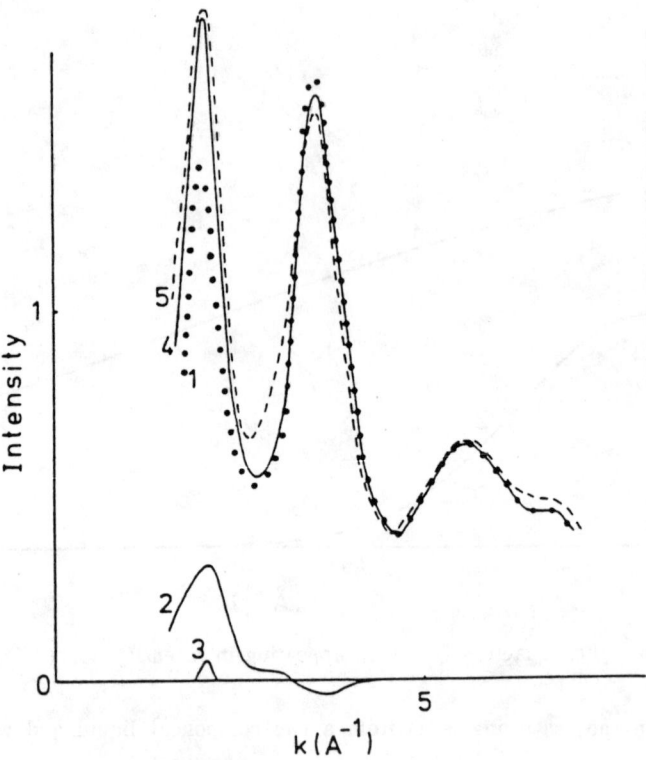

Fig.2.6 X–ray scattering intensity for amorphous silicon, using continuous random network model (1) Contribution from Si–Si correlations (2) From Si–bond centre correlations (3) Bond centre contribution (4) Total predicted intensity (5) Experiment[20] (After Stenhouse et al[17]).

Though, no doubt, some refinements of structure and of electron distribution could now be made, we emphasize in the present context that one can gain confidence in the CRN structural model (with odd membered rings[17]) and in the detailed treatment of chemical bonding. A model built from ordered units turns out to be clearly inferior to the CRN model, while a superposition of spherical blobs on the Si nuclei fails to account for the diffraction intensities. As shown, in fact, in Ref. 17, use of the bond distribution discussed above accounts rather well for the 'forbidden' 222 reflection in crystalline Si. Naturally in the crystal, however,

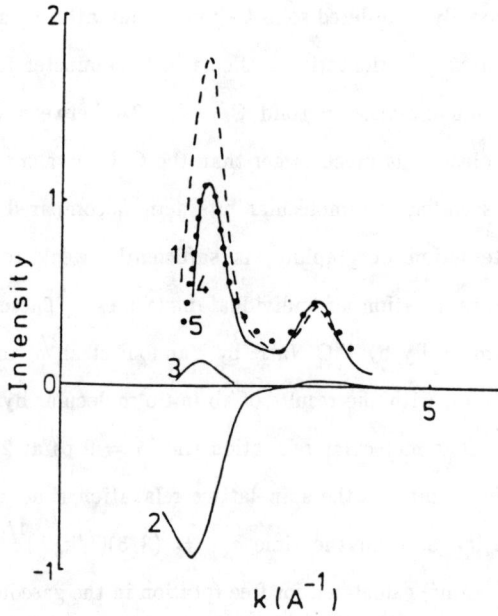

Fig.2.7   Electron scattering intensity for amorphous silicon using continuous random network model. Labelling (1) to (4) as in Figure 2.5 (5) Experiment[21] (After Stenhouse et al[17]).

modern density functional theory, combined with Bloch's theorem, leads to fully first principle electron densities, which can again be used to calculate Bragg reflection intensities[22]. There can be no doubt, therefore, that there is important information about chemical bonding in X-ray and electron diffraction intensities. However, it should be stressed that the important Car–Parrinello[23] technique (see Appendix 2.1) in which density functional theory has been combined with computer simulation of nuclear motions, has been used more recently to give a fully quantitative treatment of amorphous Si[24], and also of various phases of C[25]. We shall supplement this discussion of covalent bonding in relation to diffraction intensities by referring to studies on orientationally disordered solid C_{60} (see also Chapter 8).

2.4 Orientationally disordered solid C_{60}: neutron scattering studies.

In the context of orientational effects in buckminster fullerene, we note first that the minimum distance in solid C_{60} of ~ 3Å between two carbon atoms on adjacent C_{60} molecules is much larger than the C–C covalent bond length (average 1.4Å). The associated intermolecular interaction, compared by Hebard[26] to the interplanar interaction of graphite, is sufficiently weak to permit independent room–temperature rotation of individual molecules. These rotations were first detected experimentally by ^{13}C NMR by Yannoni et al[27] and by Tycko et al[28]. The findings accord with the results of ab initio molecular dynamics simulations of Zhang et al[29]. The molecular relaxation time $\tau = 9$ ps at 283K can be extracted from NMR experiments on the spin–lattice relaxation time. This is only about a factor of three longer than the time $\tau_{FR} = (3/5)(I/k_B T)^{1/2} \doteq 3$ ps (moment of inertia $I = 10^{-43}$ kgm^2) calculated for free rotation in the gaseous phase[30].

As the temperature T decreases, and τ increases, Heiney et al[31], by X–ray diffraction measurements, found evidence for an orientational ordering transition at a temperature ~ 250K. The transition is from a high temperature face–centred–cubic (fcc) phase to a simple–cubic (sc) phase at lower temperature.

The lower–temperature sc structure has been shown to be an orientationally–ordered phase (space group Pa $\bar{3}$) in which the equivalence of the four sites per fcc cell is lost[31]. Above the transition temperature, the molecules have random orientations uncorrelated with their neighbours and are therefore indistinguishable. For T below the transition temperature, the four molecules per fcc cell have distinguishable orientations about the four <111> crystal axes. The driving forces needed to bring about these specific orientations appear to be weak Coulomb interactions[26], since electron–rich interpentagonal double bonds are located opposite electron–deficient pentagonal centres of adjacent molecules.

It is of interest to note that the vibrational couplings responsible for these orientational effects are also revealed in Raman[32] and infrared[33] spectrocopies, which both exhibit anomalies near 250K. It is also highly relevant to record here the subsequent investigations of C_{60} orientational disorder by Blinc et al[34] and by Vashishta and co-workers[35]. The neutron studies of Soper at al[36] must also be referred to in this same context.

Having discussed at some length disorder in chemically bonded solids lacking completely ordered atomic arrangements, we want to summarize some first-principles theoretical work relating to electrons and phonons. The treatment immediately below is based largely on the studies of Stratt and coworkers.[37–40]

2.5 Classical Liquid State Theory Applied to Electron and Phonon Densities of States

Stratt et al (see also the related work of Logan and Wolynes[41] and other references given by Stratt et al) have emphasized the surprising notion that the calculation of what seems intrinsically a quantum-mechanical problem, the electronic band structure of a liquid, can be cast into a problem in classical liquid state theory.

Thus, Chen and Stratt[40] note that, regardless of the quantum-mechanical liquid-state problem of the electronic density of states, the task one has generally to face is that of diagonalizing an N x N matrix, in a liquid of N atoms, whose elements fluctuate from one liquid configuration to another. Then, what has been done in the work of Xu and Stratt[38,39] is to demonstrate that this formidable problem of diagonalizing large random matrices is mathematically identical to that of the structural theory of a classical liquid, but now with artificial internal degrees of freedom. As Chen and Stratt[40] note, the precise details of this mapping differ somewhat from one problem to another. As examples, for s-band calculations, these internal coordinates are basically one-dimensional harmonic oscillators (Stratt and Xu[38]) whereas for p-band systems the oscillators are three dimensional (Xu and Stratt[39]). Nevertheless, the resulting effective liquids turn out to be quite similar.

In fact, the interaction between the atoms in the effective liquid is always the pair potential of the real liquid, plus an additional term coupling the internal oscillators.

As shown in Appendix 2.2, the technique is to consider the averaged Green function (see Edwards[42]; also Edwards and Jones[43]; Edwards and Warner[44]). From this, the density states can be extracted almost immediately.

Of course, it has to be said that the liquid–state problem that results is itself formidable. Nevertheless, an array of approximate techniques exist, the simplest of which is the so–called mean spherical approximation (MSA) : see also Chapter 3 below. This approximation, according to Chen and Stratt[40], embodies the minimum level of sophistication necessary to yield a qualitative understanding of the electronic band structure. Thus, the asysmmetric band shape characteristic of a liquid environment is predicted for s bands (Stratt and Xu[38]): p bands have also been studied subsequently at the MSA level by Chen and Stratt[40]. The approach certainly looks promising, though improved liquid state techniques will probably be needed to transcend results obtained by earlier, admittedly more ad hoc, methods.

We turn from the brief discussion of a route to electronic densities of states to a treatment of phonons.

2.5.1 Phonons from artificial internal degrees of freedom

Xu and Stratt[39] have also utilized the same very general framework to discuss phonons. This follows the interesting simulations of Seeley and Keyes[45] of the 'phonon spectrum' of Lennard–Jones fluids.

To conclude this section it will be useful to summarize the findings of Seeley and Keyes as follows. They apply harmonic normal–mode analysis to Lennard–Jones (LJ) liquids. The configurationally–averaged density of vibrational states is obtained via numerical eigenanalysis of the force–constant matrices appropriate to an ensemble of liquid configurations. These configurations are generated by computer simulation. These workers include the contribution of unstable modes, which appears to play an important role in their analysis. Their findings are somewhat related to the model of Zwanzig[46], to be presented in Chapter 5 below, also based on a normal–mode analysis. As an example of the work of Seeley and

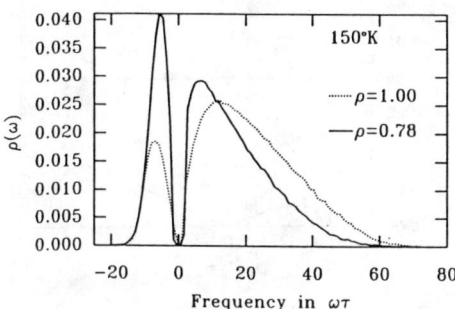

Fig.2.8   Densities of vibrational states for a Lennard–Jones liquid, at T = 150K. Solid line, $\rho\sigma^3 = 0.78$ (normal liquid); dotted line, $\rho\sigma^3 = 1.00$ (super cooled liquid).

Keyes, Fig.2.8 shows the density of vibrational states at 150K, the solid line for the normal liquid and the dotted line for the supercooled state.

Returning now to the study of Xu and Stratt[39] they exploit the analogy between p bands and phonons in liquids. Their 'density of states' is now the distribution of eigenvalues of the dynamical matrix. Fig.2.9 shows the instantaneous normal mode spectrum they thereby obtain. They divide the spectrum into two parts: one with real frequencies and one with imaginary frequencies; these are termed stable and unstable modes respectively by Seeley and Heyes. The reader interested in further technical details is referred to Appendix 2.2 and to the original papers.

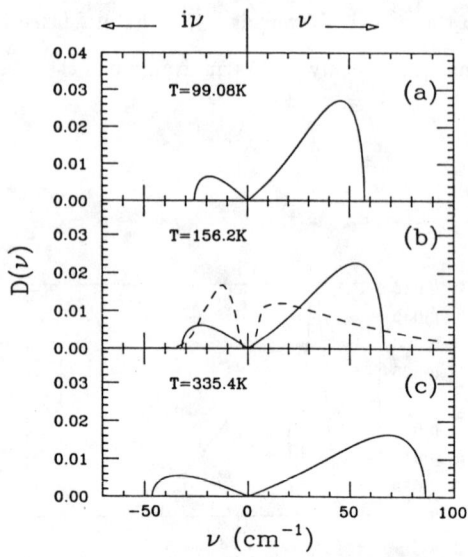

Fig.2.9   Average distribution of instantaneous normal mode frequencies predicted for Ar at a reduced density $\rho\sigma^3 = 0.75$ and for three different temperatures. Exact distribution is shown by a dashed curve : see also Fig.2.8.

2.6   Conclusion

We have discussed admittedly rather primitive, though very basic, properties of electron and phonon states in disordered systems. As emphasized by other workers (eg Stratt,[37]) much additional richness of this area of disordered structures comes from internal degrees of freedom when molecules are involved. Thus, molecules can have different conformations in liquids than their equilibrium structure in the vapour phase. They can therefore exhibit different vibrational, electronic and magnetic resonance spectra. In admittedly extreme examples, such as collision–induced spectra, liquids will absorb light at frequencies either forbidden to, or totally irrelevant to, the individual molecules.

Furthermore, liquids also have some fundamental characteristics, such as their overall electronic behaviour, which are inherently bulk properties (eg plasmons in liquid metals) but which cannot be predicted in their full richness (eg their dispersion and damping) from their radial distribution functions or equations of state. This latter category includes not only such detailed information as the electronic band structure, accessible to experiment, if indirectly, via photoelectron spectroscopy, but also, whether the liquid is electronically conducting, like Na, Rb (or Si!), a semiconductor, or an insulator like Ar.

It is to be hoped that the 'new look' that has come from the profitable introduction of internal degrees of freedom into otherwise featureless atomic liquids will herald not only further understanding of, say liquid transition metals, but of a wide variety of molecular systems in the liquid and amorphous states.

Appendix 2.1

A2.1.1   Molecular dynamics : Car–Parrinello technique

In the text, brief reference was made to the use of the Car–Parrinello technique in treating disordered systems. The purpose of this Appendix is to give a brief summary of this approach, following somewhat the review by Remler and Madden[47].

We note first the wide utility of molecular–dynamics computer simulation of systems for which the interactions can be described by sums of effective potentials, which depend explicitly only on nuclear positions. Problems arise, however, with such an approach in systems in which the nature of the chemical bonding round an atom, or the distribution of electron density, appreciably alters in the course of the atomic motions that are taking place during the simulation. One example of immediate interest in the present context is the interconversion between tetrahedral (sp^3) and graphite–type (sp^2) coordination in amorphous carbon. The difficulties that arise are linked with the itinerant nature of some of the valence electrons; these often cannot be viewed as tied to a particular atomic site or bond and move in a non-trivial manner as they respond to changes in nuclear positions.

In essence the Car-Parrinello[23] technique is an extended molecular dynamics (MD) method in which the electronic degrees of freedom are handled explicitly and which provides for stable atomic dynamics in a tractable computational fashion.

Conceptually, calculations on dynamic systems whose atomic coordinates, as in liquids say, are continually changing, could be performed as follows (cf Remler and Madden[47]):

(1) Start out from a plausible atomic configuration and atomic velocities.
(2) Solve the electronic states for this atomic configuration.
(3) Use the Hellmann-Feynman theorem to calculate the forces on the atoms.
(4) Move the atoms according to the classical equations of motion.
(5) Take the new atomic coordinates, return to stage (2) above etc.

In principle, the above method generates energy-conserving atomic motion on the potential-energy surface appropriate to the adiabatic (Born-Oppenheimer) electronic state. In thermal simulations the system could be equilibrated to the required temperature by scaling of the atomic velocities. Then its structural and dynamical properties could be examined, as in conventional MD. Unfortunately, attempts to carry out such a dynamical scheme prove to be unstable. Quite specifically, the atomic dynamics do not conserve energy unless a very high degree of convergence in the electronic-structure calculations is imposed. If this is not done, the electronic system behaves like a heat sink or source, energy gradually draining from or being added to the atomic system.

A2.1.2   Car-Parrinello method

The Car-Parrinello (CP) method is an extended molecular-dynamics (MD) scheme in which the electronic degrees of freedom are handled explicitly and which provides for stable atomic dynamics. Below, the emphasis, as in the account of Remler and Madden[47], will be concerned with the CP method as a way to simulate the dynamics of atoms in systems not appropriately modelled by the effective –

potential approach already referred to. Payne et al[48] and also Payne[49], have reviewed related methods used for structural calculations (total–energy minimizations).

A2.1.3 The Born–Oppenheimer surface

(a) <u>Simulated annealing</u>

Although the CP technique is a method for dynamics and thus for describing the evolution of the Born–Oppenheimer electronic state, it is convenient to consider at this point step (2) of the scheme above. One wishes to solve the static, energy–minimization problem involved in finding the electronic ground state for the initial atomic configuration. This variational energy minimization may be tackled, like any optimization problem, by a process of simulated annealing, as discussed by Kirkpatrick et al[50]; see also Remler and Madden[47].

(b) <u>Dynamic optimization</u>

The customary approach to optimization problems using simulated annealing has been to employ Monte Carlo techniques. Since, it turns out, this technique is not suited to the electronic–structure problem, Car and Parrinello proposed that the search through electronic configuration space should be implemented with molecular dynamics (MD) equations of motion.

The calculation starts at an arbitrary point in electronic configuration space, where the gradient of the electronic energy with respect to the variables specifying the electronic state is computed. The point in electronic configuration space is then moved according to the MD (and constraint) algorithms; the gradient of the electronic energy playing the role of a force. The process is repeated and the configuration 'rolls down' the energy surface, gaining kinetic energy. As this kinetic energy (KE) indicates the rate of movement through electronic configuration space and is not a physical KE, it is often termed 'fictitious' KE. After some suitable time, the fictitious KE is removed from the system, lowering its fictitious temperature, and the process begins anew. Eventually, the system reaches the minimum electronic energy and the optimization problem is solved.

A2.1.4 Parallel MD

Car and Parrinello[23] combined their fictitious optimization scheme with real MD by running these in parallel. In the same (very short) time step, the ions are moved according to the correct self-consistent forces, while the electronic variables are moved according to the energy gradients and constraints. The result was that the fictitious electronic dynamics moved the electronic configuration appropriate to the new ionic configuration.

A2.1.5 Use of density functional theory

We shall merely note here that above, all details of the electronic-structure method have been ignored, in order to stress the generality of the Car-Parrinello algorithm for extended MD schemes. In practice, the work to date has used electron density functional theory, reviewed for example in the books by Lundqvist and March[51] and by Parr and Yang[52]. In the present context, the reader may refer to the excellent summary by Remler and Madden[47]. Usually, this density functional approach is implemented in practical cases by employing pseudopotentials and plane-wave basis sets. We shall not go into further details here : rather we shall conclude this Appendix by reporting further results on amorphous Si[24], already considered in some detail in the body of this Chapter.

Briefly, the Car-Parrinello method outlined above has been used by these same workers to simulate the nuclear pair correlation function $g(r)$ of amorphous Si. Accurate structural data on tetravalent amorphous Ge is available from neutron scattering experiments. The simulation data on amorphous Si was therefore suitably scaled by Car and Parrinello and compared with this Ge data. The agreement between theory and experiment proved satisfactory. Both sets of data testify to the disappearance of the third crystalline coordination shell. The area under the first peak of the radial distribution function gives a local coordination of four in each case.

Inelastic neutron scattering data is also available to yield the vibrational spectrum of amorphous Si. Car and Parrinello therefore employed their simulation data to obtain the phonon density of states as the Fourier transform of the velocity autocorrleation function. Agreement between theory and experiment was again satisfactory.

Appendix 2.2

<u>Electronic band structure: use of liquid–state theory in diagonalizing an NxN matrix with random elements</u>

Following Xu and Stratt[39], we summarize in this Appendix the way in which the electronic energy bands in a system with a disordered structure, and specifically a liquid, can be turned into an equilibrium liquid statistical mechanical problem.

The central problem addressed is to diagonalize a Hamiltonian matrix (generally set up within the tight–binding framework). To be precise, one is interested in not simply one Hamiltonian matrix but rather that ensemble of matrices generated by the liquid state configurations. This leads to matrix elements which are random variables. The aim therefore is to convert this random–matrix problem of the electronic band structure into a liquid–state statistical mechanical calculation.

A2.2.1 <u>Mapping of energy band states into liquid problem with internal degrees of freedom.</u>

For any random matrix problem, the eigenvalues ϵ_m of any one realization of an NxN matrix $\mathbf{H}$ define the density of states. This means the density of states for a 'frozen ion configuration' $\{\mathbf{R}\}$. Thus one can write for this configuration

$$\rho_{\mathbf{R}}(E) = \frac{1}{N} \sum_{m=1}^{N} \delta\left[E - \epsilon_m(\mathbf{R})\right]. \qquad (A2.2.1)$$

What one will eventually require is the average of this quantity over the Boltzmann distribution of liquid configurations

$$D(E) = <\rho_{\mathbf{R}}(E)>, \qquad (A2.2.2)$$

but the averaging will be postponed.

Instead, let us consider the trace of the Green's function $G(E)$ for a given configuration

$$G_R(E) = \mathrm{Tr}\left[E\,\mathbf{1} - H(R)\right]^{-1}$$
$$= \sum_{m=1}^{N}\left[E - \epsilon_m(R)\right]^{-1} \quad (A2.2.3)$$

where $\mathbf{1}$ denotes the NxN unit matrix. The averaged density of states $D(E)$ in eqn (A1.1.2) can then be derived from the average of $G_R(E)$ in eqn (A2.2.3) by first modifying E by a small negative imaginary part $-i\eta$, and then taking the limit $\eta \to 0$:

$$D(E) = \lim_{\eta \to 0^+}(\pi N)^{-1}\,\mathrm{Im}\!<\!G_R(E{-}i\eta)\!>. \quad (A2.2.4)$$

However, Xu and Stratt now use the fact that the trace of the inverse of a matrix $\mathbf{M}$ can be expressed as a Gaussian integral

$$\mathrm{Tr}\mathbf{M}^{-1} = \frac{\int_{-\infty}^{\infty}dx_1\cdots\int_{-\infty}^{\infty}dx_N\, 2i\sum_{j=1}^{N} x_j^2 \exp\!\left[-i\sum_{j,k} x_j x_k M_{jk}\right]}{\int_{-\infty}^{\infty}dx_1\cdots\int_{-\infty}^{\infty}dx_N \exp\!\left[-i\sum_{j,k} x_j x_k M_{jk}\right]}$$

$$(A2.2.5)$$

Because, in applying eqn (A.2.2.5) to $G_R(E)$ in eqn (A.2.2.3), both numerator and denominator depend on the liquid configuration $\{R\}$, the resulting expression is still not suitable for averaging over the disorder. But, following Xu and Stratt[39], one can remove the denominator with the aid of the replica identity (see, for example, the use in spin glasses; Lubensky[53]) $A/B = \lim_{n \to 0} m^{n-1}$.

If we finally introduce a vector $\mathbf{x}_j = (x_j^{(1)},\ldots x_j^{(n)})$ for each atom, then the final, fully averaged, Green's function can be written[22]

$$<\!G_R(E)\!> = \lim_{n \to 0} 2iN\!\int dx_1\cdots\int dx_N (x_j^{(a)})^2$$
$$\times <\exp\!\left\{-i\sum_{j,k}\mathbf{x}_j\!\cdot\!\mathbf{x}_k\left[E\delta_{jk} - H_{jk}(R)\right]\right\}\!>. \quad (A2.2.6)$$

Here a is any one replica component ($a = 1, .., n$) while j is any one atom.

The conclusion from eqn (A2.2.6) is that, at the cost of introducing one n–dimensional vector per atom, one has transformed the calculation of the eigenvalue spectrum of a random matrix into the (admittedly still difficult!) computation of a classical ensemble average.

While eqn (A2.2.6) is the essential result of this Appendix, we note that Xu and Stratt have applied it to a tight–binding model of a liquid, using the so–called mean spherical approximation (MSA). The elegant result which emerges is that if one wants to get the energy bands in a tight–binding model with a hopping integral t(r) in some liquid with pair potentials u(r), then one should (a) envisage a new liquid with an effective pair potential

$$v_{eff}(r) = u(r) - 2ik_B Tqt(r). \qquad (A2.2.7)$$

Xu and Stratt set out a route to calculate the quantity q appearing in eqn (A2.2.7), but we refer the reader to their paper for the details of this. The second step (b) is then to calculate the excess internal energy of the 'new' liquid described by eqn (A2.2.7) over the liquid with pair potential u(r). This excess energy is then, in fact, the usual self–energy $\Sigma(E)$.

Appendix 2.3

<u>Dispersion relation for collective modes in liquids and amorphous solids</u>

Knipp[54] has pointed out a generalization of the simple classical delta function model in eqn (2.5), to include, in a manner motivated by hydrodynamic considerations, deviations of the specific heat ratio γ from unity. As emphasized by March and Paranjape[55] for numerous s–p liquid metals near freezing γ is in fact quite close to unity.

Knipp's form, valid only in the hydrodynamic regime of long–wavelength and low frequency is

$$S(k,\omega) = \gamma^{-1}S(k)[\tfrac{1}{2}\delta(\omega-\omega(k)) + (\gamma-1)\delta(\omega) + \tfrac{1}{2}\delta(\omega+\omega(k))]. \qquad (A2.3.1)$$

March and Paranjape[56] have now generalized the argument as follows. In Knipp's eqn. (A2.3.1), γ is allowed to become a suitable **k**–dependent quantity, $\gamma(k)$ say, such that $\gamma(0) \equiv \gamma$. Then it turns out that while the zeroth moment of $S(k,\omega)$ is unaltered, one finds from the second moment the result

$$\omega^2(k) = \frac{k^2 k_B T}{M} \frac{\gamma(k)}{S(k)} , \qquad (A2.3.2)$$

relating the dispersion relation $\omega(k)$ of the collective modes to $\gamma(k)$. Evidently, eqn (A2.3.2) reduces to the Feynman–like result (2.6) with the additional approximation $\gamma(k) = 1$.

Using next a generalization of the fluctuation theory result $S(0) = \rho k_B T \kappa_T$ where κ_T is the isothermal compressibility, namely

$$S(k) = \frac{\rho k_B T}{B_T(k)} \qquad (A2.3.3)$$

B_T denoting the isothermal bulk modulus, yields from eqn (A2.3.2) above:

$$\omega^2(k) = \frac{B_T(k) \, \gamma(k) k^2}{M\rho} . \qquad (A2.3.4)$$

This equation is equivalent to that given earlier by March and Paranjape[55] as $k \to 0$, since $B_T \gamma = B_S$. Indeed, it would agree with their result for all k if the k–dependent specific heat ratio $\gamma(k)$ introduced above is taken to be defined by

$$\gamma(k) = \frac{B_S(k)}{B_T(k)} \qquad (A2.3.5)$$

which seems physically plausible.

Unfortunately, at the time of writing, the dispersion relation $\omega(k)$ proposed by March and Paranjape[55] has not been brought into contact with neutron peaks observed in a measured $S(k,\omega)$, as in the experiment, for example, of Copley and Rowe[8] on liquid Rb near freezing. This is because the theory of $\omega(k)$ involves knowledge not only of the three–particle correlation $g^{(3)}$ introduced in eqn (2.1) but also the four–particle function $g^{(4)}$.

References

1) Y. Waseda, see preceding Chapter of this Volume.

2) M.D. Johnson and N.H. March, Phys. Lett., **3**, 313 (1963).

3) J. Worster and N.H. March, J.Phys.Chem.Solids, **25**, 1013 (1964).

4) F. Perrot and N.H. March, Phys. Rev. **A41**, 4521 (1990); see also Phys. Rev. **A42**, 4884 (1990).

5) See, for example, N.H. March. Liquid Metals (Pergamon: Oxford), (1968).

6) L Reatto, Phil. Mag., **A58**, 37 (1988).

7) N.H. March, Canadian J. Physics, **56**, 219 (1987).

8) J.R.D. Copley and J.M. Rowe, Phys. Rev. Lett., **32**, 49 (1974).

9) A. Rahman, Phys. Rev. Lett., **32**, 52 (1974).

10) R.P. Feynman, Phys.Rev., **94**, 262 (1954).

11) R.D. Cowan and J.G. Kirkwood, J.Chem.Phys., **29**, 264 (1958).

12) N.H. March and M.P. Tosi, Annals of Physics, **81**, 41 (1973).

13) P.A. Egelstaff, N.H. March and N.C. McGill, Canadian J.Phys., **52**, 1651 (1974).

14) P.J. Dobson, J.Phys., **C11**, L295 (1978).

15) M.W. Johnson, N.H. March, F. Perrot and A.K. Ray, Phil.Mag.B. 1994, in press.

16) S. Tamaki, Canadian J.Phys., **65**, 286 (1987).

17) B. Stenhouse, P.J. Grout, N.H. March and J. Wenzel, Phil. Mag. **36**, 129 (1977).

18) See, for instance, C.A. Coulson, Valence (Oxford: University Press) 1950.

19) J.C. Phillips, Phys.Rev., **166**, 832 (1968).

20) H. Richter and G. Breitling, Zeits.Naturforsch. **A13**, 988 (1958).

21) S.C. Moss and J.F. Graczyk, Phys.Rev.Lett. **23**, 1167 (1969).

22) See, for example, L.C. Balbás, A. Rubio, J.A. Alonso, N.H. March and G. Borstel, J.Phys.Chem. Solids, **49**, 1013 (1988) and other references there.

23) R. Car and M. Parrinello, Phys.Rev.Lett. **55**, 2471 (1985).

24) R. Car and M. Parrinello. Phys.Rev.Lett. **60**, 204 (1987).

25) G. Galli, R.M. Martin, R. Car and M. Parrinello, Phys.Rev.Lett. **62**, 555, (1989); ibid **63**, 988 (1989).

26) A.F. Hebard, Ann.Rev.Mat.Sci. **23**, 159 (1993).

27) G.S. Yannoni, R.D. Johnson, G. Meijer, D.S. Bethune and J.R. Salem, J.Phys.Chem. **95**, 9 (1991).

28) R. Tycko, R.C. Haddon, G. Dabbagh, S.H. Glarum, D.C. Douglas and A.M. Mujsce, J.Phys.Chem. **95**, 518 (1991).

29) Q.-M. Zhang, J.-Y. Yi and J. Bernholc, Phys.Rev.Lett., **66**, 2633 (1991).

30) R.D. Johnson, G.S. Yannoni, H.C. Dorn, J.R. Salem and D.S. Bethune, Science, **255**, 1235 (1992).

31) P.A. Heiney, J.E. Fischer, A.R. McGhie, W.J. Romanow, A.M. Denenstein, J.P. McCauley, A.B. Smith and D.E. Cox, Phys.Rev.Lett. **66**, 2911 (1991).

32) P.J.M. van Loosdrecht, P.H.M. van Bentum and G. Meijer, Phys.Rev.Lett. **68**, 1176 (1992).

33) L.R. Narasimham, D.N. Stonebach, A.F. Hebard, R.C. Haddon and C.K.N. Palet, Phys.Rev. **B46**, 2591 (1992).

34) R. Blinc, J. Seliser, J. Dolinšek and D. Arčon, Phys.Rev. **B49**, 4993 (1994).

35) J. Yu, L. Bi, R.K. Kalia and P. Vashishta, Phys.Rev. **B49**, 5008 (1994).

36) A.K. Soper, W.I.F. David, D.S. Sivia, T.J.S. Dennis, J.P. Hare and K. Prassides, J.Phys.Cond.Mat. **4**, 6087 (1992).

37) R.M. Stratt, Advances in Chemical Physics, **78**, 1 (1990).

38) R.M. Stratt and B.-C. Xu, Phys.Rev.Lett., **62**, 1675 (1989).

39) B.-C. Xu and R.M. Stratt, J.Chem.Phys., **91**, 5613 (1989); ibid **92**, 1923 (1990).

40) Z. Chen and R.M. Stratt, J.Chem.Phys., **94**, 1426 (1991).

41) D.E. Logan and P.G. Wolynes, J.Chem.Phys., **87**, 7199 (1987).

42) S.F. Edwards, Polymer networks : Eds. A.J. Chompff and S. Newman (Plenum: New York)(1970).

43) S.F. Edwards and R.C. Jones, J.Phys. **A9**, 1595 (1976).

44) S.F. Edwards and M. Warner, J.Phys. **A13**, 381 (1980).

45) G. Seeley and T. Keyes, J.Chem.Phys., **91**, 5581 (1989).

46) R. Zwanzig, J.Chem.Phys. **79**, 4507 (1983).

47) D.K. Remler and P.A. Madden, Molecular Physics, **70**, 921 (1990).

48) M.C. Payne, P.D. Bristowe and J.D. Joannopolous, Molec.Simulation, **4**, 79 (1989).

49) M.C. Payne, J.Phys.Cond.Mat. **1**, 2199 (1989).

50) S. Kirkpatrick, G.D. Gelatt and M.P. Vechi, Science, **220**, 671 (1983).

51) S. Lundqvist and N.H. March, Eds: Theory of the inhomogeneous electron gas (Plenum: New York) (1983).

52) R.G. Parr and W. Yang, Density functional theory of atoms and molecules (Oxford: University Press) 1989.

53) T.C. Lubensky,in Ill–condensed matter: Eds. R. Balian, R. Maynard and G. Toulouse (North–Holland: Amsterdam) 1983.

54) P. Knipp, Phys.Rev. **A41**, 4547 (1990).

55) N.H. March and B.V. Paranjape, Phys.Rev. **A35**, 5285 (1987).

56) N.H. March and B.V. Paranjape, Phys.Rev. **A41**, 4548 (1990).

CHAPTER 3

CONDENSED MATTER - LIQUID TRANSITION METALS AND ALLOYS

Toshio Itami

Faculty of Science, Department of Chemistry, Hokkaido University

Sapporo, 060, Japan

CONTENTS

3.1. Theories of liquids — basic concepts of liquids for discussions of physical properties of liquid transition metals	128
3.1.1. Radial distribution functions and structure factors	128
3.1.2. Models for structure factors and thermodynamic functions — from a view point of theories of liquids and computer simulations	131
(i) The Percus-Yevick ("PY") and hypernetted chain ("HNC") approximations	131
(ii) The Percus-Yevick ("PY") approximation for hard sphere ("HS") liquids	133
(iii) The mean spherical approximations ("MSA")	133
(A) The charged hard spheres in a uniform background ("CHSO")	136
(B) The one component hard spheres with the Yukawa tail ("HSYQ")	138
(C) Binary hard sphere mixtures of equal diameter with the Yukawa tail ("HSYB")	138
3.1.3. Liquid models based on the computer simulations	140
(i) The hard sphere model ("HS")	141
(ii) The one component plasma model ("OCP")	142
3.1.4. Realistic interionic potentials in liquid transition metals	143
(i) The generalized pseudopotential theory ("GPT")	143
(ii) Wills and Harrison type interionic ("WH") potentials for transition metals	145
(iii) Model potential approches and the BS interionic potential	146
(iv) Interionic potentials based on the tight binding Bethe lattice methods	148
(v) Interionic potentials due to the embedded-atom methods	149
(vi) Comments on interionic potentials in liquid transition metals	150

3.1.5. Refined theories of liquids-GB, WCA, RPA, ORPA, RYC, SMSA ZHC and VMHNC — 151
    (i) The thermodynamic variational method due to the Gibbs-Bogoliubov ("GB") inequality — 151
    (ii) The thermodynamic perturbation theories — 152
        (A) The Weeks, Chandler and Anderson ("WCA") perturbation theory — 152
        (B) The random phase approximation ("RPA") — 154
        (C) The optimized random phase approximation ("ORPA") — 155
        (D) The coupling of WCA and JA with ORPA ("WCA-ORPA" and "JA-ORPA") — 156
    (iii) Refined integral equation approaches — 157
        (A) The Ornstein-Zernike equation with the Rogers-Young closure ("RYC") — 157
        (B) The soft core MSA — 158
        (C) The Ornstein-Zernike equation with the Zerah and Hansen closure (ZHC) — 159
        (D) The variational modified HNC equation ("VMHNC") — 159

3.2. The structure of liquid transition metals — from a point of view of theories of liquids — 161
  3.2.1. Fitting of model structure factors to experimental structure factors of liquid transition metals — 162
    (i) Fitting by the HS model — 162
    (ii) Fitting by the OCP model — 162
    (iii) Fitting by the HSYO model — 163
  3.2.2. Analysis of structure of liquid transition metals by refined theories by liquids — 164
    (i) Preliminary attempts — 164
    (ii) Interionic potentials due to the GPT and refined theories of liquids — 164
    (iii) Wills and Harrison ("WH") type interionic potentials and refined theories of liquids — 165
        (A) The structure of liquid transition metals due to the GB thermodynamic variational theory — 165
        (B) The liquid structure factors due to the JA-OPRA — 166
        (C) The structure of liquid transition metals due to the integral equation of liquids — 167

	(iv)	Bretonnet and Silbert ("BS") interionic potentials and refined theories of liquids	168
		(A) The structure of liquid transition metals due to the GB thermodynamic variational method	168
		(B) The structure of liquid metals due to the VMHNC approximation	168
	(v)	Liquid structures from computer simulations — MD, Car-Parrinello dynamics simulations and reverse Monte Carlo method	168
	(vi)	Structures of liquid transition metal alloys	170
3.3.	Thermodynamic properties of liquid transition metals	171
3.3.1.	Thermodynamic data of liquid transition metals	171
	(i)	General trends and parabolic dependences of thermodynamic properties of liquid transition metals	171
	(ii)	Special techniques for experiments of physical properties of liquid transition metals	176
	(iii)	The relation between the hardness of interionic potentials and the compressibilities of liquid transition metals	177
3.3.2.	Theories of thermodynamic properties of liquid transition metals	180
	(i)	Thermodynamic properties based on simple model theories for liquid transition metals — OCP, HS, PYP and CHSO	180
		(A) The one component plasma ("OCP") model	181
		(B) The hard sphere ("HS") model	183
		(C) The Percus-Yevick phonon ("PYP") theory	183
		(D) Comparisons among calculations due to HS, OCP and PYP models for thermodynamic properties of liquid transition metals	185
		(E) The charged hard sphere ("CHSO") model	190
	(ii)	Refined theories of thermodynamic properties of liquid transition metals	190
		(A) Generalized pseudopotential approaches ("GPT")-GB-GPT and JA-ORPA-GPT	190
		(B) The GB thermodynamic variational method and the tight binding approach ("TBA")-GB-TBA	191
		(C) The GB thermodynamic variational method for WH type interionic potentials with HSYO and HS respectively as a reference	193
		(D) The GB thermodynamic variational method for WH interionic potentials with CHSO as a reference	197

3.3.3. Thermodynamic properties of liquid transition metal alloys		199
3.4. Electronic properties of liquid transition metals		201
3.4.1. Electrical resistivity		201
(i)	Experimental data of electrical resistivity ρ^L, of liquid transition metals	201
(ii)	The extended Ziman formula for ρ^L of liquid transition metals	204
(iii)	Numerical calculations based on the extended Ziman formula and attempts to improve it	205
(iv)	The s-d scattering model of Mott	207
(v)	Calculations of electrical resistivity on the s-d scattering model of Mott	207
(vi)	Calculations of electrical resistivity for non-clustered liquids with a model Hamiltonian for s and d states	208
(vii)	Cluster calculations for electrical resistivities of liquid transition metals	208
(viii)	The electrical resistivities of liquid transition metal alloys	210
(ix)	Comments on the electrical resistivity of liquid transition metals	212
3.4.2. Thermoelectric power		212
(i)	Theoretical expressions for the thermoelectric power	214
	(A) The case of extended Ziman formula	215
	(B) The case of s-d scattering model of Mott	215
	(C) The case of Kubo-Greenwood basis	215
(ii)	Numerical calculations of thermoelectric powers	215
3.4.3. Magnetic susceptibility		217
3.4.4. Hall coefficient		220
3.4.5. Density of states		221
3.4.6. The transition metal impurities in liquid noble metals, liquid simple metals and their liquid alloys		223
3.5. Surface properties and atomic transport properties of liquid transition metals		224
3.5.1. Surface properties and the Mn anomaly		224
3.5.2. Atomic transport properties		229
3.6. Concluding remarks		235
Acknowledgement		236
References		237

In this chapter, recent studies of the physical properties of liquid transition metals and alloys are reviewed and discussed from the fundamental point of view. At present no detailed reviews are present for thermodynamic properties of liquid transition metals. Therefore, discussions are given in detail for theoretical studies of thermodynamic properties of liquid transition metals. Rather brief reviews about liquid transition metals are given for electronic properties("EP"), atomic transport properties ("ATP"), surface properties ("SP"), experimental data of thermodynamic properties ("ETP") and experimental aspects of atomic structures (EAS). Readers who are much interested in these branches are expected to consult recent excellent reviews, for example, by Shimoji,[1] Yonezawa[2] and Cusack[3] for EP, Waseda[4] for EP and EAS, Shimoji and Itami[5] for ATP and Iida and Guthrie[6] for ETP, ATP and SP.

In the case of liquid simple (or non-tnansition) metals with s and p valence electrons, the small core approximation is valid in pseudopotentials of ionic potentials exerting on conduction electrons. Under this approximation the separation between core electrons and sp conduction electrons is strictly possible in space and in the energy state. Moreover, the effect of pseudopotential can be considered to be weak. As a result, successful predictions have been given for electronic structures, electron transport properties and thermodynamic properties of liquid simple metals and alloys based on the nearly free electron ("NFE") model. In this NFE model, perturbation theories are coupled with the perturbing "weak" ionic pseudopotentials applied to the uniform electron gas. In fact, the validity of the NFE model for liquid metals can be confirmed rigourously by many theoretical calculations of density of states ("DOS") for liquid simple metals and alloys, which have been performed by the Green function method of Edwards[7] or more refined versions of it.[8-11] Calculated results of density of states are in most cases very close to that of the free electron model near the Fermi energy E_F.

However, in the case of liquid transition metals with partially filled d bands, the small core approximation breaks down between broad sp conduction states and narrow d states because of the presence of s-d hybridization effects; the overlaping and the mixing of d bands with sp bands occur near the Fermi energy E_F. In this situation, the scattering of s electrons in conduction bands by ionic cores becomes considerably strong, which can be described by the resonance scattering scheme. Moreover, some of d electrons may contribute to the electonic conduction in these transition metals and alloys; it is not always easy to determine the number of conduction electrons.[12] Simple "weak" perturbation techniques, which have been powerful for liquid simple metals, are not appropriate to liquid transition metals. To describe the physical properties of liquid transition metals, roles of d electrons must be deeply taken into account. In this respect, essential features described in this chapter for liquid transition metals are frequently common to heavy alkali metals, divalent metals, alkali earth metals and noble metals under high pressures or atmospheric conditions because of the existence of d bands close to the E_F. In this review, discussions are given for liquid 3d, 4d and 5d transition metals with a partially filled d shell. Liquid noble metals and liquid lanthanide metals with a partially filled f shell are inferred only in the important case for the understanding of "typical" liquid transition metals. An excellent review by van Zytveld[13] is available for physical properties of liquid lanthanoid metals.

It is almost impossible for the author to illuminate all studies of a plenty amount in this field, particularly studies in metallurgy. If necessary, readers are expected to consult the textbooks and reviews in metallurgy.[14,15] Many books and reviews in physical and chemical fields are also available as the references for physical properties of liquid transition metals and alloys.[16-20] Recent proceedings [21] of international conference on liquid (and amorphous) metals are valuable for the survey of recent studies in this field.

3.1 Theories of liquids - basic concepts of liquids for discussions of physical properties of liquid transition metals

3.1.1 Radial distribution functions and structure factors

For single component liquids with a number density n, the radial distribution function, g(r), is defined so that $n^2 g(r)$ may represent the joint probability of finding simultaneously one atom at the position $\mathbf{r}'$ and another atom at the position $\mathbf{r}''$; r is the absolute value of vector $\mathbf{r}'' - \mathbf{r}'$. The $n^2 g(r)$ corresponds to the two body distribution function in the statistical mechanics of liquids. The radial distribution function is usually the function of relative distance r only and can be employed to describe sufficiently the structure of single component liquids. This g(r) is related to the structure factor, S(q), as follows:

$$S(q) = 1 + n \int \{g(r) - 1\} \exp(i\mathbf{q}\cdot\mathbf{r}) d^3 r. \qquad (3\text{-}1\text{-}1)$$

In the case of binary liquids, three radial distribution functions, $g_{11}(r)$, $g_{12}(r)$ and $g_{22}(r)$ are required to describe the structure of liquids. The $n^2 x_\alpha x_\beta g_{\alpha\beta}(r)$ represents the joint probability of finding simultaneously one particle of species α at the position r'' and another particle of species β at the position r' in liquid mixtures with the distance $r=|r'' - r'|$. That is, the $n^2 x_\alpha x_\beta g_{\alpha\beta}(r)$ corresponds to the two body distribution function between species α and β in binary liquid mixtures(α, β=1 or 2). The partial structure factor of Ashcroft and Langreth [22] type ("AL type"), $S_{\alpha\beta}(q)$, is related to the radial distribution function between α and β species, $g_{\alpha\beta}(r)$, as follows: [1, 22]

$$S_{\alpha\beta}(q) = \delta_{\alpha\beta} + \sqrt{x_\alpha x_\beta}\{a_{\alpha\beta}(q) - 1\}, \quad (3\text{-}1\text{-}2)$$

$$a_{\alpha\beta}(q) = 1 + n\int \{g_{\alpha\beta}(r) - 1\}\exp(iq.r)d^3r. \quad (3\text{-}1\text{-}3)$$

In these equations, $\delta_{\alpha\beta}$ is the Kronecker delta and X_α is the atomic fraction of species α. The $a_{\alpha\beta}(q)$ is the so called partial structure factor of Faber - Ziman [23] type ("FZ type").

In addition to these AL and FZ types, partial structure factors of Bhatia - Thornton [24] type ("BT type"), $S_{NN}(q)$, $S_{NC}(q)$ and $S_{CC}(q)$, can be also adopted for discussions of structure of binary liquids. The BT type partial structure factors are related to the AL and FZ types as follows: [4]

$$S_{NN}(q) = x_1 S_{11}(q) + 2\sqrt{x_1 x_2}\, S_{12}(q) + x_2 S_{22}(q), \quad (3\text{-}1\text{-}4a)$$

$$= x_1^2 a_{11}(q) + 2x_1 x_2 a_{12}(q) + x_2^2 a_{22}(q). \quad (3\text{-}1\text{-}4b)$$

$$S_{NC}(q) = x_1 x_2\left[\{S_{11}(q) - S_{22}(q)\} + (x_2 - x_1)\frac{S_{12}(q)}{\sqrt{x_1 x_2}}\right], \quad (3\text{-}1\text{-}5a)$$

$$= x_1 x_2 [x_1\{a_{11}(q) - a_{12}(q)\} - x_2\{a_{22}(q) - a_{12}(q)\}]. \quad (3\text{-}1\text{-}5b)$$

$$S_{CC}(q) = x_1 x_2 \{x_2 S_{11}(q) - 2\sqrt{x_1 x_2}\, S_{12}(q) + x_1 S_{22}(q)\}, \quad (3\text{-}1\text{-}6a)$$

$$= x_1 x_2 [1 + x_1 x_2 \{a_{11}(q) + a_{22}(q) - 2a_{12}(q)\}]. \quad (3\text{-}1\text{-}6b)$$

The BT partial structure factors, $S_{NN}(q)$, $S_{NC}(q)$ and $S_{CC}(q)$ correspond to correlation functions of density– density, density– concentration and concentration– concentration respectively.

Under the two body approximation, the structure of binary liquids are described sufficiently by only single set among these three kinds of structure factors $\{S_{11}(q), S_{12}(q), S_{22}(q)\}$, $\{a_{11}(q), a_{12}(q), a_{22}(q)\}$ and $\{S_{NN}(q), S_{NC}(q), S_{CC}(q)\}$.

The S(q) are obtained experimentally from the scattered intensity of the X-ray or neutron diffraction patterns for single component liquids. In the case of binary liquids, the $S_{\alpha\beta}(q)$ or $a_{\alpha\beta}(q)$ for certain composition can be obtained experimentally in principle in terms of total scattered intensities of neutron diffractions of more than three different isotope enrichments ("isotope enrichment technique"). It can be also obtained approximately with the assumption of concentration independent $a_{\alpha\beta}(q)$ by using total scattered intensities of neutrons or X rays for binary liquid mixtures of more than three different concentrations. The information of $S_{CC}(q)$ can be extracted exclusively from the total scattered intensity of neutrons for "zero alloys" whose average scattering length of neutrons is zero. These experimental techniques were described in detail in the book by Waseda.[4]

Contrary to these diffraction experiments for real liquids, computer simulations, such as the molecular dynamics ("MD") and the Monte-Carlo ("MC") method, are also important to investigate the structure of liquids with difficult experimental conditions or to provide ideal experiments for the verification of the correctness of the theory of liquids concerned or the interionic (or interatomic) potential employed in it. The interionic potential must be given beforehand and the periodic boundary condition is applied to remove the effect of the boundary due to a finite (rather small) number of particles of assembly considered.

In the MC method integrals required for evaluations of ensemble average is replaced by the summation of many configurations of assembly of a few hundred (or thousand) particles, which are produced on fast computers by using important sampling techniques.[25] On the other hand, in the MD method a few hundred (or thousand) particles are forced to move on fast computers under the control of Newton's law of motion with a given interatomic potential. The equilibrium properties are calculated as time averages. The non-quilibrium properties are calculated in terms of time correlation functions in the "equilibrium" MD or directly calculated by the "non- equilibrium " MD.[25]

Recently a different kind of simulation technique, the Car-Parrinello method[26], has been developed. This method depends on the density functional theory and the energy of system corresponds to the minimum of energy functional $E(\{\Psi_J, R_K\})$ with the variation of set of one electron wave function, $\{\Psi_j\}$ and the set of position of ions, $\{R_K\}$. The minimization for energy functional $E(\{\Psi_J, R_K\})$ reduces to the problem to solve two Newtonian equations of motion for $\{\Psi_J\}$ and $\{R_K\}$. Therefore it is possible to simulate motions of ions in disordered system only with the knowledge of pseudopotentials of ion for electrons, that is dynamic simulations can be performed without the knowledge of interionic potentials with many approximations.

3.1.2 Models for structure factors and thermodynamic functions - from a view point of theories of liquids and computer simulations

Recently various theoretical methods for determinations of $S(q)$ and $S_{\alpha\beta}(q)$ have been developed extensively. For example, they are given by solving integral equations of liquids,[1,3,4,27,38] such as, the Born-Green ("BG"), the Percus-Yevick ("PY") and the hypernetted chain ("HNC") equations, the mean spherical approximation ("MSA"), the Ornstein-Zernike ("OZ") equation with the Rogers-Young closure ("RYC"),[28] soft core mean spherical approximation ("SMSA"),[29] the OZ equation with the Zerah and Hansen closure("ZHC").[30,31] In addition to these integral equation approaches, the variational method of Gibbs-Bogoliubov ("GB") [32], thermodynamic perturbation theories and the variational modified HNC equation ("VMHNC")[33] can be also employed. One series of thermodynamic perturbation theory was started by Weeks, Chandler and Anderson ("WCA"),[34,35] followed by the random phase approximation ("RPA"),[36,37] and the optimized random phase approximation ("ORPA"),[36,37] as described in Sec. 3.1.5.

In these liquid theories, the number density, n, and the interionic potential, $\phi(r)$, must be given beforehand as input informations. In this section discussions are given only for typical models of liquids employed for discussions of liquid transition metals; in these models artifitial interionic potentials are assumed. More sophisticated theories of liquids such as, GB, WCA, JA, ORPA, RYC, SMSA, ZHC, VMHNC are described in Sec. 3.1.5, for which realistic interionic potentials are employed.

(i) The Percus-Yevick ("PY") and hypernetted chain ("HNC") approximations

The PY and HNC equations correspond respectively to one special case of the well known Ornstein - Zernike ("OZ") equation. The OZ equation itself is strictly correct for liquids with pair interactions because it can be derived from the exact cluster expansion of radial distribution functions. It is written explicitly as follows:

$$h(r_{12}) = C(r_{12}) + n \int C(r_{13}) h(r_{23}) d^3 r_3. \qquad (3\text{-}1\text{-}7)$$

In this equation $h(r_{12})$ is the total correlation function, that is $g(r_{12})-1$ and $C(r_{12})$ is the direct correlation function. The r_{ij} indicates the distance between the position of particle i and that of particle j.

The OZ equation (3-1-7) proves to be the same form as the HNC equation if following form of C(r) is adopted,

$$C(r) = h(r) - \ln g(r) - \frac{\phi(r)}{k_B T}, \qquad (3\text{-}1\text{-}8a)$$

or equivalently

$$g(r) = \exp\left\{-\frac{\phi(r)}{k_B T}\right\} \exp\{h(r)-c(r)\}, \quad \text{for HNC}, \quad (3\text{-}1\text{-}8b)$$

where k_B is the Boltsmann's constant and T is the absolute temperature.
In the case of PY equation, the following form of C(r) is adopted in the OZ equation,

$$C(r) = g(r)\left[1 - \exp\left\{\frac{\phi(r)}{k_B T}\right\}\right]. \qquad (3\text{-}1\text{-}9a)$$

This equation can be written also in a different form as

$$g(r) = \exp\left\{-\frac{\phi(r)}{k_B T}\right\}(1+h(r)-C(r)), \quad \text{for PY}. \qquad (3\text{-}1\text{-}9b)$$

Linearized limit of Eqn. (3-1-8b) ("HNC") corresponds to the PY case, Eqn. (3-1-9b). It is known that HNC well suites to liquids with long range potentials, such as coulombic potentials and PY to liquids with harshly repulsive potentials, such as hard sphere potentials.[38]

The closure relations for the HNC and PY approximations, (3-1-8a)–(3-1-9b), can be rewritten in terms of the Bridge function, $\overline{B}(r)$, which represents the contribution of "elementary graph" in the terminology of diagramatic analysis of density expansion of radial distribution function, g(r). The exact expression for this density expansion of g(r) is written as follows:

$$g(r) = \exp\left(-\frac{\phi(r)}{k_B T} + N(r) - \overline{B}(r)\right), \qquad (3\text{-}1\text{-}10)$$

where N(r) is the contribution of "nodal graph" and can be related to C(r) and h(r) as N(r)=h(r) - C(r). Then, the exact closure relation is written formally as

$$C(r) = h(r) - \{\ln y(r) + \overline{B}(r)\}, \qquad (3\text{-}1\text{-}11)$$

where y(r) is defined as $y(r) \equiv g(r) \exp(\phi(r)/k_B T)$. The closure relations for $\overline{B}(r)$ of the HNC and PY approximations can be written respectively as follows:

$$\overline{B}(r)=0. \qquad \text{for HNC,} \quad (3\text{-}1\text{-}12)$$

$$\overline{B}(r)=y(r) - \ln y(r) - 1. \qquad \text{for PY.} \quad (3\text{-}1\text{-}13)$$

(ii) The Percus-Yevick ("PY") approximation for hard sphere ("HS") liquids

The simplest assumption for interionic potentials is the hard sphere potential. Particles with impenetrable diameters behave like free particles except for complete repulsions at the moment of contact, $r=\sigma$ (σ: the hard sphere diameter), that is

$$\phi(r) = 0 \qquad \text{for } r>\sigma, \qquad (3\text{-}1\text{-}14a)$$
$$= +\infty \qquad \text{for } r<\sigma. \qquad (3\text{-}1\text{-}14b)$$

The importance of this hard sphere potential is derived from the fact that the structure of liquids is determined primarily by the repulsive part of interatomic potentials.[34,39] Moreover, the "exact" analytic solution of the PY equation has been obtained in the case of this hard sphere potential for single component liquids [40,41] and liquid mixtures.[42] For the hard sphere potential, the pressure predicted by the PY equation is in better agreement with results of MD for hard spheres compared with calculated results by the BG and HNC equation.

The "exact" analytic solution of the PY equation for hard spheres can be written as follows:

$$C(r) = a + b \left(\frac{r}{\sigma}\right) + c \left(\frac{r}{\sigma}\right)^3 \qquad \text{for } r>\sigma, \quad (3\text{-}1\text{-}15a)$$
$$= 0 \qquad \text{for } r<\sigma. \quad (3\text{-}1\text{-}15b)$$

Coefficients a, b and c are given as a function of packing fraction ξ [1], which is defined as

$$\xi = \frac{1}{6}\pi n \sigma^3. \qquad (3\text{-}1\text{-}16)$$

The analytic form of structure factor $S(q)$ can be obtained by inserting the Fourier transform of $C(r)$ (Eqns. (3-1-15a) and (3-1-15b)) into following equation,

$$S(q) = \frac{1}{1 - nC(q)}, \qquad (3\text{-}1\text{-}17)$$

which corresponds to the Fourier transform of the OZ equation (3-1-7) and $C(q)$ is the

Fourier transform of C(r).

The application of this HS structure factor to predictions of structure of liquid metals was successfully made by Ashcroft and Leckner,[43] who found that the calculated S(q) with $\xi=0.45$ well reproduces the experimental S(q) of many liquid simple metals at the melting point.

The OZ equation for binary liquids takes the form as

$$h_{\alpha\beta}(r) = C_{\alpha\beta}(r) + \sum_\gamma n_\gamma \int C_{\alpha\gamma}(r') h_{\gamma\beta}(|r-r'|) d^3r', \quad (3\text{-}1\text{-}18)$$

where $h_{\alpha\beta}(r)$ ($\equiv g_{\alpha\beta}(r)-1$) and $C_{\alpha\beta}(r)$ indicates the total and direct correlation functions between α and β species pair respectively; n_γ indicates the number density of species γ. Note that Eqn. (3-1-18) is written as the representation of three independent integral equations for cases $(\alpha=1,\beta=1)$, $(\alpha=1,\beta=2)$ and $(\alpha=2,\beta=2)$.

The Fourier transform of the OZ equation (3-1-18) leads to following equation,

$$S_{\alpha\beta}(q) = \frac{\{\delta_{\alpha\beta} - (-1)^{\alpha+\beta}\sqrt{n_\alpha n_\beta}\, C_{\alpha\beta}(q)\}}{D(q)}, \quad (3\text{-}1\text{-}19)$$

with

$$D(q) = \{1 - n_1 C_{11}(q)\}\{1 - n_2 C_{22}(q)\} - n_1 n_2 C_{12}(q)^2. \quad (3\text{-}1\text{-}20)$$

In these equations, $C_{\alpha\beta}(q)$ is the Fourier transform of direct correlation function $C_{\alpha\beta}(r)$.

The PY equation for binary mixtures corresponds to the OZ equation (Eqn. (3-1-18)) with the "PY closure" described as follows:

$$C_{\alpha\beta}(r) = g_{\alpha\beta}(r)\left[1 - \exp\left\{\frac{\phi_{\alpha\beta}(r)}{k_B T}\right\}\right]. \quad (3\text{-}1\text{-}21)$$

As already described, the PY equation for hard sphere mixtures has been solved exactly by Lebowitz;[42] the explicit form of $C_{\alpha\beta}(r)$ (analogous to Eqns. (3-1-15a) and (3-1-15b)) and $C_{\alpha\beta}(q)$ are given, for example, in Shimoji,[1] Waseda,[4] Ashcroft and Langreth[22] and Enderby and North.[44] Applications of this "exact" analytic solution of hard sphere mixtures were performed successfully to predict the structure of liquid alloys by Ashcroft and Langreth[22] and Enderby and North.[44]

The PY equation for hard sphere mixtures corresponds to the case

$$C_{\alpha\beta}(r) = 0 \quad \text{for} \quad r > \sigma_{\alpha\beta}, \qquad (3\text{-}1\text{-}22a)$$

$$h_{\alpha\beta}(r) = -1 \quad \text{for} \quad r < \sigma_{\alpha\beta}, \qquad (3\text{-}1\text{-}22b)$$

in the OZ equation; $\sigma_{\alpha\alpha}$ (or $\sigma_{\beta\beta}$) indicates the hard sphere diameter of species α (or β) and $\sigma_{\alpha\beta} = (\sigma_{\alpha\alpha} + \sigma_{\beta\beta})/2$. This indicates that the PY equation corresponds to one extreme case of mean spherical approximation (zero limit of attractve part of potential), as desribed in the next part.

The expressions of the Helmholtz free energy and the entropy for hard spheres are concisely summarized in Refs.1 and 3 both for single component and binary cases.

(iii) The mean spherical approximations ("MSA")

In the OZ equation, C(r) is frequently approximated everywhere in the r-space of attractive region of interionic potentials by its asymptotic form $C(r) = -\phi(r)/k_BT$ at $r \to \infty$. This approximation is called as the random phase approximation ("RPA").

In the so called mean spherical approximation ("MSA"), this RPA is adoptd for the attractive part of interatomic potential together with the hard sphere ("HS") approximation for the repulsive part. For this MSA the OZ equation must be combined with two restrictions, described below.

$$C(r) = -\frac{\phi(r)}{k_BT} \quad \text{for} \quad r > \sigma, \qquad (3\text{-}1\text{-}23a)$$

$$h(r) = -1 \quad \text{for} \quad r < \sigma. \qquad (3\text{-}1\text{-}23b)$$

These two restrictions are called as the closure relations for the MSA.

The MSA can be shown to be closely related to the ORPA by the diagram analysis of radial distribution functions,[45] as mentioned in Sec.3.1.5 (ii) (C).

Analytical solutions for structures and the thermodynamic functions, such as the the Helmholtz free energy, are available on this MSA for various model cases, for example, for one component hard spheres and hard sphere mixtures as described above, for charged hard spheres in an oppositely charged uniform background ("CHSO"),[46-48] one component hard spheres with the Yukawa tail ($\propto \exp(-\alpha r)/r$)[49] and binary hard sphere mixtures of equal diameters with the Yukawa tail.[50-53] The latter two models with the Yukawa tail have been frequently called as the hard sphere Yukawa model ("HSY") in the studies of structure of liquid metals and alloys. Here the notation "HSYO" is employed for this one component hard spheres with the Yukawa tail; the notation "HSYB" is employed for binary hard sphere mixtures of equal diameters with the Yukawa tail.

In addition to these structural models employed for discussions of liquid transition metals, analytical solutions of MSA are available, for example, for charged hard sphere mixtures with equal diameters,[54] those with different diameters [55] and hard sphere mixtures with the Yukawa tail with different diameters.[56] The MSA was solved exactly for hard sphere mixtures of equal diameters with the Yukawa tail in a uniform neutralizing background charge.[57] MSA was solved also for the combination of dipole-dipole interaction($\propto r^{-3}$) and Yukawa tail.[58] Only CHSO, HSYO and HSYB are discussed in the next part.

(A) The charged hard spheres in a uniform background ("CHSO")

In this CHSO model, charged hard spheres (with diameter σ and charge Ze; e: the electron charge) are interacted with each other in terms of coulombic interactions for $r > \sigma$, which are immersed in an oppositely charged uniform background. The explicit form of $\phi(r)$ is as follows:

$$\phi(r) = \frac{(Ze)^2}{r} \quad \text{for} \quad r > \sigma, \quad (3\text{-}1\text{-}24a)$$

$$= +\infty \quad \text{for} \quad r < \sigma. \quad (3\text{-}1\text{-}24b)$$

The solution of MSA for this potential has been obtained [46] as a function of $x = r/\sigma$, as follows:

$$C(x) = -\frac{\Gamma^*}{x} \quad \text{for} \quad x > 1, \quad (3\text{-}1\text{-}25a)$$

$$= A + Bx + Cx^2 + Dx^3 + Ex^5 \quad \text{for} \quad x < 1. \quad (3\text{-}1\text{-}25b)$$

Coefficients A, B, C, D and E are expressed as a function of two parameters, packing fraction ξ defined in Eqn.(3-1-16) and the parameter Γ^*, which is defined as $(Ze)^2/k_B T\sigma$. The Γ^* is closely related to the so called plasma parameter Γ in the OCP model described in Sec.3.1.3, as follows;

$$\Gamma = \frac{(Ze)^2}{\hat{a} k_B T} = \frac{\sigma}{\hat{a}} \Gamma^* . \quad (3\text{-}1\text{-}26)$$

In this equation, Z is the valence of point ion and $\hat{a}$ is the sphere radius of point ion, that is $\hat{a} = (3/(4\pi n))^{1/3}$. Hereafter, Γ is employed preferentially for conveniences. Inserting the Fourier transform of C(r) into Eqn. (3-1-17), the analytical expression for C(q) is obtained.[47,48] Analytic expressions for this model are available also for thermodynamic functions, such as the internal energy $U^{46)}$ and entropy $S,^{59,60)}$ for which subscript CHSO is added here.

$$U_{CHSO}(\xi,\Gamma) = - \frac{k_B T}{4\pi\sigma^3}\left[\left(1+\xi - \frac{1}{5}\xi^2\right)\theta^2 + \theta(1+2\xi)\left\{1 - \left[1+\frac{2(1-\xi)^3\theta}{(1+2\xi)^2}\right]^{1/2}\right\}\right],$$

(3-1-27)

$$S_{CHSO}(\xi,\Gamma) = S_{gas} + \Delta S_{CHSO}(\xi,\Gamma). \quad (3\text{-}1\text{-}28)$$

In Eqn.(3-1-27), $\theta = (24\xi\Gamma)^{1/2}$. The S_{gas} is the entropy of ideal gas as follows:

$$S_{gas} = Nk_B\left[\ln\left\langle\left(\frac{V}{N}\right)\left(\frac{2\pi Mk_B T}{h^2}\right)^{3/2}\right\rangle + \frac{5}{2}\right]. \quad (3\text{--}1\text{--}29)$$

$\Delta S_{CHSO}(\xi,\Gamma)$ indicates the excess entropy relative to S_{gas} and is written as the sum of the hard sphere term and the charging term, as follows:

$$\Delta S_{CHSO}(\xi,\Gamma) = Nk_B\frac{\xi(1-\xi)}{(1-\xi)^2} + \Delta S_{CHR}(\xi,\Gamma),$$

(3-1-30)

The charging term, ΔS_{CHR}, can be derived[60] by using $\Gamma(\partial/\partial\Gamma)_{Ze,n} = -T(\partial/\partial T)_{Ze,n}$ as follows:

$$\Delta S_{CHR}(\xi,\Gamma) = -Nk_B\left[B(\xi)\Gamma^{1/2}\left\{1+(1+C(\xi)\Gamma^{1/2})^{1/2}\right\}\right.$$
$$\left. - \frac{4B(\xi)}{3C(\xi)}\left\{(1+C(\xi)\Gamma^{1/2})^{3/2}-1\right\}\right], \quad (3\text{-}1\text{-}31)$$

where $B(\xi) = -(\sqrt{3})(1+2\xi)/12\xi^{2/3}$ and $C(\xi) = (2\sqrt{12})(1-\xi)^3\xi^{1/3}/(1+\xi)^2$. For practical applications of this CHSO, an additional term has been introduced as a correction derived from the presence of thermodynamic inconsistency.[60]

The Helmholtz free energy for the CHSO model, F_{CHSO}, can be expressed by the combination of Eqn. (3-1-27) and Eqn. (3-1-28). The slightly different form of F_{CHSO} have been given by Bretonnet et al.[48] from the integration of Eqn.(3-1-27) with respect to T, as follows:

$$\frac{F_{CHSO}}{k_B T} = \frac{\eta(4-\eta)}{(1-\eta)^2} - \frac{\theta^2}{24\eta}(1+\eta-\frac{1}{5}\eta^2)$$

$$-\frac{(1+2\eta)^3}{36\eta(1-\eta)^3}\left[1-\left\{1+\frac{2(1-\eta)^3}{(1+2\eta)^2}\theta\right\}^{3/2}\right] - \frac{(1+2\eta)}{12\eta}\theta. \quad (3\text{-}1\text{-}32)$$

(B) The one component hard spheres with the Yukawa tail ("HSYO")

The explicit form of interionic potential for one component hard spheres with Yukawa tail ("HSYO") is as follows:

$$\phi(r) = \varepsilon\sigma\frac{\exp\{-k_S(r-\sigma)\}}{r} \quad \text{for} \quad r>\sigma, \quad (3\text{-}1\text{-}33a)$$

$$= +\infty \quad \text{for} \quad r<\sigma. \quad (3\text{-}1\text{-}33b)$$

In this equation, σ is the hard sphere diameter, ε is the strength of Yukawa interaction at the hard sphere contact and k_s is the inverse of screening length. The k_s represents the steepness (or softness) of repulsive part in the interionic potential for $\varepsilon > 0$. The analytical form of S(q) is available for MSA for this interionic potential,[49,50] as already described.

Both the HSYO and the PY solutions for hard spheres are frequently employed for the discussions of structure of single component liquids. Roles of essential feature of the interionic potential on liquid structures can be conveniently tested with their analytical solutions for S(q).

In addition, analytic expressions are available for thermodynamic functions, F, U and S of HSYO model though expressions for them are a little complicated [61]

(C) Binary hard sphere mixtures of equal diameter with the Yukawa tail ("HSYB")

In some binary alloys, chemical short range ordering ("CSRO") effects are substantial to explain their particular features of structures and physical properties. These CSRO effects are considered to be induced by the transfer of electronic charges from an atom of one component to that of another component in liquid mixtures. Therefore such a liquid mixture shows sometimes fully or partially ionic characters. The MSA provides a very good model also for this kinds of (partially) ionic liquids. In this model, two kinds of particles with equal hard core diameter σ interact with each other in terms of the screened coulombic potentials (Yukawa type) at $r > \sigma$ (σ: the diameter of hard cores). The interionic potential between a particle of species α and that of species β, $\phi_{\alpha\beta}(r)$, can be written as follows:

$$\phi_{\alpha\beta}(r) = Q_\alpha Q_\beta \frac{\exp\{-k_S(r-\sigma)\}}{r} \quad \text{for } r > \sigma, \quad (3\text{-}1\text{-}34a)$$

$$= +\infty \quad \text{for } r < \sigma. \quad (3\text{-}1\text{-}34b)$$

In these equations, k_s is the screening constant and Q_α is the charge on the hard spheres of species α. The OZ equation (3-1-18) for this interionic potential has been solved exactly under the restriction of a charge neutrality condition,[50-53]

$$x_1 Q_1 + x_2 Q_2 = 0. \quad (3\text{-}1\text{-}35)$$

The direct correlation functions of Bhatia–Thornton type, $C_{NN}(r)$, $C_{NC}(r)$ and $C_{CC}(r)$ can be introduced conveniently as follows:

$$C_{NN}(r) \equiv x_1^2 \, C_{11}(r) + 2x_1 x_2 \, C_{12}(r) + x_2^2 \, C_{22}(r). \quad (3\text{-}1\text{-}36a)$$

$$C_{CC}(r) \equiv x_1 x_2 \{C_{11}(r) + C_{22}(r) - 2C_{12}(r)\}. \quad (3\text{-}1\text{-}36b)$$

$$C_{NC}(r) \equiv x_1 \{C_{11}(r) - C_{12}(r)\} - x_2 \{C_{22}(r) - C_{12}(r)\}. \quad (3\text{-}1\text{-}36c)$$

The total correlation functions of Bhatia - Thornton type, $h_{NN}(r)$, $h_{NC}(r)$ and $h_{CC}(r)$ can be defined similarly. Also for interionic potentials, $\phi_{NN}(r)$, $\phi_{NC}(r)$ and $\phi_{CC}(r)$ are defined similarly. The closure relations for the MSA for binary mixtures must be written in three sets of pair equations similar to Eqns. (3-1-23a) and (3-1-23b) for three α and β pairs. Then, three independent OZ integral equations (3-1-18) coupled with these closure relations of MSA for α and β pairs can be transformed into another set of three independent OZ integral equations for $\{h_{NN}(r), C_{NN}(r)\}$, $\{h_{NC}(r), C_{NC}(r)\}$ and $\{h_{CC}(r), C_{CC}(r)\}$. The solution of the equation for $\{h_{NC}(r), C_{NC}(r)\}$ is trivial, that is, $h_{NC}(r) = C_{NC}(r) = 0$. That for $\{h_{NN}(r), C_{NN}(r)\}$ becomes completely the same form as the PY equations for single component hard spheres (Eqns.(3-1-7), (3-1-9a) or (3-1-9b), (3-1-14a) and (3-1-14b)) because of $\phi_{NN}(r) = 0$. That for $\{h_{CC}(r), C_{CC}(r)\}$ takes the following form ($n(=n_1+n_2)$: the number density):

$$h_{CC}(r) = C_{CC}(r) + n \int h_{CC}(r') C_{CC}(|r-r'|) d^3 r'. \quad (3\text{-}1\text{-}37)$$

with

$$C_{CC}(r) = -x_1 x_2 (Q_1 - Q_2)^2 \frac{\exp\{-k_S(r-\sigma)\}}{k_B T \, r} \quad \text{for } r > \sigma \quad (3\text{-}1\text{-}38a)$$

$$h_{CC}(r) = 0 \quad \text{for } r < \sigma \quad (3\text{-}1\text{-}38b)$$

The Eqns. (3-1-37), (3-1-38a) and (3-1-38b) have been already solved exactly by Waisman [50] and the analytical expression is available for $C_{CC}(r)$, which contains the packing fraction ξ, the screening constant k_s and the strength of the ordering potential at the hard sphere contact ε ($= x_1 x_2 (Q_1 - Q_2)^2 \exp(-k_s \sigma / k_B T)$) as the characteristic parameters. Thus, partial structure factors of this liquid can be evaluated by the following equations, into which analytic forms for $C_{CC}(q)$ [53] and $C_{NN}(q)$ [43] must be inserted.

$$S_{CC}(q) = \frac{x_1 x_2}{1 - n C_{CC}(q)} \quad (3\text{-}1\text{-}39)$$

$$S_{NN}(q) = \frac{1}{1 - n C_{NN}(q)} \quad (3\text{-}1\text{-}40)$$

$$S_{NC}(q) = 0. \quad (3\text{-}1\text{-}41)$$

The Helmholtz free energy for this HSYB liquid [52,53] is expressed as

$$F_{HSYB} = \frac{3}{2} k_B T + \Delta E_{ORD} - T(S_{HS} + \Delta S_{ORD}). \quad (3\text{-}1\text{-}42)$$

This equation indicates that the Helmholtz free energy for this system can be considered to be the sum of that of hard sphere liquids $3/2 k_B T - T S_{HS}$ and the ordering free energy $\Delta E_{ORD} - T \Delta S_{ORD}$; S_{HS} is the entropy of hard spheres and can be estimated by, for example, Eqns. (3-1-47) and (3-1-48). ΔS_{ORD} and ΔE_{ORD} mean respectively the entropy and the internal energy of ordering induced by oppositely charged hard spheres ("charge ordering"). The most important point of this HSYB model liquids is present in the fact that this liquid has completely analytric solutions for structures, the Helmholtz free energy and the entropy.[52,53]

3.1.3 Liquid models based on the computer simulations

Realistic interionic potentials in liquid metals, particularly in liquid transition metals, are complicated to obtain the solution when it is inserted into integral equations of liquids. Therefore, as described in Sec. 3.1.5, various refined treatments of liquids have been proposed to take account of this realistic potentials approximately, such as variational methods, perturbation theories and the OZ equation with various closure relations. For such theories, thermodynamic and structural properties must be known exactly for reference liquids. In Sec.3.1.2, some such models have been already described on the HS and MSA approximations; analytical solutions are exactly given for them with artificial interionic potentials. Results of computer simulations have been also employed as a model reference system for liquid (transiton) metals. Such examples are the hard sphere ("HS") model and the one component plasma ("OCP") model; in the former case particles interact with each other in terms of hard sphere potentials (Eqns.(3-1-14a) and (3-1-14b)) and in the latter case the coulombic potential is assumed for the interaction of particles.

(i) The hard sphere model ("HS")

Alder and Wainwright[62] performed the MD for hard spheres. The density (or packing franction ξ) dependence of pressure P obtained by them is close to the pressure equation, $P_{HS,P}$ and the compressibility equation, $P_{HS,C}$, given as the solutions of PY equation for hard spheres.[1,40,41] It is to be noted that the PY equation for hard spheres provides two kinds of pressure equation, $P_{HS,P}$ and $P_{HS,C}$, due to the inconsistency of PY theory itself. In fact, the pressure equation obtained by averaging these pressures as $(2P_{HS,P} + P_{HS,C})/3$ is found to express quite well the "experimental" pressure given by the MD simulation due to Alder and Wainwright. The explicit form of this is written as follows:

$$P = nk_BT \frac{(1 + \xi + \xi^2 - \xi^3)}{(1 - \xi)^3}. \qquad (3\text{-}1\text{-}43)$$

Eqn.(3-1-43) is called as the Carnahan-Stirling equation which was derived originally by Carnahan and Stirling.[63]

The Helmholtz free energy for this system can be easily obtained by integrating P by the volume of this system and is found to be the sum of the Helmholtz free energy for the ideal gas, F_{gas} and the contribution of packing, $\Delta F_{HS}(\xi)$, as follows:[64]

$$F_{HS}(\xi) = F_{gas} + \Delta F_{HS}(\xi). \qquad (3\text{-}1\text{-}44)$$

$$F_{gas} = -Nk_BT \ln\left\{\left(\frac{2\pi Mk_BT}{h^2}\right)^{3/2} \frac{V}{N}\right\} - Nk_BT. \qquad (3\text{-}1\text{-}45)$$

$$\Delta F_{HS}(\xi) = Nk_BT\left\{\frac{(3-2\xi)}{(1-\xi)^2} - 3\right\}. \tag{3-1-46}$$

In these equations N is the number of particles with mass M in a volume V and h is the Planck's constant.

The entropy of HS model is easily written as follows:

$$S_{HS}(\xi) = S_{gas} + \Delta S_{HS}(\xi). \tag{3-1-47}$$

In this equation the entropy of ideal gas, S_{gas}, is already given in Eqn.(3-1-29). The $\Delta S_{HS}(\xi)$ means the excess entropy relative to S_{gas}, as expressed as follows:

$$\Delta S_{HS}(\xi) = Nk_B\frac{\xi(3\xi - 4)}{(1 - \xi)^2}. \tag{3-1-48}$$

Applications of $F_{HS}(\xi)$ given in Eqn.(3-1-44) are shown in Secs.3.2 and 3.3 for structures and thermodynamics of liquid transition metals.

(ii) The one component plasma model ("OCP")

In the one component plasma ("OCP") model, point ions of charge Ze with number density n interact with each other by the coulombic potential in a uniform background of opposite charge. Hansen[65] performed the MC simulation for this system. He obtained the radial distribution function g(r), the Helmholtz free energy F, etc. for this system as a function of the plasma parameter Γ, which is given in Eqn.(3-1-26).

The Helmholtz free energy of OCP system can be written as

$$F_{OCP}(\Gamma) = F_{gas} + \Delta F_{OCP}(\Gamma). \tag{3-1-49}$$

In this equation, $\Delta F_{OCP}(\Gamma)$ means the excess Helmholtz free energy relative to the F_{gas}, which can be expressed by a very simple interpolation formula[65,66] of Γ reproducing the result of computer simulation. The explicit form of $\Delta F_{OCP}(\Gamma)$ is as follows:

$$\Delta F_{OCP}(\Gamma) = Nk_BT(A\Gamma+B\Gamma^{1/4}+C\ln\Gamma+D), \tag{3-1-50}$$

where A=-0.896434, B=3.447408, C=-0.555130 and D=-2.995976.

The entropy of OCP model can be written as follows:

$$S_{OCP}(\Gamma) = S_{gas} + \Delta S_{OCP}(\Gamma) . \qquad (3\text{-}1\text{-}51)$$

$$\Delta S_{OCP}(\Gamma) = Nk_B \left\{ -\frac{3}{4}B\Gamma^{1/4} - C(\ln\Gamma - 1) - D \right\} . \qquad (3\text{-}1\text{-}52)$$

This OCP model corresponds to the extreme case of "soft" repulsive potential compared with the HS model of "hard core" repulsion. The calculations due to this model [64] is described in Sec.3.3.2(i)(A) for thermodynamics of liquid transition metals.

3.1.4 Realistic interionic potentials in liquid transition metals

In Secs.3.1.2 and 3.1.3, liquid theories have been discussed for artificial interionic potentials. However, informations of accurate realistic interionic potentials are required to predict the structures and thermodynamic properties of liquid transition metals based on the computer simulations and liquid theories. Recently interionic potentials in liquid transition metals themselves have been formulated rigourously from the refined electron theories of metals such as augmented plane wave ("APW") methods,[67] linearized APW methods ("LAPW")[68], effective medium theories, [69] etc. Here discussions are given only for interionic potentials employed practically for liquid transition metals, which are based on the nearly free electron ("NFE") pseudopotential theories, Bethe lattice tight binding approaches and embedded atom methods.

(i) The generalized pseudopotential theory ("GPT")

The realistic interionic potential $\phi(r)$ can be given also for (liquid) transition metals from the NFE pseudopotential theories similarly to the case of simple liquid metals [1] It can be extracted from the structure dependent energy in the second order perturbation theory for the cohesion of (liquid) transition metals based on the generalized (or transition metal) pseudopotential theory of Harrison [70] and Moriarty.[71] Unfortunately the small core approximation breaks down for the d states, that is the atomic d states are not the exact eigenstates of metallic states. In other words, the potential seen by d states in metallic states, v, differs with the amount of $-\delta v$ from that in atomic states, v^a, as $v = v^a - \delta v$. Such effects can be discussed by the hybridizing operator Δ,

$$\Delta |d\rangle = \delta v |d\rangle - \langle d| \delta v |d\rangle |d\rangle , \qquad (3\text{-}1\text{-}53)$$

where |d⟩ means the atomic d states. For the metallic hamiltonian H it can be written as follows:

$$H|d\rangle = (E_d - \Delta)|d\rangle .\qquad(3\text{-}1\text{-}54)$$

In this equation, E_d represents the expectation value of metallic hamiltonian H for atomic $|d\rangle$ states. The similar equation can be written for the core states $|c\rangle$. However, it is possible to assume that the core states do not vary appreciably when atoms are combined to form a metal, that is δv (or Δ) is zero for the core states $|c\rangle$. Under these circumstances the generalized pseudopotential can be defined based on the OPW formalism with the basis set composed of $|k\rangle$ (plane wave like states), atomic $|d\rangle$ and $|c\rangle$.

The generalized pseudopotential $w(k,q)$ can be written by using the hybridizing operator Δ, as follows:

$$\begin{aligned}w(k,q) =& \langle k+q| \left\{ v + \sum_{\alpha=c,d} (E_k-E_\alpha) \right\} |\alpha\rangle \langle \alpha|k\rangle \\ &+ \sum_d \{ \langle k+q|d\rangle \langle d|\Delta|k\rangle + \langle k+q|\Delta|d\rangle \langle d|k\rangle \} \\ &+ \sum_d \frac{\langle k+q|\Delta|d\rangle \langle d|\Delta|k\rangle}{E_k-E_d} ,\end{aligned}\qquad(3\text{-}1\text{-}55)$$

where E_k is the kinetic energy of free electron. In the limit of $\Delta=0$ the generalized pseudopotential reduces to the simple metal pseudopotential with both $|d\rangle$ and $|c\rangle$ as core states, which is formally expressed by the first term.

In spite of this complicated form of $w(k,q)$, the expression of interionic potential $\phi(r)$ can be extracted formally from the structure dependent term in the second order perturbation theory for cohesive energies of transition metals, as follows:

$$\phi(r) = \frac{(Z^*e)^2}{r} - \frac{2(Z^*e)^2}{\pi} \int F_N(q) \frac{\sin(qr)}{qr} dq + v_{ol}(r) .\qquad(3\text{-}1\text{-}56)$$

In this equation, Z^* is the effective valence corrected by the orthogonalization hole, that is, Z^* electons among sp and d electrons are considered to be extending over transition metals in single band. $F_N(q)$ is the so called normalized energy wavenumber characteristic and $v_{ol}(r)$ is the overlap potential due to the overlapping of d shells in metallic states; $v_{ol}(r)$ should be zero if d shells are considered to be in an isolated state. Explicit forms for Z^*, $F_N(q)$ and $v_{ol}(r)$ are given by Moriarty [71] and Regnaut et al., [72] though expressions for these are extremely complicated and lengthy.

Recently quite similar form of $\phi(r)$ to Eqn.(3-1-56) was obtained by Moriarty[73] from the reformulation of generalized pseudopotential theory for transition metals based on the density functional theory.

The s-d hybridization effects induce the effective charge Z^* in Eqn.(3-1-56) to be larger than the formal valence Z due to sp electrons. In the limit $\Delta \to 0$ and $v_{ol}(r) \to 0$, the interionic potential given by Eqn.(3-1-56) takes the same form as that for simple metals with valence Z. For liquid noble metals, interionic potentials due to Eqn.(3-1-56) have been calculated by Joarder and Gopala Rao,[74] and Regnant et al.,[72,75,76] as described later.

The first principle calculations of GPT are extremely complicated and more tractable interionic potentials and pseudopotentials are desired for the calculation of physical properties of (liquid) transition metals. Pseudopotentials have been studied and employed for a long time and still now various pseudopotential formalisms have been proposed for tansition metals.[77] Among these pseudopotentials the Vanderbilt ultrasoft pseudopotential (last paper in Ref.77) is expected to be applied for the Car-Parrinello-dynamics simulations for liquid transition metals.[78,79]

(ii) Wills and Harrison type interionic ("WH") potentials for transition metals

Unfortunately numerical calculations for $\phi(r)$ are too much complicated, as described above. Wills and Harrison[80] presented a simple theory of cohesion of transition metals from which a simple form of interionic potential $\phi(r)$ can be extracted. In this theory the nearly free electron ("NFE") theory of simple metals is extended to the theory for transition metals including the effects of d bands. The separation is adopted between free electron like states and d like states. The free electron like states can be treated in terms of the NFE theory for simple metals with the empty core pseudopotential of Ashcroft type,[1,3] whose bare potential is zero in the inside region of core and coulombic in the ouside region of it; this simple model potential is quite successful for simple metals. The d-d interactions are calculated by the atomic sphere approximation of Anderson,[81] which is the modification of muffin tin picture with two approximations, overlapping atomic Wigner-Seits spheres and zero wave vector of electrons in the interstitial regions; by the latter approximation calculations turn out to be insensitive to the interstitial regions between muffin tin spheres. The d-d interaction on this ASA are combined with the Friedel model of density of states[82] for d bands, that is the "rectangular model" for density of states for d bands is assumed. The effects of s-d hybridization are taken into account by considering the relative occupancies of s and d states. The resultant interionic potential ("WH" potential) is composed of three parts; the contribution of s electrons represented by an effective screened coulombic potential for an empty core pseudopotential ("NFE term"), $\phi_s(r)$, the attractive contribution of d electrons derived from the d band width term ($\sim r^{-5}$), $\phi_b(r)$, and repulsive one derived from the shift of the d band center ($\sim r^{-8}$), $\phi_{rep}(r)$. Recently this type of interionic potentials has been derived from the combination of NFE theories for s electrons and a tight binding Bethe lattice method for d electrons, which is described as "HKH" potential in part (iv) in this section. This type of interionic potential has been adopted with some refinements on the electron screening functions, the density of states ("DOS") for d bands, etc. for theoretical calculations of S(q) and thermodynamic properties and for the MD of liquid 3d transition metals.[59,61,83-87] A typical form of WH potential is given in Hartree energy unit as follows:

$$\phi(r) = \phi_s(r) + \phi_b(r) + \phi_{rep}(r) , \qquad (3\text{-}1\text{-}57)$$

$$\phi_s(r) = \frac{Z_s^2}{r}\left\{1 - \frac{2}{\pi}\int (1 - \frac{1}{\varepsilon(q)})\cos^2(qR_c)\frac{\sin(qr)}{q}dq\right\} \cdot (3\text{-}1\text{-}58)$$

$$\phi_b(r) = - Z_d(1 - \frac{Z_d}{10})\left(\frac{12}{N_c}\right)^{\frac{1}{2}}\left(\frac{28.06}{\pi}\right)\frac{R_d^3}{r^5} . \qquad (3\text{-}1\text{-}59)$$

$$\phi_{rep}(r) = Z_d\left(\frac{225}{\pi^2}\right)\frac{R_d^6}{r^8} . \qquad (3\text{-}1\text{-}60)$$

In these equations, Z_s is the effective number of s electrons, $\varepsilon(k,q)$ the dielectric screening function, r_c the radius of empty core for model potentials. Z_d, R_d and N_c are respectively the number of d electrons per atom, the d state radius and the number of nearest neighbour coordination.

(iii) Model potential approaches and the BS interionic potential

As described above, numerical calculations of GPT are not always easy because of its complicated and lengthy form. Only for noble metals, w(k,q) has been calculated by Eqn. (3-1-55).[71] At present applications of model pseudopotentials to transition metals, which are very convenient for practical applications to simple metals, have been also limitted to noble metals.[88-91] Borchi and De Gennaro[83] proposed a simple form of pseudopotential for noble metals. The bare pseudopotential contains a core repulsion term, C_1 (positive sign), an attractive s-d hybridization term, $- C_2$ (negative sign) and coulombic attractive

$-Ze^2/r$. The positive constant parameters, C_1 and C_2, are determined so that the form factor (the Fourier transform of this bare pseudopotential), may reproduce numerical values of gen eralized pseudopotentials for noble metals; numerical values themselves are given by Moriarty [71] based on the OPW calculation of Eqn. (3-1-55). Dagen[89] also presented the model potential of transition metals, whose bare potential is constructed by combining the Heine-Abarenkov like model potential (simple metal part) and the resonant type potential (d part).

Recently Bretonnet and Silbert [92] proposed a new form of model potentials for transition metals. The contribution of NFE band due to s and p states is considered by an empty core type pseudopotential and that of d band ("s-d mixing") is taken into accounts by the first two terms of a Dirichlet series sum of short range exponential functions, which is based on the distorted plane wave method for deducing potential interactions from elastic scattering phase shifts.[93] The explicit form of bare pseudopotential is as follows:

$$w(r) = \sum_{n=1}^{2} B_n \exp\left(-\frac{r}{n\bar{a}}\right) \quad \text{for } r<R_c \quad (3\text{-}1\text{-}61a)$$

$$= -\frac{Z_s e^2}{r} \quad \text{for } r>R_c \quad (3\text{-}1\text{-}61b)$$

where $\bar{a}$, B_1 and B_2 are constant parameters. R_c and Z_s are respectively the empty core radius and the effective number of valence electrons per atom. For practical applications the continuity condition was applied at $r=R_c$ for $w(r)$ and its first derivative.

Interionic potentials are calculated similarly to the case of simple metals on the pseudopotential NFE theory. That is

$$\phi_s(r) = \frac{(Z_s e)^2}{r}\left\{1 - \frac{2}{\pi}\int F_N(q) \frac{\sin(qr)}{q} dq\right\}, \quad (3\text{-}1\text{-}62)$$

where $F_N(q)$ is the normalized energy wavenumber characteristic,

$$F_N(q) = \left(\frac{q^2}{4\pi e^2 Z_s}\frac{V}{N}\right)^2 |w_0(q)|^2 \left\{1 - \frac{1}{\epsilon(q)}\middle|\frac{1}{1-G(q)}\right\}. \quad (3\text{-}1\text{-}63)$$

In this equation $w_0(q)$ is the unscreened form factor, $\epsilon(q)$ and $G(q)$ are respectively the dielectric screening function and the local field function which is derived from the exchange and correlation effects of electrons. The $w_0(q)$ corresponding to $w(r)$ given in Eqns. (3-1-61a) and (3-1-61b) can be given from the Fourier transform of $w(r)$ (see Eqn. (5)

in Ref.92). Hereafter this type of $\phi(r)$ is called as the BS interionic potential.

The depth of attractive well of the "BS" interionic potential is shallower than that of WH interionic potential and it was employed to predict the structure of liquid 3d transition metals.[94, 95]

(iv) Interionic potentials based on the tight binding Bethe lattice methods

Quite recently Do Phuong et al.[96] calculated the pair interionic potentials for liquid 3d transition metals based on the tight binding scalar cluster Bethe latice method.[99] According to this approach an interionic potential can be written as the sum of attractive contribution, $\phi_{bond}(r)$, and repulsive one, $\phi_{rep}(r)$. The former results from the covalent bonding energy on the density functional theory and is expressed by the bond order [97]. The latter results from the repulsion of electron clouds and is expressed semiempirically by a sum of power forms of distance r. A pair potential from this approach can be written explicitly as follows:

$$\phi(r) = \phi_{bond}(r) + \phi_{rep}(r) , \qquad (3\text{-}1\text{-}64)$$

$$\phi_{bond}(r) = \sum_{\alpha,\beta} t_{\alpha,\beta}(r)\theta_{\beta\alpha} , \qquad (3\text{-}1\text{-}65)$$

$$\phi_{rep}(r) = \frac{C_s}{r^4} + \frac{C_d}{r^{15}} . \qquad (3\text{-}1\text{-}66)$$

In the attractive contribution $\phi_{bond}(r)$, the $t_{\alpha,\beta}(r)$ is the transfer integral between state α on one atom and state β on another atom, whose dependence on distance r between these two atoms can be approximated by the Harrison's power law dependence.[98] That is, explicit r dependence is assumed to be $t_{\alpha,\beta}(r) \propto r^{-(l+l'+1)}$, where l and l' are respectively the number corresponding to angular momentum quantum numbers of state α and state β; l=l'=0.5 for s and p states and l=l'=2 for d states. The $\theta_{\alpha\beta}$ is the configurational average of bond order in liquid structures, which is related to configurational average of the local Green function, $\langle G^{i,j}_{\alpha,\beta} \rangle_{conf}$, between state α on site i and state β on site j, as follows:

$$\theta_{\alpha,\beta} = -\frac{2}{\pi}\int_0^{E_F} \text{Im}\langle G^{i,j}_{\alpha,\beta} \rangle_{conf} dE . \qquad (3\text{-}1\text{-}67)$$

The configurational average $\langle \text{---} \rangle_{conf}$ can be calculated by the scalar cluster Bethe lattice method.[99] In this approach the tight binding approximation is applied to a cluster of a few atoms, whose surface is surrounded by dangling bonds. The effect of remaining atoms are considered as the Bethe lattice with a coordination number Z, which is attached to these dangling bonds. Z=12 is assumed for the Bethe lattice for liquid 3d transition metals.[94,95]

In the repulsive contribution $\phi_{rep}(r)$, the contribution of s electrons is expressed by the minus 4th power of r and that of d electrons is by the minus 15th power of r. Coefficients of these terms were determined so that the volume and the bulk modulus may be reproduced theoretically.

This type of interionic potential was applied to molecular dynamic simulations to predict liquid structures of 3d transition metals by Do Phuong et al.[96] Hereafter this type of interionic potential is called as "DPPNM" interionic potential.

Hafner and collaborators [84,86,87] developed theories of interionic potentials in which the NFE interionic potential for s electrons, $\phi_s(r)$, is coupled with the attractive d band contribution, $\phi_{bond}^{d,d}(r)$, derived from bond energies due to the tight binding Bethe lattice method for d electrons, as follows:

$$\phi(r) = \phi_s(r) + \phi_{bond}^{d,d}(r) + \frac{C_d}{r^8}. \quad (3\text{-}1\text{-}68)$$

As for the repulsive part, only d contribution, the third term in the right hand side, is explicitly taken into account in Eqn. (3-1-68) because the contribution of s electrons is included in the NFE term. Hereafter this type of interionic potential is caled as "HKH" interionic potential. This interionic potential takes essentially the same form as the WH interionic potential within the second moment approximations for bond energies.[87,106] Therefore, in this review, this HKH and WH interionic potentials are treated as the WH type interionic potential in a same category.

(v) Interionic potentials due to the embedded-atom methods

In the embedded-atom method[100] the dominant contribution of energy of metals is considered to be the energy to embed an atom into the local electron density due to the remaining atoms of the system, which is supplemented by a short-range core-core repulsion $\phi_{rep}(r_{ij})$. The former energy is obtained by summing the embedding energy, $F(\rho_i)$, with resprect to atom i in the system; ρ_i is the (local) total electron density at atom i due to the host, that is the rest of atoms in this system. Therefore, the total energy of the system, E_{tot}, can be written as

$$E_{tot} = NF(\rho_i) + \frac{1}{2} \sum_{i,j(\neq i)} \phi_{rep}(r_{ij}). \quad (3\text{-}1\text{-}69a)$$

From this theory [101] a pair potential can be written as follows:

$$\phi(r) = \phi_{rep}(r) + 2F'(\bar{\rho}) \rho^a(r) + F''(\bar{\rho}) \rho^a(r)^2, \qquad (3\text{-}1\text{-}69b)$$

where F' and F" are respectively the first and the second derivative of the embedding energy of some atom with respect to electron density evaluated at average host electron density $\bar{\rho}$; $\rho^a(r)$ is the atomic electron density of some atom at the distance r from the nucleus. This interionic potential has been employed for predicting atomic structures of liquid Ni, Cu, Ag and Au. [101]

(vi) Comments on interionic potentials in liquid transition metals

As described above many interionic potentials have been proposed. Up to date the atomic structure of liquid transition metals has been discussed within two body approximations from these pair interactions described above though at least for solid transition metals and alloys the importance of many body potentials has been stressed.[97,102] This simplification is supported partly from the fact that interionic potentials and radial distribution functions in liquids depend only on the absolute value of relative distance r and an angular dependence can be neglected as a first approximation("spherically symmetric"). Another reason is that liquids have been traditionally forced to be treated in liquid theories within "effective" two body potentials because of difficulty of liquid theories themselves for many body potentials.

Apart from the problem of many body potential we must compare these various pair interionic potentials described above. At present applications of interionic potentials due to embedded-atom method have been limited to liquid transition metals with more than half filled d shells and liquid noble metals. Hausleitner and Hafner[103] pointed out that this type of interionic potentials encounters serious difficulties when applied to the transition metals with a half filled d band. The HKH interionic potential is almost identical to the WH interionic potential [87,103] on the second moment approximation for the bond energy. The DPPNM interionic potential have a nearly same depth of attractive tail as the WH interionic potential. These depths of DPPNM and WH interionic potentials are far deeper than the depth of BS interionic potential. It is shown that the Helmholtz free energy, F, derived from liquid thoeries due to the BS interionic potential [48] does not show a W shape dependence on 3d shell filling, which is required to explain the Mn anomaly as discussed in Sec.3.5.1. On the other hand the F due to the WH interionic potential [61,104] shows such a W shape behaviour as described in Secs.3.3.2(ii)(C), Fig.2 and 3.5.1. Therefore, the BS interionic potential seems to have some problem for the prediction of thermodynamic properties of liquid transition metals though it is very suitable to predict their liquid structures.

3.1.5 Refined theories of liquids - GB, WCA, RPA, ORPA, RYC, SMSA, ZHC and VMHNC

Descriptions of structures and thermodynamic properties of liquids have been refined in recent theories of liquids so that the detailes of realistic interionic potentials may be included correctly. Those are thermodynamic variational methods, perturbation theories and integral equation approaches combined with certain closure relation. In the integral equation approaches, hybrid closure relations have been employed recently to remove the thermodynamic inconsistencies involved in many usual theories of liquids, as described below.

(i) The thermodynamic variational method due to the Gibbs-Bogoliubov ("GB") inequality

Variational calculations of structures and thermodynamics of liquids are performed mostly based on the well-known Gibbs - Bogoliubov (GB) inequality,[32]

$$F \le F_R + <\Phi - \Phi_R>_R . \qquad (3\text{-}1\text{-}70)$$

In this equation F and Φ are respectively the Helmholtz free energy and the potential energy of real liquids considered; quantities with subscript R correspond to those of reference liquids; $<\text{----}>_R$ means the statistical average in the canonical ensemble of reference system with the weighing factor $\exp(-\Phi_R/k_BT)$. This inequality indicates that the trial function $F^V = F_R + <\Phi-\Phi_R>_R$ becomes closest to the "true" F when the F^V is minimized with the variation of structure parameters of reference liquids; consecquently the structure of reference system also becomes closest to "true" liquids when F^V is minimized.

The GB inequality can be written for the reference system with a set of vaiational structure parameter $\{z_i\}$, as follows:

$$F \le F^V \equiv \frac{3}{2}k_BT - TS_R(\{z_i\}) + N\, u_{eg}(n) - TS_{el} + \frac{1}{2}nN \int g_R(r;\,\{z_i\})\phi(r;\,n)d^3r.$$

$$(3\text{-}1\text{-}71)$$

In this equation, $S_R(\{z_i\})$ indicates the entropy of reference liquid. The term $Nu_{eg}(n)$ is the structure independent (density dependent) part of cohesive energy, which is important for metallic liquids (see Eqn.(3-3-4)). In the case of liquid metals and alloys, it is derived from a uniform electron gas contribution in the second order perturbation theory due to the pseudopotential formalism. The S_{el} indicates the contribution of entropy of electron gas as a Fermion assembly, which is important for transition metals (see Eqn. (3-3-5a)). In the case of non metallic liquids, the terms, $Nu_{eg}(n)$ and S_{el}, can be omitted. The pair potential $\phi(r;n)$ depends slightly on the density n for liquid metals and alloys due to the screening effect of electron gas.

The set of variational parameters is selected so that F^V may be minimized. Examples of such variational parameters are σ or ξ for one component HS model, $\{\sigma_1, \sigma_2 ---\}$ or $\{\xi_1, \xi_2 ---\}$ for HS model of hard sphere mixtures, Γ for OCP model, σ or ξ and Γ for CHSO model, σ, k_s and ε for the HSYO and HSYB models.

The thermodynamic quantities can be calculated at the minimized F^V, F^*. For example, the entropy, S, can be written as follows:

$$S = -\left(\frac{\partial F}{\partial T}\right)_V = -\left(\frac{\partial F^V}{\partial T}\right)_{V,z^*} - \sum_i \left(\frac{\partial F^V}{\partial z_i}\right)_{V,T} \left(\frac{\partial z_i}{\partial T}\right)_V . \qquad (3\text{-}1\text{-}72a)$$

$$= -\left(\frac{\partial F^V}{\partial T}\right)_{V,z^*} . \qquad (3\text{-}1\text{-}72b)$$

The z^* represents the set of variational parameter $\{z_i\}$ which gives the minimum of F^V. The second term in Eqn. (3-1-72a) vanishes because of this minimization condition.[105,106] Therefore the entropy S of real liquids is the same as that of the reference liquids with $\{z_i\}=z^*$ on this GB variational approximation.

(ii) The thermodynamic perturbation theories

(A) <u>The Weeks, Chandler and Anderson ("WCA") perturbation theory</u>

It is found that the structure of liquids is predominantly determined by the repulsive core potentials, as can be seen from the fact that considerably good predictions are possible for the gross feature of the experimental S(q) of liquid metals near the melting point in terms of the solution of the PY equation for hard spheres.[43] Weeks, Chandler and Andersen[34,36] developed the perturbation theory of liquids, which forcuses on the roles of repulsive part of interionic potentials. The Helmholtz free energy of liquids is expanded as a function of the deviation of repulsive part in the interionic potential, $\phi(r)$, from the hard sphere potential,

$\phi_{HS}(r)$ (see also the review by Anderson et al.[36] and Kumaravadivel and Evans[35]).

For this purpose, $\phi(r)$ is divided into two parts, $\phi_0(r)$ and $\phi_1(r)$, as follows:

$$\phi(r) = \phi_0(r) + \phi_1(r) . \tag{3-1-73}$$

$$\phi_0(r) = \phi(r) - \phi(r_0) \quad \text{for } r \leq r_0 , \tag{3-1-74a}$$

$$= 0 \quad \text{for } r \geq r_0 . \tag{3-1-74b}$$

$$\phi_1(r) = \phi(r_0) \quad \text{for } r \leq r_0 , \tag{3-1-75a}$$

$$= \phi(r) \quad \text{for } r \geq r_0 . \tag{3-1-75b}$$

In these equations r_0 is the position of the minimum of $\phi(r)$.

The explicit expansion parameter is the so called blip function, $B(r)$, defined by

$$B(r) = y_{HS}(r)\left[\exp\left\{-\frac{\phi_0(r)}{k_BT}\right\} - \exp\left\{-\frac{\phi_{HS}(r)}{k_BT}\right\}\right]. \tag{3-1-76}$$

In this equation, $y_{HS}(r)$ is defined as follows:

$$y_{HS}(r) = \exp\left\{\frac{\phi_{HS}(r)}{k_BT}\right\} g_{HS}(r) . \tag{3-1-77}$$

Hereafter, subscript "HS" is added to quantities of hard sphere system. If a hard sphere diameter is chosen as

$$\int B(r)d^3r = 0 , \tag{3-1-78}$$

the Helmholtz free energy F and the radial distribution function g(r) can be well approximated as

$$F_{WCA} = F_{HS}(\xi) + Nu_g(n) - TS_{el} + \frac{1}{2}nN \int y_{HS}(r)\exp\left\{-\frac{\phi_0(r;n)}{k_BT}\right\} \phi_1(r,n)d^3r . \tag{3-1-79}$$

$$g_{WCA}(r) \cong y_{HS}(r)\exp(-\frac{\phi_0(r;n)}{k_BT}) . \tag{3-1-80}$$

Therefore, the structure factor S(q) on WCA approximation can be written as

$$S_{WCA}(q) = S_{HS}(q) + n \int \exp(-iqr)\bar{B}(r)d^3r, \quad (3\text{-}1\text{-}81a)$$

$$= S_{HS}(q) + nB(q), \quad (3\text{-}1\text{-}81b)$$

$$= \frac{1}{1 - nC_{HS}(q)} + nB(q). \quad (3\text{-}1\text{-}81c)$$

This equation indicates that the deviation of $S(q)$ from the hard sphere model, $S_{HS}(q)$, is expressed by the Fourier transform of $B(r)$. The perturbation theory described here is called as the WCA approximation and in Eqns.(3-1-79)-(3-1-81c), the subscript "WCA" is added to F, $g(r)$, and $S(q)$. The WCA provides the rigorous condition (Eqn.(3-1-78)) for the determination of hard sphere diameter σ.

The slightly different form of $S(q)$ has been presented to express the effects of deviation of repulsive part from the hard core repulsion in the interionic potentials by Jacobs and Andersen,[107] as follows:

$$S_{JA}(q) = \frac{1}{1 - nC_{HS}(q) - nB(q)}. \quad (3\text{-}1\text{-}82)$$

In the case of $S_{WCA}(q)$ (Eqn.(3-1-81c)), only a lower order in the expansion of $g(r)$ is taken into account. In the $S_{JA}(q)$ it contains the partial sum up to higher order terms, as can be seen from the presence of $B(q)$ in the denominator of Eqn.(3-1-82) for JA. In recent studies, JA is preferred to WCA on applications of this perturbation theory for predictions of $S(q)$ of liquid transition metals, as described in Sec.3.2.

(B) The random phase approximation ("RPA")

In the WCA and JA approximations, detailed considerations were added for roles of the repulsive part in the interionic potentials. The attractive part must be also deeply taken into account. The random phase approximation ("RPA")[37] is one of methods for this purpose as already described in Sec.3.1.2. On this RPA, the direct correlation function $C(r)$ is approximated to be

$$C(r) = C_{HS}(r) - \frac{\phi_1(r)}{k_B T}, \quad (3\text{-}1\text{-}83)$$

where $C_{HS}(r)$ is the direct correlation function for hard spheres and $\phi_1(r)$ is given by Eqns. (3-1-75a) and (3-1-75b). According to the OZ equation, $S(q)$ and $g(r)$ on this RPA can be written as

$$S_{RPA}(q) = S_{HS}(q)\left\{1 + \frac{n\phi_1(q)S_{HS}(q)}{k_BT}\right\}^{-1}, \qquad (3\text{-}1\text{-}84a)$$

$$= \left\{1 - nC_{HS}(q) + \frac{n\phi_1(q)}{k_BT}\right\}^{-1}. \qquad (3\text{-}1\text{-}84b)$$

$$g_{RPA}(r) = g_{HS}(r) - \frac{1}{(2\pi)^3}\int \frac{\{S_{HS}(q)\}^2 v(q)}{k_BT} \exp(iq\cdot r) d^3q. \qquad (3\text{-}1\text{-}85)$$

In Eqn. (3-1-85), $v(q)$ is defined as

$$v(q) \equiv \phi_1(q)\left\{1 + \frac{n\phi_1(q)S_{HS}(q)}{k_BT}\right\}^{-1}. \qquad (3\text{-}1\text{-}86)$$

The Helmholtz free energy due to the perturbation theory on this RPA approcximation is given as

$$F_{RPA} = F_{HS}(\xi) + Nu_g(n) - TS_{el} + F_{HTL} + F^C_{RPA}. \qquad (3\text{-}1\text{-}87)$$

$$F_{HTL} = \frac{n^2 V}{2k_BT}\int g_{HS}(r)\phi_1(r) d^3r. \qquad (3\text{-}1\text{-}88)$$

$$F^C_{RPA} = -\frac{V}{2(2\pi)^3}\int\left[\frac{n\phi_1(q)S_{HS}(q)}{k_BT} - \ln\left\{1 + \frac{n\phi_1(q)S_{HS}(q)}{k_BT}\right\}\right] d^3q. \qquad (3\text{-}1\text{-}89)$$

The second term F_{HTL} in Eqn. (3-1-87), which represents the direct contribution of attractive potential to the free energy, is called as the "high temperature limit". This naming is derived from the hard sphere approximation for structures of trial liquids, which is valid at the "high temperature limit" condition. This term corresponds to the last term in the right hand side of Eqn. (3-1-79). The last term, F^C_{RPA}, in Eqn. (3-1-87) indicates the correction of "random phase approximation". This corresponds to terms containig two and more than two $\phi_1(r)$ bonds and $h_{HS}(r)$ bonds in the diagram expansion of free energy for the system with interionic potential, $\phi_{HS}(r) + \phi_1(r)$ ($\phi_{HS}(r)$: the hard sphere potential).

(C) The optimized random phase approximation ("ORPA")

It is well known in this field that the RPA has an unphysical feature, that is $g_{RPA}(r)$ has a non zero value for $r<\sigma$ in spite of hard core assumption, which can be easily seen by the existence of the second term in Eqn. (3-1-85). In order to avoid this inconsistency, Anderson et al.[36,37] developed the method called as the optimized random phase approximation ("ORPA"). In this ORPA, the form of $\phi_1(r)$ for $r<\sigma$ ($< r_0$) given in equations (3-1-75a) is abandoned and is variationally varied so that the g(r) may show the correct behavior, that is g(r)=0 for $r<\sigma$. The optimized potential $\phi_1^*(r)$ for $r<\sigma$ is determined so that

$$\frac{\delta F_{RPA}^C(\phi_1(r))}{\delta \phi_1(r)} = 0 \quad \text{for } r<\sigma \quad (3-1-90)$$

The disappearance of second term in Eqn. (3-1-85) and as a result g(r)=0 for $r<\sigma$ can be easily checked by Eqns. (3-1-85), (3-1-86), (3-1-89) and (3-1-90).

The optimized potential for Eqns. (3-1-74a)-(3-1-75b) turns out to be as follows:

$$\phi_0(r) = \phi(r) - \phi_1^*(r) \quad \text{for } r \leq \sigma, \quad (3\text{-}1\text{-}91a)$$

$$= \phi(r) - \phi(r_0) \quad \text{for } \sigma \leq r \leq r_0, \quad (3\text{-}1\text{-}91b)$$

$$= 0 \quad \text{for } r \geq r_0. \quad (3\text{-}1\text{-}91c)$$

$$\phi_1(r) = \phi_1^*(r) \quad \text{for } r \leq \sigma, \quad (3\text{-}1\text{-}92a)$$

$$= \phi(r_0) \quad \text{for } \sigma \leq r \leq r_0, \quad (3\text{-}1\text{-}92b)$$

$$= \phi(r) \quad \text{for } r \geq r_0. \quad (3\text{-}1\text{-}92c)$$

Thus, S(q), g(r) and F for the ORPA approximation can be given respectively as the same form as Eqns. (3-1-84a), (3-1-84b), (3-1-85) and (3-1-87) with the optimized $\phi_1^*(r)$ as $\phi_1(r)$.

(D) The coupling of WCA and JA with ORPA ("WCA-ORPA" and "JA-ORPA")

The treatment of ORPA neglects effects of softnesses or deviations from the hard core repulsions of interionic potentials. Therefore, when such deviations are taken into account together with the ORPA, following expressions are obtained for F, g(r) and S(q) with the subscript "WCA-O" respectively.

$$F_{WCA-O} = F_{HS}(\xi) + N u_g(n) - TS_{el} + F_{HTL} + F_{ORPA}^C. \quad (3\text{-}1\text{-}93)$$

$$g_{WCA-O}(r) = g_{WCA}(r) - \frac{1}{(2\pi)^3} \int \frac{\{S_{HS}(q)\}^2 v(q)}{k_B T} \exp(i q \cdot r) d^3 q. \quad (3\text{-}1\text{-}94)$$

$$S_{WCA-O}(q) = \left\{1 - nC_{HS}(q) + \frac{n\phi_1(q)}{k_BT}\right\}^{-1} + nB(q) \ . \qquad (3\text{-}1\text{-}95)$$

If $S_{JA}(q)$ is adopted in place of $S_{WCA}(q)$ together with the ORPA, $S(q)$ can be written (with the subscript "JA-O") as follows:

$$S_{JA-O}(q) = \left\{1 - nC_{HS}(q) + \frac{n\phi_1(q)}{k_BT} - nB(q)\right\}^{-1} . \qquad (3\text{-}1\text{-}96)$$

In these equations, the attractive potential $\phi_1(r)$ for $r<\sigma$ must satisfy the stationally condition Eqn.(3-1-90).

(iii) Refined integral equation approaches

(A) <u>The Ornstein-Zernike equation with the Rogers and Young closure ("RYC")</u>

As is well known, many liquid theories, such as PY, HNC, MSA, etc., contain some thermodyamic inconsistency in themselves. For obtaining the pressure, there are two routes in liquid theories. In one route, it can be obtained by the so called pressure (or virial) equation,

$$p = nk_BT - \frac{n^2}{6}\int r\frac{\partial \phi(r)}{\partial r}g(r)d^3r \ . \qquad (3\text{-}1\text{-}97)$$

In another route, the pressure can be obtained by integrating with respect to the density the compressibility equation,

$$\frac{1}{k_BT}\left(\frac{\partial p}{\partial n}\right) = 1 - n\int C(r)d^3r \ , \qquad (3\text{-}1\text{-}98)$$

or equivalently,

$$k_BT\left(\frac{\partial n}{\partial p}\right) = 1 + n\int h(r)d^3r \ . \qquad (3\text{-}1\text{-}99)$$

Because of thermodynamic inconsistency, the pressures obtained by these two routes do not coincide with each other, as can be typically shown by the fact that there are two pressure equations, $P_{HS,C}$ and $P_{HS,P}$ as analytical solutions of PY equation for hard spheres (Sec. 3.1.3).

It is formally possible to express both PY and HNC equations by single equation with one parameter f(r), as follows:

$$g(r) = \exp(-\phi(r)/k_BT)(1 + \frac{\exp[\{(h(r) - c(r)\}f(r)] - 1}{f(r)}), \quad (3\text{-}1\text{-}100)$$

where $f(r) = 1 - \exp(-\alpha r)$, $\alpha > 0$.[28,31] In the limit of $r \to 0$, this equation reduces to the PY closure (Eqn.(3-1-9b)); in the limit of $r \to \infty$, this reduces to the HNC closure (Eqn.(3-1-8b)). The parameter α is selected so that the physical quantities obtained by these two routes, such as the isothermal compressibility, may coincide with each other.[28,31] This hybrid form of closure relation between PY and HNC is expected to incorpolate both the efficiency of HNC for long range potential and that of PY for short range repulsive potentials, as described in Sec.3.1.2(i). The OZ equation (Eqn.(3-1-7)) coupled with Eqn.(3-1-100) (Rogers-Youg closure[28]) is called here as the "RYC". This approximation yields excellent results for inverse power potentials. Unfortunately it is not promising for potentials with attractive well such as Lennard-Jones potentials.

(B) The soft core MSA

In Sec. 3-1-2 (iii), the mean spherical approximations ("MSA") have been introduced for various important model liquids with hard core repulsive potentials. In real liquids the attractive part of interionic potential has no such singuralities. Therefore, MSA must be improved for the case with continuous core repulsions in interionic potentials. For this purpose, a new closure relation is introduced into the OZ equation.[29] This theory is called as the soft core MSA ("SMSA") and its closure relation is written as follows:

$$g(r) = \exp\left\{-\frac{\phi_0(r)}{k_BT}\right\}\left\{1+h(r) - c(r) - \frac{\phi_1(r)}{k_BT}\right\}. \quad (3\text{-}1\text{-}101a)$$

or equivalently

$$c(r) = \left[1 - \exp\left\{\frac{\phi_0(r)}{k_BT}\right\}\right]g(r) - \frac{\phi_1(r)}{k_BT}. \quad (3\text{-}1\text{-}101b)$$

In this approximation, $\phi_0(r)$ and $\phi_1(r)$ are respectively the attractive part and the repulsive part of interatomic potential, $\phi(r)$; separations of $\phi(r)$ into $\phi_0(r)$ and $\phi_1(r)$ are performed as in Eqns.(3-1-74a), (3-1-74b), (31-75a) and (3-1-75b). Eqns.(3-1-101a) and (3-1-101b) are derived from the coupling of PY equation among $C_0(r)$, $g_0(r)$ and $\phi_0(r)$ with the HNC equation among $\delta C(r)$ (=$C(r)-C_0(r)$), $\delta g(r)$ (=$g(r)-g_0(r)$) and $\phi_1(r)$. $C_0(r)$ and $g_0(r)$ are respectively the direct correlation function and the radial distribution function given as the solution of PY equation for $\phi_0(r)$. Note that $C(r) = -\phi_1(r) / k_B T$ for $r > r_0$. For the Lennard-Jones potential, the g(r) obtained by SMSA is much more accurate than that obtained by PY and HNC.

(C) The Ornstein-Zernike equation with the Zerah and Hansen closure ("ZHC")

To improve inaccuracies of RYC for potentials with an attractive well such as the Lennard-Jones potential, a new closure relation was introduced into the OZ equation by Zerah and Hansen.[31]

$$g(r) = \exp\left\{-\frac{\phi_0}{k_B T}\right\}\left[1 + \frac{\exp\left\{f(r)\left[h(r) - c(r) - \frac{\phi_1}{k_B T}\right]\right\} - 1}{f(r)}\right]. \quad (3\text{-}1\text{-}102)$$

In this equation f(r) takes the same form as that for the case of RYC and is determined similarly to the RYC case. This relation interpolates continuously between SMSA at small r and HNC at large r. Quite succesful predictions are realized for thermodynamics and structure facors of liquids with realistic potentials including the cases of liquid simple metals.

(D) The variational modified HNC equation ("VMHNC")

The OZ equation, Eqn.(3-1-7), can be solved under the closure relation, Eqn.(3-1-10), with the universality condition for the bridge function, $\overline{B}(r)$, which was proposed by Rosenfeld and Ashcroft[108]. As a result $\overline{B}(r)$ can be taken to be same as the $\overline{B}_{PY}(r;\xi)$, which represents the bridge function of the PY approximation for hard spheres. Explicitly this universality condition can be written as follows:

$$\overline{B}(r) = \overline{B}_{PY}(r;\xi) = y_{HS}(r;\xi) - \ln y_{HS}(r;\xi) - 1. \quad (3\text{-}1\text{-}103)$$

The $\overline{B}_{PY}(r;\xi)$ can be written as follows:

$$\overline{B}_{PY}(r;\xi) = \{-C_{HS}(r;\xi)\} - \ln\{-C_{HS}(r;\xi)\} - 1, \quad \text{for } r<\sigma. \quad (3\text{-}1\text{-}104a)$$

$$= g_{HS}(r;\xi) - \ln g_{HS}(r;\xi) - 1, \quad \text{for } r>\sigma. \quad (3\text{-}1\text{-}104b)$$

For $\overline{B}_{PY}(r;\xi)$, $y_{HS}(r;\xi)$, $g_{HS}(r;\xi)$ and $C_{HS}(r;\xi)$ the packing fraction, ξ, is explicitly written as a variable to express it to be treated as a parameter. The ξ is determined, for example, by fitting the compressibility from the long wavelength limit of S(q) to that from the virial equation, as described in Sec.3.1.5 (iii) (A). This theory of liquids is called as the modified HNC equation("MHNC"). It is to be noted that in the limit of $\xi \rightarrow 0$ $\overline{B}(r)$ tends to be zero. This means that in this limit the MHNC approximation reduces to the HNC approximation, which is known to be correct in the low density limit.

Rosenfeld [33)] presented the variational method for solving the MHNC equation, which is called as the variational MHNC ("VMHNC") approximation. In this method, the parameter ξ is determined so that the reduced Helmholtz free energy functional $f^{VMHNC}(T, n; \xi) = F^{VMH}(T, n; \xi) / Nk_BT$ may be stationary with respect to variations of the variational parameter, ξ. That is,

$$\left(\frac{\partial f^{VMHNC}(T,n;\xi)}{\partial \xi}\right)_{T,n} = 0. \qquad (3\text{-}1\text{-}105)$$

The reduced Helmholtz free energy $f^{VMHNC}(T, n; \xi)$ is given as follows:

$$f^{VMHNC}(T,n;\xi) = f^{MHNC}(T,n;\xi) - \Delta_V(T,n;\xi) . \qquad (3\text{-}1\text{-}106)$$

In this equation $f^{MHNC}(T, n; \xi)$ is the MHNC free energy functional[33)], which is written as follows:

$$f^{MHNC}(T,n;\xi) = \frac{1}{2}\int dx\, g_{HS}(x;T,n;\xi)\left[\frac{\phi(x/n^{1/3})}{k_BT} + \overline{B}_{PY}(x;T,n;\xi)\right]$$

$$- \frac{1}{2}\int dx\, [\frac{1}{2}h_{HS}(x;T,n;\xi)^2 + h_{HS}(x;T,n;\xi)$$

$$- g_{HS}(x;T,n;\xi) \ln g_{HS}(x;T,n;\xi)]$$

$$- \frac{1}{2}(2\pi)^{-3}\int dk\left[\ln\left\{1 + \tilde{h}_{HS}(x;T,n;\xi)\right\} - \tilde{h}_{HS}(x;T,n;\xi)\right] .$$

$$(3\text{-}1\text{-}107)$$

The $\phi(x/n^{1/3})$ is an interionic potential and x is the reduced distance as $x=r\, n^{1/3}$. The symbol "~" indicates the Fourier transform. $\Delta_V(T, n; \xi)$ is given as follows:

$$\Delta_V(T,n;\xi) = \frac{1}{2}\int_0^\xi d\xi'\int dx\, g_{HS}(x;T,n;\xi')\frac{\partial \overline{B}_{PY}(x;T,n;\xi')}{\partial \xi'} - \delta_V(T,n;\xi') .$$

$$(3\text{-}1\text{-}108)$$

In this equation $\delta_V(T,n;\xi)$ is written as follows:

$$\delta_V(T,n;\xi) = f_{CS}(T,n;\xi) - f_{PYV}(T,n;\xi) , \qquad (3\text{-}1\text{-}109)$$

where $f_{CS}(T,n;\xi)$ and $f_{PYV}(T,n;\xi)$ are respectively the reduced Helmholtz free energy due to the Carnahan-Stirling equation and that due to the virial pressure equation of hard spheres.

Based on the universality condition for the bridge function, B(r), similar liquid theories to MHNC described above have been developed by Rosenfeld and Ashcroft,[108] Lado et al.,[109] etc.

3.2 The structure of liquid transition metals - from a point of view of theories of liquids

The first systematic study of structures of liquid transition metals was performed by Waseda and Tamaki [110] by means of the X-ray diffraction technique. Since then, extensive studies have been performed for many liquid transition metals and alloys and even for liquid lanthanoid and actinoid metals. Readers interested in this field can consult the book by Waseda.[4]

For a few liquid transition metals both X-ray and neutron diffraction experiments have been performed under difficult experimental conditions due to high reactivities of samples with containers and atmospheres at high temperatures. The height of first peak of S(q) obtained by recent neutron diffraction experments seems to be considerably higher than that of S(q) obtained by the X-ray diffraction experiments [4] for liquid Cu [111], Ni [112], La, Ce and Pr.[113] However, the position of the first peak is in accord considerably well with each other regardless of experimental methods, X-ray or neutron diffractions. Recently these differences between S(q) obtained by X-ray diffractions and that by neutron diffractions have been discussed from a view point of electron - electron and electron - ion correlations in liquid metals.[114 -117] Therefore, it is very important to advance the understanding of these differences experimentally and theoretically.

In this section discussions are performed mostly based on the results of X-ray diffractions due to Waseda [4], by which the systematic trends of the structure of liquid transition metals are expected to be extracted without being suffered from errors due to the difference of experimental methods and conditions. Only essential features of structure of liquid transition metals are discussed from theories of liquids, which is substantial for the theoretical understanding of thermodynamics of liquid transition metals presented in Sec.3.3.

3.2.1 Fitting of model structure factors to experimental structure factors of liquid transition metals

(i) Fitting by the HS model

The structure of liquid metals has been classified into three types, "liquid Al", "liquid Zn" and "liquid Sn" types.[4] The structure factors of liquid transition metals do not show anomalous features such as an asymmetry of the first peak ("liquid Zn type") and a small hump on the high angle side of the first peak ("liquid Sn type"). The structure of liquid transition metals looks like "liquid Al type". In fact, except for the case of Ti and V, structure factors of liquid transition metals are well described by the hard sphere structure factor, which is given as the solution of PY equation for hard spheres. Values of packing fraction ξ are found to be 0.42–0.47 (in many cases 0.44–0.45) to reproduce the experimental structure factors.[4,110] Therefore, the structure factors of liquid transition metals seem to be rather simple. Nevertheless, effects of presence of d electrons appear on their structure. The oscillation of the second peak increases from liquid Ti to liquid Ni in the 3d transition metal series. Waseda and Tamaki[110] explained this tendency by the increase of the hardness of the repulsive part in the interionic potential with the increase of electron filling in 3d shell. As desribed above, disagreements are found for liquid Ti and V between expFrerimental $S(q)$ and the calculated $S(q)$ due to the hard sphere model. These disagreements are speculated to be caused by the partial overlap of one atom with another in these liquid transition metals with almost empty d shells.[4,110] Such systematic tendency is found to be present from liquid Ce to liquid Yb in the Lanthanide series with 4f filling[4,13]

(ii) Fitting by the OCP model

In the case of hard sphere model, singularly steep repulsive potentials are considered. In real liquids repulsions are considered to be far more softer than the hard sphere model. Such a "softness" was incorporated into calculations of $S(q)$ in an extreme manner by Khanna and Cyrot–Lackmann.[118] They found that the $S(q)$ of OCP model presented by the MC simulations of Hansen[65] is able to reproduce very well the experimental $S(q)$ of liquid transition metals if values of plasma parameter Γ are suitably chosen. Chosen Γ is 100 - 120 for liquid 3d transition metals and 80-110 for liquid Lanthanide metals. In the OCP model, point ions in a uniform background of opposite charge interact with each other in terms of long range repulsion of r^{-1} ("coulombic") form, which corresponds to a limitting case for soft interionic potentials. They showed that the predicted S(q) for liquid Eu and Ce by the OCP model also gives a better agreement with experimental S(q) than that by the HS model. With the increase of temperature the first peak of S(q) lowers for liquid Lanthanide metals; this experimental tendency was also explained by the decrease of Γ typically in the case of liquid La; this lowering is caused by the explicit temperature dependence of Γ (see the Eqn.(3-1-26)) and the decrease of density n with the increase of T.

(iii) Fitting by the HSYO model

Recently Meyer et al.[119,120] theoretically tried to confirm the increasing tendency of the hardness in the repulsive part of the interionic potentials with the increase of electron filling in the d and f shells for liquid transition and Lanthanoid metals. This tendency was speculated from the experimental S(q) for these liquid metals by Waseda and Tamaki,[110] as described before. They employed the solution of MSA for hard spheres with the Yukawa tail in the repulsive part ("HSYO"); the explicit form of potential is given in Eqns.(3-1-33a) and (3-1-33b). We note that the steepness (or hardness) of repulsion increases with the increase of softness parameter k_s for $\varepsilon > 0$. The parameters, k_s, ε and σ, were determined so that the calculated S(q) due to HSYO model was able to reproduce very well the experimental S(q) in a sense of least square method.

For liquid 3d transition metal series, the softness parameter k_s as a whole appears to increase with the increase of electron filling in the 3d shell. In addition, the value of $\varepsilon = \phi(\sigma)$, the strength of Yukawa (repulsive) interaction at the hard sphere contact, decreases from liquid V to liquid Cu. The results for liquid Lanthanide metals seem to have the same trend as those for liquid 3d transiton metals though it is not so clear. In other words, the steepness of repulsions in interionic potentials increases, which is shown particularly by the increase of k_s, with the increase of electron filling in the 3d (or 4f) shell. This relation is considered to be consistent with the speculation [110, 121] that the tendency of penetration of sp electrons into core regions is inhibited increasingly with the increase of electron filling in the 3d (or 4f) shell. This tendency is once more discussed in Sec.3.3 with the relation to the tendency of compressibility of liquid transition and lanthanide metals

3.2.2 Analysis of structure of liquid transition metals by refined theories of liquids

(i) Preliminary attempts

Eder et al.[111,112] tried to take into account of roles of attractive potential to reproduce theoretically the experimental S(q) of liquid Ni and Cu, which was obtained by them by means of the neutron diffraction experiments. The Yukawa type attractive potential was taken into account in Eqns.(3-1-66a) and (3-1-66b) for the random phase approximation ("RPA"). Calculated S(q) is in better agreement with experiments than the HS structure factor for both liquid Ni and Cu.

Mitra[122] evaluated the interionic potentials of liquid Cu and Ni by the combination of MSA, fully thermodynamically consistent ("FTC") integral equation of Brennan[123] and MD. The trial form of interionic potential, $\phi(r)$, was given by the MSA and then it was improved by the combination of FTC and MD. Obtained $\phi(r)$ for liquid Cu and Ni became slightly positive after the deep minimum in r–space ("Friedel Oscillation"). The intrerionic potential of liquid Cu is shown to be harder than that of liquid Ni.

Mayer et al.[124] tried to fit a slightly modified form of structure factor on WCA appoximation to the experimental S(q) of many liquid metals including liquid transition metals and lanthanide metals. The determined diameter of hard spheres decreases monotonously from liquid V to liquid Cu in 3d series. This tendency may suggest the trend of softness (or hardness) of repulsive part in interionic potentials of liquid 3d transition metal series described in Sec.3.2.1(iii). That is, in spite of increase of electron number the hard sphere radius decreases from V to Cu in 3d series. This may indicates the increase of steepness or hardness of repulsive part in the interionic potential from V to Cu.

(ii) Interionic potentials due to the GPT and refined theories of liquids

The existence of s conduction electrons and d electrons was not taken into account explicitly on the prediction of S(q) by theories of liquids described above. In these discussions, for example, the " softness parameters " k_s in the repulsive part in the interionic potential is simply speculated to be related to the electron filling in d and f shells. However, from the theoretical point of view structures of liquids should be given, in principle, from theories of liquids with the knowledge of interionic potential, $\phi(r)$. For this purpose the interionic potentials in transition metals can be formally given by , for example, the generalized pseudopotential theory ("GPT"), as described in Sec.3.1.4(i).

Unfortunately, practical calculations of $\phi(r)$ based on this GPT have been limited only to noble metals and alloys because of their complications.

Joarder and Gopala Rao[74] calculated the interionic potential of liquid noble metals to predict their thermodynamic properties, which is described in Sec.3.3. The Borchi-De Gennaro[88] type model potential was employed in place of exact OPW pseudopotential. For $v_{ol}(r)$ in Eqn.(3-1-56), the analytic form given by Moriety[71] was used.

Regnaut et al.[72,75,76] calculated theoretically the interionic potential of liquid noble metals by Eqn.(3-1-56). They introduced it into the ORPA theory of liquids to calculate theoretically the S(q) of liquid noble metals. In these studies calculations were performed by means of both the parametrized version of the generalized pseudopotential and the resonant model pseudopotential ("RMP") of Dagen.[89] Calculated S(q) is generally in good agreement with experiments for liquid Cu, Ag and Au, particularly in the case of GPT.

Moreover, these studies indicate that the s-d hybridization and the overlap potential are essential to predict liquid structures, entropies and resistivities of liquid noble metals judgeing from the poor agreement between experiments and calculations for $\Delta=0$ and $V_{ol}=0$ ("simple metal" limit). In the OPW case, the packing fraction ξ determined is shown to be a little sensitive to the simplification for the calculation of OPW; $\xi=0.5$ in Regnaut et al.[75] and $\xi=0.46$ in Regnaut et al.[72,76] at the same temperature near the melting point of liquid Cu. The value of $\xi=0.59$ for liquid Au is a little too large.[75]

(iii) Wills and Harrison ("WH") type interionic potentials and refined theories of liquids

As described in Sec.3.1.4(ii), the WH interionic potential, a simplified version of GPT, can be employed without any serious technical difficulties for theoretical calculations of structures and thermodynamic properties based on liquid theories. The HKH interionic potential has essentially the same form as this WH interionic potential and theoretical calculations due to both potentials are discussed here.

(A) The structure of liquid transition metals due to the GB thermodynamic variational theory

Hausleitner and Hafner[61] applied the thermodynamic variational theory based on the Gibbs-Bogoliubov inequality to predict the structures and thermodynamic properties of liquid 3d transition metals. The reference system was selected to be the HSYO and HS liquids. The HS reference system corresponds to the case $\varepsilon=0$ for the HSYO reference. The parameters for WH potentials were adopted from informations of solid states and the zero pressure cndition was applied for determinations of empty core radius. As for the valence of s electrons $Z_s = 1.5$ is assumed. The simple metal part, $\phi_s(r)$, shows the Friedel oscillation though it is masked by a deep attractive interaction of the sum, $\phi_b(r)+\phi_{r\ ep}(r)$. The particular feature of WH potential for liquid transition metals is a "soft" repulsive part and a "deep" attractive well.

The predicted structures due to HSYO model show a better agreement with experiments than those due to HS model. This fact confirms the soft charactors of interionic potentials in liquid transition metals. For Zr and the middle element in 3d series (Cr, Mn, Fe), agreements were excellent. However, calculations provide not so good results for early (Ti, V) and end (Co, Ni) transition metals.

Bari et al.[59] tried the GB variational method with both the HS model and CHSO model as a refernce system for calculations of structures and thermodynamic properties of liquid Fe, Co and Ni. For a CHSO reference system only Γ was treated as the variational parameter and ξ was taken to be the same as that for the case of HS reference system. Calculated structure factors due to the CHSO reference system are in better agreement with experiments than calculations due to the HS reference system.

In the WH interionic potentials, Friedel oscillations in the simple metal part are masked in the rather deep well of interionic potentials due to the d states, as described above. This is completely different from a conclusion of Mitra,[122] which was deduced from the combination of FTC integral equations and MD simulations as described above; the Friedel oscillations appeared in the tail part of interionic potential for liquid Ni. In addition another forms of interionic potentials have been proposed by Bretonnet and Silbert [92] and Do Phuong et al.[96] Therefore, in the next stage of study, it is very important to confirm the accuracy of interionic potentials for transition metals, which has been discussed in Sec.3.1.4. Anyway, these studies indicate that liquids with "soft" ionic potentials ("HSYO" and "CHSO") are prefered as reference liquids to liquids with "hard" interionic potentials ("HS") to obtain the accurate theoretical prediction of structures and thermodynamics of liquid transition metals.

(B) The liquid structure factors due to the JA-ORPA

Russier et al. [83] employed the WH interionic potential in the JA-ORPA for calculations of S(q). Calculated S(q) for liquid Ni is in excellent areement with experiments. However, for liquid Fe, Mn and Co, agreements are not so good and for liquid Ti and V even a large shift is found in the high q region of S(q). They pointed out the difficulty of prediction of S(q) for liquid Ti and V with nearly empty d shells in terms of simple theories such as the HS, OCP and JA-ORPA.

Regnaut [85] also calculated the structure of liquid 3d transition metals by WCA and JA-ORPA, for which WH potentials were employed. Calculated results from WCA is similar to those of GB variational calculations with HSYO reference system though both results are not so sufficient for descriptions of S(q) of liquid transition metals. They found that with an appropriate choice of empty core parameter r_C, calculated results of S(q) proved to be in excellent agreement with experiments for liquid Ni and Co. However, for liquid Sc no such success was obtained with any choice of r_C. They explained these results as described below. The structure of liquid transition metals are very sensitive to the tail of WH potentials, which is caused by the delicate cancellation between positive $\phi_s(r)$ and negative $\phi_b(r)+\phi_{rep}(r)$. It may be speculated that the choice of r_C works for liquid Ni and Co with sufficient d charactors and it does not work for liquid Sc because of less d band contributions on interionic potentials for such transition metals with almost empty d shells.

Hausleitner et al.[84] also adopted interionic potentials of WH type in some modified form in the JA-ORPA for calculating the structure of liquid transition metals. In place of the Friedel rectangular model for the DOS of d bands, which is required in the $\phi_b(r)$ in Eqn.(3-1-51), they employed the semi-elliptic form of DOS, which is derived from the Bethe lattice for a monoatomic system with only nearest neibour interactions and large cordination numbers ("HKH" potential). This choice was performed for conveniences for applications to binary alloys. Slightly lower values were adopted as Zs compared with Zs =1.5 in GB-HSYO calculations[61] described before. Obtained conclusions are in good agreement with those of Regnaut et al.[83,85]; the S(q) predicted by JA-ORPA is in poor agreement with experiments for liquid transition metals with half filled d bands, such as liquid Cr and Mn, though good agreements are obtained for liquid Fe, Co and Ni with more than half filled d bands. It should be noted that conclusions of JA-ORPA are quite contradictory to conclusions from GB-HSYO described in (A) in this section except for an agreement for the case of liquid Fe.

(C) The structure of liquid transition metals due to the integral equations of liquids

Hausleitner et al.[84] employed slightly modified WH interionic potentials ("HKH" potential) in three theories of liquids, SMSA, HNC and ZHC as input interionic potentials. Among these three liquid theories, ZHC gives a best agreement with experiments for S(q) of liquid Ni, though the first peak of S(q) is higher than experiments. They concluded that all existing integral equations fail to describe the S(q) of liquid transition metals except for those with nearly filled d bands when WH type inteionic potentials are inserted.

(iv) Bretonnet and Silbert ("BS") interionic potentials and refined theories of liquids

(A) <u>The structure of liquid transition metals due to the GB thermodynamic variational method</u>

Bretonnet et al.[48] performed the GB thermodynamic variational calculation of liquid 3d transition metals with the CHSO model as a reference liquid. The BS interionic potential was adopted as a interionic potential in liquid 3d transition metals. Calculated results of S(q) are in fair agreement with experiments except for liquid Ti and V. In this calculation predicted S(q) in the low q region is fitted to experimental one and adopted BS interionic potential with $Z_s=2$ differs considerably from the WH interionic potential; the former has a very shallow attractive well and the latter has a deep one.

(B) <u>The structure of liquid transition metals due to the VMHNC approximation</u>

Bhuiyan et al.[94,95] applied the variational MHNC approximation to the prediction of S(q) of liquid 3d transition metals. They employed the BS interionic potential as a input potential with $Z_s=1.5$ for liquid Ti and Ni, 1.7 for liquid Sc and 1.4 for other liquid 3d transition metals. The depth of attractive well for this BS interionic potential is far shallower than that of the WH interionic potential; by the deep WH potential almost all theories of liquids have been found to fail to reproduce the structure of liquid transition metals, as described above. As a result of shallow BS interionic potential, their calculated results were in fair agreement with experiments except for liquid Ti and V. Therefore the BS interionic potential is recommended for the prediction of structure of liquid transition metals in terms of liquid theories. However there seems to remain some problem for the prediction of the Helmholtz free energy, which has been already pointed out in Sec.3.1.4.

(v) Liquid structures from computer simulations–MD, Car-Parrinello dynamic simulations and reverse Monte Carlo methods

As desribed above, no good predictions have been obtained for S(q) of liquid transition metals from the coupling of WH interionic potentials with refined theories of liquids, such as JA-ORPA, HNC and ZHC. For predictions of structures and thermodynamics of liquid transition metals from liquid theories, it must be investigated in detail whether input WH potentials are incorrect or theories of liquids themselves are incorrect for such interionic potentials. For this purpose, MD simulations [84] were performed to obtain the $S(q)$ of liquid 3d and 4d transition metals, in which WH type interionic potentials are employed as input informations. Obtained $S(q)$ is in good agreement with experiments except for liquid Ti and V with nealy empty 3d shells. For 4d trasition metals MD simulations are successful for the prediction of $S(q)$ even for liquid Y and Zr, which is isoelectronic to liquid Ti and V in 3d series from the aspect of nearly empty d shell filling. Therefoe, it was concluded that all existing liquid theories are incorrect for liquids with "soft" and "deep" interionic potentials like WH type.

Recently Do Phong et al.[96] presented a quite excellent prediction of $S(q)$ of liquid 3d transition metals in terms of MD simulation, in which employed interionic potentials were calculated by the tight binding scalar cluster Bethe lattice method[99], that is DPPNM interionic potentials. In the case of this MD obtained $S(q)$ is considerably satisfactory even for liquid Sc, Ti and V. It is to be noted that two calculated quantities in the liquid state, the atomic volume and the bulk modulus, were fitted to experimental ones to determine the two parameters, C_s and C_d, in the DPPNM interionic potentials given in Eqn.(3-1-66).

Hausleitner and Hafner [103] performed the MD of liquid 3d and 4d metals, in which the core radius R_c was determined so that the $S(q)$ due to MD simulations may well reproduce the experimental $S(q)$.

Foiles [125] applied the interionic potentials due to the embedded atom method (Eqn. (3-1-69b)) to the modified HNC equation due to Lado et al.[109] Obtained $S(q)$ and $g(r)$ are considerably satisfactory for liquid Cu, Ni, Ag and Au though it has been pointed out [103] that the applications of the embedded atom method itself are difficult for typical transition metals with half filled d bands. Foiles concluded that a pair potential approximation, Eqn.(3-1-69b), is considerably good judging from a close agreement between the results of MHNC due to Lado[109] and those of computer simulations with full energies of embedded-atom method (Eqn. (3-1-69a)), though the zero pressure condition was not fulfilled in this simulation for liquid transition metals.

The Car-Parrinello ("CP") molecular simulation is a very attractive tool for studies of liquid metals and semiconductors. In this method electronic structures are determined by the density functional local density approximation and ionic forces are directly determined from the electronic structure of the system independently from any empirical parameters. As for liquid transition metals the scheme for this method has been presented,[78,79] in which ultrasoft Vanderbilt pseudopotential has been employed.[77] Obtained results of g(r) for liquid Cu is considerably good. It is desired that the CP molecular simulations are performed for many liquid transition metals.

Recently the reverse Monte Carlo ("RMC") technique[126] has been applied to many liquid metals and alloys.[127] Atomic configurations in liquids are determined so that the calculated scattered intensity of X rays or neutrons for postulated atomic arrangements may reproduce the experimental scattered intensity of liquids. The knowledge of atomic form factors or atomic scattering length must be given beforehand. How et al.[127] have performed the RMC for various liquids including liquid transition metals.

(vi) Structures of liquid transition metal alloys

The structures of liquid transition metal alloys have been studied extensively by Waseda and coworkers. Readers interested in this field, particularly in the characteristic short range order effects in liquid transition metal alloys, can consult the book by Waseda.[4] Here we only list recent experimental studies for structures of liquid transition metal alloys, that is Lemarchand et al.[128] for liquid Ni-V alloys, Andonov et al.[129] for liquid Pd - Si, Ji-CHEN Li et al. for liquid Mn-Sn and Mn-Cu [130] and Wagner and Boldrick for liquid Pb-Pd.[131]

At present not so many noteworthy researches have been presented for theoretical studies of structures of liquid transition metal alloys. Joarder et al.[132] applied the GB thermodynamic variational method to liquid noble metal alloys with hard sphere mixtures as a reference system; interionic potentials were calculated from the GPT, in which the Borchi-De Gennaro pseudopotential was employed. Pasturel and Hafner [53] presented a theory of chemical short range order in liquid transition metal alloys, in which the tight binding bond model for internal energies is coupled with the GB thermodynamic variational theory with HSYB as a reference system. According to this theory, the charge transfer between d bands can be treated in some self consistent manner between electronic energies and structures of liquids with charge orderings. Hausleitner and Hafner[86] also presented the formalism of hybridized NFE tight binding bond approach for binary disordered transition metals alloys, which corresponds essentially to the extension of application of WH interionic potential in liquid transition metals[84] to binary cases. Latter two theories have been applied only to metallic glasses.

3.3 Thermodynamic properties of liquid transiton metals

3.3.1 Thermodynamic data of liquid transition metals

(i) General trends and parabolic dependences of thermodynamic properties of liquid transition metals

In table 3.1 are summarized the density, ρ, the thermal expansion coefficient, α_p, and the molar volume, V_M, of liquid transition metals. The molar volume changes systematically with the variation of d shell filling in the atomic state. It shows a minimum at d^6-d^8 configutation, for example, Ni ($3d^84s^2$) in 3d series, Ru ($4d^75s^1$) and Rh ($4d^85s^1$) in 4d series and Os ($5d^66s^2$) in 5d series.

The V_M of first elements (Y and La) shows a far larger value than that of last elements (Cu, Ag and Au) in 3d, 4d and 5d series. This tendency may be explained by the difference of screening effects, that is, much more sufficient screening effects from the nuclear charge are expected for outer d and s electrons in first elements than in last elements in the same d series. In these systematic (parabolic) trends, the exception is the V_M of liquid Mn, whose anomalous behaviour is inferred in Sec.3.5. Originally for solid transition metals many physical quantities have been found to show such systematic trends with the variation of electron occupations in d (or f) shells. Also for liquid transition metals, such systematic trends have been already reported for various physical properties shown in many figures by Wilson,[14] Siegel [133] and Iida and Guthrie.[6]

Siegel [133] has shown the maximum of surface tension and viscosity in the liquid state, melting temperature and sublimation enthalpy and the minimum of atomic volume in the liquid state at the half filled d shells for 3d, 4d, and 5d metals except for a few exceptions, such as the physical properties described above for liquid Mn. He proposed that the d electron bonding persists from the solid to the liquid on melting for transition metals. Here it is to be noted that such a trend can be seen also for the specific heat capacity at constant volume, C_V, as shown in table 3.2. The maximum of C_V is observed for liquid transition metals in the middle of 3d, 4d and 5d series. These facts imply that binding energies in liquid transition metals are largest for nearly half filled d shells with a few exceptions, such as the case of Mn ("Mn anomaly").

According to the Richard's law the entropy of fusion is approximately R (R: gas constant, 8.309 J K^{-1} mol^{-1}). Most of liquid transition metals follow this law within 90% ~ 118% except for Ni, Y and La, as shown in table3.2; it was considered by Grimvall.[134] that

Table 3.1 The density ρ_m, the thermal expansion coefficient α_p, and the molar volume V_M, and the volume change % (vc%) on melting, $\{(V_M-V_S)/V_S\}\times 100$ (V_S: the molar volume of solid), at the melting point of liquid transition metals. ECA: the electronic configuration in atomic states.

ECA	T_m /K	ρ_m /10^3Kg·m^{-3}	$-\dfrac{1}{\rho_m}\dfrac{d\rho_m}{dT}(\equiv\alpha_p)$ /10^{-4}K^{-1}	V_M /10^{-6}m^3·mol^{-1}	vc /%	source
Sc 3d4s^2						
Ti 3d^{2}4s^2	1941	4.11	1.7	11.6		A
	1945	4.13	—	11.6		L
	1943	4.155	non-linear	11.52		SK
	\multicolumn{6}{l}{$V_M/V_0=1.093+1.575\times 10^{-4}(T-T_m)+5.671\times 10^{-9}(T-T_m)^2$}
	\multicolumn{6}{l}{for 1943K(T_m)~5100K ; V_0:V_M at room Temp.}
V 3d^{3}4s^2	2173	5.75	—	8.86		A
	2190	5.55	—	9.18		L
	2175	5.565	non-linear	9.15		SK
	\multicolumn{6}{l}{$V_M/V_0=1.098+9.595\times 10^{-5}(T-T_m)+8.943\times 10^{-9}(T-T_m)^2$}
	\multicolumn{6}{l}{for 2175K(T_m)~6600K ; V_0:V_M at room Temp.}
Cr 3d^{5}4s	2146	6.46	—	8.05		A
	2130	6.31	—	8.24		L
Mn 3d^{5}4s^2	1518	5.73	1.27	9.59		A
Fe 3d^{6}4s^2	1810	7.01	1.19	7.97		A
	1809	7.024	0.89	7.95	3.55±0.6	L
Co 3d^{7}4s^2	1768	7.67	1.60	7.68		A
	1767	7.73	0.93	7.62	5.8±1	L
Ni 3d^{8}4s^2	1726	7.77	1.42	7.55		A
	1726	7.79	0.84	7.53	5.4±0.7	L
Cu 3d^{10}4s	1356	7.99	1.00	7.95		A
	1358	7.937	0.96	8.01	5.2	L

Table 3.1 continued.

ECA	T_m /K	ρ_m /10^3Kg·m^{-3}	$-\frac{1}{\rho_m}\frac{d\rho_m}{dT}(\equiv\alpha_p)$ /10^{-4}K^{-1}	V_M /10^{-6}m^3·mol^{-1}	vc /%	source
Y 4d5s^2	1782	~4.15	—	21.4		L
Zr 4d^{2}5s^2	2125	6.06	—	15.1		A
	2125	5.5	—	16.6		L
	2125	5.8	—	15.7		L
	2125	6.09	—	15.0		L
Nb 4d^{4}5s	2741	7.68	—	12.1		A
	2750	7.83	—	11.9		L
	2750	7.57	—	12.3		L
Mo 4d^{5}5s	2883	(9.35)	—	(10.3)		A
	2892	9.04	—	10.6		L
	2890	9.10	non-linear	10.5		SK
	\multicolumn{6}{l}{$V_M/V_0=1.1199+9.912\times10^{-5}(T-T_m)+2.194\times10^{-9}(T-T_m)^2$}	
	\multicolumn{6}{l}{for 2890K(T_m)~7000K ; V_0:V_M at room Temp.}	
Tc 4d^{5}5s^2	2473	—	—	—		—
Ru 4d^{7}5s	2523	(10.9)	—	(9.28)		A
Rh 4d^{8}5s	2239	(11.1)	—	(9.27)		A
	2236	10.7	—	9.62		L
Pd 4d^{10}	1825	10.52	1.21	10.11		A
	1827	10.49	—	10.14	5.91	L
		10.29	—	10.34		SK
Ag 4d^{10}5s	1234	9.33	0.97	11.6		A
	1234	9.32	non-linear	11.6	3.70	L
	\multicolumn{6}{l}{$V_M=11.56+8.92\times10^{-4}(T-T_m)+1.56\times10^{-7}(T-T_m)^2$}	
	\multicolumn{6}{l}{for 1234K(T_m)~1773K}	

Table 3.1 continued.

ECA	T_m /K	ρ_m /10^3Kg·m^{-3}	$-\frac{1}{\rho_m}\frac{d\rho_m}{dT}(\equiv \alpha_p)$ /10^{-4}K^{-1}	V_M /10^{-6}m^3·mol^{-1}	vc /%	source
La 5d6s^2	1193	5.95	0.40	23.3		A
Hf 5d^{2}6s^2	2495	(12.0)	—	(14.9)		A
	2504	11.1	—	16.1		L
	2504	12.0	—	14.9		L
	2504	11.5	—	15.5		L
	2504	11.97	—	14.9		L
Ta 5d^{3}6s^2	3269	(15.0)	—	(12.1)		A
		14.43	0.90	12.5		L
W 5d^{4}6s^2	3683	(17.5)	—	(10.5)		A
	3660	16.26	—	11.30		L
	3680	16.37	non-linear	11.23		SK

$V_M/V_0 = 1.18 + 6.20 \times 10^{-5}(T-T_m) + 3.23 \times 10^{-8}(T-T_m)^2$

for 3680K(T_m)~7500K ; V_0:V_M at room Temp.

ECA	T_m /K	ρ_m /10^3Kg·m^{-3}	$-\frac{1}{\rho_m}\frac{d\rho_m}{dT}(\equiv \alpha_p)$ /10^{-4}K^{-1}	V_M /10^{-6}m^3·mol^{-1}	vc /%	source
Re 5d^{5}6s^2	3453	(18.7)	—	(9.69)		A
Os 5d^{6}6s^2	3283	(20.1)	—	(9.46)		A
Ir 5d^{7}6s^2	2727	(20.0)	—	(9.61)		A
	2720	19.39	—	9.91		L
Pt 5d^{9}6s	2042	18.91	1.53	10.32	6.63*	A, L*
Au 5d^{10}6s	1336	17.29	0.70	11.39		A
	1337	17.31	0.78	11.38	5.78	L

Data sources

   A :Allen,B.C. , "Liquid Metals : Chemistry and Physics" (ed by Beer,S.Z.) , Marcel Dekker , New York , Ch.4(1972).

   L :Lucas,L.D. Tech l'Ing. , 7,Form. M65(1984).

   SK:Seydel,U. and Kitzel,W. , J. Phys. F:Metal Phys. , 9.L153(1979).

Table 3.2 The atomic mass M_A, the melting temperature T_m, the specific heat capacity at constant pressure Cp, the enthalpy H, the entropy S, and the entropy of fusion ΔS_m of liquid 3d,4d,5d transition metals.

	M_A	T_m /K	C_p /J·mol^{-1}·K^{-1}	H /kJ·mol^{-1}	S /J·mol^{-1}·K^{-1}	ΔS_m /J·mol^{-1}·K^{-1}	source
Sc	44.96	1812	44.22	68.29	100.12	7.778	(S)
Ti	47.88	1933	35.6	73.09	99.06	9.632	(I)
V	50.94	2175	41.84	85.20	100.37	9.62	(S)
Cr	52.00	2130	39.3	83.80	95.40	7.950	(S)
Mn	54.94	1517	46.02	61.89	99.65	7.950	(S)
Fe	55.85	1809	46.02	72.94	100.25	7.632	(S)
Co	58.93	1768	40.50	70.05	99.19	9.159	(S)
Ni	58.69	1726	43.10	64.89	95.98	10.12	(S)
Cu	63.55	1357	31.38	42.52	83.81	9.778	(S)
Y	88.91	1799	39.79	64.22	108.12	6.335	(S)
Zr	91.22	2125	33.47	80.80	110.57	9.845	(I)
Nb	92.91	2740	33.47	99.19	108.45	9.619	(I)
Mo	95.94	2892	56.21	117.11	107.47	9.623	(S)
Tc	(99)	2473	41.84	96.33	107.32	9.623	(I)
Ru	101.1	2523	41.84	95.46	100.13	9.623	(S)
Rh	102.9	2233	41.84	86.39	103.68	9.623	(S)
Pd	106.4	1825	34.73	63.89	100.19	9.623	(S)
Ag	107.9	1234	33.47	37.73	90.91	9.155	(S)
La	138.9	1193	34.31	36.99	106.34	5.192	(S)
Hf	178.5	2500	33.47	104.06	122.38	9.623	(S)
Ta	180.9	3287	41.84	126.38	120.96	9.623	(S)
W	183.8	3680	35.56	153.72	118.46	9.169	(I)
Re	186.2	3453	41.84	140.20	121.36	9.623	(S)
Os	190.2	3300	35.98	123.26	110.45	9.623	(I)
Ir	192.2	2176	41.84	106.36	111.93	9.623	(S)
Pt	195.1	2043	34.73	72.96	107.31	9.627	(I)
Au	197.0	1336	30.96	41.59	97.78	9.393	(S)

(I) I.Barin and O.Knacke,
"Thermochemical Properties of Inorganic Substances" (1973,Springer-Verlag).
(S) I.Barin, O.Knacke and O.Kubaschewski,
"Thermochemical Properties of Inorganic Substances" (1977,Springer-Verlag).

many metals including simple metals follow the Richard's law within this range of reliability. Therefore, he concluded that the order of spin as a cause of paramagnetism in solid transition metals just below the T_m persists on melting judging from the absence of contribution of electron spin to the entropy change on melting.

(ii) Special techniques for experiments of physical properties of liquid transition metals

From the experimental point of view, there remains considerable inaccuracies in the data of physical properties of liquid transition metals due to high experimental temperatures, high reactivities of liquid transition metals with atmospheres and container materials. This situation can be seen easily from the fact that, even for the melting temperature T_m of same liquid transition metals, different values are reported depending on data sources (see tables 3.1 and 3.2).

Particularly it is very important for improvements of accuracy of measurements to prevent the samples from contaminations due to the reactions with containers. In this respect, containerless measurements with inert (or vacuum) atmospheres are very important for highly reactive liquid transition metals. There are many methods for realizing the containerless experimental technique. Typical methods are as follows:

(1) levitation method.[135]
(2) fast resistive pulse heating method.[136]

The first technique, the levitation technique, can be realized by, for example, the repulsion of electromagnetic fields against conducting samples, operations of static electric field on slightly charged samples, a levitation by the pressure of ultrasonic sound wave, etc. This levitation technique turns out to be a powerful tool for determinations of surface tensions and is discussed once more in Sec. 3.5. By means of this tecnique, many physical properties can be measured, for example, surface tensions, densities, viscosities, heats of mixing, etc.

The second technique, the fast resistive pulse heating method, can be realized by resistive heating of conducting samples with a wire or foil shape in a very short period. The pulse of electric current required is obtained by the capacitance discharge. Samples were heated up to extremely high temperatures during a very short time of micro seconds with a speed of 10^{10} Ks^{-1}. This method can be utilized in principle for determinations of many thermodynamic data at high temperatures of solid and liquid materials up to vaporizations. In fact, many reports have been published, for example, electrical resistivities (liquid Ti, V, Mo and V up to 7500K),[137] densities and thermal expansions of volume (liquid Ti, V, Mo, Pd and W up to 7500K),[138] thermal expansions (liquid Al, Cu, Mo,Ta and W),[139] critical data of liquid Mo[140] and the heat of fusion, thermal expansions and electrical resistivities of Mo and Pd in liquid and solid states.[141] The detail of this fast resistive heating method is described in the review written by Gathers.[136]

In spite of very difficult experimental conditions, some researchers have succeeded also in obtaining the velocity of sound of liquid transition metals at high temperature conditions, as shown in table 3.3. Usually, to protect the transducer from high temperature conditions, a buffer rod is inserted between transducers and liquid transition metal samples. Sapphire rods were employed as buffer rods by Shiraishi et al.[142] for liquid Fe, Co and Ni. In the case of Casas et al[143] for liquid Ti, V and Cr, containerless measurements were realized by heating the middle part of solid sample bar by a induction coil heating; the melted part works as a liquid sample and the solid bar works as a buffer rod. The velocity of sound was measured for liquid Au-Co alloys by the pulse echo overlap method.[144]

(iii) The relation between the hardness of interionic potentials and the compressibilities of liquid transition metals

In table 3.3 are also shown the adiabatic compressibilities, κ_s's, of liquid transition metals, which are evaluated by the well known formula, $\kappa_s = (nv^2)^{-1}$, between κ_s, number density, n, and the velocity of sound, v. In addition in this table the isothermal compressibility, κ_T, is shown. The data, κ_T (1), are derived from the velocity of sound for liquid Fe, Co and Ni. The data, κ_T (2), are derived from the long wavelength limit of X ray structure factor, S(0), by the relation $S(0)=nk_BT\kappa_T$. For discussions below for liquid transition metals, the κ_s is employed.

One important point is that the values of κ_s of liquid transition metals are very small compared with those of simple liquid metals; for example, the κ_s of liquid alkali metals is a few times or a few ten times larger than the κ_s of liquid Fe. This fact also corresponds to the implication that the cohesion of transition metals tends to be strong because of the existence of unfilled d band and as a result the κ_s of them tends to be small. Anyway, compared with liquid simple metals such as alkali metals, liquid transition metals with unfilled d bands are "hard" from a thermodynamic point of view. However, the "softness" (or "hardness") in interionic potentials plays an important role for systematic variations of physical properties of liquid transition metals with the variation of d shell filling.

In Fig.3.1 the adiabatic compressibility of liquid transition metals is plotted as a function of "softness parameter k_s" in the Yukawa repulsive tail determined by Meyer et al [119,120,] by

Table 3.3 The velocity of sound v and its temperature coefficient $-(dv/dt)$ at the melting point of liquid transition metals together with the estimated adiabatic compressibility κ_S and the isothermal compressibility κ_T. (1):κ_T from the velocity of sound ; (2):κ_T from the long wavelength limit of $S(q)$, $S(0)=nk_BT\kappa_T$.

	T_m/K	$v/m \cdot s^{-1}$	$-(dv/dT)/m \cdot s^{-1} \cdot K^{-1}$	sources	$\kappa_S/10^{-11}N^{-1} \cdot m^{-2}$	$\kappa_T/10^{-11}N^{-1} \cdot m^{-2}$ (1)	(2)
Sc	–	–	–	–	–	–	3.64
Ti	1945	4407	–	CKS	1.24	–	1.40
V	2163	4742	–	CKS	0.77	–	1.31
Cr	2130	4298	–	CKS	0.84	–	1.10
Fe	1808	4400	–	FKP	0.80	1.04	1.05
		3917	–	KSO	0.93	–	–
		3912	–	KMS	0.93	–	–
		3983±27	1.0021	ST	0.90	–	–
		4052	–	CKS	0.87	–	–
Co	1765	4090	–	SB	0.78	0.97	0.96
		4033±20	0.5325	ST	0.80	–	–
Ni	1728	4045	–	SKO	0.79	0.98	1.03
		4036±5	0.3501	ST	0.79	–	–
Cu	1356	3440	0.50	FKP	1.06	–	1.45
		3460	0.46	BR	1.05	–	–
		3485±8	0.524	SKO	1.03	–	–
Pd	1825	–	–	–	–	–	1.32
Ag	1234	2770	0.47	FKP	1.47	–	1.86
		2710	0.41	BR	1.46	–	–
		2810±10	0.336	TSTS	1.35	–	–
La	1193	2023±5	0.078±0.004	MC	4.11	–	4.29

BR : Beyer,R.T.and Ring,E.M.," Liquid Metals : Chemistry and Phisics " (ed. S.Z.Beer). Ch.9, Marcel Dekker , New York.
CKS : Casas,J. , Keita,N.M. and Steinemann,S.G. , Phys. Chem. Liqids,14, 155(1984).
FKP : Fillipov,S.I. , Kazakov,N.B. and Pronin,L.A. , Izv. VUZ Chern. Met.,9,8(1966).
KMS : Keita,N.M., Morita,H. and Steinemann,S.G. , Helv. Phys. Acta,55,153(1982).
KSO : Kats,Ya L. , Sokolov,L.N. and Okorokov,G.N. , Izv. Akad. Nauk.SSSR Met.,5,57(1978).
MC : McAlister,S.P. and Crozier,E.D. , Solid State Comm. , 40,435(1981).
SB : Steeb,S. and Bek,R. , Z. Naturforsch.,31A,1348(1976).
SKO : Sokolov,L.N. , Katz,Ya L. and Okorokov,G.N.,Izv. Akad. Nauk SSSR Met.,4,62(1977).
ST : Shiraishi,Y. and Tsu,Y. , The 140th Committee , the Japan Society for the Promotion of Sicience (JSPS) , Rep. No.129 , Dec. 1982
TSTS: Tsu,Y. , Suenaga,H. , Takano,K. and Shiraishi,Y., Trans. JIM,23,1(1982).
(1) Tsu,H. , Takano,K. and Shiraishi,Y. , Bull. Res. Inst. Met. SENKEN , Tohoku Univ. , 41,1(1985). see also Itami,T. and Shimoji,M. , J. Phys. F:Metal Phys. , 14,L15(1984).
(2) Waseda,Y. and Ueno,S. , Sci. Rep. Res. Inst. Tohoku University,34A,1(1987).

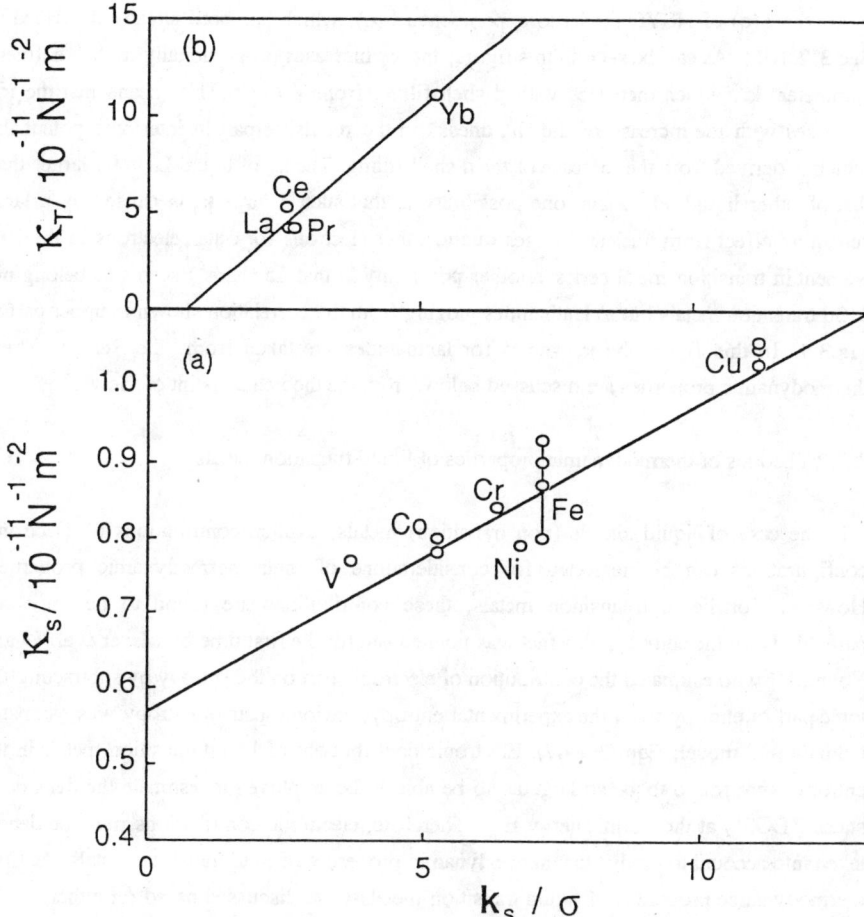

Figure 3.1 The relation between the compressibility and the softness parameter k_s (in unit of hard sphere diameter σ) in Yukawa type repulsive tail.[A,B]

(a) Liquid 3d transition metals (K_s: adiabatic compressibility).

(b) Liquid lanthanide metals (K_T: isothermal compressibility).

A: Meyer,A. , Silbert,M. and Young,W.H. , Zeit. für Phys. Chemie Neue Folge,156,S519(1988).

B: Meyer,A. , Silbert,M. and Young,W.H. , Phys. Chem. Liquids, 19,97(1989).

fitting the S(q) of HSYO model to experimental S(q), which has been already described in Sec.3.2.1(iii). As can be seen in this figure, the κ_s increases proportionally to the "softness parameter" k_s, which increases with d shell filling from V to Cu. This means that the κ_s increases with the increase of the "hardness" in the repulsive part in interionic potentials, which is derived from the increase of the d shell filling. The κ_s of liquid La is far larger than that of other liquid 3d metals; one possibility is that such a large κ_s is caused by a large screening effect from nuclear charges due to other electrons for outer electrons in the first element in transition metal series; another possibility is that La ($5d6s^2$) seems to belong not to 3d transition metals but to lanthanides judging from the correlation shown in upper part in Fig.3.1. In this figure the κ_T values for lanthanides are taken from the Ref.13. These thermodynamic properties are discussed below from the theoretical point of view.

3.3.2 Theories of thermodynamic properties of liquid transition metals

In the case of liquid simple (non transition) metals, explicit contributions of electronic configurations can be neglected for considerations of their thermodynamic properties. However, for liquid trnasition metals, these contributions are found to be important particularly on the entropy; this fact was pointed out for the first time by Meyer et al.[145] and Young,[146] who estimated the contribution of electronic part on the entropy by subtracting the ionic part of entropy from the experimental entropy; the ionic patr of entropy was estimated from the HS model, Eqn.(3-1-47). Electronic contributions of liquid transition metals in the entropy were found to be so large as to be able to be employed to estimate the density of states ("DOS") at the Fermi energy E_F. Therefore, electronic contributions must be deeply taken into account to predict the thermodynamic properties of liuid transition metals. At first, thermodynamic properties of liquid transition metals were discussed based on rather simple models.[64,147-149] Recently, discussions have been given for thermodynamics of liquid transition metals by introducing refind theories of liquids, such as the GB thermodynamic variational method, etc., which is discussed in part (ii) in this section 3.3.2.

(i) Thermodynamic properties based on simple model theories for liquid transiton metals - OCP, HS, PYP and CHSO

Accoding to Itami and Shimoji,[64] the Helmholtz free energy of liquid transition metals with a volume V can be given simply by

$$F=F_R+U_g(V,T), \qquad (3\text{-}3\text{-}1)$$

where F_R is the Helmholtz free energy of a model reference liquid capable of describing the structure of system of interest; $U_g(V,T)$ is a function of the volume V and the temperature T representing the difference between the free energy of the reference liquid and that of the real liqiud. $U_g(V,T)$ includes the effect of conduction electrons and d electrons. At a normal pressure, the pressure P, given by $P = -(dF/dV)_T$, is so small that

$$P = P_R - \left(\frac{dU_g}{dT}\right)_T = 0, \qquad (3\text{-}3\text{-}2)$$

where P_R is defined to be $P_R = -(dF_R/dV)_T$.

(A) The one component plasma ("OCP") model

If the one component plasma (OCP) model is adopted as the reference liquid,

$$U_{g,OCP}(V,T) = N\left\{u_{eg} + \frac{B_{OCP}}{\hat{a}^6} - \frac{1}{6}\pi^2 k_B^2 T^2 v_a N(E_F)\right\}. \qquad (3\text{-}3\text{-}3)$$

In this equation (in unit of Rydberg), v_a is the volume per atom, V/N, and a is the Wigner-Seitz radius, that is $\hat{a} = (3v_a/4\pi)^{1/3}$. The first term in the brace means the energy of electron gas containing the kinetic, exchange and correlation energies, as follows:

$$u_{eg} = \frac{2.21 Z^{5/3}}{\hat{a}^2} - \frac{0.916 Z^{4/3}}{\hat{a}} + Z(0.031 \ln\cdot\frac{\hat{a}}{Z^{1/3}} - 0.115). \qquad (3\text{-}3\text{-}4)$$

The third term in the brace of Eqn. (3-3-3) indicates the contribution of electron assembly as Fermions at the non zero temperature T; this term is important in the case of liquid transition metals because of the large values of the density of states per unit volume, $N(E_F)$, for d bands at the Fermi energy, E_F, and negligible in the case of simple (non transition) liquid metals.[145] The second term represents the contribution from the remaining effects of ionic cores (in which d electrons are involved). The s-d hybridization energy of solid transition metals can be expressed as a sum of terms which are inversely propotional to the sixth power of the lattice constant on the atomic sphere approximation ("ASA").[150] The form of the second term is taken to be analogous to this ASA approximation and B_{OCP} is a parameter to be determined.

After some rutine manipulations of thermodynamics, expressions for the entropy, S_{OCP}, the heat capacity at constant volume, $C_{V,OCP}$ and the isothermal compressibility, $\kappa_{T,OCP}$ are obtained.[64] The particular feature of thermodynamics is that the contribution of electron assembly appears for physical quantities. In the case of entropy, the following electronic term appears.

$$S_{el} = \frac{1}{3}N\pi^2 k_B^2 T v_a N(E_F) . \qquad (3\text{-}3\text{-}5)$$

Various thermodynamic quantities can be written on the OCP model, as follows:

$$S_{OCP} = Nk_B\left\{\ln(\frac{V}{N})(\frac{2\pi Mk_B T}{h^2})^{3/2} + \frac{5}{2}\right\}$$

$$+ \frac{1}{3}N\pi^2 k_B^2 T v_a N(E_F) + Nk_B\left\{-\frac{3}{4}B\Gamma^{1/4} - C(\ln\Gamma - 1) - D\right\} . \qquad (3\text{-}3\text{-}6)$$

$$C_{V,OCP} = Nk_B\left\{\frac{3}{2} + \frac{3}{16}B\Gamma^{1/4} + C + \frac{1}{3}\pi^2 k_B T v_a N(E_F)\right\} . \qquad (3\text{-}3\text{-}7)$$

$$\frac{1}{\kappa_T^{OCP}} = \frac{Nk_B}{V}(1 + \frac{4}{9}A\Gamma + \frac{13}{144}B\Gamma^{1/4} + \frac{1}{3}C)$$

$$+ v_a\left(\frac{\partial^2(U_{g,OCP}(V,T)/N)}{\partial v_a^2}\right)_T . \qquad (3\text{-}3\text{-}8)$$

In Eqn. (3-3-6), the first term indicates the entropy of ideal gas, S_{gas}, expressed by Eqn. (3-1-29); the second term is the contribution of electron assembly as Fermions, S_{el} given in Eqn.(3-3-5). The third term is the excess entropy of classical particles due to the OCP model relative to the ideal gas (Eqn.(3-1-52)).

In Eqn. (3-3-6) for the entropy, $(dF/d\Gamma)_{T, V} = 0$ is assumed for the Γ value which can describe the structure of real liquids, as has been done in the spirit of the GB inequality by Young.[106] Up to date, studies of the density of states, N(E), for (liquid) transition metals have been an interesting and important field in the solid state physics. However, no available data of N(E) are present for the purpose of realistic calculations of thermodynamic properties described here. Therefore, the simplest assumption of N(E) is adopted, that is a rectangular model; the N(E) of d bands can be approximated as the rectangular form with the width, W, as follows:

$$N(E)_{rec} = \frac{10}{Wv_a} . \qquad (3\text{-}3\text{-}9)$$

(B) <u>The hard sphere ("HS") model</u>

If an assembly of hard spheres is taken as a reference system, $U_g(V,T)$ for HS model can be written as

$$U_{g,HS}(V,T) = N\left\{u_{eg} - \frac{1.8Z^2}{a} + \frac{B_{HS}}{a^6} - \frac{1}{6}\pi^2 k_B^2 T^2 v_a N(E_F)\right\}. \quad (3\text{-}3\text{-}10)$$

The second term in the brace of Eqn. (3-3-10) is explicitely written as the electrostatic energy of ions immersed in a uniform neutralizing background of conduction electrons, which is included in ΔF_{OCP} for the case of OCP model. The third and fourth terms have the same meaning as the Eqn. (3-3-3) for the OCP model; B_{HS} a parameter to be determined. Thermodynamic quantities for this HS model can be written as follows:[64]

$$S_{HS} = Nk_B\left\{\ln(\frac{V}{N})(\frac{2\pi M k_B T}{h^2})^{3/2} + \frac{5}{2}\right\}$$

$$+ \frac{1}{3}N\pi^2 k_B^2 T v_a N(E_F) + Nk_B\frac{\xi(3\xi - 4)}{(1 - \xi)^2}. \quad (3\text{-}3\text{-}11)$$

$$C_{V,HS} = Nk_B\left\{\frac{3}{2} - \frac{2(2 - \xi)}{(1 - \xi)^3}T(\frac{\partial \xi}{\partial T})_V\right\} + \frac{1}{3}N\pi^2 k_B^2 T v_a N(E_F). \quad (3\text{-}3\text{-}12)$$

$$\frac{1}{\kappa_T^{OCP}} = \frac{Nk_B}{V}\frac{2\xi(4 - \xi)+(1 - \xi)^4}{(1 - \xi)^4} + v_a\left\{\frac{\partial^2(U_{g,HS}(V,T)/N)}{\partial v_a^2}\right\}_T. \quad (3\text{-}3\text{-}13)$$

In Eqn.(3-3-11), $(dF/d\xi)_{T,V}=0$ (analogous to $(dF/d\Gamma)_{T,V}=0$ in OCP model) is also assumed for the ξ value which can well describe the structure of real liquids, as was done by Edwards and Jarzynski.[105] $N(E_F)$ can be estimated by the rectangular model given in Eqn.(3-3-9).

(C) <u>The Precus - Yevik phonon ("PYP") theory</u>

The similar simple approach, analogous to Eqns. (3-3-6)-(3-3-8) for the OCP model and Eqns. (3-2-11) -(3-3-13) for the HS model, have been developed for thermodynamics of liquid transition metals based on the Percus-Yevick phonon ("PYP") theory of liquids.[148,149] In this approach, atomic motions in liquids can be described by collective motions, represented by 3N independent normal modes similar to phonons in solids.[151,152]

Taking this PYP theory as the reference system, the free energy of liquids on this model can be written as follows:

$$F_R = F_0 + F_1 + \cdots \tag{3-3-14}$$

The first term indicates the Helmholtz free energy of assembly for 3N indepndent phonons and the second term is the leading term of corrections due to anharmonic contributions. Following the routine procedure of thermodynamics, the entropy, S, on this PYP approximation can be expressed as follows ($N(E_F)$ can be estimated by the rectangular model):

$$S_{PYP} = S_0 + S_1 + \frac{1}{3} N\pi^2 k_B^2 T N(E_F). \tag{3-3-15}$$

S_0 is the independent phonon part of the entropy and S_1 is a leading order of corrections due to the anharmonic part of phonons. S_0 and S_1 are written explicitly as follows:

$$S_0 = \frac{Nk_B}{2\pi^2 n} \int_0^{q_0} q^2 \left\{ \frac{x}{e^x - 1} - \ln(1 - e^{-x}) \right\} dq . \tag{3-3-16}$$

$$S_1 = \frac{Nk_B}{128\pi^4 n^2} \int_0^{q_0} \frac{dq}{q} \int_0^{q_0} \frac{dq'}{q'} \left\{ (q^2 + q'^2)\Phi_1 - 2(q^2 + q'^2)\Phi_3 + \Phi_5 \right\}. \tag{3-3-17}$$

In these equations, q_0 is the Debye cut-off wavenumber multiplied by $3^{1/2}$, that is $q_0 = (18\pi^2 n)^{1/3}$; n is the number density of ions and $x = \hbar\omega(q)/k_B T$; the $\omega(q)$ is the zeroth order phonon frequancy and its dispersion relation is assumed to be written in terms of structure factor $S(q)$ in the long wave length part as follows:[153]

$$\omega(q) = q \sqrt{\frac{k_B T}{MS(q)}} . \tag{3-3-18}$$

This relation well reproduces the dispersion relation of the collective motion in liquid Rb observed by the MD simulations due to Rahman[154] and neutron diffraction experiments by Copley and Rowe.[155]

In Eqn.(3-3-17), Φ_n is defined as follows:

$$\Phi_n = \int_{|q-q'|}^{q+q'} S(q)\, q^n\, dq , \text{ for } n=1,3,5. \tag{3-3-19}$$

The specific heat at constant volume and the isothermal compressibility on this PYP approximation can be written as follows:

$$C_V^{PYP} = 3Nk_B + \frac{1}{3}N\pi^2 k_B^2 TN(E_F). \qquad (3\text{-}3\text{-}20)$$

$$\frac{1}{\kappa_T^{PYP}} = nk_BT\left\{1 - \frac{9}{2}\int_0^1 d(\frac{q}{q_0})\frac{(q/q_0)^2}{a(q/q_0)} n\left(\frac{\partial a(q/q_0)}{\partial n}\right)_T\right\}$$

$$+v_a\left\{\frac{\partial^2(U_{g,PY}(V,T)/N)}{\partial v_a^2}\right\}_T \cdot (3\text{-}3\text{-}21)$$

The $U_{g,PYP}$ has the same form as $U_{g,HS}$ of Eqn.(3-3-10), in which B_{PYP} should be writtten in place of B_{HS}. To obtain the expression for C_V^{PYP}, the contribution of S_1 can be neglected and that of S_0, $T(\partial S_0/\partial T)_V$, is approximated to be $3Nk_B$.

(D) Comparisons among calculations due to HS, OCP and PYP models for thermodynamic properties of liquid transition metals

Calculations for these model theories described above have been performed under the zero pressure condition, P = 0, by which the coefficients, B_{OCP}, B_{HS}, and B_{PYP} were determined.[64,149] The value of ξ required for calculations was taken from the ξ which was determined so that the S(q) of HS model may reproduce the experimental S(q).[4,156] For the isothermal compressibility for OCP and HS models, calculations are performed for valence Z=1 and Z=Z_T; the latter value was estimated from the value of plasma parameter Γ obtained by fitting the S(q) of OCP model to the experimental S(q).[118] The band width W was taken from Varma and Wilson.[157]
The Z_T is slightly larger than 1, that is 1.22 - 1.52 for 3d transition metals, as shown in table 3.4. As for the valence of liquid transition metals, some arbitrariness is present on applications of the extended Ziman formula for electrical resistivity as discussed in Sec.3.4. Esposito et al.[12] presented a rigorous theoretical method for determinations of effective valence Z*. Z*=1.21 has been obtained for liquid Fe and Cu and Z* = 1 for liquid Co and Ni. Moriety[158] evaluated Z* to be 1.5 for liquid Cu from the density functional theory of the generalized pseudopotential. Moruzz[159] also presented the valence of s type electrons as Z_s =1.29 ~1.43 for solid transition metals from the self consistent calculations. In contrast to

Table 3.4 The entropy S (in unit of Nk_B) of Liquid 3d transition Metals. Z_Γ is the valence caluculated from the plasma parameter Γ; S_{el}^{rec} is S_{el} calculated using the rectangular model; S_{el}^{AY} is S_{el} calculated using the theoretical $N(E_f)$ of Asano and Yonezawa (1980); S_{el}^{solid} is S_{el} calculated using the $N(E_f)$ obtained from solid band calculations (see Meyer et al. 1976); experimental values S_{exp} are referred to table 3.2. Theoretically calculated results are taken from Itami and Shimoji (1984) for the hard sphere(HS) and OCP model. Calculationa are taken from Itoh et al. (1986) for the PY phonon (PYP) theory.

	T(K)	Γ	Z_Γ	S_{OCP}	ξ	S_{HS}	$q_0(\text{Å}^{-1})$	S_{PYP}	S_{exp}	S_{el}^{rec}	S_{el}^{AY}	S_{el}^{solid}
Sc	1833	100	1.42	11.87	0.43	11.48	—	—	12.05	0.92	—	—
Ti	1973	110	1.47	11.55	0.437	11.06	2.10	11.17	12.01	0.81	—	0.49,1.18
V	2173	110	1.49	11.61	0.436	11.24	2.24	11.31	12.08	0.83	—	1.23
Cr	2173	120	1.52	11.34	0.445	10.95	2.35	11.07	11.58	0.80	0.75	~0.38
Mn	1533	120	1.30	10.87	0.449	10.41	2.27	10.63	12.05	0.67	—	—
Fe	1833	110	1.33	11.39	0.438	10.98	2.38	11.08	12.11	0.90	—	1.82
Co	1823	110	1.32	11.49	0.444	10.98	2.41	11.19	12.08	0.97	—	1.21-1.82
Ni	1773	110	1.30	11.56	0.442	11.08	2.41	11.29	11.69	1.09	1.51	1.92
Cu	1423	120	1.22	10.26	0.45	9.68	—	—	10.27	0.095	—	—

Itami,T. and Shimoji,M., J. Phys. F:Metal Phys., 14,L15(1984).
Itoh et al.: Itoh,H., Yokoyama,I. and Waseda,Y., J. Phys. F:Metal Phys.,16,L113(1986).
Tsu,Y. and Takano,K. 88th Spring Conf.,2-4 April, 88 (Sendai : Japan Inst. Met.),86(1981).

the case of transition metals, the Z_T is estimated to be smaller than unity for liquid simple alkali metals by Mon et al.[160] Thus estimated Z_T (>1) for the OCP model of liqud transition metals indicates the effect of s-d hybridization effects.

The calculated results are considerably good for all these three models for $Z=Z_T$, as shown in tables 3.4, 3.5, and 3.6, particularly for the case of S due to OCP, C_V due to PYP and OCP and κ_T due to HS (dσ/dT=0) and OCP. In addition, according to the study of Itoh et al.[149] calculated C_V for liquid transition metals by OCP is in better agreement with experments than that by PYP.

As a conclusion the OCP is the best model among these three models for predictions of thermodynamic properties of liquid transition metals. This fact reflects the importance of "soft" charactor in interionic potentials for liquid transition metals. The HS model was employed for describing the structure and thermodynamic properties of liquid transition metals by Tamaki and Waseda.[147] The entropy calculations based on the HS model were performed by Asano and Yonezawa[161] to verify the theoretically calculated DOS. The OCP model has been successfully applied also to liquid Lanthanide metals by Yokoyama and Naito.[162]

As described in Sec.3.2 it is possible for OCP and HS models to refine their treatments by means of advanced theoretical techniques of liquid theories such as ORPA, GB variational method, etc. It seems to be not so promising to improve the PYP theory beyond the present limitations because of difficulty for obtaining dispersion relations of phonons in liquids theoretically or experimentally from the knowledge of interionic potentials. On this model, contributions of phonon like collective motion are assumed for the free energy; this point is completely different from the HS and OCP models where motions of independent particles are essentially considered. The PYP model contains a little ambiguous assumptions, such as the existence of cut-off frequency of phonons, assumed dispersion relations for phonon frequency and slightly complicated calculations, for which "HS" model structure factors are inserted as structure informations. Anyway, it is very interesting and important to clarify the collective behaviour of particle motions in liquid (transition) metals from the microscopic structure investigations.[5]

Recently Venkatesh[162a] calculated the transverse and the longitudinal phonon dispersion relations and the elastic constant of liquid Pt and Pd. The former dispersion relation was evaluated based on the Takeno and Goda's equation,[162b] in which the WH interionic potential and theoretically obtained radial distribution function were inserted. The latter elastic constant was calculated from three methods, from the Schofield equation[162c], from the phonon frequency and from the compressibility estimated from the long wave length

Table 3.5 The heat capacity at constant volume C_V (in unit of Nk_B) of liquid 3d transition metals.

The experimental value C_V^{exp} of Fe, Co and Ni are caluculated from the relation $C_V^{exp} = C_P^{exp} \kappa_S^{exp}/\kappa_T^{exp}$; κ_S^{exp} and κ_T^{exp} indicate the experimental values (Tsu and Takano 1981) of adiababatic and isothermal compressibility respectively ; C_P^{exp} can be referred to table 3.2. Theoretically calculated results are taken from Itami and Shimoji (1984) for the hard sphere (HS) and OCP model. Calculations are taken from Itoh et al. (1986) for the PY phonon (PYP) theory.

	T(K)	C_V^{el}	C_V^{OCP}	$C_V^{HS}(\partial\xi/\partial T=0)$	$C_V^{HS}(\partial\xi/\partial T\neq 0)$	C_V^{PYP}	C_V^{exp}
Sc	1833	0.92	3.91	2.42	4.75†	—	4.39‡
Ti	1973	0.81	3.85	2.31	4.90†	3.92	3.53‡
V	2173	0.83	3.87	2.33	5.17†	3.91	4.15‡
Cr	2173	0.80	3.88	2.30	5.26†	3.94	3.90‡
Mn	1533	0.67	3.75	2.17	4.33†	3.78	4.56‡
Fe	1833	0.90	3.94	2.40	4.84	4.08	4.38‡
Co	1823	0.97	4.01	2.47	4.93	4.19	4.00
Ni	1773	1.09	4.13	2.59	4.99	4.33	4.46

†: Calculated using $(\partial\xi/\partial T)_V = -7.5\times 10^{-5} K^{-1}$, that is, the average values of liquid Fe, Co and Ni.

‡: Derived from C_P^{exp} multiplied by the average values of $\kappa_S^{exp}/\kappa_T^{exp}$ of liquid Fe, Co and Ni, i.e. 0.824.

Itami,T. and Shimoji,M., J. Phys. F:Metal Phys. ,14,L15(1984).

Itoh et al. : Itoh,H., Yokoyama,I. and Waseda,Y., J. Phys. F:Metal Phys.,16,L113(1986).

Tsu,Y. and Takano,K. 88th Spring Conf.,2-4 April, 88 (Sendai : Japan Inst. Met.),86(1981).

Table 3.6 The isothermal compressibility κ_T (10^{-11} m$^2\cdot$N^{-1}) of liquid 3d metals.

The κ_T^{exp} are referred to Tsu and Takano (1981). Theoretically caluculated results are taken from Itami and Shimoji (1984) for the hard sphere (HS) and OCP model. Calculations are taken from Itoh et al. (1986) for the PY phonon (PYP) theory.

	T(K)	κ_T^{OCP}(Z=1)	κ_T^{OCP}(Z=Z$_\Gamma$)	κ_T^{HS}(Z=1)	κ_T^{HS}(Z=Z$_\Gamma$)	κ_T^{PYP}(Z=Z$_\Gamma$)	κ_T^{exp}
Sc	1833	2.16	2.31	3.32	1.91	—	—
Ti	1973	1.41	1.56	2.33	1.28	1.69	—
V	2173	1.07	1.23	1.87	1.01	1.23	—
Cr	2173	0.87	1.01	1.56	0.82	1.10	—
Mn	1533	1.37	1.52	1.84	1.23	1.63	—
Fe	1833	1.11	1.26	1.59	1.02	1.37	1.04
Co	1823	1.09	1.23	1.50	0.98	1.35	0.97
Ni	1773	1.11	1.26	1.51	1.00	1.37	0.98

Itami,T. and Shimoji,M., J. Phys. F:Metal Phys. ,14,L15(1984).

Itoh et al. : Itoh,H., Yokoyama,I. and Waseda,Y., J. Phys. F:Metal Phys., 16,L113(1986).

Tsu,Y. and Takano,K. 88th Spring Conf.,2-4 April , 88 (Sendai : Japan Inst. Met.),86(1981).

limit of calculated structure factors. Elastic constants obtained from these three methods agree with each other considerably well though many assuptions are included, such as the squre well potentials for calculations of structures, the adoption of effective mass in place of atomic mass, etc.

(E) The charged hard sphere ("CHSO") model

Recently the charged hard sphere ("CHSO") model was applied to predict thermodynamic properties of liquid transition metals and rare earth metals in addtion to liquid alkali metals.[60] Similarly to the case of OCP model, Eqn.(3-3-3) was adopted as a density dependent energy, $U_{g,CHSO}(V,T)$, in whch a parameter B_{OCP} is replaced by B_{CHSO}. On this CHSO approximation the Helmholtz free energy can be written as follows:

$$F_{CHSO} = U_{CHSO}(\xi, \Gamma) - TS_{CHSO}(\xi, \Gamma) + U_{g,CHSO}(V,T) . \qquad (3\text{-}3\text{-}22)$$

A parameter B_{CHSO} was determined by a similar condition to the zero pressure one. The calculated results are inferior to the OCP[64] for the S and C_V, as can be seen in table II in Ref.60. The structure parameters required for calculations were taken from the fitted values of Γ and ξ, which were determined so that the respective model structure $S(q)$ could reproduce the experimental $S(q)$, that is Γ by the OCP and ξ by the HS structure factors.[4,118,156] Therefore, there remains some question whether these independently determined parameters are surely able to reproduce the experimental $S(q)$ or not when these two parameters are inserted into the $S(q)$ of CHSO itself.

(ii) Refined theories of thermodynamic properties of liquid transition metals

(A) Generalized pseudopotential approaches ("GPT") - GB-GPT and JA-ORPA-GPT

In Sec.3.2, perturbation theories of liquids with WH type interionic potentials are not always successful to predict structures of liquid transition metals because of their "soft" repulsions and "deep" attractive wells. Therefore, up to date, applications of such thermodynamic perturbation theories, WCA-ORPA and JA-ORPA, have been limited to liquid noble metals and never applied to describe thermodynamic properties of typical liquid transition metals, such as liquid Cr, Fe, Mn, etc.

Joarder and Gopala Rao[74] applied GB thermodynamic variational calculations with a hard sphere reference system to predict the entropy, S, the heat capacity at constant volume, C_V, the thermal pressure coefficient, γ_V, the thermal expansion coefficient, α_P, and the isothermal compressibility, κ_T, of liquid noble metals. The Borchi-De Gennaro [88] model potential was adopted. Calculated results ("GB-GPT") are in rather good agreement with experiments.

Also for liquid noble metals, Rignaut et al.[72,75,76] calculated S and C_V by using the GPT of transition metals due to Moriaty,[71] for which some simplifications were tried for numerical calculations of form factors. They calculated the interionic potential theoretically by Eqn.(3-1-56). The structure factor of liquid noble metals was obtained by JA-ORPA in perturbation theories of liquids ("JA-ORPA-GPT"). The S and C_V were calculated in terms of the free energy given by this ORPA approximation. The explicit expression for the entropy is given as follows:

$$S = S_{el} + k_B \frac{\partial}{\partial T} \left(F_{HS} + \frac{2\pi}{V_M} \int g_{HS}(r) \phi_{OP}(r) r^2 dr \right)_V$$

$$+ k_B \frac{\partial}{\partial T} \left[\frac{V_M T}{4\pi^2} \int \left\{ 1 - \frac{S_{HS}(q)}{S_{ORPA}(q)} + \ln\left(\frac{S_{HS}(q)}{S_{ORPA}(q)}\right) \right\} q^2 dq \right]_V . \quad (3\text{-}3\text{-}23)$$

In this equation, F_{HS}, $g_{HS}(r)$, and $S_{HS}(q)$ are the Helmholtz free energy, the radial distribution function and the structure factor of hard sphere liquids respectively. S_{el} is the electronic contribution of entropy given in Eqn.(3-3-5).

The $\phi_{op}(r)$ is the optimized interionic potential on the OPRA sence, which is determined in the range $r<\sigma$ by "staionary condition" Eqn.(3-1-90) for a free energy functional $F(\phi(r))$. $S_{ORPA}(q)$ corresponds to the structure factor $S(q)$ on the ORPA approximation in the case of $B(q) = 0$ (see Eqns.(3-1-84b) and (3-1-95))

In addition to the GPT, Regnaut et al.[75] employed the Dagen's resonance potential[89] and calculated results for S and C_V of liquid noble metals are in good accordance with experiments equally to the case of GPT. The contribution of the hybrdizing operator Δ, which takes account of the effect of s-d hybridization, is important to improve the agreement between theories and experments.

(B) The GB thermodynamic variational method and the tight binding approach ("TBA") - GB-TBA

In Secs.3.2 and 3.3.2(i) and (ii)(A), the starting point of most of discussions for structures and thermodynamics of liquid transition metals has been put on the NFE theory. However, there are many systematic, typically parabolic, dependence of physical properties on d shell filling for liquid transition metals, as described in Sec.3.3.1(i). These trends imply the parabolic dependence of cohesive energy, E_c, as a function of d shell filling with a minimum for half filled d band metals.

Therefore, Aryasetiawan et al.[163] presented the theory of thermodynamics of liquid transition metals, in which the tight binding model ("TBA") for electron theories was adopted as the starting point of discussions. They coupled the TBA with the variational mehtod of the Gibbs - Bogoliubov with the hard sphere reference system. According to Ducastelle,[164] the total potential energy of system can be written by the TBA scheme as follows:

$$\Phi(\{r_i\}) = \sum \left\{ -10 \left(\frac{\mu_i}{2\pi}\right)^{1/2} \exp\left(-\frac{E_F^2}{2\mu_i}\right) \right\} + \frac{1}{2} \sum_{i,j(i\neq j)} \phi_{rep}(r_{ij}) . \quad (3\text{-}3\text{-}24)$$

In this equation, the first term is derived from the sum of occupied d electron eigenvalues and a gaussian form is assumed for the density of states for site i; E_F is the Fermi energy. The second term represents the pair wise repulsive contribution of remaining effects to energy. The second moment μ_i, which represents the width of partial density of states $n_i(E)$, is written by the hopping integrals $\beta(r_{ij})$ as follows:

$$\mu_i = \sum_{j(\neq i)} \beta^2(r_{ij}) . \quad (3\text{-}3\text{-}25)$$

Both for the $\beta(r_{ij})$ and the pair wise repulsive potential $\phi_{rep}(r_{ij})$, the Born - Meyer form was simply assumed, that is $\beta(r_{ij}) = \beta_0 \exp(-qr_{ij})$ and $\phi_{rep}(r_{ij}) = C_0 \exp(-pr_{ij})$.

The GB inequality for this TBA system with the HS reference system can be written as folllows:

$$F \leq F^V(\xi; n, T) = \frac{3}{2} N k_B T - \frac{10N}{\sqrt{2\pi}} \sqrt{\langle \mu \rangle_{HS}} \exp\left(-\frac{E_F^2}{2 \langle \mu \rangle_{HS}}\right)$$

$$+ \frac{n}{2} N \int dr \, g_{HS}(r) \phi_{rep}(r) - T S_{HS} - \frac{\pi^2}{3} N \langle n(E_F) \rangle_{HS} (k_B T)^2 . \quad (3\text{-}3\text{-}26)$$

The last term in the right hand side of Eqn.(3-3-26) represents the ad hoc term to give the contribution of electrons to the entropy. In this equation $g_{HS}(r)$ and S_{HS} are the radial distribution function and the entropy for hard spheres respectively; $<n(E_F)>_{HS}$ is the densty of states at E_F for HS reference system and the ensemble average of second moment is given as $<\mu>_{HS} = n \int dr g_{HS}(r) \beta^2(r)$. The gaussian form was assumed for the $N_d(E)$. $\beta(d)$ was determined by fitting the cohesive energy of solid at T=0 to experimental one. Required parameters were taken with some approximations.

The free energy and the ξ was determined variationally based on Eqn.(3-3-26) as a function of density n. The equilibrium number density per volume was determined by the so called zero pressure condition from the relation between the free energy and the density. Calculated free energies and densities are in good agreement with experiments for 3d, 4d, and 5d transition metals in spite of many assumptions in it, for examples, very simple form of $\beta(r)$ and $\phi_{rep}(r)$ and the number of d electrons per atom were referred to the solid state data; approximations for the ensemble average with the reference structure of first term in Eqn.(3-3-24); only energies of d elecrons are taken into account and the sd hybridization energy and the energy of sp electrons are put aside; the entropy of tightly binding electrons was estimated by the expression of free Fermions. The ξ determined by the GB inequality became considerably smaller (0.22 - 0.42) than ξ (~0.46) determined by fitting of S(q) due to the HS model to the experimental S(q) as described in Sec.3.1. It is to be noted that the parabolic dependence of binding energy with Mn anomaly ("W shape") in 3d series was repoduced only when $\beta(d)$ was determined by fitting the cohesive energy at T=0. According to Sato et al.,[165] determinations of number density by the zero pressure condition in liquid alloys are rather insensitive to the detailed shape of the entropy term and depend predominantly on the gross features of energy terms in the free energy of liquids. In this respect, it is desired that more sensitive entropies, specific heats at constant volume and compressibilities are also determined together with the volume by the thoery of Aryasetiawan et al.[163]

(C) The GB thermodynamic variational method for WH type interionic potentials with HSYO and HS respectively as a reference

Hard spheres with the Yukawa tail ("HSYO") was adopted by Hausleitner and Hafner[61] as a reference in the GB thermodynamic variational theory to predict the thermodynamic properties of liquid transition metals. The GB ineqaulity can be written as

$$F \leq F^V(\sigma;\varepsilon,k_B) = N\left\{\frac{3}{2}k_BT + u_{eg} + E_{BS} + E_{ES} + E_d - (S_{HSYO} + S_{el})\right\} \cdot (3\text{-}3\text{-}27)$$

In this equation, the energy and the entropy of electron gas, u_{eg} and S_{el} are given by Eqns.(3-3-4) and (3-3-5) respectively. The band structure energy E_{BS} and the electrostatic energy E_{ES} for the HSYO reference structure are respectively written explicitely as follows:

$$E_{BS} = (2\pi^2 n)^{-1} \int_0^\infty F(q) \, S_{HSYO}(q;\sigma,\varepsilon,k_S) \, q^2 dq \, . \qquad (3\text{-}3\text{-}28)$$

$$E_{ES} = \frac{2Z_S^2}{\pi} \int_0^\infty (S_{HSYO}(q;\sigma,\varepsilon,k_S) - 1) dq \, . \qquad (3\text{-}3\text{-}29)$$

In Eqn.(3-3-28), $F(q)$ is the wave number characteristic and following form was adopted.

$$F(q) = \frac{v_a q^2 |w_b(q)|^2}{8\pi e^2} \left[\frac{1}{\varepsilon_p(q)} - 1 \right] . \qquad (3\text{-}3\text{-}30)$$

where $w_b(q)$ is the the pseudopotential form factor for a bare potential $w_b(r)$ and $\varepsilon_p(q)$ is the dielectric screening function for electron gas. The d electron part, E_d, is expressed by using the d part of WH interionic potential (Eqns.(3-1-59) and (3-1-60)) as follows:

$$E_d = \frac{1}{2} n \int g_{HSYO}(r;\sigma,\varepsilon,k_S)(\phi_b(r) + \phi_{rep}(r)) \, . \qquad (3\text{-}3\text{-}31)$$

In table 3.7 are shown numerical results of calculations for thermodynamic properties due to this theory. $Z_s=1.5$ was assumed throughout. The minimum condition of $F^V(\sigma, \varepsilon, k_s)$ was searched with respects to the variation of σ, ε, and k_s. The calculated entropy on this theory is equally good or better than the OCP model [64] except for the case of liquid Ni and Cr. In the case of liquid Mn, the HSYO calculation is best though the degree of agreements are insufficient.

The variationally determined ξ shows a minimum at Mn in liquid 3d transition metal series and except for Ti, the softness parameter, k_s, decreases with the increase of 3d shell filling. These trends are consistent with the speculations that the binbding is strongest for the half filled d shell and with the increase of d shell filling the interionic potential becomes harder.

It should be mentioned that the HSYO model posseses somewhat curious properties of penetrable hard spheres ("non additivity"), that is the uniform negative background charge of elecrons penetrates even into the inside of hard spheres.

In Fig.3.2 are shown the Helmholtz free energy, F, the internal energy, U, and the entropy of liquid 3d transition metals determined variationally by the GB inequality by Hausleitner

Table 3.7 The predicted entropy S (in unit of Nk_B) for liquid transition metals by variational calculations with one component hard spheres with the Yukawa tail (HSYO) as a reference system[IH]. The values given in the parentheses refer to hard-sphere variational calculations (with $\varepsilon=0$).

	T(K)	Ω(au)	ξ	ε(mRyd)	k_S(au^{-1})	S_{HSYO}^{HH}	experiment	S_{OCP}^{IS}	S_{IIS}^{IS}	S_{PYP}^{IYW}
Ti	1973	130.51	0.362 (0.392)	58	1.36	12.42	12.01	11.55	11.06	11.17
V	2173	106.43	0.374 (0.419)	92	1.12	12.15	12.08	11.61	11.24	11.31
Cr	2173	92.93	0.380 (0.416)	87	1.22	12.00	11.58	11.34	10.95	11.07
Mn	1533	103.06	0.404 (0.436)	50	1.54	11.14	12.05	10.87	10.41	10.63
Fe	1823	89.29	0.389 (0.411)	43	2.00	11.93	12.11	11.39	10.98	11.08
Co	1823	85.85	0.370 (0.386)	30	3.04	12.44	12.08	11.49	10.98	11.19
Ni	1773	85.24	0.351 (0.372)	28	3.37	12.76	11.69	11.56	11.08	11.29
Zr	2173	172.40	0.428 (0.473)	133	0.90	12.54	13.24	—	—	—

HH :Hausleitner,C. and Hafner,J. , J. Phys. F:Metal Phys. , 18,1025(1988).
IS :Itami,T. and Shimoji,M. , J. Phys. F:Metal Phys. ,14,L15,(1984).
IYW:Itoh,H.,Yokoyama,Y. , J. Phys. F:Metal Phys. ,16,L113,(1986) .

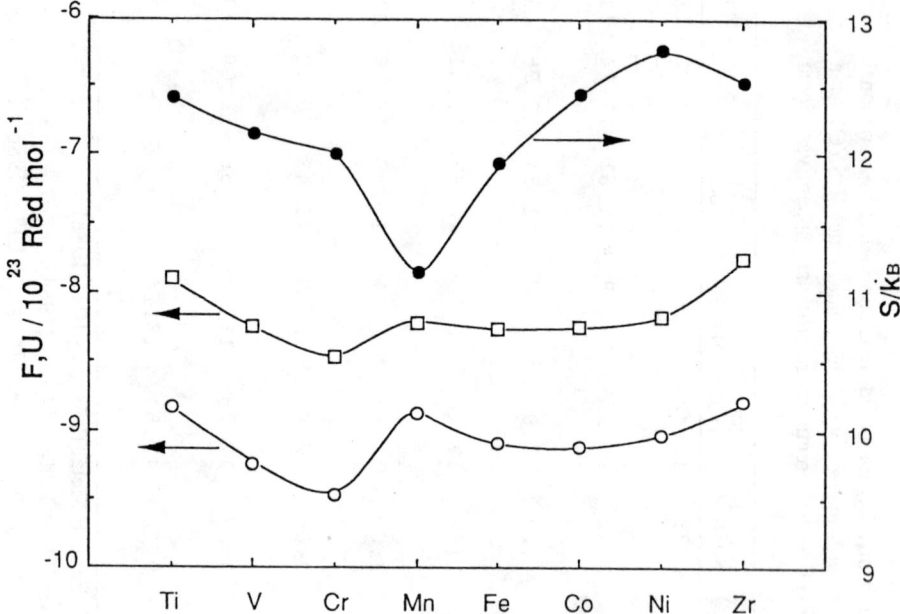

Figure 3.2 The variation of the Helmholtz free energy F(○), the internal energy U (□), and the entropy S (●) of liquid 3d transition metal series due to theoretical calculations bared on the Gibbs-Bogoliubov method with HSYO as a reference system.[†]

†: Hausleitner,C. and Hafner,J. , J. Phys. F:Metal Phys.,18,1013(1988).

and Hafner.[61] As can be seen in this figure, the nearly parabolic binding energy with the anomaly of liquid Mn ("W shape") case can be reproduced also from this NFE point of view. This success may be probably derived from the process of introducing the experimental volume to this calculation. This situation is very similar to the GB-TBA, in which the fitting procedure of cohesive energy at T=o seems to be essential to obtain such trends of free energy.

The case of $\varepsilon=0$ for GB-HSYO corresponds to the Gibbs-Bogoliubov variational case with the hard sphere model as a reference liquid, that is GB-HS. Hausleitner and Hafner[61] performed the GB-HS with WH interionic potentials for liquid transition metals. Bretonnet and Derouiche[104] also performed similar calculations with WH interionic potentials for liquid 3d transition metals. The Helmholtz free energy obtained by these studies shows nearly "W shape" dependence on d shell filling, which may be consistent with the Mn anomaly, as discussed in Secs.3.1.4, 3.3.1 and 3.5.

(D) The GB thermodynamic variational method for WH type interionic potentials with CHSO as a reference

The charged hard sphere model (CHSO) was also applied as a reference of GB variational method[59] for calculations of thermodynamic properties of liquid Fe, Co and Ni. The GB inequality can be written for the CHSO reference system as follows:

$$F \leq F^V = F_{CHSO} + \frac{n}{2} u_1(o;n) + \frac{1}{2} \int \frac{d^3q}{(2\pi)^3} u_1(q;n) \left(S_{CHSO}(q) - 1 \right) + E_d \cdot \quad (3\text{-}3\text{-}32)$$

In this equation, F_{CHSO} is given by Eqn.(3-3-22); $S_{CHSO}(q)$ is the structure factor of CHSO model liquid; E_d is defined as Eqn.(3-3-31); $u_1(q,n)=\phi_s(q) - 4\pi(Z_s e)^2/q^2$ where $\phi_s(q)$ is the Fourier transform of s part of interionic potential of WH type. The parameter ξ was variationally changed only for the case of hard sphere reference($Z_s=0$ or $\Gamma=0$). For variational calculations of the CHSO reference, the ξ was set to be same as that for the HS reference system and only Γ was variationally changed as a variational parameter. Introduction of coulombic tail into the HS model improved the calculated results, as can be seen in table 3.8. However, agreements were excellent only for the heat capacity at the constant volume of liquid Co. This CHSO model also corresponds to non addtive model for hard spheres and allows the penetration of negative uniform charge of electrons into the inside of hard spheres.

Table 3.8 The predicted entropy S and the heat capacity at constant volume C_V for liquid Fe, Co and Ni by the thermodynamic variational methods with the charged hard sphere (CHSO) and hard sphere (HS) references[BDJ] (in unit of Nk_B).

	T/K	V_M/a.u.	ξ	Γ	r_C	S_{HS}	S_{CHSO}	S^{exp}	C_V^{CHSO}	C_V^{exp}
Fe	1823	89.29	0.438	118	1.61	11.13	11.21	12.11	3.86	4.83
			0.451	138	1.66	10.90	11.08		3.70	
			0.420	99	1.56	11.44	11.40		4.03	
Co	1823	85.85	0.448	133	1.48	11.21	11.36	12.44	4.02	4.00
			0.462	150	1.52	10.94	11.25		3.74	
			0.437	116	1.44	11.40	11.49		4.07	
Ni	1773	85.24	0.440	122	1.36	11.12	11.21	11.69	3.87	4.46
			0.452	136	1.40	10.90	11.10		3.71	
			0.431	108	1.32	11.27	11.31		3.95	

BDJ: Bari,A., Das,T. and Joarder,R.N., J. Non-Crist. Solids. 136,173(1991).

3.3.3 Thermodynamic properties of liquid transition metal alloys

Most theoretical studies of thermodynamic properties of liquid transition metal alloys have been performed from very simple view points of TBA for d bands, in which almost no attentions have been paied for liquid structures. So far as the author knows, from a point of view of liquid theories, thermodynamic properties have been studied only for liquid noble metals alloys (Ag-Au, Au-Cu, Ag-Cu) by Joarder et al.[132] They calculated the excess entropy and the heat of mixing of these liquid systems. The GB variational method was employed for determinations of the optimum hard sphere diameters σ_i (i=1,2) in the hard sphere reference system. The simple model potential of the Borchi-De Gennaro [88] was employed for evaluations of interionic potentials. Calculated entropies are in fairly good agreement with experiments. However, calculated heat of mixing was in poor agreements with experiment even if the overlap contributions are taken into account. This poor prediction of the heat of mixing by the pseudopotential theory is common to the case of simple metals [166] and may be derived from the sensitivity of such calculations to potential parameters, overlap correction and dielectric screening functions.

Tight binding methods are successful for predicting the concentration dependence of thermodynamic properties of liquid transition metals, which shows interesting behaviors at the intermediate concentration range. For example, largely negative enthalpies and entropies of mixing (ΔH_M and ΔS_M) are found at certain Si concentration for liquid Fe-, Co- and Ni-Si alloys[167] and liquid Co- and Ni-Al alloys.[168] In the case of liquid Pd-Si alloys, minima of the entropy of mixing are present at two Si concentrations, $X_{Si}=0.25$ and 0.55.[169] Hicter and Cyro-Lackmann and collaborators[170-172] have succeeded in explaining these behaviors theoretically by introducing the idea of critical polyvalent metal concentration X_p^* in the tight binding scheme for d-bands, which is implicitly or explicitly pressumed. The idea for their discussion is as follows: for low concentrations of polyvalent metal, such as Si or Al, all sp electrons on a Si or Al atom transfer into d bands, which is formed by transition metal atoms; for $X_p < X_p^*$ only d like electrons are present; at $X_p = X_p^*$ a complete filling of d bands is accomplished; for $X_p > X_p^*$ all conduction electrons are sp electrons though their number per atom is decreased as $Z_{eff} = (X_p - X_p^*)Z_p$ (Z_p: the valence of polyvalent metals). The entropy of mixing can be considered to be composed of contributions of conduction electrons, S_{el}, magnetic terms, S_{mag}, derived from d band electrons and ionic contributions, S_{conf}. For $X_p < X_p^*$, S_{el} and S_{mag} decrease linearly as $S_{el} = S_{el0}(1 - X_p/X_p^*)$ and $S_{mag} = S_{mag0}(1 - X_p/X_p^*)$ respectively with the increase of d electrons, which is caused by the addition of sp metals or the increase of X_P. S_{el0} and S_{mag0} indicate the electronic and magnetic entropy of the transition metal of interest in the pure state respectively.[170] The entropy of pure liquid transition metals S_{el0} was estimated from the theoretical calculation of the density of states, $N(E_F)$, for d states based on the hard sphere and Lennard-Jones relaxed models respectively.[173,174] For $X_P > X_P^*$, the S_{el} was evaluated by the well known formula Eqn. (3-3-5a), for which values of $N(E_F)$ were taken from solid band structure calculations for these systems. The magnetic entropy, S_{mag0}, was taken from the data of Lytton.[175] The contribution of S_{conf} was neglected for simplicity.

In this way, Khanna et al.[175a] succeeded in reproducing qualitatively the largely negative ΔS_M for liquid Fe-, Co- and Ni-Si alloys. Moreover, Pasturel et al.[171] also succeeded in explaining the largely negative ΔH_M in addition to the negative behavior of ΔS_M for these liquid transition metal alloys. The method adopted by them is essentially same as described above. However, the DOS was calculated in their work directly by the tight binding method.[176]

The presence of two minima in ΔS_M for liquid Pd-Si alloys has been also successfully explained by Pasturel et al.[172] by taking into account of ΔS_{conf} by the hard sphere model; the minimum position of ΔS_{conf} on the isotherm is situated at $X_{Si} = 0.5$ differently from at $X_{Si} = 0.25$ for that of ΔS_{el}. Their conclusions are as follows: both electronic and ionic configurational (or packing) effects are responsible for "anomalies" in the entropy, the minima of these two effects may arise at neibouring compositions for liquid Fe-, (Co-), Ni-Si alloys and those may arise at distinctly different compositions each other for ΔS_M in liquid Pd-Si alloys with two minima.

Similar discussions due to the d band filling were extended successfully to the prediction of negative ΔH_M for solid transition metal-In, -Ga and -Ge alloys.[177,178] In addition, a similar approach was made for evaluations of the partial excess entropy of transition metal at a infinite dilution in liquid polyvalent metals, such as Ni, Fe, Co and Pd in liquid Si. The calculated results are in fair agreement with experimental data and the predicted values are given for systems Ni, Co, Fe and Pd in liquid Al, Ni in liquid Sn, Ni and Pd in liquid Ge and Pt in liquid Si.[179,180]

As mentioned in Sec.3.2, Pasturel and Hafner[53] formulated the self consistent theory of thermodynamics with d charge transfer in liquid and amorphous transition metal alloys from the coupling of GB variational method with the TBA approximation for internal energies. Unfortunately, its application have been performed only to metallic glasses.

Quite recently Pasturel et al.[181] presented the first principles determination of the Ni-Al phase diagram, in which th configurational free energy of liquid alloy phases is replaced by the free energy of FCC structure due to the tetrahedron approximation in the cluster variation method.

3.4 Electronic properties of liquid transition metals

3.4.1 Electrical resistivity

(i) Experimental data of electrical resistivity, ρ^L, of liquid transition metals

In table 3.9 experimental data of electrical resistivity, ρ^L, of liquid transition metals are summarized together with various theoretical calculations, which have been reported so far. Particular features of ρ^L of liquid transition metals are summarized as follows:

(1) Values of ρ^L at the melting point of liquid transition metals change in the range from 83 µΩcm to 199.5µΩcm.
(2) Values of ρ^L show a maximum for Mn with the half filled 3d shell with the variation of 3d shell filling in the atomic state.
(3) The value of ρ^L for liquid Ti shows an anomalously large in the trend of 3d transition metal series.
(4) The increase of resistivity on melting is as large as 12~27% of solid resistivity value, ρ^S, for liquid V, Cr, Co, Mo and for some liquid transition metals it is larger than 40% of ρ^S, for example, 40% for Pt, 44% for Ni and 70% for Pd.
(5) The increase of resistivity on melting is smaller (1 ~ 10%) for liquid Ti, Mn, Fe, La and W.

The point (2) indicates that d electrons or d states play important roles in the electronic conduction process though the magnitude of ρ^L is not so large compared with values of simple sp metals. From the points (4) and (5), it is known that the resistivity change on melting is very small for transition metals with a half filled d shell in the atomic state and large for those with a nearly filled d shell in it; for example, only 3% of solid values for Mn with $3d^54s^2$ and 70% for Pd with $4d^{10}$ in the atomic state. The change of resistivity on melting was discussed by Dupree et al. [182) based on the theoretical analysis of ρ^L for liquid transition metals due to Ten Bosch and Bennemann. [183) According to their interpretation, there are two kinds of contributions to the ρ^L of liquid transition metals, d contributions and

Table 3.9 Electrical resistivities of liquid transition metals at the T_m (in unit of $\mu\Omega$cm).

	experiments			calculations							
	ρ^L	$\dfrac{\rho^L-\rho^S}{\rho^S}$	source	EGL (1)	(2)	DEGK	B1 (1)	(2)	LEB	HWJS	BLET
Sc	–	–	–	–	–	–	–	–	–	–	–
Ti	167	0.01	VZ2	–	–	–	–	–	–	350	138~469
V	135	0.20	VZ2	–	–	–	–	–	–	280	–
Cr	150	0.27	VZ1	–	–	–	–	–	–	120	–
Mn	199.5	0.03	OO	–	–	–	–	–	–	188	–
Fe	138	0.08	VZ2	372	276	196	–	–	–	182	–
Co	115	0.18	VZ2	–	–	–	–	–	–	83.3	–
Ni	83	0.44	VZ2	227	106	78	27.2	56.3	–	54.9	–
Zr	–	–	–	–	–	–	–	–	–	235	–
Mo	97	0.12	SF	–	–	–	–	–	–	–	–
Pd	83	0.70	DVZE	–	–	–	46.6	62	83~218	51.8	–
La	135	0.06	GHK	–	–	–	–	–	–	–	–
W	135	0.1	SF	–	–	–	–	–	–	–	–
Pt	83	0.4	VZ2	–	–	–	–	–	–	94.6	–

	calculatons												
	DJ (1)	(2)	WJT	EEG	DB (1)	(2)	B2	GP (1)	(2)	F	BBH1	BBH2	B3	OFK
Sc	–	–	–	–	–	–	–	–	–	–	–	–	–
Ti	–	–	–	–	–	–	627	–	–	–	–	–	–
V	385	452	–	–	–	–	455	–	–	–	–	118	–
Cr	115	276	–	–	–	–	204	–	–	–	106	130	–
Mn	172	499	–	–	–	–	131	–	–	–	124	166	–
Fe	174	360	–	1130	–	–	476	130	1173	129	180	139	136	–
Co	71	167	–	329	–	–	178	61	187	87	–	–	–
Ni	102	175	–	74	–	–	52	41	57	67	–	–	–	67
Zr	–	–	–	–	–	–	–	–	–	–	–	–	–
Mo	–	–	–	–	–	–	–	–	–	–	–	–	–
Pd	84	90	–	–	–	–	–	36	–	–	–	–	–
La	–	–	165	–	133	158	–	–	–	–	151	–	–
W	–	–	–	–	–	–	–	–	–	–	–	–	–
Pt	146	200	–	–	–	–	–	62	–	–	–	–	–

ρ^L: the electrical resistivity ρ of liquid at the T_m ; ρ^S: the ρ of solid at the T_m.
Experiments
- DVZE :Dupree,B.C. , Van Zytveld,J.B. and Enderby,J.E. , J. Phys. F:Metal Phys., 5,L200(1975).
- GHK :Güntherodt,H.-J. , Hauser,E. and Künzi,H.U. , Inst. Phys. Conf. Ser. , No.30,324(1977).
- OO :Okada,T. and Ohno,S. , J. Phys. Soc. Jpn. , 55,599(1986).
- SF :Seydel,U and Fucke,W. , J. Phys. F:Metal Phys. , 10,L203(1980).
- VZ1 :Van Zytveld,J.B. , J. de Physique , C8,503(1980).
- VZ2 :Van Zytveld,J.B. , J. Non Cryst. Solids , 61&62,1085(1984).

Calculations
- B1 :Brown,J.S. , J. Phys. F:Metal Phys. , 3,1003(1973).
- BLET :Brown,J.S. , Lopez-Escobar,A.H.M. and Todd,J.R. , J. Phys. F:Metal Phys. , 8,1703(1978).
- B2 :Brown,J.S. , J. Phys. F:Metal Phys. , 11.2099(1981).
- B3 :Ballentine,L.E. , "Rapidly Quenched Metals" (ed. by Steeb,S. and Warlimont,H.),Elsevier Science Pubblishers B.V. , 981,(1985).
- BBH1 :Ballentine,L.E. , Bose,S.K. and Hammerberg,J.E. , J. Non Cryst. Solids , 61&62,1195(1984).
- BBH2 :Bose,S.K. , Ballentine,L.E. and Hammerberg,J.E. , J. Phys. F:Metal Phys. , 13,2089(1983).
- DB :Delley,B. and Beck,H. , J. Phys. F:Metal Phys. , 9,517(1979).
- DEGK :Dreirach,O. , Evans,R. , Güntherodt,H.-J. and Künzi,H.-U., J. Phys. F:Metal Phys. , 2,709(1972).
- DJ :Dunleavy,H.N. and Jones,W. , J. Phys. F:Metal Phys. , 8,1477(1978).
- EEG :Esposito,E. , Ehrenreich,H. and Gelatt,Jr.,C.D. , Phys. Rev., 18B,3918,(1978).
- EGL :Evans,R. , Greenwood,D.A. ans Loyd,P. , Phys. Lett. , 25A,57(1971).
- F :Fujiwara,T. , J. Phys. F:Metal Phys. , 9,2011(1979)
- GP :Gorecke,J. and Popielawski,J. , J. Phys. F:Metal Phys. , 13,2107(1983).
- HWJS :Hirata,K. , Waseda,Y. , Jain,A. and Srivastava,R. , Phys. F:Metal Phys. , 7,419(1977)
- LEB :Lopez-Eskoba,A.H. and Brown,J.S. , Phil. Mag. ,35,1609,(1977).
- OFK :Ostermier,H. , Fembacher,W. and Krey,U. , Zeit. für Physikalische Chemie , Neue Folge ,157,S489(1988).
- WJT :Waseda,Y. , Jain,A. and Tamaki,S. , J. Phys. F:Metal Phys. , 8,123(1978).

s contributions. The former is insensitive to the melting and the latter is sensitive to it. The ρ^L of liquid transition metals becomes larger with the increase of unfilling of d bands due to d contributions. Implicitly it is assumed that transition metal atoms with unfilled d shells form solids and liquids with an unfilled d band and those with fully filled d shells form solids and liquids with a fully filled d band. Therefore, for liquid transition metals with unfilled d bands the resistivity is large and it changes only a little on melting. This explains very well the important roles of d bands on the resistivity though the conduction mechanism itself in liquid transiton metals has not yet been established, as described in the following discussions.

For Cr, a comparably large change of resistivity on melting can be seen in spite of large resistivity. This may be derived partly from the slightly large change of DOS due to the structure change from the BCC below T_m to the FCC like in the liquid state, as speculated as one possibility by Asano and Yonezawa.[161] They found from their theoretical calculations that the DOS of liquid transition metals resembles that of FCC solids and differs considerably from that of BCC solids. The structure of liquid transition metals is closer to the FCC structure from the fact that the number of nearest neighbours ("NNN") obtained from the X ray structure analysis,[4] 10 ~11, is close to the NNN of FCC solids, 12. However, from this point of view, the very small change of ρ^L is difficult to be understood for liquid Fe with a similar structure change on melting from the BCC to the FCC like. In fact, the change of magnetic susceptibility on melting is found to be large probably due to this change of DOS, as described in Sec.3.4.3.

The large value and small change on melting (1%) of resistivity for liquid Ti ($3d^2 4s^2$) (point (3)) seem to be mysterious only from the interpretation based on d shell filling.

(ii) The extended Ziman formula for ρ^L of liquid transition metals

As described above, the ρ^L of liquid transition metals is not so large compared with that of liquid simple metals, which is successfully explained by the nearly free electron model of Ziman.[184] Evans et al.[185,186] presented the so called "extended Ziman formula" based on the force-force correlation function formalism of electrical conductivity. This extended Ziman formula takes the form which can be obtained in the Ziman formula simply by replacing the weak pseudopotential matrix element with the single site, on-shell t-matrix $t(\mathbf{k},\mathbf{k}+\mathbf{q})$, which is defined by the muffin tin ("MT") approximation. The explicit form is as follows:

$$\rho^L = \frac{3\pi V_a}{|e|^2 (h/2\pi) v_F^2} \langle |T(\mathbf{k}_F, \mathbf{q}+\mathbf{k}_F)|^2 \rangle, \qquad (3\text{-}4\text{-}1)$$

where v_a is the atomic volume, v_F the Fermi velocity, k_F the Fermi wave vector. <----> indicates the integral described below.

$$\langle |T(k,K')|^2 \rangle \equiv 4\int_0^1 |T(k,K')|^2 \left(\frac{q}{2k_F}\right)^3 d\left(\frac{q}{2k_F}\right). \qquad (3\text{-}4\text{-}2)$$

In Eqn.(3-4-2)

$$|T(k,K')|^2 \equiv S(q)|t(k,K')|^2, \qquad (3\text{-}4\text{-}3)$$

where $q = |k_F - k'_F|$ and $S(q)$ is the structure factor, as defined in Eqn.(3-1-1). The $t(k,k')$ is the muffin tin t-matrix element, which represents the intensity of scattering center (transition metal ion) for scattering events of conduction electron from a momentum state k to another k'. The $t(k,k')$ is related to the phase shift $\eta_l(E)$ as follows:

$$t(k,k') = -\frac{h^3 n}{4\pi^2 m\sqrt{2mE}} \sum (2l+1) \sin\eta_l(E) \exp(i\eta_l(E)) P_l(\cos\theta), \qquad (3\text{-}4\text{-}4)$$

where $P_l(\cos\theta)$ is the Legendre polynominal of l-th order as a function of angle θ between k and k'. The phase shift $\eta_l(E)$ is calculated as a function of energy E by matching at the muffin tin ("MT") radius R_{MT} the solution of the radial Schrodinger equation for the non-overlapping muffin tin potential inside the MT spheres to the free space solution outside of MT spheres. The form of latter free space solution takes the linear combination of spherical Bessel function and Neumann function. The non-overlapping MT potential required is set to be constant in the range $R_{MT} < r < R_{WS}$ (R_{WS}: the radius of the Wigner Seitz sphere) and spherically symmetric in the range $r < R_{MT}$. This spherical potential can be constructed by the combination of exchange potential and coulombic potential due to the overlap of both atomic potentials and electron charge density of nearest neighbours on the central atom.[12,187,188]

(iii) Numerical calculations based on the extended Ziman formula and attempts to improve it

Numerical calculations of ρ^L for liquid transition metals were performed based on Eqn. (3-4-1) by many authors (see EGL (1) and (2), DEGK, B1(1) and (2), LEB, HWJS, BLET and WJT in table 3.9). Calculated results were generally good except for the case of liquid Ti and V as can be seen in the case of HWJS. In spite of these rather remarkable successes, the extended Ziman theory has been criticized [2,3,189] for the weak foundation from the theoretical point of view, such as the complete neglection of multiple scattering effects for considerably strong muffin tin t-matrix, arbitrarinesses of input parameters for theoretical calculations of ρ^L. In fact, there are many arbitrarinesses in input parameters, such as the valence Z, k_F and the Fermi energy E_F, and in the form of exchange potentials included in the MT potential. For example, Z was taken to be 2 by Evans et al.[185] (see EGL in table 3.9) and 1 by Dreirach et al.[190] (DEGK) for liquid Fe and Ni, 1 for liquid transition metals by Hirata et al.[191] (HWJS), 3 by Delly and Beck[192] (DB) and Waseda et al.[193] (WJT) for liquid La. Even 0.36 and 0.46 were adopted for liquid Pd and Ni respectively by Brown[194] (B1).

Several forms of exchange potential were tried by Lopez-Escobar and Brown[195] (LEB) for liquid Pd and by Brown et al.[196] (BLET) for liquid Ti. The calculated ρ^L due to the Slator exchange gave a close agreement with experiments for liquid Pd; on the other hand a good agreement with experiments was obtained for the calculated ρ^L of liquid Ti due to the Kohn-Sham-Gaspar exchange and $3d^34s^1$ electron configulation assumed in the atomic state. Esposito et al.[12] performed the self consistent calculation of ρ^L of liquid transition metals and amorphous transition-metaloid glasses in which the Fermi energy E_F and effective valence Z^* was determined rigorously in a self consistent manner; E_F was determined so that all s and d eletrons are contained in the integrated density of states up to E_F; effective valence Z^* was determined from the integration of free electron density of states up to E_F. However, calculated results for liquid Co and Fe were too large (EEG); Z^* was determined to be 1.21 for Liquid Fe, 0.99 for liquid Co and 0.95 for liquid Ni.

Some attempts were performed to improve the extended Ziman theory. Brown[197] took account of corrections for the effect of short mean path of conduction electrons based on the theory of Ferraz and March,[198] which incorpolates the bluring of Fermi surface consistently with the finite mean free path. In the case of liquid Fe, the mean free path was as extremely short as 0.8 multiplied by the muffin tin radius, R_{MT}. Therefore, the multiple scattering corrections improved considerably the result of single site approximation, EEG, in table 3.9, though the calculated result for liquid Fe was still far larger than experimental values (B2). Another attempt was done by Dunleavy and Jones[199] in which multiple scattering effects were taken into accounts on quasi crystalline approximation ("QCA"), by which some same kind of higher order terms are systematically taken up to an infinite order in the expansion of scattering matrix. The results (DJ(2)) were improved considerably compared with the single site t-matrix case (DJ(1)) though the calculated result for liquid V remained to be insufficient. Goreck and Popielawski[200] estimated numerically the contribution of triplet correlation function to the ρ^L for many liquid metals based on the self consistent quasi crystalline approximation ("SQCA"). In table 3.9 "GP(1)" indicates the calculated results with triplet contributions and the calculations within the extended Ziman approximation are given as "GP(2)"; for liquid Cu, Ag, Ca and Hg the contribution of higher order correlations may be small; for liquid Pt, Ni, Fe and Co calculated resistivities are an order of experimental values; the contribution of triplet term decreases considerably the calculated values based on the single site t-matrix approximation. However, for light transition metals, Ti, V and Cr, contributions of triplet term are as large as that of two body term and it was concluded that the NFE model for the electron transport is not appropriate for these light transition metals.

(iv) The s-d scattering model of Mott

Mott [201] presented another formula of ρ^L on the s-d scattering model. In this theory, the different mean free path, L, was assumed respectively for s electrons (L_s) and d electrons (L_d), which was suggested for solid transition metals.[202,203] Because of $L_s \gg L_d$, ρ^L is determined by the transition of s electrons on one d site to another d site. The explicit form of ρ^L on this model is as follows:

$$\rho^L = \frac{m}{Z_s |e|^2} \left(\frac{N}{V}\right)\left(\frac{4\pi^2}{h}\right) |\gamma_{sd}|^2 N_d(E_F) . \qquad (3\text{-}4\text{-}5)$$

In this equation, Z_s (=1) is the valence of s electrons, $N_d(E_F)$ is the DOS of d states; $|\gamma_{sd}|$ is the coupling constant for the s-d scattering. This explains rather structure insensitive resistivities for transition metals, which can be seen from the smaller change of experimental resistivity on melting and smaller temparature dependence of experimental ρ^L for transition metals compared with those for simple metals.

(v) Calculations of electrical resistivity on the s-d scattering model of Mott

Fujiwara [204] calculated the resistivity of liquid Fe, Co and Ni based on Eqn.(3-4-5). The liquid structure was realized on computers by the MD simulation. Under this simulated structure, the DOS required for Eqn.(3-4-5) was calculated by the recursion method[205] on the TBA. For this numerical calculation the appropriate forms of atomic orbitals and hopping integrals were introduced. Calculated results are considerably good as can be seen in F in table 3.9.

(vi) Calculations of electrical resistivity for non-clustered liquids with a model Hamiltonian for s and d states

Ten Bosch and Bennemann [183] discussed the contribution of d electrons to resistivity based on the Kubo-Greenwood formula [1,3] by using the model Hamiltonian containing s and d states. The obtained conclusion is that the contribution of d states becomes comparable order of magnitude as that of s states for a half filllled d band and it becomes smaller for a more than half filled or a more than half empty d band. Thus they explained the maximum resistivity of liquid Mn in 3d series qualitatively.

Asano Yonezawa [161] also calculated the ρ^L based on the Kubo-Greenwood formula with some simplifications due to the decoupling of the Green functions. The ρ^L was calculated as a function of energy for liquid transition metals with muffin tin potentials and HS structure factors. The calculated results with the Ni potential agree well with experimental values for liquid Ni, Fe and Mn. These results also explain the maximum tendency of ρ^L with a half filled d shell qualitatively.

These calculations were performed for an infinitely disordered array of s and d states. This point is different from cluster calculations of finite size, which were performed by Fujiwara,[204] as described above and Ballentine and collaborators [206-209] discussed in the next part.

(vii) Cluster calculations for electrical resistivities of liquid transition metals

Based on the recursion method for the conductivity on the tight binding approximation ("TBA"), Ballentine and collaborators [206-209] have developed the calculation of ρ^L for a cluster with liquid structures of a few hundred particles realized by the MC method on computers.

The expression of ρ^L adopted for numerical calculations by them is written in the Butcher's form [210], as follows:

$$\rho^L = \frac{1}{|e|^2}\left(\frac{V}{N}\right)\frac{1}{N(E_F)D(E_F)}. \qquad (3\text{-}4\text{-}6)$$

In this equation the $N(E_F)$ is the DOS at the E_F and $D(E_F)$ is the energy dependent diffusivity at the E_F. The $D(E_F)$ is defined in terms of diffusivity function, as follows:

$$D(E_F) = \lim_{\varepsilon \to 0} D(E_F, E_F; \varepsilon). \qquad (3\text{-}4\text{-}7)$$

The diffusivity function has the form,

$$D(E,E';\varepsilon) = -\frac{h}{2\pi}\text{Im}\{\langle E_n|v_x G(E'+i\varepsilon)v_x|E_n\rangle\}_{E_n=E}, \quad (3\text{-}4\text{-}8)$$

where v_x is the component of the velocity operator in the direction of the current, $v_x = i[H,X]/(h/2\pi)$. The notation $\{\text{---}\}_{E_n=E}$ means an average over all energy eigenstates $|E_n\rangle$ on the shell of energy E. Eqn. (3-4-8) has a quite similar form to the expression of DOS when the $v_x|E_n\rangle$ is considered to be some kind of eigenstates. Therefore evaluations of it can be possible by the recursion method,[205] which is recently found to be very powerful for the calculation of DOS on TBA scheme.

Fundamentally this formula for the conductivity (or resistivity) is written in terms of the Green functions and is almost equivalent to the rigorous Kubo-Greenwood theory, which corresponds to the quite general linear response theory of conductivity for isotropically disordered materials with many scattering potential centers and with electrons described by one electron wave functions.[3]

One merit of this method is that s and d electrons are equally treated and no assumptions about electrons in a conduction band are required. This point differs completely from the case of extended Ziman theory which includes many assumptions about them. Ballentine and collaborators [208,209] calculated by means of this method the ρ^L of liquid La, Cr, Mn and Fe and obtained results are considerably good (BBH2 and B3 in table 3.9). According to their analysis, the contribution of s-d cross term in ρ^L is very small (less than 1% of total ρ^L) compared with that of s-s and d-d terms and the d states dominate the conductivity because of high partial density of states of d states though the diffusivity itself of d electrons is less than that of s electrons. Similar calculations were performed successfully for ρ^L of liquid Ni by Ostermeier et al[211] (OFK in table 3.9). At a first look obtained results were a little contradictory to that of Ballentine et al.[208-209] In the calculation of OFK the contribution of s-d hybridization is dominant for ρ^L of liquid Ni compared with s-s and d-d contributions. However, similar dominant contributions of s bands are reported for liquid Ni by Ballentine.[208,209] He concluded that the diffraction approach is applicable to liquid metals whose E_F lies outside of d bands and the LCAO approach is successful for liquid metals whose E_F lies inside of d band such as liquid Cr, Mn and Fe. Liquid Ni, with a rather long mean free path and an almost occupied d band, is a marginal case of the validity of cluster calculation of ρ^L based on the LCAO, which requires rather short mean free path due to the limitation of finite size of clusters.

(viii) The electrical resistivities of liquid transition metal alloys

There are two kinds of concentration dependence of ρ^L for liquid transition metal alloys.[20] In liquid transition metal-noble metal alloys, such as Ni-Au and Fe-Au systems, the parabolic concentration dependence can be seen similarly to the Nordheim rule in solid solutions and the temperature coefficients of ρ^L are kept to be positive. In liquid transition metal-polyvalent metal alloys, such as liquid Fe-Ge and Ni-Ge systems, negative temperature coefficients of ρ^L appear in the middle concentration range though a similar parabolic concentration dependence are found for ρ^L itself. The concentration of maximum of ρ^L does not always accord with that of minimum of the temperature coefficient of ρ^L.

It is to be noted that the concentration range for negative temperature coefficients of ρ^L is much wider for liquid Ni-, Co-, Fe- and Mn-Ge alloys than for liquid Cu-Ge alloys. This feature has been explained by the mechanism of transfer of sp electrons in Ge to d states in transition metal atoms,[20,212] as described below. As is well known for simple liquid metals the negative temperature coefficient of ρ^L appears when the $2k_F$ (k_F: the Fermi wave number) becomes nearly equal to the q_0, which is defined as the wave number of the first peak position of partial structure factors $a_{ij}(q)$. Roughly speaking this condition corresponds to effective valence $Z^* \approx 2$.[1] In transition metal-Ge alloys, it may be considered that some part of sp electrons derived from the Ge atom is removed from the conduction band and is transfered into d states of transition metal atom; transition metal atoms act as a monovalent metals sharing only one electron to the conduction band. It needs much more concentrated Ge atoms for liquid transition metal-Ge alloys to establish the condition $Z^* \approx 2$ than for liquid Ge mixtures alloyed with Cu, which has no room in the 3d shell to accomodate electrons transfered from Ge. This explains wider concentration range of negative temperature coefficient for liquid transition metal-Ge alloys than for liquid Cu-Ge alloys.

Therefore the Ziman type formula of ρ^L may work for liquid transition metal alloys. The extended Ziman formula, Eqn. (3-4-1), can be easily extended into the case of liquid transition metal alloys. The factor $|T(k,k')|^2 = S(q)t(k,k')^2$ in Eqn. (3-4-1) must be replaced for liquid transition metal alloys as follows:

$$|T(k,k')|^2 \equiv x_1|t_1(k,k')|^2(1-x_1+x_1 a_{11}(q)) + x_2|t_2(k,k')|^2(1-x_2+x_2 a_{22}(q))$$
$$+ x_1 x_2 (t_1(k,k')^* t_2(k,k') + t_1(k,k') t_2(k,k')^*)(a_{12}(q)-1), \quad (3\text{-}4\text{-}9)$$

where x_k and $t_k(k,k')$ are the atomic fraction and the muffin tin t-matrix of component k respectively; $a_{ik}(q)$ is the partial structure factor of FZ type.

The resistivity of liquid transition metal alloys has been successfully explained by the extended Ziman formula by Dreirach et al.[190] and Hirata et al.[191] However, Dupree et al.[213] were forced to take an unphysically low value of 0.2 as the valence Z in the extended Ziman formula to reproduce the measured ρ^L and thermoelectric power, Q, of liquid Co-Ni alloys. Newport et al.[214] also reported experiments of ρ^L and Q of liquid Ag-Pd alloys and their results supported the s-d scattering model of Mott as the conduction mechanism of electrons compared with the extended Ziman theory. It is desired that the cluster calculations on the LCAO are performed also for liquid transition metal alloys though computations for them may be much more time consuming.

Experimental studies of ρ^L for liquid transition metal alloys have been introduced in previous reviews and papers for liquid Co-Sn and Mn-Ge,[191] Fe-Au, Fe-Ge, Ni-Au, Ni-Ge and Co-Ge [20,212] and Ni-Sn.[190] Besides these studies, the author notices that experimental studies have been performed for following liquid systems containing transition metals ("TM"); Cr-Te [215] Mn-Bi,[216] Mn-Te,[215,217] Fe-Co, Fe-Ni and Ni-Cu,[218] Fe-Ni,[219] NI-Ge,[220] Ni-Sb,[221] Ni-Te,[222] Fe-Ni-B-Si,[223] Pd-Bi,[224] Pd-Te[225] and 3d TM-Te.[226] Xu and van der Lugt [224] reported the ρ^L of liquid Pd-Bi alloys up to 60 at.% Pd. At 30 at.% Pd ρ^L shows a maximum of 188 $\mu\Omega$cm and the temperature coefficient of ρ^L shows a minimum. This minimum behaviour of temperature coefficient is quite anomalous, that is the very narrow minimum with a nearly zero value which appears in the broad concentration range of comparably large positive temperature coefficients. The maximum of ρ^L itself is speculated to be derived from the accordance of Fermi energy with the pseudo gap of DOS, proposed by Mott [3,227]; their discussions were given based on the solid band structure calculations.

Okada and Ohno[217] showed distinctive metal-nonmetal transition behaviours in liquid Mn-Te by measurements of σ (inverse of ρ^L). With the addition of Mn the σ decreases steeply and a sharp minimum of conductivity appears at a stoichiometry of Mn-Te. Similar decrease of σ was found for the addition of Ti, V and Cr in 3d series though its tendency is weaker than the case of Mn.[226] On the other hand the increase of σ was found for the addition of Fe, Co, Ni[226] and Pd.[225] The minimum conductivity of liquid stoichiometric MnTe is 21.7 ohm^{-1} cm^{-1} ($4.6\times10^4 \mu\Omega$cm) rather in the range of insulaters. Such a metal to non-metal transition is concluded to be related to the formation of Mn^{2+} and Te^{2-} ions in liquid structures at stoichiometric MnTe.[217]

Okada and Ohno[226] discussed this difference of the effect of addition of transition metal elements based on the strong scattering model of Mott.,[227] which assumes the diffusive motion of electrons with a short mean free path of an order of interionic spacing in the disordered system with very strong scattering potentials. According to this model the electrical conductivity σ is proportional to the square of DOS at the Fermi energy, $N(E_F)$, that is $\sigma \propto N(E_F)^2$. In the case of addition of Co and Ni, d states of these elements are degenerate and the position of E_F lies in these degenerate d states; $N(E_F)$ is large at E_F and as a result σ is large. On the other hand the d state in liquid Te-Mn splits into up spin state below E_F and down spin state above E_F. In addition the formation of Mn^{2+} and Te^{2-} is present. $N(E_F)$ for both s and d bands is small at E_F and as a result σ is small. As for the concentration dependence of σ for liquid Mn-Te, Laundy et al.[215] discussed from the Mott-Hubbard model of magnetic insulators.[228]

(ix) Comments on the electrical resistivity of liquid transition metals

Results due to cluster calculations are in general in better agreement with experiments compared with the case of extended Ziman theory. It is to be remembered that the Kubo-Greenwood formula (in the Bucher's form) for the conductivity in cluster calculations is far more general than the extended Ziman formula. In addition s and d electrons are treated in eqaul weight without any approximations. Therefore, it is very intersting to perform such cluster calculations also for lighter liquid transition metals such as Ti and V, for which the extended Ziman formula has never given good results in most cases. The possibility of insufficient screening from nuclear charge for outer electrons in liquid Ti was mentioned in Sec.3.3 with the relation to the large atomic volume. For these metals with large atomic volumes, a localization tendency of electrons may increase due to the narrowing of d and s bands. In this respect measurements of the ρ^L for liquid states are expected for Sc, which is known to have the far larger atomic volume than the other 3d transition metals in the solid state.[229] The density functional calculation of cohesive energy for solid states are very successful even for Ti and V,[229] in which the exchange and correlation of electrons have been taken into account in detail. Therefore it is very interesting and important to clarify the roles of s-s, d-d and s-d transitions and electron correlations for these lighter transition metals.

3.4.2 Thermoelectric power

Experimental data of thermoelectric power Q of liquid transition metals are summarized in table 3.10 together with the calculated values based on various theories. Particular features of experimental data are as follows:

Table 3.10 Thermoelectric power Q (in unit of μVK^{-1}) and thermoelectric power parameter X (no dimension) of liquid transition metals.

	experiments			calculations	
	Q^L/X	Q^S	sources	Q^L/X	sources
Fe	-4±2/0.7	-1±1	E	-48/8.4	EGL
				-43/7.5	EGL,HWJS
				-7/1.2	F
Co	-4±2/0.7	-10±2	E	-10.9/1.9	HWJS
	-3±2/0.53			-12/2.2	F
				0.7/-0.13	KCLD
				-5/0.90	KCLD
Ni	-38±3/7.8	-44±3	HE	-38/7.8	EGL
				-33/6.8	EGL
				-11/2.3	HWJS
				-52/10.7	F
				-25/5.1	KCL
				45/-9.2	OFK
Pd	-41±3/6.1	-55	E	-78/11.6	DVZE
				-74/11.0	DVZE
				17/-2.5	HWJS
La	-7.5±1.0/1.06	-7.4	VZ	-2.4/0.34	WJT
				-7±1/0.99	BH

Experiments
E :Enderby,J.E. , Inst. Phys. Conf. Ser. , No.39,214(1978).
HE :Howe,R.A. and Enderby,J.E. , J. Phys. F:Metal Phys. , 3,L12(1973).
VZ :Van Zytveld,J. , "Handbook on the Physical Chemistry of Rare Earths ,vol.12,357(1989).

Calculations
BH :Ballentine,L.E. and Hammerberg,J.E. , Can. J. Phys. , 62,692(1984).
DVZE:Dupree,B.C. , Van Zytveld,J.B. and Enderby,J.B. , J. Phys. F:Metal Phys. 5,L200,(1975).
EGL :Evans,R. , Greenwood,D.A. ans Loyd,P. , Phys. Lett. , 25A,57(1971).
F :Fujiwara,T. , J. Phys. F:Metal Phys. , 9,2011(1979).
HWJS:Hirata,K. , Waseda,Y. , Jain,A. and Srivastava,R. , J. Phys. F:Metal Phys. , 7,419(1977).
KCL :Khanna,S.N. and Cyrot-Lackmann,F. , Phyl.Mag.,38B,197(1978).
KCLD:Khanna,S.N.,Cyrot-Lackmann,F. and Desjonquéres,M.C. , J. Phys. F:Metal Phys.,9,79(1979).
OFK :Ostermier,H. , Fembacher,W. and Krey,U. , Zeit. für Physikalische Chemie , Neue Folge ,157,S489(1988).
WJT :Waseda,Y. , Jain,A. and Tamaki,S. , J. Phys. F:Metal Phys. 8,125(1978).

(1) In small changes of Q on melting for liquid transition metals, they are comparably large for liquid Fe and Co and comparably small for liquid Pd and Ni.
(2) Large negative values for liquid Ni and Pd and small negative values for liquid Fe, Co and La.

As for point (1) it is to be noted that the comparably large change of Q on melting is found for liquid Fe and Co with the small change of resistivity and the large change of magnetic susceptibility, χ_M, on melting. On the other hand the comparably small change of Q on melting are found for Ni and Pd with the large change of resistivity and the small change of χ_M on melting. As for point (2), it is to be noted that the negatively small Q are found for liquid Fe and Co with the high resistivity and the large χ_M. On the other hand the comparably largely negative Q are found for Ni and Pd with the low resistivity and the small χ_M.

Though the mechanisms of Q and χ_M are discussed later both in this section and in Sec.3.4.3, following comments are simply presented here. Point (2) may be related to the fact that the conduction of electrons in Fe and Co is dominated by a d band, whose resistivity is high and which contributes less negatively to Q and paramagnetically to χ_M. On the other hand the conduction of electrons in liquid Ni and Pd is dominated by a s electron, whose resistivity is low, whose Q is largely negative and which contributes less paramagnetically to χ_M. It is difficult to explain the point (1) completely. However, as for the change of Q and χ_M on melting it may correspond to the small change of DOS on melting for Pd and Ni and comparably large change of DOS for Fe and Co. The change of resistivity may be explained as in the same manner as Dupree et al.,[182] as already described in Sec.3.4.1.

(i) Theoretical expression for the thermoelectric power

As is well known, the thermoelectric power Q is closely related to the electrical resistivity as follows:

$$Q = -\frac{\pi^2 k_B^2 T}{3|e|E_F} X, \qquad (3\text{-}4\text{-}10)$$

$$X = -\left(\frac{\partial \ln \rho^L(E)}{\partial \ln E}\right)_{E_F} \qquad (3\text{-}4\text{-}11)$$

Therefore three kinds of theoretical formula are available for the Q, depending on pictures for conduction mechanisms of electrons in liquid transition metals.

(A) The case of extended Ziman formula

The thermoelectric parameter X can be written as

$$X = 3 - 2\tilde{q} - \frac{1}{2}\tilde{r}. \qquad (3\text{-}4\text{-}12)$$

In this equation q and r with symbol "~" mean that

$$\tilde{q} = \frac{T(k_F, -k_F)}{\langle T(k_F, k_F+q)\rangle}, \qquad (3\text{-}4\text{-}13)$$

$$\tilde{r} = \frac{k_F \langle \partial T(k, k+q)/\partial k \rangle k_F}{\langle T(k_F, k_F+q)\rangle}. \qquad (3\text{-}4\text{-}14)$$

(B) The case of s-d scattering model of Mott

The Q for the s-d scattering model of Mott can be written as follows:

$$Q = Q_1 + Q_2, \qquad (3\text{-}4\text{-}15)$$

$$Q_1 = -\frac{3}{2}\left(\frac{\pi^2 k_B^2 T}{3|e|E_F}\right), \qquad (3\text{-}4\text{-}15a)$$

$$Q_2 = \left(\frac{\pi^2 k_B^2 T}{3|e|E_F}\right)\left(\frac{\partial \ln N_d(E)}{\partial \ln E}\right)_{E_F}. \qquad (3\text{-}4\text{-}15b)$$

(C) The case for the Kubo-Greenwood basis

For Eqn.(3-4-6) based on the Kubo-Greenwood basis, the Q is written as follows:

$$Q = -\frac{2\pi^2}{3}\frac{k_B^2 T}{|e|}\left(\frac{\partial}{\partial E'}\ln D(E_F, E')\right)_{E_F}. \qquad (3\text{-}4\text{-}16)$$

In this equation $D(E,E') = \lim_{\varepsilon \to 0} D(E, E'; \varepsilon)$.

(ii) Numerical calculations of thermoelectric powers

The Q was calculated based on the extended Ziman formula by Eqns. (3-4-10) and (3-4-12), whose results are shown with symbol, EGL, HWJS, DVZE and WJT in 3.10.

There are two sets of calculations due to this extended Ziman formula in the case of EGL for liquid Ni and in the case of DVZE for liquid Pd; in the former case the difference is present in the value of k_F used and in the latter case it is present in the structure factor used. Taking into account of the sensitive dependence of Q in Eqns. (3-4-13) and (3-4-14) on the phase shift and its energy dependence, calculated results are rather in good agreement with experiments except for the case of EGL for liquid Fe and HWJS for liquid Fe and Pd. On the other hand, cluster calculations with the symbol BH, F, KCLD and KCL provide as a whole reasonable values though calculations are very sensitive to the energy dependence of $D(E_F,E')$ or DOS at the E_F. In the case of OFK a comparable order with an opposite sign was reported for liquid Ni near T_m. However, details of calculations are not clear from their paper. In the case of BH, Q was calculated by Eqn. (3-4-16). In the cases of KCL and KCLD, calculations were performed based on Eqn. (3-4-15); these calculated data were derived from the DOS obtained by the coupling of hard sphere network model ("geometrical model") of 8800 particles for atomic structures with the momemt method[230] for DOS, which will be mentioned once more in Sec. 3.4.5. Two data for KCLD are derived from the difference of structure used. Slightly better results for cluster calculations may be derived from the absence of errors for structure informations and the more general basis of the Kubo-Greenwood formula for ρ^L and Q.

Measurements of Q of liquid transition metal alloys were performed by Dupree et al.[213] for liquid Co-Ni, by Newport et al. for liquid Ni-Te[222] and liquid Ag-Pd[214] and by van Zytveld[231] for liquid Cu-Ni.

Those experimental studies do not always support the extended Ziman formula. Sometimes Mott formula is supported. Therefore roles of d and s states must be analyzed in detail.

Ballentine[209] speculated the trends of Q with the variation of electron filling in the 3d shell. Due to the energy dependence of diffusivity, $D(E,E')$, the contribution of s electrons to Q is always negative because of less than half filled s band for all transition metals and that of d electron to Q changes from negative to positive as the d band is filled. Early transition metals should have negative Q; large Q in liquid Ni is derived from the dominant contribution of s band; for liquid Mn a positive contribution of d bands dominates to predict a positive Q; small values of Q for liquid Fe, Cr and Co may be derived from the cancellation of s and d band contributions.

3.4.3 Magnetic susceptibility

The magnetic susceptibility, χ_M, of liquid transition metals was measured, for example, by Nakagawa[232] and Urbain and Übelacker.[233] Experimental data of magnetic susceptibility, χ_M, of liquid TM are summarized in table 3.11. Particular features are as follows:

(1) The change of χ_M on melting is large for Fe and Co and small for Mn and Ni.
(2) The magnitude of χ_M is a few ten or hundred times larger than that of liqud simple metals.
(3) The temperature coefficient is slightly negative except for a small positive coefficient of liquid Mn.
(4) In the 3d series, the χ_M increases from liquid Cr to liquid Fe with the filling of 3d states and it shows the maximum for liquid Co and then it decreases from liquid Ni to liquid Cu.

The tendencies (2) and (4) may be explained by the change of DOS at the Fermi energy, E_F, $N(E_F)$, with the electron filling in d bands. In fact Asano and Yonezawa[161] and Jank et al.[87] estimated the $N(E_F)$ by using experimental χ_M of liquid transition metals from an expression of χ_M on the Hartree-Fock approximation, which is expressed as follows:

$$\chi_M = 2N_V\mu_B^2 \left\{ \frac{N(E_f)}{1 - J_{eff} N(E_f)} \right\}. \qquad (3\text{-}4\text{-}17)$$

In this equation N_V is the Avogadro number and J_{eff} is the effective exchange interaction factor. The value of J_{eff} is taken to be , for example, 0.05.[87,161] Estimated $N(E_F)$ by them, particularly Jank et al.,[87] was in close agreement with theoretically calculated DOS at the E_F. Therefore, this large magnitude of χ_M for liquid transition metals (point (2)) and their small temperature coefficients (point (3)) are derived from the fact that the large $N(E_F)$ of d states for solid transition metals is kept stably also in liquid states, as speculated by Grimvall.[134]

A large change of χ_M for Fe (point (1)) may be explained by the change of DOS on melting. Due to theoretical calculations of DOS by Asano and Yonezawa;[161] the DOS of liquid transition metals is very similar to that of FCC solids and Fe takes the BCC structure just below the T_m. The importance of similar paramagnetic spin susceptibility on the change of χ_M on melting was also pointed out by Adachi and Eisaka.[234] The slightly large change of χ_M on melting of liquid Co may be probably derived from the slight change of DOS due to the structure change on melting from the HCP below T_m to the FCC like in the liquid state.

Table 3.11 Magnetic susceptibility χ_M^L (in unit of 10^{-4} cgs emu mol^{-1})[1], its temperature coefficient $d\chi_M^L/dT$ and Hall coefficient R_H^L (in unit of 10^{-11} m^{3}A^{-1}s^{-1})[2] for liquid transition metals ; data of solid state are also shown with the subscript "s".

	χ_M^L	$d\chi_M^L/dT$	χ_M^S	R_H^L	R_H^S	$(R_H/|R_0|)^{exp}$	$(R_H/|R_0|)^{calc}$ [3]
Cr	~3.1	—	—	—	—	—	—
Mn	6.7	VS (>0)	6.4	~11	—	—	—
Fe	14.1	VS (<0)	17.4	38	—	4.5	-2.1
Co	30.8	VS (<0)	33.0	13	14	1.5	-0.6
Ni	3.7	VS (<0)	3.7	-11	-9	-1.5	-1.9
La	1.2	0	—	6.2	5.0	—	—

R_0 corresponds to the free electron value of Hall coefficient with valence Z=1 , that is R_0 is $-1/n_s|e|$ (n_s : the electron number density ; |e| : the absolute value of electron charge). $|R_0|$ is the absolute value of R_0.

VS means " very small".

1) Dupree,R and Seymour,E.F.W. , "Liquid Metals:Chemistry and Physics" (ed. by Beer,S.Z.),Ch.11,Marcel Dekker(1972).
2) Künzi H.U. and Güntherodt H.-J. , "The Hall Effect and Its Applications" (ed. by Chien,C.L. and Westgate,C.R.),201(1980).
3) Ballentine,L.E. "The Hall Effect and Its Applications" (ed. by Chien,C.L. and Westgate,C.R.),201(1980).

As for χ_M of liquid (transition) metals and alloys, excellent reviews have been already published by Dupree and Seymour,[235] Busch and Güntherodt[20] and March and Sayers.[236]

The concentration dependence of χ_M for liquid transition metal ("TM")-simple metal ("SM") alloys is classified into two groups; in one group (group A) positive deviations from the linear law of χ_M are found in the TM rich side and negative or slightly positive deviations are found in the SM rich side. In the other group (group B) negative deviations are found in the whole concentration range. The group A includes liquid Fe-Au and -Ge alloys. The group B includes liquid Ni-Au, -Cu, -Zn, -Ga and -Ge alloys and liquid Co-Au ,-Zn, -Ga and -Ge alloys. These tendencies are explained by Büsch et al.,[20,212] as follows. The DOS of liquid transition metals is considered to have a double peak structure, which was concluded from the theoretical calculation of DOS (see the Refs.20,212 and 237) and has been surely confirmed by recent calculations of DOS, as described in Sec.3.4.5. The E_F is situated in the low energy side of the high energy peak for group A and it is situated over high energy peak for group B. In group A an addition of SM increases the DOS at E_F with the increase of E_F. Roughly speaking the χ_M is proportional to the DOS at E_F and as a result it is enhanced. On the other side, in group B an addition of SM reduces the DOS at E_F with the increase of E_F and as a result the χ_M is depressed. These tendencies were explained by Bass[238] and Bennemann[239] in a theoretically refined manner though lattice models with or without vacancies were employed in place of realistic liquid structures.

Frequently the temperature dependence of χ_M for liquid transition metal alloys shows curious behaviours. The minimum of χ_M was found in the temperature dependence curve for liquid Fe-Ge alloys in the Ge rich side[237] For liquid Fe-B[240] and Co-Y[241] alloys, two different Curie-Weiss behaviours were reported in the liquid states. Particular short range order effects were proposed in the low temperature range above T_m.

Okada et al.[242] studied the χ_M of liquid transition metal-Te alloys. The d states of Cr, Mn and Fe ions split into two sub bands of spin up and spin down in these liquid alloys judging from the fact of Curie-Weiss behaviour of χ_M. On the other hand liquid Ti-Te and Ni-Te alloys are non magnetic and only in the high V and Co concentrations the splitting of d band appears for liquid V-Te and Co-Te alloys respectively.

The existence of liquid ferromagnetism was theoretically predicted by Gubanov.[243] The most promizing candidates are liquid Co alloys because of highest paramagnetic Curie temperature, Θ_p, of Co among transition metals. Since ferromagnetic beghaviors of liquid Au-Co alloys were reported by Busch and Güntherodt,[244] many investigations[245-249] were performed to confirm the existence of ferromagnetism in liquids. For example, Nakagawa[245,249] measured the χ_M of liquid Co-Pd, Co-Au, many eutectic binary Co alloys with low melting points and Co-B-P alloys. For all these liquid Co alloys no ferromagnetism was found, that is $\Theta_P < T_m$ and only for amorphous Co-B-P alloys the ferromagnetism was found. However, Kraeft and Alexander[248] reported the strong evidence for the existence of spontaneous magnetization in the temperature range some 17 degrees above the T_m.

Besides studies described above, the χ_M of liquid transition metal alloys was studied in following papers, Nakagawa[232] for V-Fe, Cr-Mn, -Fe, -Co, -Ni, Mn-Fe, -Co, -Ni, Fe-Co, -Ni, Co-Ni and Briane[250] for Fe-Ni.

3.4.4 Hall coefficient

Experimental data of Hall coefficient, R_H, for liquid transition metals are summarized in table 3.11. Up to date only for liquid Ni, Co and La measurements of R_H have been performed and for liquid Mn and Fe data have been obtained as the extraporated value of Ge and Au alloys in the zero concentration limit of Ge and Au respectively. A positive R_H has been observed for liquid Mn, Fe, Co and La.

Busch and Guntherodt[20] speculated from the simple effective mass theory that positive R_H is possible in the case of large $dN(E)/dE$ at the E_F. Recent theoretical calculations of DOS may be helpful for explanation of positive R_H, for example, the energy derivative of DOS at E_F is positive for liquid Fe, Mn due to Jank et al.[87] and also positive for liquid La due to Ballentine.[208,209] Unfortunately slightly negative derivatives were obtained for liquid Co by Jank et al.[87] and for liquid Fe by Bose et al.[206] It seems to be difficult to obtain the accurate energy derivative of DOS purely from present theoretical calculations with many assumptions and simplifications. The possibility of positive R_H is speculated to appear when the group velocity, dE/dk is negative.[251,252] Nguyen-Manh et al.[253] showed that the sign of dE/dk may accord with that of $dN(E)/dE$ for strong s-d hybridisation system. However, negative dE/dk was not supported by the theoretical calculation of spectrum density for liquid Fe due to the recursion calculation on the TBA by Bose et al.[206] It is not plausible that this positive R_H may be derived from "holes in bands" of liquid transition metals, as in the case of semiconductors. The R_H of liquid transition metals is no longer the measure of carrier density.

Apart from these simple views, some researchers tried to find another origin for the positive R_H. Ten Bosch[254] showed that the spin-orbit coupling causes the change of electronic charge density in the presence of the magnetic field and the current. The correction for this charge redistribution always leads to an increase of NFE Hall coefficient. Szabo[255] tried to explain this problem only from the effect of electron-ion interactions. Ballentine[256] proposed the importance of different mechanism for the origin of deviation of R_H from the NFE value. The skew scattering in the spin-orbit interaction was forcussed; in the skew scattering the direct and inverse collision probabilities are considered to be unequal each other, that is $W(\mathbf{k},\mathbf{k}') \neq W(\mathbf{k}',\mathbf{k})$. The collision probability $W(\mathbf{k},\mathbf{k}')$ was calculated from the phase shift analysis based on the muffin tin scheme. Calculations well reproduces the large departure of R_H from the free electron values for liquid polyvalent Tl, Pb and Bi. However, for liquid transition metals, only the negative R_H of liquid Ni has been reproduced by theoretical calculations and an opposite negative sign to experiments was obtained for R_H of liquid Fe, Co.

Thus, to solve the origin of positive R_H is a challenging problem in the future study of liquid metals. Readers much interested in this problem can consult excellent reviews for experiments due to Künzi and Güntherodt [257] and for theoretical aspects due to Busch and Güuntherod [20] and due to Ballentine. [256]

The R_H was measured for liquid transition metal alloys such as liquid Ge-TM (TM=Fe, Mn, Co, Cu and Ni)[20] and liquid Ce-Cu. [258] In liquid Ge-Ni alloys, the NFE model works only in the dilute Ni concentration range, in which Ni can be considered to be a monovalent metal. [20] Studies of Hall coefficients have been popular rather in the field of amorphous metals and alloys. [259a,259b] Readers much inerested in this field must consult also the excellent review[259b] of R_H for amorphous metals and alloys.

3.4.5 Density of states

Experimental studies of the DOS for liquid transition metals have been performed only in a few cases. Garg and Källne [260] measured the K emissions (valence band→1s) of liquid Cu and Fe. Hague [261] also measured the L3 emissions (valence band→$2p_{3/2}$) and absorption ($2p_{3/2}$→ valence band) of liquid Fe, Co, Ni. Another techniques for studying the valence band, UPS and XPS, were applied to liquid Cu by Williams and Norris. [262] These methods have been applied to studies of DOS for liquid Pd-Si and Bi-Pd alloys by Oelhafen and collaborators. [263] They adopted the wire cleaning technique for obtaining clean surfaces on liquid metal samples. This technique enables measurements of photoelectron spectroscopy for liquid metals at high temperatures near the melting point with low evaporation rates of samples. The liquid alloy systems, Au-Si, Au-Sn and Ag-Cu-Ge, were also studied by their group. [264] Dose et al. [265] studied the DOS of liquid Ni by means of Auger appearance potential spectroscopy. It is not always straightforward to obtain the DOS from these experiments because, for example, transition probabilities of electrons between initial and final states for each process are not always known exactly. The positron annihilation experiment was reported with the expectation for obtainig the informations of electron states and trapped sites.[266,267]

The DOS is given also from theoretical calculations. The muffin tin potential model is the one starting point for the studies of DOS for liquid transition metals. Asano and Yonezawa[161] investigated the DOS of liquid transition metals by means of the KKR method for the Geen function of non-overlapping muffin-tin potentials. Calculations were performed on the three kinds of single site approximations, QCA, IY("Ishida and Yonezawa") and EMA("effective medium approximation"), which differ each other with respect to the pick of higher order terms systematically in the expansion of Green functions. The EMA is found to be the best approximation among these three approximations. The calculated DOS for liquid Ni resembles the calculated DOS of FCC solid.

Another starting point for the theoretical DOS is the tight binding approximation, TBA. Such TBA calculations of DOS were performed by Cyrot-Lackmann and collaborators for liquid Ni[173] and Co.[268] The liquid structures were made by a hard sphere network model (geometrical model) of 8800 particles and based on this structure the DOS was calculated by the moment method[176] on TBA for liquid Co and Ni. Fujiwara[204] calculated the DOS of liquid Fe by coupling of MD with the recursion method on the TBA. Ballentine[207,208] also calculated the DOS of liquid La by coupling of MC with the recursion method on the TBA; liquid structures were given from the MC and the DOS was calculated for the cluster of 365 particles with a free surface. Bose et al.[206] adopted the similar approach to calculate the DOS of liquid Fe. Ostermeijer and Fembacher[211] also performed the calculation of DOS for liquid Ni in a similar manner.

Recently Jank et al.[87] performed the theoretical calculation of DOS for liquid 3d and 4d transition metals. Liquid structures were given by the MD with the WH type interionic potential. The DOS of liquid transition metals were calculated by the supercell technique of linear muffin tin orbital ("LTMO") method with 65 particles; the result of MD were inserted as the structure information of unit cell and the periodic boundary condition was applied.

Condsiderably common results were obtained by these theoretical calculations of DOS. The shape of DOS in the liquid states is far smoother than that of solid and its height is far taller than that of sp bands. The Fermi energy lies inside the d bands. The splitting of d bands can be seen from two peaks in the extreme case and shoulders in other cases. Calculated results of DOS are considerably reliable; calculated values of DOS at E_F for liquid 3d transition metals are in extremely good agreement with the estimations of DOS at E_F from their experimental χ_M, which can be found impressively in the work of Jank et al.[87]

Similar comparisons of the DOS were performed also for liquid 3d transition metals by Do Phuong et al.,[96ced by intra-atomic exchange interactions. From the similar point of view, transition metal impurities in liquid noble metals and liquid simple metals have been studied. For example, Gardner and Flynn [269] measured the χ_M and the Knight shift of liquid Cu. Similar experiments were performed for liquid Al containing transition metal impurities.[270] The χ_M of liquid Cu containing Cr, Mn, Fe and Co was found to be paramagnetically large with a Curie-Weiss type temperature dependence and the number of unpaired electrons are estimated, which are derived from the difference of occupation numbers in splitted d bands with partial overlap.[269,271] Such unpaired electrons are absent in liquid Al host. This situation, the existence in Cu and the nonexistence in Al of lacalized moments due to transition metal impurities, is completely similar between solid solutions and liquid solutions. Several studies were performed for investigations of the magnetic-non magnetic transition in liquid transition metal alloys, such as liquid Cu-Al alloys.[272-275] Since Tamaki [276] studied the localized impurity states in liquid In, Sn, Bi and Sb metals, such studies have been extended to various liquid metal and alloy systems including liquid semiconductors. As for transition metal impurities in liquid noble metals and simple metals including Te, a remarkablly detailed review has been presented by March and Seyers.[236] Readers much intersted in this field can consult this reviw. Here only a list is presented for liquid systems in which localized magnetic moments are found or speculated to be present; Cr, Mn, Fe and Co in liquid Cu, Mn in liquid Sn, Cr, Mn in liquid Te, Mn and Cr in liquid Sb, Co, Mn and Fe in liquid Cu-Al, Mn in liuid Cu-Ga, Mn in liquid Cu-Sn, Mn in liquid Se-Te, Cr, Mn, Fe in liquid In-Te. Following list is available for further studies. In this list are given solvents, corresponding impurities and measured physical quantities and impurities are given in parenthesis.

Liquid Cu solvent : χ_M and ^{63}Cu Knight shift (3d TM),[269] ^{63}Cu Knight shift (3d TM) and χ_M (Mn),[271] χ_M(3d TM),[272] electrotransport (3d TM).[273]

Liquid Au solvent : Knight shift and relaxation time (Co).[274]

Liquid Zn solvent : χ_M(3d TM).[272]

Liquid Al solvent : χ_M and ^{27}Al Knight shift (3d TM),[270]
^{27}Al Knight shift and χ_M (Cr and Mn),[275] χ_M, ρ^L and Q(Mn)[277]

Liquid Ga solvent: χ_M (3d TM).[272]

Liquid In solvent: χ_M (Mn),[276] χ_M (3d TM).[278,279]

Liquid Ge solvent : χ_M(3d TM).[272]

Liquid Sn solvent : χ_M (Mn),[276] ρ^L and Q(Ni, Co and Te),[280] χ_M (3d TM),[281] χ_M (Mn)[282], χ_M (Co),[283] χ_M (3d TM).[278,279]

Liquid Sb solvent : χ_M(Mn),[276] χ_M(3d TM),[284] ρ^L(3d TM),[285] χ_M(3d TM).[278,279]

Liquid Bi solvent : ρ^L(Ni),[286] χ_M (Mn),[276] χ_M, ^{55}Mn and ^{209}Bi Knight shift and relaxation time(Mn).[287]

Liquid Te solvent : ρ^L(3d TM),[288] χ_M(3d TM),[278,279,289,291] ρ^L(3d and 4d TM).[290]

Liquid Al-Cu solvent : χ_M(Fe, Mn),[293] χ_M(Fe),[296] χ_M(Co),[294] Co Knight shift(Co).[295]

Liquid Al-Zn solvent : χ_M(Mn).[292]   Liquid As-Ge solvent : χ_M (3d TM).[272]

Liquid Cu-Ga solvent : χ_M(Mn).[297]   Liquid Cu-Sn solvent : χ_M(Mn).[298]

Liquid Cu-Zn solvent : χ_M(3d TM).[272]   Liquid Ga-Ge solvent : χ_M(3d TM).[272]

Liquid Ga-Ge solvent :χ_M(3d TM).[272]   Liquid Ga-Zn solvent : χ_M(3d TM).[272]

Liquid In-Te solvent : ρ^L(3d TM),[299,300] χ_M(3d TM).[301]

Liquid Se-Te solvent : χ_M(3d TM),[302,303] ρ^L(3d TM)[304]

3.5 Surface properties and atomic transport properties of liquid transition metals

3.5.1 Surface properties and the Mn anomaly

Up to date many experimental data of surface tension, γ, have been accumulated even for liquid transition metals at very high temperatures, as shown in table 3.12. Systematic trends can be seen in 3d, 4d and 5d series; the γ shows a maximum for a middle element in a parabolic dependence with the variation of d shell filling of electrons in the same d series. These tendencies for γ can be more easily seen in Fig. 10 in Allen's review.[305] Exceptions are smaller values of γ for liquid Mn and Tc than this parabolic dependence. These exceptions correspond to the slight anomaly of V_M for liquid Mn, which was mentioned in Sec.3.3.1. Roughly speaking the γ is a measure of cohesion energy. It is well known that

Table 3.12 The surface tention γ and its temperature coefficient $d\gamma/dT$ at the melting temperature T_m of liquid transition metals.

	T_m/K	γ/mN·m^{-1}	$\dfrac{d\gamma}{dT}$ /mN·m^{-1}·K^{-1}		data sources
Sc	1812	954±12	-0.124		VZ
Ti	1941	1650	(-0.26)		A
	1945	1650			L
V	2173	1950	(-0.31)		A
	2190	1950			L
Cr	2146	1700	(-0.32)		A
	2130	1700±50			L
	2130	1510	-0.243		L
	2148	1780	-0.57		CNMM
Mn	1518	1090	-0.2		A
	1517	1220	-0.35		L
Fe	1810	1872	-0.49		A
	1809	1806	-0.42	-0.41(calc)	L
		1978	-0.49		NOMM
Co	1768	1873	-0.49		A
	1767	1884	-0.37	-0.41(calc)	L
		1993	-0.57		NOMM
Ni	1726	1778	-0.38		A
	1726	1782	-0.31	-0.41(calc)	L
		1845	-0.43		N
Cu	1356	1360	-0.21	-0.40(calc)	A
	1358	1286±20	-0.19±0.03	-0.27(calc)	L
		1372	-0.31		NOMM
Y		871±15	-0.086		VZ
	1782	610			L
Zr	2125	1480	(-0.20)		A
Nb	2741	1900	(-0.24)		A
	2750	1900	—		L
Mo	2883	2250	(-0.30)		A
	2892	2250	—		L
Tc	2473	1935±15			L
Ru	2523	2250	(-0.31)		A
Rh	2239	2000	(-0.30)		A
	2236	1915	-0.664		L
Pd	1825	1500	(-0.22)		A
	1827	1500	—		L
		1475	-0.279		L
Ag	1234	903	-0.16		A
	1234	926	-0.15		L

Table 3.12   continued.

	T_m/K	γ/mN·m^{-1}	$\dfrac{d\gamma}{dT}$ /mN·m^{-1}·K^{-1}	data sources
La	1193	720	−0.32	A
	1193	700	−0.078	L
Hf	2495	1630	(−0.21)	A
	2504	1630		L
Ta	3269	2150	(−0.25)	A
		2360		L
W	3683	2500	(−0.29)	A
	3660	2500, 2316		L
Re	3453	2700	(−0.34)	A
Os	3283	2500	(−0.33)	A
	3300	2500		L
Ir	2727	2250	(−0.31)	A
Pt	2042	1800	(−0.17)	A
	2045	1746	−0.307	L
		1699(2073K)		L
Au	1336	1140	−0.52	A
	1337.4	1138	−0.19	L

Values with (calc) is theoretical calculations due to the hard sphere model (Itami, T. and Shimoji, M., J. Phys F : Metal Phys, $\underline{9}$, L15, (1979)).

A : Allen, B.C., "Liquid Metals : Chemistry and Physics", (ed. by Beer, S.Z.), Marcel Dekker, New York, Ch.4 (1972).

CNMM: Chung, W.B., Nogi, K., Miller, W.A. and McLean, A., Mater. Trans. JIM, $\underline{33}$, 753(1992).

L : Lucas, L.D., Tech. l'Ing., $\underline{7}$, Form.M67(1984).

NOMM: Nogi, K., Ogino, K., McLean, A. and Miller, W.A., Met. Trans., $\underline{17B}$, 163(1986).

VZ : compiled data by van Zytveld, J. "Handbook on the Physics and Chemistry of Rare Earths" (ed. by Gschneider, K.A.Jr and Eyring, L.), Ch.85, Elsevier Science Publishers B.V., 357(1989).

the cohesive energy, E_C, of rectangular DOS for d bands shows the parabolic dependence on d shell filling as follows:

$$E_C = \frac{1}{2} Z_d \left(1 - \frac{Z_d}{10}\right) W. \qquad (3\text{-}5\text{-}1)$$

In this equation Z_d and W are respectively the valence(or number) of d electrons and the band width of d bands for the rectangular model of DOS. Besides this simple model, similar parabolic dependences of γ were obtained for solid 3d and 4d transition metals from the moment method on the TBA.[306] Thus, a maximum of γ is predicted for liquid Mn ($3d^5 4s^2$; Z_d=5) in 3d series from a theoretical point of view; this prediction differs from experiments.

As for the origin of this famous anomaly no clear explanations can be found. Sayers[307] concluded that the electron-electron interaction on the cohesion is responsible to a lower cohesive energy of liquid Mn than that of Friedel model given in Eqn.(3-5-1). In the calculations for thermodynamic properties of liquid transition metals in Sec.3.3, it is to be noted that both the GB-TBA[163] and GB-HSYO with WH potential[61] give the nealy parabolic shape with a negativre curvature of Helmholtz free energy or internal energy with an anomaly of slight hump for Mn case("W shape" dependence on d shell filling), as already shown in Fig.3.2 for GB-HSYO. Similar dependence of F can be seen in GB-HS with WH potential.[61,104] These facts indicate that these model calculations of thermodynamics can reproduce the essential feature of cohesion of liquid transition metals. However, these facts do not solve the problem of Mn anomaly because those minima in F and U may be brought by input informations taken from the solid state experimental data, such as the cohesive energy at T=0 (GB-TBA[163]) and the atomic volume (GB-HSYO[61]), as described before. Here it is to be noted that the density functional calculation due to Moruzzi et al.[229] gives a parabolic dependence of cohesive energy with an anomalous depression for Mn case ("inverse W shape") in 3d series. In this calculation the exchange and the correlation of electrons is taken into account deeply. Note that the parabolic dependence of γ was obtained on the TBA calculation with no electron correlations.[306] Thus, the Mn anomaly may be surely related to electron-electron interactions, as pointed out by Sayers[307] though it is difficult to obtain the vivid picture for this anomaly.

It is well known that the γ at the melting point, T_m, of liquid metals, γ_m, is proportional to the factor ($T_m/V_M^{2/3}$), as can be seen typically in Fig.9 in the review by Allen.[305] However, it should be remembered that there remains considerable inaccuracies in these experimental data. Recently levitation techniques due to electromagnetic fields under an innert gas atmosphere have been adopted for measurements of γ for liquid Fe, Co, Ni and Cu by Nogi et al.[308-310] Similar measurements were performed for liquid Fe and Ni by Sauerland et al.[311] The γ can be obtained experimentally from the knowledge of frequency for surface vibrations of liquid levitated sample with a droplet shape under the electromagnetic field. They found that the difference, Δγ, increases with the increase of oxygen affinity of element: Δγ is defined as the difference between γ obtained by their levitation technique and the mean values of previous data of γ obtained by conventional methods, such as a sessile drop method; in these conventional methods liquid samples are necessarily kept in contact with container walls or supporting plates. This fact indicates that the previous data of γ for highly reactive "liquid transition metals" are suffered from the contaminations due to containers and supporting plates at very high experimental temperatures. Another merit of this levitation method is that this needs no density data and only the mass of droplet sample is required together with the frequency of vibration. The containerless technique is very important for the accurate determination of γ. Floating of large amount of samples by this technique is easier under the microgravity than on earth or under a gravity condition. In future "space age", data of accurate γ are expected to be established.

As for liquid simple metals, considerable progresses have been obtained for the theoretical analysis of surface properties based on an inhomogeneous electron theory [312-314] and a pseudo neutral atom approximation.[315] So far the auther knows, no such microscopic theories have been presented for surface properties of liquid transition metals. Therefore, at present theoretical approaches remain to be in a rather crude or phenomenological stage. Brown and March[316] discussed the thickness of surface layer L based on the density gradient theory for inhomogeneous electron gas in the surface layer. By considering only the contribution of kinetic energy of electron gas, a simple relation was derived for L and the product of isothermal compressibility κ_T and γ.

$$L \approx \kappa_T \gamma . \qquad (3\text{-}5\text{-}2)$$

This relation was tested for many substances including liquid transition metals and liquid rare earth metals.[317-319] Estimated L is nearly 0.2 Å for liquid Cu, Ag, Mn, Fe, Co and Ni and 0.37-0.48 Å for liquid alkali metals. The surface of liquid transition metals is sharper than that of liquid alkali metals (and other simple metals). For the further advance of understanding, contributions of potential energy must be taken into account particularly for liquid heavier metals.

For the concentration dependence of liquid alloys, Bhatia and March[320] derived an expression for the concentration dependence of γ from the extension of t he Cahn-Hilliard's concentration fluctuation theory,[321] as follows:

$$\gamma \approx \frac{1}{\kappa_T}\left(1+\frac{\delta^2 S_{CC}(0)}{nk_B T \kappa_T}\right). \qquad (3\text{-}5\text{-}3)$$

In this equation, δ is the so called size factor and $S_{CC}(o)$ is the long wave length limit of Bhatia-Thornton type structure factor. They discussed the surface enrichment of solute atoms in many liquid metal solvents. As for liquid Fe, the enrichment of solute Cu and Sn is predicted in accordance with experiments roughly from the fact, $\kappa_T^{Fe} < \kappa_T^{Cu}$ and $\kappa_T^{Fe} < \kappa_T^{Sn}$, where κ_T^i indicates the κ_T of solute i.

The predictions of the temperature coefficient of γ, $d\gamma/dT$, were performed for liquid Fe, Co and Ni, in addition to liquid simple metals, based on the hard sphere model.[322] Agreements between theory and experiments are good and the relation between $d\gamma/dT$ and the surface entropy is stressed.

From the industrial interests and purposes, the γ of liquid transition metal alloys has been studied, particularly for liquid Fe containing non-metallic elements, such as C, N, O and S. The effect of addition of C does not change the γ of pure Fe; that of N decreases the γ slightly. On the other hand the addition of O and S decreases respectively the γ of liquid Fe drastically and at a few % of O and S the γ shows a saturated values due to the formation of FeO and FeS respectively on the surface layer. Similar studies were reported for liquid Cu and Ag. These problems have been reviewed in detail in Refs. 6 and 14.

The auther can not find so many studies about surface properties of liquid transition metals and alloys in recent journals in fields of physics and chemistry. The experiments of γ have been reported for liquid Fe-P [323], Ni-10at.%Ti,[311] and liquid Co-S, Co-C and Co-C-S systems.[324] Russier et al.[83] have presented the analysis of work functions for liquid transition metals by employing the WH interionic potentials.

3.5.2 Atomic transport properties

As for the atomic transport in liquid metals, many theories are presented, as can be seen in Refs. 1 and 5. However, good agreements with experiments are obtained only for the hard sphere ("HS") model of self diffusion coefficient D and viscosity η. In this HS model, so called "back scattering corrections" are important, which are derived from the effect of surrounding atoms to prevent a central atom from advancing forwardly out of cage formed by them ("cage effect"). Theoretically the self diffusion coefficient, D_{HS}, for hard spheres can be given by the molecular kinetic theories of Enskog [1,5], in which unfortunately this back scattering effect is completely neglected. Therefore, cage effects can be estimated numerically by comparing the D_{HS} due to the Enskog theory with that due to molecular dynamics of hard spheres, in which the back scattering is "experimentally" realized. Recently Speedy [325] compiled the data of D_{HS} of MD simulations due to various data sources. He presented an excellent interpolating formula for representing D_{HS} obtained by MD simulations of hard spheres in the various density range by various reseachers. According to this analysis, the D_{HS} for hard spheres is given as follows:

$$D_{HS} = \left(\frac{\pi D_0}{6\xi}\right)\left(1 - \frac{6\xi}{1.09\pi}\right)\left[1+\left(\frac{6\xi}{\pi}\right)^2\left\{0.4 - 0.83\left(\frac{6\xi}{\pi}\right)^2\right\}\right], \quad (3\text{-}5\text{-}4)$$

where $D_0 = (3/8)\sigma(k_B T/\pi M)^{1/2}$ (M: mass of hard spheres), σ the hard sphere diameter, ξ the packing fraction.

The viscosity of hard spheres, η_{HS}, can be calculated by the Stokes equation with slip boundary conditions,

$$\eta_{HS} = \frac{k_B T}{2\pi\sigma D_{HS}}, \quad (3\text{-}5\text{-}5)$$

whose validity was confirmed rigorously by computer simulations of hard spheres.[326]

If Eqn. (3-5-4) is written as $D_{HS} = C_{BS} D_{ENS}$ in terms of the Enskog formula, D_{ENS}, the back scattering factor C_{HS} can be obtained as follows:

$$C_{BS} = \frac{(1-\xi/2)^2}{(1-\xi)^3}\left(1 - \frac{6\xi}{1.09\pi}\right)\left[1+\left(\frac{6\xi}{\pi}\right)^2\left\{0.4 - 0.83\left(\frac{6\xi}{\pi}\right)^2\right\}\right], \quad (3\text{-}5\text{-}6)$$

In this equation, the folowing equation is employed as D_{ENS}.

$$D_{ENS} = D_0\left\{\frac{6\xi\, g_{HS}(\sigma)}{\pi}\right\}^{-1} = \frac{3\sigma}{8}\left(\frac{k_B T}{\pi M}\right)^{1/2}\left\{\frac{6\xi\, g_{HS}(\sigma)}{\pi}\right\}^{-1}, \quad (3\text{-}5\text{-}7)$$

where $g_{HS}(\sigma)$ $(=(1-\xi/2)/(1-\xi)^3)$ is the radial distribution function of hard sphere system at the hard sphere contact.

If the temperature dependence of σ, $\sigma(T)$, is required to predict D and η in the wide temperature range, the form (3-5-8) proposed by Protopapas et al.[327] can be employed.

$$\sigma(T) / \sigma(T_m) = 1.126\left\{1 - 0.112(T / T_m)^{1/2}\right\}. \quad (3\text{-}5\text{-}8)$$

From the theoretical point of view, it is very important for the HS model to clarify the meaning of back scattering factor and the temperature dependence of hard spheres; those have been determined rather "experimentally" by means of computer simulations.

At present there are almost no data for the D of liquid transition metals. The experimental data of η are shown in table 3.13. In Fig.3.3, $\eta_m V_M^{2/3}/T_m^{1/2}$ is plotted as a function of $M_A^{1/2}$. The dotted line corresponds to the theoretical relation based on the HS model with $\xi=0.46$ at T_m. The proportional constant of this dotted straight line is 5.37×10^{-2}, which is in good agreement with the value 5.7×10^{-2} determined empirically by Andrade.[328] Here it must be remembered that experimental values of η in literatures are scattered in a wide range due to difficult experimental conditions with many causes of large errors; such a situation can be seen typically in the book by Iida and Guihrrier.[6] In table 3.13 experimental values of η are adopted from the table of Lucas,[329] which seems to list lower values in the widely scattered values of experimental η in literatures.

Therefore, it is very important to improve the reliability for experimental data of η. The levitation technique due to electromagnetic fields may be expected also for accurate determinations of η because of absence of contaminations of samples due to containers and atmospheres. The η can be experimentally determined from the damping of surface vibrations of liquid sample with a droplet shape.[330]

The estimated η and D due to Eqns.(3-5-4) and (3-5-5) are shown respectively in table 3.13 for $\xi=0.46$ at T_m. The agreement of calculated η with experiments is not so bad (calculated η is 60-120% of experimental values). Unfortunately no reliable experimental data are given for the D of liquid transition metals. Calculated results in table 3.13 can be expected to be utilized as an order estimation of D. Recently measurements of D under the microgravity is found to improve much the accuracy of experimental data by Frohberg et al.[331] The microgravity condition provides ideal conditions for D measurements; liquid dynamics due to the gravity, which spoils experiments of D on earth, is suppressed under the microgravity. On measurements of η, a levitation technique, which is free from contaminations of samples, is easier under microgravity conditions than under 1 G conditions. Space experiments under the microgravity are extremely expected for accurate determination of D and η.

Electrotransport experiments were performed for liquid Cu containig 3d transition metal impurities.[273] The drag coefficient, P_i, was presented for solute V, Cr, Mn, Fe, Co and Ni in liquid Cu. The P_i is defined as the ratio of the flux of the atoms or ions of species i to the electron flux divided by the composition. Species with a positive P_i will move along with the conduction electrons towards the anode; those with a negative P_i move towards the cathod oppositely to the direction of movement of conduction electrons. The sign of P_i is positive

Table 3.13 Experimental and theoretical values of viscosity η and self diffusion coefficient D of liquid transition metals. The η is conventionally expressed here as $\eta=A\exp(E/RT)$.

	T_m /K	η_{exp} /10^{-3}Pa·s	A /10^{-3}Pa·s	E /kJ·mol^{-1}	η_{calc} /10^{-3}Pa·s	$D_{calc}(D_{exp})$ /10^{-9}m^2·s^{-1}
Sc						
Ti	1945	2.2			2.9	5.7
V	2190	2.4			3.8	5.3
Cr	2130	~4.85			4.1	5.0
Mn	1517	5.26	0.4786	30.259	3.2	4.4
Fe	1809	5.03	0.3162	41.606	3.9	4.5
Co	1767	4.17	0.3293	37.346	4.0	4.3
Ni	1728	4.61	0.490	32.217	4.0	4.2
Cu	1358	3.05	0.6389	17.636	3.6	3.6(3.98)
Y					2.5	4.9
Zr	2125	5.45(2138) 3.50(2133)			3.6	4.7
Nb					4.7	4.9
Mo					5.5	4.7
Tc						
Ru					5.6	4.1
Rh					5.3	3.9
Pd					4.6	3.5
Ag					3.5	3.0(2.55) (2.56)
La	1193	3.13	0.2317	26.255	2.5	3.3
Hf	2504	5.0			5.4	3.6
Ta					7.2	3.8
W					8.4	3.9
Re					8.5	3.6
Os					8.7	3.5
Ir					7.9	3.1
Pt	2045	6.74	1.527	25.263	6.6	2.8
Au	1337.4	4.74	1.133	15.900	5.0	2.3

D_{exp} is taken from table 3.1 in "Atomic Transport in Liquid Metals"(Shimoji,M. and Itami,T., Trans. Tech. Publication,Switzerland-Germany-UK-USA, 1986) η_{exp} is taken from the table given by Lucas,L.D.(Tech. l'Ing., 7,Form. M66(1984)).

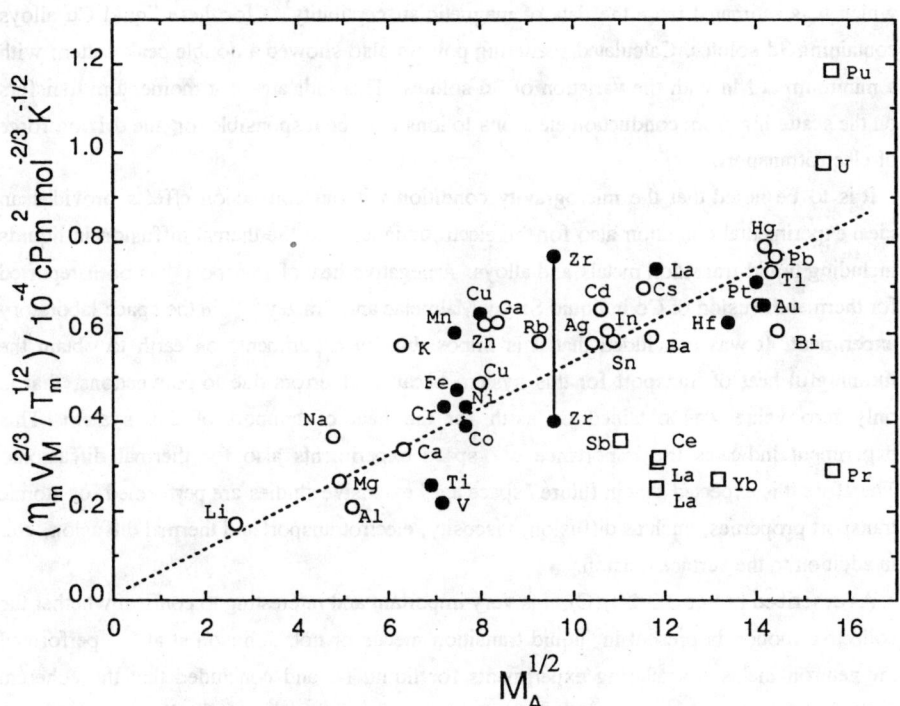

Figure 3.3 Relation between $\eta_m V_M^{2/3}/T_m^{1/2}$ and $M_A^{1/2}$ for liquid metals. The straight line corresponds to the hard sphere model with $\xi=0.46$ at the melting point T_m.

in liquid Cu for all solutes studied in that work. Double peaks with the minimum at Mn are found on the solute dependence from V to Ni. This tendency is quite similar to the resistivity increase due to the addition of 3d transition metal impurities in solid Cu.[332] They also estimated the scattering power of solute transition metals for conduction electrons in liquid Cu. The phase shift η_2 for l=2 required was evaluated from the number of d electrons, which was estimated from the data of magnetic susceptibility[269] for these liquid Cu alloys containing 3d solutes. Calculated scattering powers also showed a double peak pattern with a minimum at Mn with the variation of 3d solutes. This indicates that momentum transfers on the scattering from conduction electrons to ions may be responsible for the driving force of electrotransport.

It is to be noted that the microgravity condition without convection effects provides an ideal experimental condition also for the electrotransport and the themal diffusion in liquids including liquid transition metals and alloys. A negative heat of transport has been reported for thermal diffusion of Co in liquid Sn by Malmejac and Praizey[332a] in the space laboratory experiment. It was concluded that it is impossible by experiments on earth to obtain the meaningful heat of transport for this system because of errors due to convections, that is only zero value was obtained on earth for the heat of tranport of this system. This experiment indicates the importance of space experiments also for thermal diffusions. Therefore it is expected that in future "space age" extensive studies are performed for atomic transport properties, such as diffusion, viscosity, electrotransport and thermal diffusion, etc. in addition to the surface tension.

As described in Sec.3.3.2(i)(C), it is very important and interesting to confirm whether the collective motion is present in liquid transition metals or not. Johnson et al.[333] performed the neutron inelastic scattering experiments for liquid Ni and concluded that the coherent scattering data provides some preliminary evidences for the existence of collective modes.

Ono[334] presented the experimental data of impurity diffusion coefficient, D_i, of various elements (Si, P, S, V, Cr, Mn, Co, Ni and Cu) in carbon saturated iron. A good correlation is found between D_i and the standard free energy of solution for impurity elements dissolved in molten carbon saturated iron. Different correlations were found between the group of non metallic elements and the group of metallic elements.

The η was measured for liquid Fe-Ni alloys by Adachi et al.[335] and results of them were compared with the theory of excess viscosity based on the hard sphere model.[336] The η of liquid Fe-C, Fe-P and Fe-P-C eutectic liquid alloys was measured by Nishi et al.[337] by means of a capillary viscometer, with particular interests in the glass formations. Yamazaki et al. reported the η of liquid Al-Ni-La and Al-Ni-Mm (Mm:Mischmetal) alloys.[338]

Egry et al.[339] reported that the ratio of surface tension $\gamma(Nm^{-1})$ to viscosity η (mPa s), γ/η, is constant (0.82-1.02) for liquid Fe,Co,Ni,Cu,Ag and Au at the melting point. Discussions were given based on the Fowler eqution for γ and the Born-Green formula for η. It is to be noted that both theories themselves are not always supported to be reliable.[1,5]

Theoretical calculations of atomic transport properties of liquid Cu was performed by Mujibar and Rahman[340] based on the standard statistical theory of self diffusin and vicosity of liquids,[1,5] in which contributions of hard, soft and cross terms of interionic potentials were taken into account based on the generalized pseudopotential theory ("GPT") described in Sec.3.1.4(i). Agreements between theory and experiments are considerably good. The reliability of this calculations depends on that of interionic potential due to the GPT because of sensitivity of calculations to interionic potentials employed; these calculations requires the second derivatives with respect to the distance, r, of interionic potentials. It is to be noted that similar calculations are not so successful for liquid simple metals.[1,5] However, this may indicate the possibility that liquid Cu and transition metals are more adaptable to theories of atomic transport with the hard sphere repulsions than simple liquid metals; the repulsive parts of interionic potentials in liquid Cu and transition metals are "harder" than those in liquid simple metals, such as liquid alkali metals, judging from the data of isothermal compressibility, κ_T (or adiabatic compressibility κ_S.). The Car Parrinello dynamic simulations are expected also for liquid transition metals because of their freedom from the assumption about interionic potentials. Such calculations were performed for the selfdiffusion coefficient of liquid Cu by Pasquarello et al.[78] by using the ultrasoft Vanderbilt pseudopotential. It is desired that such simulations of atomic transport properties are extended to various liquid transition metals.

3.6 Concluding remarks

The physical properties of liquid transition metals have been reviewed in this article both from experimental and theoretical aspects. Considerable progresses have been obtained for the theoretical analysis of electronic properties, structures and thermodynamic properties of liquid transition metals. The effects of d states are taken into account indirectly in terms of "softness" of ionic core and s-d hybridization term and directly in terms of WH type (NFE) interionic potentials and interionic potentials based on the tight binding scheme containing explicitly transfer integrals between s and d states. The model theories of liquids are valid to describe the atomic structures and thermodynamic properties of liquid transition metals. The

WH type interionic potentials can be successfully employed for such model theories of liquids and computer simulations for structures of liquid transition metals. Cluster calculations of tight binding basis are successful for calculations of density of states, electrical resistivities and thermoelectric powers for typical transition metals, Fe, Co and Cr in the liquid state. The supercell technique coupled with the LMTO is also powerful to study the density of states of liquid transition metals. However, in almost all cases, theoretical predictions are unsuccessful for liquid light transition metals with nearly empty d shells, such as Ti and V. The study of light liquid transition metals is a future subject for theoretical studies. Thermodynamic perturbation theories and many integral equation approaches coupled with the WH type interionic potentials are not successful for the prediction of structures and thermodynamic properties of liquid transition metals with unfilled d shells. Liquid theories must be developed for such liquids with soft cores and deeply attracive long range wells.

Coupling of BS interionic potentials with VMHNC equation is better for predicting structures of liquid transition metals than that of WH interionic potentials with various liquid theories. For structures of even liquid Ti a good agreement is obtained between experiments and the MD simulations if employed interionic potentials are due to the tight binding Bethe lattice method("DPPNM" interionic potential). It is very intresting to study the applicability of new two interionic potentials("BS" and "DPPNM") to the prediction of thermodynamics of liquid transition metals. It is also very important to confirm the accuracy of various interionic potentials. In this respect it is to be noted that the Car-Parrinello method has just started for liquid transition metals, by which simulations are free from the assumptions and arbitrarinesses of interionic potentials. Positive Hall coefficients of liquid Mn, Fe and Co are remained to be a future challenging problem.

From experimental aspects, accuracies of data are still now insufficient and, in addition to this, almost no data have been reported for the diffusion coefficient of liquid transition metals. New techniques must be introduced, such as containerless measurements, pulse heating methods and measurements under microgravity and their combinations.

Acknowledgement

The author is much appreciated with Professor Emeritus M. Shimoji for guiding him in this branch of science. He also thanks Dr. S. K. Srivastava for providing him the chance of writing this review. He is much appreciated with Prof. N. H. March and Prof. Y. Waseda for their encouragements during the course of writing this article.

1) Simoji,M., "Liquid Metals ", Academic Press, London-New York-San Francisco, 1977.
2) Yonezawa,F, J. de Physique, 41, C8-447, 1980.
3) Cusack,N.E., "The Physics of Structurally Disordered Matter:An Introduction", Adam Hilger, Bristol and Philadelphia, 1987.
4) Waseda,Y., "The structure of Non-Crystalline Materials:Liquids and Amorphous Solids", McGraw - Hill International Book Company, 1980.
5) Shimoji,M. and Itami,T., "Atomic Transport in Liquid Metals", Trans Tech Publications, Switzerland - Germany - UK - USA, 1986.
6) Iida,T. and Guthrie,R.I.L., "The physical Properties of Liquid Metals", Clarendon Press, Oxford, 1988.
7) Edwards, S.F., Proc.Roy.Soc., A267, 518, 1962.
8) Ballentine, L.E., Can.J.Phys., 44, 2533, 1966; Adv.Chem.Phys. 31, 263, 1975.
9) Itami,T. and Shimoji,M., Phil.Mag., 25, 229, 1972.
10) Wagner,N.C., Phys.Stat.Sol., 115, 9, 1983.
11) Van Oosten,A.B. and Geertsma,W., Physica B&C, 133, 55, 1985.
12) Esposito,E., Ehrenreich,H. and Gelatt,C.D., Phys.Rev., B18, 3913, 1978.
13) Van Zytveld,J., Handbook on the Physics and Chemistry of Rare Earths, 12, 357, 1989.
14) Wilson,J.R., Met.Rev., 10, 381, 1965.
15) Richardson,F.D., "Physical Chemistry of Metals in Metallurgy", Academic Press., London - New York, 1974; Turkdogan E.T. "Physical Chemistry of High Temperature Technology", Academic Press., 1980.
16) Faber,T.E., "Introduction to the Theories of Liquid Metals", Univ. Press, Cambridge, 1972.
17) Beer,S.Z.(ed.), "Liquid Metals", Marcel Dekker, New York, 1972.
18) March,N.H. and Tosi,M.P. "Atomic Dynamics in Liquids", The Macmillian Press LTD., 1976.
19) Ziman,J.M., "Models of Disorder", Cambridge Univ. Press., London - New York - Melbourne, 1979.
20) Busch,G. and Güntherodt,H-J., Solid State Physics (ed. Ehrenreich,H., Seitz,F. and Turnbull,D., Academic Press., New York), 29, 235 1974.
21) Adams,P.A., Davies,H.A. and Epstein,S.G.(ed.), "The Properties of Liquid Metals", Tayler & Francis Ltd, 1967; Takeuchi,S.(ed.), "The Properties of Liquid Metals", Tayler & Francis Ltd, 1973; Evans,R. and Greenwood,D.A.(ed.), "Liquid Metals, 1976",

Conference Series Number 30, The Institute of Physics, Bristol and London; Cyrot-Lackmann,F. and Desre,P.(ed.), "Fourth International conference on Liquid and Amorphous Metals", Les Editions de Physique, 1980 (J.de Physique, 41, C-8,1980); Wagner,C.N.J. and Johnson,W.L.(ed.),"Liquid and Amorphous Metals V", North Holland, 1984; Gässer,W., Hensel,F. and Lüscher,E.(ed.), "Liquid and Amorphous Metals LAM 6", R.Oldenboug Verlag, München, 1987; Endo,H.(ed.), "Liquid and Amorphous Metals VII", North Holland, 1990; Hafner,J.(ed.), "Liquid and Amorphous Metals VIII", North Holland, 1992.

22) Achcroft,N.W. and Langreth,D.C., Phys.Rev., 156, 685, 1967; ibid., 159, 685, 1967.
23) Faber,T.E. and Ziman,J.M., Phil.Mag., 11, 153, 1965.
24) Bhatia,A.B. and Thornton,D.E., Phys.Rev., B2, 3004, 1970.
25) Allan,M.P. and Tildesley, D.J. "Computer Simulation of Liquids", Oxford Press, London, 1989; Ciccotti,G., in "Liquids, freezing and glass transition"(ed. Hansen,J.P., Levesque,D. and Zinn-Justin,J.), Part ÎI, p.943, North Holland, 1991.
26) Car,R. and Parrinello,M., Phys.Rev.Lett., 55, 2471, 1985.
27) Rice,S.A. and Gray,P, "Statistical Mechanics of Liquids - An Introduction to the Theory of Equilibrium and Non-Equilibrium Phenomena", Interscience Publishers, 1965; Croxton,C., "Liquid State Physics - A Statistical Mechanical Introduction", Cambridge Univ Press, 1974.
28) Rogers,F.J. and Young,D.A., Phys.Rev., A30, 999, 1984.
29) Madden,W.A. and Rice,S.A., J.Chem.Phys., 72, 4208, 1980; see also Chihara,J., Prog.Theor.Phys., 50, 1156, 1973.
30) Hansen,J.P and Zerah,G., Phys.Lett., 108A, 277, 1983.
31) Zerah,G. and Hansen,J.-P., J.Chem.Phys., 84, 2336, 1986.
32) Ishihara,A., J.Phys.A, 1, 539, 1968; Lukens,T. and Jones,R., J.Phys.A, 1, 29, 1968.
33) Rosenfeld,Y., J.Stat.Phys., 42, 437, 1986; Gonzalez,L.E., Gonzalez,D.J. and Silbert,M., Phys.Rev., 45A, 3803, 1992.
34) Weeks,J.D., Chandler,D. and Anderson,H.C., J.Chem.Phys., 54, 5237, 1971.
35) Kumaravadivel,R. and Evans,R., J.,J.Phys.C:Solid State Phys., 9, 3877, 1976.
36) Anderson,H.C., Chandler,D. and Weeks,J.D., Adv.Chem.Phys., 34, 105, 1976.
37) Anderson,H.C., Chandler,D. and Weeks,J.D., J.Chem.Phys., 56, 3812, 1972.
38) Hansen,J.P. and McDonald,I.R., "Theory of Simple Liquids", Academic Press, London - New York - San Francisco, 1976.
39) Widom,B., J.Chem.phys., 39, 2808, 1963; Barker,J.A. and Henderson,D., J.Chem.Phys., 47, 2856, 1967; Mansoori,G.A. and Canfield,F.B., J.Chem.Phys., 51,

4958, 1969.
40) Thiele,E., J.Chem.Phys., $\underline{39}$ 474, 1963.
41) Wertheim,M.S., Phys.Rev.Lett., $\underline{10}$, 321, 1963.
42) Lebowitz,J.L, Phys.Rev., $\underline{A133}$, 895, 1964.
43) Ashcroft,N.W. and Leckner, J., Phys.Rev., $\underline{145}$, 83, 1966.
44) Enderby ,J.E. and North D.M., Phys.Chem.Liquids, $\underline{1}$, 1, 1968.
45) Anderson,H.C. and Chandler,D., J.Chem.Phys., $\underline{57}$, 1918, 1972.
46) Palmer, R.G. and Weeks,J.D., J.Chem.Phys., $\underline{58}$, 4171, 1973.
47) Singh,H.B. and Holz, A., Phys.Rev., $\underline{A28}$, 1108, 1983.
48) Bretonnet,J.L., Bhyiyan,G.M., and Silbert, M., J.Phys.: Condens.Matt., $\underline{4}$, 5359, 1992.
49) Hayter,J.B. and Penfold,J., Mol.Phys., $\underline{42}$, 109, 1981; Waisman,E., Mol.Phys., $\underline{25}$, 45, 1973.
50) Waisman,E., J.Chem.Phys., $\underline{59}$, 495, 1973.
51) Copestake,A.P., Evans,R. and Telo da Gama, M.M., J. de Physique, $\underline{41}$, C8-145, 1980; Copestake,A.P and Evans,R., "Lecture Notes in Physics" (ed. Bennemann, K.H. and Quitmann,D., Springer, Berlin), $\underline{172}$, 86, 1982;
Copestake,A.P, Evans,R., Ruppersberg,H. and Schirmacher,W.W., J.Phys.F:Metal Phys., $\underline{13}$, 1993, 1983.
52) Hafner,J., Pasturel,A. and Hicter,P., J.Phys.F:Metal Phys., $\underline{14}$, 1137, 1984; ibid., $\underline{14}$, 2279, 1984.
53) Pasturel,A. and Hafner,J., Phys.Rev., $\underline{B32}$, 5009, 1985; ibid., $\underline{B34}$, 8357, 1986.
54) Waisman,E. and Lebowitz,J.L., J.Chem.Phys., $\underline{56}$, 3093, 1972.
55) Hiroike,K., Mol.Phys., $\underline{33}$, 1195, 1977.
56) Ginoza,M., J.Phys.Soc.Jpn., $\underline{54}$, 2783, 1985; $\underline{55}$,1782,1986.
57) Ginoza,M., J.Phys.Soc.Jpn., $\underline{56}$, 5, 1987.
58) Høye,J.S., Lebowitz,J.L. and Stell,G., J.Chem.Phys., $\underline{61}$, 3253, 1974.
59) Bari,A., Das,T. and Joarder,R.N., J.Non Cryst.Solids, $\underline{136}$, 173, 1991.
60) Joarder,R.N. and Das,T., Phys.Scr., $\underline{37}$, 762, 1988.
61) Hausleitner,C. and Hafner,J., J.Phys.F:Metal Phys. $\underline{18}$, 1013, 1988.
62) Alder,B.J. and Wainwright,T.E., J.Chem.Phys., $\underline{33}$, 1439, 1960.
63) Carnahan,N.F. and Starling,K.E., J.Chem.Phys., $\underline{51}$, 635, 1969; ibid., $\underline{53}$, 600, 1970.
64) Itami,T. and Shimoji,M., J.Phys.F:Metal Phys., $\underline{14}$, L15, 1984.
65) Hansen,L.P., Phys.Rev., $\underline{A8}$, 3096, 1973.
66) Galam,S and Hansen,J.P., Phys.Rev., $\underline{A14}$, 816, 1976.

67) Soler,J.M. and A.R.Williams, Phys.Rev., 42B, 9728, 1990.
68) Yu,R., Singh,D. and Krakauer,H., Phys.Rev., 43B, 6411, 1991.
69) Jacobson,K.W., Nørskov,J.K. and Puska,M.J., Phys.Rev., B35, 7423, 1987.
70) Harrison,W., Phys.Rev., 181, 1036, 1969.
71) Moriarty,J.A., Phys.Rev., B1, 1363, 1970; ibid., B6, 1239, 1972.
72) Regnaut,C, Fusco,E. and Badiali,J.P., Phys.Rev., B31, 771, 1985.
73) Moriarty,J.A., Phys.Rev.38B, 3199, 1988; see also ibid, Phys.Rev., B16, 2537, 1977 and B26, 1754, 1982.
74) Joarder,R.N. and Gopara Rao,R.V., Phys.Stat.Sol., B109, 137, 1982.
75) Regnaut,C, Fusco,E. and Badiali,J.P, Phys.Stat.Sol., 120, 373, 1983.
76) Regnaut,C, Fusco,E. Rosinberg, M.-L. and Badiali, J.Non Cryst. Solids, 61&62, 207, 1984.
77) Bylander,D.M. and Kleinman,L., Phys.Rev., 46B, 9837, 1992; see also Starkloff,Th. and Joannopoulos,J.D., Phys.Rev., 16, 5212, 1977; Harris,J. and Jones,R.O., Phys.Rev.Lett., 41, 191, 1978; Hamann,D.R., Sclüter and Chiang,C., Phys.Rev. Lett., 43, 1494, 1979; Kleinman,L. and Bylander,D.M., Phys.Rev.Lett., 48, 1425, 1982; Allan,D.C. and Teter,M., Phys.Rev.Lett., 59, 1136, 1987; Blöchl, P., Phys.Rev., B41, 5414, 1990; Vanderbilt,D., Phys.Rev., B41, 7892, 1990.
78) Pasquarello A., Laasonen,K., Car, R., Changyol,L. and Vanderbilt,D., Phys.Rev.Lett., 69,1992.
79) Laasonen,K., Pasquarello A., Car, R., Changyol,L. and Vanderbilt,D., Phys.Rev., B47, 10142, 1993.
80) Wills,J.M. and Harrison,W., Phys.Rev., B28, 4363, 1983.
81) Anderson,O.K., Skiver,H.L. and Nohl,H., Pure & Appl.Chem., 52, 93, 1979.
82) Friedel,J., The physics of Metals:1.Electrons, (ed. Ziman,J.M, Cambridge At the Univ.Press.), 340,1969.
83) Russier,V., Regnaut,C. and Badiali,J.P., Zeit. für Phys. Chemie Neue Folge, 156, S489, 1988.
84) Hausleitner,C, Kahl,G. and Hafner,J., J.Phys.:Condens.Matter, 3, 1589, 1989.
85) Regnaut,C., Z.Phys.B-Condensed Matter, 76, 179, 1989.
86) Hausleitner,C. and Hafner,J., J.Phys.:Condens.Matter, 2, 6651, 1990.
87) Jank,W., Hausleitner,C. and Hafner,J., J.Phys.:Condens.Matter, 3, 4477, 1991.
88) Borchi,E. and De Gennaro,S., Phys.Lett.., A32, 301, 1970.
89) Dagen,L., J.Phys.F:Metal Phys., 6, 1801, 1976; ibid., 7, 1167, 1977.
90) Idrees,M., Khwaja,F.A. and Razmi,M.K.S., Solid State Comm., 41, 469, 1982.

91) Ermakov,Yu A. and Denisenko,G.A., J.Phys.F:Metal Phys., 11, 2085, 1981.
92) Bretonnet,J.L. and Silbert,M., Phys.Chem.Liquids, 24, 169, 1992.
93) Swan,P. and Pearce,W.A., Nuclear Phys., 79, 77, 1966; Swan,P., Nuclear Phys., A90, 436, 1967.
94) Bhuiyan,G.M., Bretonnet,J.L., Gonzalez,L.E. and Silbert,M., J.Phys.:Condens.Matt., 4, 7651, 1992.
95) Bhuiyan,G.M., Bretonnet,J.L. and Silbert,M., J.Non-Cryst. Solids, 156-158, 145, 1993.
96) Do Phuong,L., Pasturel,A. and Nguyen Manh,D., J.Phys.F:Condens.Matt., 5, 1901, 1993.
97) Pettifor,D.G., "Many-Atom Interactions in Solids"(ed. Nieminen,R.M., Puska,M.J. and Manninen,M.J., Berlin, Springer, p.64, 1991; Pettifor,D.G.and Aoki, M., Phil. Trans. R. Soc., A334, 439, 1991;.Pettifor,D.G., Phys.Rev.Lett., 63, 2480, 1989.
98) Harrison,W.A., "Electronic Structure and the Properties of Solids", San Francisco, Freeman, p.503, 1980.
99) Mayou,D., Nguyen Manh,D., Pasturel,A. and Cyrot Lackmann,F., Phys.Rev., B33, ,3384, 1986.
100) Daw, M.S. and Baskes,Phys.Rev.Lett., 50, 1285, 1983; ibid, Phys.Rev., B29, 6443, 1984.
101) Foiles,S.M., Phys.Rev., B32, 3409, 1985.
102) Moriarty,J.A., Phys.Rev., 42A, 1609, 1990.
103) Hausleitner,Ch and Hafner,J., Phys.Rev., 45B, 115, 1992.
104) Bretonnet,J.L. and Derouiche,A., Phys.Rev., 43B, 8924, 1991.
105) Edward,D.J. and Jarzynski,J., J.Phys.C:Solid State Phys., 5, 1745, 1972.
106) Young,W.H., J.Phys.F:Metal Phys., 12, L19, 1982.
107) Jacobs,R.E. and Andersen,H.C., Chem.Phys., 10, 73, 1975.
108) Rosenfeld,Y. and Ashcroft,N.W., Phys.Rev., A20, 1208, 1979.
109) Lado,F., Foiles,S.M. and Ashcroft,N.W., phys.Rev., A28, 2374, 1983.
110) Waseda,Y. and Tamaki,S., Phil.Mag., 32, 273, 1975.
111) Eder,O.J., Erdpresser,E., Kunsch,B., Stiller,H. and Suda,M., J.Phys.F:Metal Phys., 10, 183,1980.
112) Eder,O.J., Erdpresser,E., Kunsch,B., Stiller,H., Suda,M and Weinzier,P., J.Phys.F: Metal Phys., 9, 1215, 1979.
113) Rudin,H., Milhouse,A.H., Fisher,P. and Meier,G., Inst.Phys.Conf.Ser., No.30, 241, 1977.

114) Egelstaff.,P.A., March,N.H. and Mcgill,N.C., Can.J.Phys., 52, 1651, 1974.
115) Dobson,J.B., J.Phys.C:Solid State Phys., 11, L295, 1978.
116) Takeda,S., Tamaki,S. and Waseda,Y., J.Phys.Soc.Jpn., 54, 2552, 1985.
117) Chihara,J., J.Phys.F:Metal Phys., 17, 295, 1987.
118) Khanna,S.N. and Cyrot-Lackmann,F., J.de Physique, 40, L45, 1979.
119) Meyer,A., Silbert,M. and Young ,W.H., Zeit.für.Phys.Chemie Neue Folge, 156, S.519, 1988.
120) Meyer,A., Silbert,M. and Young,W.H., Phys.Chem.Liquids, 19, 97., 1989.
121) Waseda,Y., Inst.Phys.Conf.Ser., No.30, 230, 1977.
122) Mitra,S.K., Inst.Phys.Conf.Ser. No.30, 146, 1976.
123) Brennan,M., Ph D Thesis, Reading University,1974.
124) Meyer,A., Silbert,M. and Young,W.H., Phys.Chem.Liquids, 13, 293, 1984.
125) Foiles,S.M., Phys.Rev., B32, 3409, 1985.
126) McGreevy,R.L. and Pusztai,L., Mol.Simulations, 1, 359, 1988.
127) Howe,M.A.,McGreevy,R.L. and Pusztai,L.,Borzsak,I, phys.Chem.Liquids, 25, 205, 1993.
128) Lemarchand,J.L., Bletry, J. and Desre, P., J. de Physique, 41, C8-163, 1980.
129) Andonov,P., Bellissent-Funel,M.-C., Bellissent,R. and Tourand,G., J.Phys.F:Metal. Phys., 12, .. 2757, 1982. 128)Li Ji-Chen, Cowlan,N. and He Fenglai,F., Phys.Chem.Liquids, 18, 31, 1988.
130) Li,Ji-Chen,Cowlam,N. and HE Fenglai,F, Phys.Chem.Liquids, 18, 31, 1988.
131) Wagner,C.N.J., Boldrick,M.S., J.Non-Cryst.Solids, 156-158, 38, 1993.
132) Joarder,R.N., Palchaudhuri,S. and Gopara Rao,R.V., Phys.Rev., B30, 4417, 1984.
133) Siegel,E., Phys.Chem.Liquids, 5, 29, 1976.
134) Grimvall,G., Inst.Phys.Conf.Ser., No.30, 241, 1977.
135) Herlach,D.M., Ann.Rev.Mater.Sci., 21, 23, 1991.
136) Gathers,G.R., Rept.Prog.Phys., 49, 341, 1986.
137) Seydel,U. and Fucke,W., J.Phys.F:Metal Phys., 10, L203, 1980.
138) Seydel,U and Kizel,W., J.Phys.F:Metal Phys., 9, L153, 1979.
139) Ivanov,V.V., Lebedev,S.V. and Savvatimskii,A.I., J.Phys.F:Metal Phys., 14, 1641, 1984.
140) Seydel,U. and Fucke,W., J.Phys.F:Metal Phys., 8, L157, 1978.
141) Seydel,U. and Fucke,W., J.Phys.F:Metal Phys., 8, 1397, 1978.
142) Shiraishi,Y. and Tsu,Y., The 140th Committee, The Japan Society of Promotion of Science (JSPS), Rep.No.129, Dec. 1982.

143) Casas,J., Keita,M.M. and Steinemann,S.G., Phys.Chem.Liquids, 14, 155, 1984.
144) Bek,R., Phys.Chem.Liquids, 6, 113, 1977.
145) Meyer,A., Stott,M.J. and Young,W.H., Philos.Mag., 33, 381, 1976.
146) Young,W.H., Ber.Bunsen-Gesellschaft, 80, 749, 1976.
147) Tamaki,S. and Waseda,Y., J.Phys.F:Metal Phys., 6, L89, 1976.
148) Yokayama,I., Ohkoshi,I. and Satoh,T., J.Phys.F:Metal Phys., 13, 729, 1983.
149) Ito,H., Yokoyama,I. and Waseda,Y., J.Phys.F:Metal Phys., 16, L113, 1986.
150) NørskoV,J.K., Phys.Rev., B26, 2875, 1982.
151) Gray,P., Yokoyama,I. and Young,W.H., J.Phys.F:Metal Phys., 10, 197, 1980.
152) Percus,J.K. and Yevick,G.J., Phys.Rev., 110, 1, 1958.
153) Egelstaff,P.A., Rept.Prog.Phys., 29, 333, 1966.
154) Rahman,A., Phys.Rev.Lett., 32, 52, 1974; Phys.Rev., A9,1667,1974.
155) Copley,J.R., and Rowe,J.M., Phys.Rev.Lett., 32, 49, 1974; Phys.Rev., A9, 1656, 1974.
156) Meyer,A.,Stott,M.J. and Young,W.H., Phil.Mag., 33, 381, 1976.
157) Varma,C.M. and Wilson,A.J., Phys.Rev., B22, 3795, 1980.
158) Moriety,J.A., Phys.Rev., B26, 1774, 1982.
159) Moruzzi,V.L., PhD thesis, Tecnische Universität Wien, 1985; see also Ref.65.
160) Mon,K.K., Gann,R: and Stroud,D., Phys.Rev., A24, 2145, 1981.
161) Asano,S. and Yonezawa,F., J.Phys.F:Metal Phys., 10,75, 1980.
162) Yokoyama,I. and Naito,S., Zeit .für Phys.Chemie, 156, S469, 1986.
162a) Venkatesh,R., Phys.Stat.Sol., 176, 91, 1993.
162b) Takeno S. and Goda,M., Prog.Theor.Phys., 45, 331, 1971.
162c) Schofield,P., Proc.Phys.Soc., 88, 149, 1966.
163) Aryasetiawan,F., Silbert,M. and Stott,M.J., J.Phys.F:Metal Phys., 16, 1419, 1986.
164) Ducastelle,F., J.Physique, 31, 1055, 1970.
165) Sato,T., Itami,T. and Shimoji,M., J.Phys.Soc.Jpn., 51, 2493, 1982.
166) Umar,I.H., Meyer,A., Watabe,M. and Young,W.H., J.Phys.F:Metal Phys., 4, 1691, 1974.
167) Schwerdtfeger,K. and Engell,H.J., Arch.Eisenhuttenw., 35, 533, 1964; Trans. Met. Soc. AIME, 233, 1327, 1964.
168) Petrushevskiy,M.S., Esin,Yu O., Gel'd,P.V. and Sandakov,V.M., Izv, Vuz. Tsvet. Metal., 1, 37, 1972; Izv.An.S.S.S.R. Metally, 1, 37, 1972; Izv.An.S.S.R. Metally, 6, 193, 1972.
169) Bergman,C., Chastel,R., Gillbert,M., Castanet,R. and Mathieu,J.C., J.de

Physique, 41, C8-591, 1980..
170) Khanna,S.N., Cyrot-Lackmann,F. and Hicter,P., J.Chem.Phys., 73, 4636, 1980; J.de Physique, 41, C8, 582, 1980.
171) Pasturel,A., Hicter,P., Mayou,D. and Cyrot-Lackmann,F., Scr.Met., 17, 841, 1983.
172) Pasturel,A., Colinet,C. and Hicter,P., J.Phys.F:Metal Phys., 15, L81, 1985.
173) Khanna,S.N., Cyrot-Lackmann,F., Phil.Mag., B38, 197, 1978.
174) Khanna,S.N., Cyrot-Lackmann,F., Phys.Rev., B21, 1412, 1980.
175) Lytton,L., J.Appl.Phys., 35, 2397, 1964.
175a) Khanna,S.N., Cyrot-Lackmann,F.and Hicter,P., J.Chem.Phys., 73, 4636,1980.
176) Gaspard,J.P. and Cyrot-Lackmann,F., J.Phys.C:Solid state Phys., 6, 3077, 1973.
177) Pasturel,A., Hicter,P., and Cyrot-Lackmann,F., J.Less Common Metals, 86, 181, 1982.
178) Pasturel,A., Hicter,P., and Cyrot-Lackmann,F., Solid State Comm., 48, 561, 1983.
179) Pasturel,A., Hicter,P., and Cyrot-Lackmann,F., Solid State Comm., 92, 105, 1983.
180) Colinet,C., Bessoud,A., Pasturel,A., Hicter,P., Physica, B133, 103, 1985.
181) Pasturel,A., Colinet,C., Taxton,A.T. and van Schilfgaarde, M., J.Phys.: Condens. Matt., 4, 945, 1992.
182) Dupree,B., Van Zytveld,J.B. and Enderby,J.E., J.Phys.F:Metal Phys., 5, L200, 1975.
183) Ten Bosch,A. and Bennemann,K.H., J.Phys.F:Metal Phys., 5, 1333, 1975.
184) Ziman,J.M., Phil.Mag., 6, 1013, 1961; Advan.Phys., 16, 551, 1967.
185) Evans,R., Greenwood, D.A. and Lloyd,P., Phys.Lett., A35, 57,1971.
186) Evans,R., Gyorffy,B.L., Szabo,N. and Ziman,J.M., "The Properties of Liquid Metals" (ed.Takeuchi,S.), Wiley, New york,219,1973.
187) Mattheiss,L.F., Phys.Rev., A133, 1399, 1964.
188) Mukhopadhyay,G., Jain,A. and Ratti,V.K., Solid State Comm., 13, 1623, 1973.
189) Greig D. and Morgan,G.J., Phil.Mag., 27, 929, 1973.
190) Dreirach,O.D., Evans,R., Güntherodt,H.-J. and Künzi,H.-U., J.Phys.F:Metal Phys., 2, 709, 1972.
191) Hirata,K., Waseda,Y., Jain,A. and Srivastava,R., J.Phys.F:Metal Phys., 7, 419, 1977.
192) Delley,B. and Beck,H., J.Phys.F:Metal Phys., 9, 517, 1979.
193) Waseda,Y., Jain,A. Tamaki,S., J.Phys.F:Metal Phys., 8, 125, 1978.
194) Brown,J.S.,J.Phys.F:Metal Phys., 3, 1003, 1973.
195) Lopez-Escobar,A.H. and Brown,J.S., Phil.Mag., 35, 1609, 1977.

196) Brown,J.S., Lopez-Escobar,A.H. and Todd,J.R., J.Phys.F:Metal Phys., 8, 1703, 1978.
197) Brown,J.S., J.Phys.F:Metal Phys., 11, 2099, 1981.
198) Ferratz,A. and March,N.H., Phys.Chem.Liquids, 8, 271, 1979.
199) Dunleavy,H.N. and Jones,W., J.Phys.F:Metal Phys., 8, 1477, 1978.
200) Goreck,J. and Popielawski,J., J.Phys.F:Metal Phys.,13, 2107, 1983.
201) Mott,N.F., Phil.Mag., 26, 1249, 1972.
202) Mott,N.F., Proc.Roy.Soc., A153, 699, 1936.
203) Mott,N.F. and Jones,H., "The Theory of the Properties of Metals and Alloys", Oxford University Press., London, 1936.
204) Fujiwara,T., J.Phys.F:Metal Phys., 9, 2011, 1979.
205) Haydock,R., Solid State Physics (ed.Ehrenreich,H., Seitz,F. and Turnbull,D., Academic Press., New York), 35, 215, 1980; Heine,V.,ibid., 38, 1, 1980.
206) Bose,S.K.,Ballentine,L.E. and Hammerberg,J.E., J.Phys.F:Metal Phys, 13, 2089, 1983.
207) Ballentine,L.E., Can.J.Phys., 62, 692, 1984.
208) Ballentine,L.E., J.Non.Cryst.Phys., 61&62, 1195,1984.
209) Ballentine,L.E., "Rapidly Quenched Metals"(ed.Steeb,S. and Warlimont,H, Elsevier Science Publishers), 981, 1985.
210) Bucher,P.N., J.Phys., C5, 3164, 1972.
211) Ostermeijer,H.,Fembacher,W. and Krey,U., Zeit.Phys.Chemie, 157, S489, 1988.
212) Büsch,G., Güntherodt,H.-J., Künzi,H.-U. and Meijer,H.A.,"The Properties of Liquid Metals" (ed.Takeuchi,S., Wiley, New York), 263,1973.
213) Dupree,B.C., Enderby,J.E., Newport,R.J., van Zytveld,J.B., Inst.Phys. Conf.Ser., No.30, 337, 1977.
214) Newport,R.J., Dupree,B.C., Enderby,J.E., and Howe,R.A.,J.Phys.F:Metal Phys., 11, 2539, 1981.
215) Laundy,D., Enderby,J.E., Gay,M. and Barnes,A., Phil.Mag., B48, 1981, L29.
216) Kefif,B.,Halim,H.,Ghemaz,EL and Gasser,J.G., J. Non Cryst.Solids, 117&118, 387, 1990.
217) Okada,T. and Ohno,S., J.Phys.Soc.Jpn., 55, 599, 1983.
218) Kita,Y. and Morita,Z., J. Non.Cryst.Solids, 61&62, 1079, 1984.
219) Hirayama,K., Kuwano,N. and Ono,Y., Tetsu to Hagane, 56, S91, 1970.
220) Benazzi,N.,Gasser,J.G. and Kleim,R., Phys.Chem.Liquids, 24, 177, 1992.
221) Benazzi,N.,Gasser,J.G. and Terzieff,P., J. Non Cryst.Solids, 117&118,391, 1990.

222) Newport,R.J., Howe,R.A. and Enderby,J.E., J.Phys.C:Solid St.Phys., 15, 4635, 1982.

223) Tschumi,A., Laubscher,T., Jecker,R., Schpfer,E., Künzi,H.-U. and Güntherodt H.-J., J.Non.Cryst.Solids, 61&62, 1091, 1984.

224) Xu,R., de Groot,R.A., and van der Lugt,W., J.Phys.:Condens.Matt., 4, 2389, 1992.

225) Okada,T., Kakinuma,F and Ohno,S., J.Phys.Soc.Jpn., 52, 3526, 1983.

226) Okada,T. and Ohno,S., J.Phys.Soc.Jpn., 56, 1092, 1986; Zeit für Phys.Chem. Neue Folge, 157, 675, 1988.

227) Cutler,M.,"Liquid Semiconductors", Academic Press, New York, 1977.

228) Wilson,J.A., Adv.Phys., 21, 143, 1972.

229) Moruzzi,V.L.,Janak,J.F. and Williams,A.R., "Calculated Electronic Properties of Metals", Pergamon Press Inc, 1978.

230) Cyrot-Lackmann,F. and Khanna,S.N.,"Excitations in Disordered Systems"(ed. Thorpe,M.F.,Plenum Press), 59,1982; Heine,V., Solid State Physics (ed.Ehrenreich,H., Seitz,F. and Turnbull,D., Academic Press.; New York), 38, 1, 1980.

231) Van Zytveld,J.B., J.Non Cryst.Solids, 117&118, 437, 1990.

232) Nakagawa,Y., J.Phys.Soc.Jpn., 11, 855, 1956.

233) Urbain,G. and Übelacker,E., "The Properties of Liquid Metals" (ed.Takeuchi,S., Wiley, New York), 429, 1973.

234) Adachi,K. and Eisaka,T., "The Properties of Liquid Metals" (ed.Takeuchi,S., Wiley, New York), 313,1973.

235) Dupree,R. and Seymour,F.W., "Liquid Metals" (ed.Beer,S.Z, Marcel Dekker, New York), 461, 1972.

236) March,N.H. and Seyers, C.M.,Adv. in Phys., 28, 1, 1979.

237) Busch,G.,Güntherodt,H.-J.,Künzi, H.-U.and Meijer,H.A.,Ten Bosch,A. and Zimmer-mann,A., "The Properties of Liquid Metals" (ed.Takeuchi,S.), Wiley, New york, 277,1973.

238) Bass,R., J.Phys.F:Metal Phys., 4, 1256, 1974.

239) Bennemann,K.H., J.Phys.F:Metal Phys., 6, 43, 1976.

240) Weiss,W. and Alexander,H., J.Phys.F:Metal Phys., 17, 1983, 1987.

241) Kamp,P.G. and Methfessel,S., Zeit.für Phys.Chemie, 156, S575,1988.

242) Okada,T., Ohno,S. and Iida,M., J.Non Cryst.Solids, 117&118, 367, 1990.

243) Gubanov,A.I., Soviet Phys.,Solid St., 2,468, 1960.

244) Busch,G. Güntherodt,H.-J.Phys.Rev., A27, 110, 1968.

245) Nakagawa,Y., Phys.Lett., 28, 494, 1969.
246) Wachtel,E. and Kopp,W.U., Phys.Lett., A29, 164, 1969.
247) Menth,A. and Bagley,B.G., Appl.Phys.Lett., 15, 67, 1969.
248) Kraeft,B. and Alexander,H., Phys.Kondens.Matter, 16, 281, 1973.
249) Nakagawa,Y., Yamaguchi,K., Mizoguchi,T., "The Properties of Liquid Metals" (ed.Takeuchi,S., Wiley, New York), 307,1973.
250) Briane,M., C.r.hebd.Séanc.Acad.Sci.,Paris, 276, 139, 1973; ibid, 277, 695,1973.
251) Morgan,G.J. and Weir,G.F., Phil.Mag., 47, 177, 1983.
252) Gallagher,B.L., Greig,D., Howson,M.A. and Croxton, A.A.M., J.Phys.F:Metal Phys., 13, 2331, 1983.
253) Nguyen-Manh,D.,Mayou,D.,Morgan,G.J. and Pasturel,A, J.Phys.F:Met.Phys., 17, 999, 1987.
254) Ten Bosch,A., Phys.Kondens.Materie, 16, 289, 1973.
255) Szabo,N., J.Phys.C:Solid State Phys., 5, L241, 1972.
256) Ballentine,L.E., "The Hall effect and its applications"(ed.Chien,C.L. and Westgate,C.R., Plenum Press, New York and London), 201,1980.
257) Künzi, H.-U, and Güntherodt, H.-J., "The Hall effect and its applications" (ed. Chien, C.L. and Westgate,C.R., Plenum Press, New York and London), 215,1980.
258) Ivkov,J. and Babic,E., J.Non Cryst.Solids, 156-158, 307, 1993.
259a) Ivkov,J. and Babic,E., J.Phys.F:Condens.Matt., 2, 3891, 1990.
259b) Howson,M.A. and Gallaghter,B.L., Phys.Rept., 170, 265, 1988.
260) Garg,K.B. and Källne,E., Phys.Stat. Soli., B70, K121, 1975.
261) Hague,C,F., Inst.Phys.Conf.Ser., No.30, Chap.2, Part1, 360, 1977.
262) Williams G.P. and Norris, C., J.Phys.F:Metal Phys., 4, L175, 1974.
263) Pflugi,A., Indlekofer,G. and Oelhafen,P., J.Non Cryst.Solids, 117&118, 336, 1990.
264) Indlekofer,G., Oelhafen,P., Güntherodt,H.-J., Hague,C.F. and Mariot,J.M., Zeit.füf Phys.Chemie Neue Folge, 157, 575, 1988; Indlekofer,G., Pflugi,A., Oelhafen,P., Güntherodt,H.-J., Häussler,P.,Boyen,H.-G. and Baumann,F., J.Mater. Sci. Engineering, 99, 257, 1988; Oelhafen,P, Indlekofer,G. and Pflugi,A., J.Non Cryst.Solids, 117&118, 267, 1990.
265) Dose,V., Drube,R. and Härl,A., Solid State Comm., 57, 273, 1986.
266) Tsuji,K., Endo,H., Kita,Y., Ueda,M. and Morita,Z., Inst.Phys.Conf.Ser., No.30, 367, 1977.
267) Fluss,M.J., Smedskjaer,L.C.,Chakraborty,B and Chason,M.K., J.Phys.F:Metal Phys., 13, 817, 1983.

268) Khanna,S.N., Cyrot-Lackmann,F,Desjonéres, J.Phys.F:Metal Phys., 9, 79, 1979.
269) Gardner,J.A. and Flynn,C.P., Phil.Mag., 15, 1233, 1967.
270) Flynn,C.P, Rigney,D.A. and Gardner,J.A., Phil.Mag., 15, 1255, 1967.
271) Gardner,J.A. and Flynn,C.P., Phys.Rev.Lett., 17, 579, 1966.
272) Peters,J.J. and Flynn,C.P., Phys.Rev., B6, 3343, 1972.
273) Lakshmanan,T.S. and Rigney,D.A., Mater.Sci.Engineering, 12, 285, 1973.
274) Dupree,R., Walstedt,R.e. and Warren, W.W., Bull.Am.Phys.Soc., 21, 328, 1976.
275) How,R.A., Rigney,D.A. and Flynn,C.P., Phys.Rev., B6, 3358, 1972.
276) Tamaki,S. and Takeuchi,S., J.Phys.Soc.Jpn., 22, 1042, 1967.
277) Terzieff,P., Auchet,J.and Bretonnet,J.L., J.Phys.F:Condens.Matt., 5. 1777, 1993.
278) Ohno,S., J.Phys.Soc.Jpn., 53, 1459, 1984.
279) Ohno,S., J.Non Cryst. Silids, 61&62, 1341, 1984.
280) Tamaki,S., J.Phys.Soc.Jpn., 25, 1596, 1968.
281) Tamaki,S., J.Phys.Soc.Jpn., 25, 1602, 1968.
282) Collings,E.W., Solid St. Comm., 8, 381, 1970; J.Phys., Paris C1, 516, 1971.
283) Gardner,J.A. and Ardary,C., Solid St. Comm., 19, 143, 1976.
284) Tamaki,S., J.Phys.Soc.Jpn., 25, 379, 1968.
285) Ohno,S., Okazaki,H. and Tamaki,S., J.Phys.Soc.Jpn., 35, 1060, 1973.
286) Tamaki,S., J.Phys.Soc.Jpn., 22, 865, 1967.
287) Dupree,R., Walstedt,R.E. and DiSalvo,F.J., Phys.Rev., 19B, 4444, 1979.
288) Takeda,S., Ohno,S. and Tamaki,S., J.Phys.Soc.Jpn., 40, 113, 1976.
289) Ohno,S. Nomoto,T. and Tamaki,S., J.Phys.Soc.Jpn., 40, 72, 1976.
290) Ohno,S., J.Phys.Soc.Jpn., 42, 194, 1977.
291) Ohno,S. and Harada,S., J.Phys.Soc.Jpn., 49, 189, 1980.
292) Romer,O. and Wachtel,E., Z.Metallk., 62, 871, 1971.
293) Grüber,O.F. and Gardner,J.A., Phys.Rev., B4, 3994, 1971.
294) Bessel,J. and Gardner,J.A., Phys.Rev., B8, 4303, 1973.
295) Ritter,A.L., Bremer,J.C. and Gardner,J.A., Phys.Rev., B10, 3246, 1974.
296) Wachtel, E.and Pantasis,A., Z.Metallk., 66, 172, 1975.
297) Gardner,J.A.,Zollner,R.and Sotier,S., Phys.Rev., B12, 5245, 1975.
298) Grünther,H.H.,Wachtel,E. and Gerold,V., Physica, B80, 473, 1975.
299) Ohno,S., Okada,T. and Togashi,M., J.Phys.Soc.Jpn., 56, 3616, 1987.
300) Ohno,S., Zeit. für Phys.Chemie Neue Folge, 157, 669, 1988.
301) Togashi,M., Okada,T and Ohno,S., J.Phys.Soc.Jpn., 56, 3609, 1987.
302) Ohno,S., J.Phys.Soc.Jpn., 50, 1934, 1981.

303) Tamaki,S. and Ohno,S., Inst.Phys.Conf.Ser., No.30, 474, 1977.
304) Ohno,S., J.Phys.Soc.Jpn., 55, 295, 1986.
305)) Allen,B.C., "Liquid Metals"(ed.Beer,S.Z.), Marcel Dekker, New York, 161, 1972.
306) Cyrot-Lackmann,F., Sur.Sci., 15, 535, 1969.
307) Sayers,C.M., J.Phys.F:Metal Phys., 7, 1157, 1977.
308) Nogi,K., Ogino,K, McLean,A., Miller,W.A., Met.Trans., B17, 163, 1986.
309) Nogi,K., Chung,W.B., McLean,A., Miller,W.A., Material Trans. JIM., 32, 164, 1991.
310) Nogi,K., Chung,W.B., McLean,A., Miller,W.A., Material Trans. JIM., 33, 753, 1992.
311) Sauerland,S.,Lohöfer,G.L.and Egly,I., J.Non Cryst.Solids, 156-158, 830-832, 1993.
312) Lang,N.D. and Kohn,W., Phys.Rev., B3, 1215, 1971; Lang,N.D., Solid State Physics(ed. Ehrenreich,H., Seitz,F and Turnbull,D.,Academic Press, New York), 28, 225, 1973.
313) Allen,J.W. and Rice,S.A., J.Phys.Chem., 67, 5105, 1977.
314) Evans,R. and Hasegawa,M., J.Phys.C:Solid St. Phys., 14, 5225, 1981; Hasegawa,M. and Watabe,M., J.Phys.C:Solid St. Phys., 15, 353, 1982.
315) Evans,R., J.Phys.C:Solid St. Phys., 7, 2808 1974; Kumaravadivel,R. and Evans,R., J.Phys.C:Solid St. Phys., 8, 793, 1975; Kumaravadivel,R. and Evans,R., J.Phys.C:Solid St. Phys., 9, 1891, 1976.
316) Brown,R.C. and March, N.H.,J.Phys., C5, L363, 1973.
317) Egelstaff, P.A. and Widom,B., J.Chem.Phys., 53, 2667, 1970.
318) Alonso,J.A. and Silbert,M., Phys.Chem.Liquids, 17, 209, 1987.
319) Alonso,J.A., Phys.Chem.Liquids, 21, 257, 1990.
320) Bhatia,A.B. and March,N.H., J.Chem.Phys., 68, 1999, 4651, 1978.
321) Cahn,J.W. and Hilliard,J.E., J.Chem.Phys., 28, 258, 1958.
322) Itami,T. and Shimoji,M., J.Phys.F:Metal Phys., 9, L15, 1979.
323) Dragomier,I., "The Properties of Liquid Metals" (ed.Takeuchi,S., Wiley, New York), 507, 1973.
324) Wang,J.,Wang,H. and Bian,M., Zeit. für Phys.Chemie Neue Folge, S156, 599, 1988.
325) Speedy,R.J., Mol.Phys., 66, 577, 1989.
326) Alder,B.J., Gass,D.M. and Wainwright,T.E., Phys.Rev.Lett., 53, 3813, 1970.
327) Protopapas,P., Anderson,H.C. and Parlee,A.D., J.Phys.Chem., 59, 15, 1973.
328) Andrade,E.N. da C., Phil.Mag., 17, 497, 1934.
329) Lucas,L-D, Techniques de L'ingénieur, Form. M66, 1, 1984.

330) Egry,I., Feuerbacher, B., Lohöfer,G. and Neuhaus,P., Proc. VIIth Euro.Sympo.on Mater. and Fluid Sci., in Microgravity, Oxford, UK, 10-15, Sept., 1989, ESA SP-295(Jan.),1990.
331) Frohberg,G., "Material Sciences in Space: A Contribution to the Scientific Basis of Space Processing" (ed.Feuerbacher,B., Hamacher,H. and Naumann,R.J., Springer-Verlag, Berlin Heiderberg New York Tokyo), 425, 1986.
332) Friedel,J., Suppl. to Al Volume VIII, Serie X, Del Nuovo Cimento, 7, 287, 1958.
332a) Malmejac,Y. and Praizey,J.P.,see the Ref.331.
333) Johnson,M.W., McDoy,B. and March,N.H., Phys.Chem.Liquids, 6, 243, 1977.
334) Ono,Y., "The Properties of Liquid Metals" (ed.Takeuchi,S., Wiley, New york) ,543, 1973.
335) Adachi,A., Morita,Z., Ogino,Y and Ueda,M., "The Properties of Liquid Metals" (ed.Takeuchi,S.), Wiley, New york ,561, 1973.
336) Morita,Z, Iida,T. and Ueda,M., Inst.Phys.Conf.Ser., No.30, 600, 1977.
337) Nishi,Y., Watanabe,H., Suzuki,K., Masumoto,T., J.de Physique, 41, C8-359, 1980.
338) Yamazaki,T, Kanatani,S., Ogino,Y.and Inoue,A.,J.Non Cryst.Solids, 156-158, 1993.
339) Egly,I., Lohöfer,G.L. and Sauerland,S., J.Non Cryst.Solids, 156-158, 830-832, 1993.
340) Mujibar Rahman,S.M. and Lutful Bari Bhuiyan, Phys.Rev., B33, 7243, 1986.

CHAPTER 4
CONDENSED MATTER - NON TRANSITION
LIQUID METALS AND ALLOYS

S.K. Srivastava
School of Physics, D.A. University, Indore, India

CONTENTS

4.1. Introduction	252
4.2. Different approaches	253
(a) Structural approach	253
(b) Model potential approach	256
(i) Pseudopotential theory of metals	256
(ii) Model pseudopotentials	260
(c) Phase-shift analysis approach	263
(d) Computer simulation approach	264
(e) Other approaches	265
4.3. Electronic states (Electronic structure) and electronic properties	266
(a) Dispersion of electrons and electronic band structure	267
(b) Electrical-transport properties of metals in liquid phase	268
(c) Electrical-transport properties of metals in solid phase	273
(d) Thermal conductivity	275
(e) Magnetic properties	276
(i) Susceptibility	276
(ii) Knight shift	277
(iii) Hall effect	281
4.4. Interatomic or Intermolecular Properties	283
(a) Effective ion-ion interaction or pair potential	286
(b) Crystal structure energy	289
(c) Dispersion of phonons	292
(i) Solid phonons	292
(ii) Liquid phonons	294
4.5. Thermodynamic properties	296
(a) Heat capacity	296
(b) Gruneisen parameter	299
(c) Thermal expansion	301
(d) Self-diffusion coefficient	302
(e) Compressibility	303
(f) Entropy	304
(g) Viscosity	307
(h) Surface tension	308
References	309

4.1 INTRODUCTION

The condensed matter researches essentially deal with the structure and properties of different states of matter[1]. Metals are conducting solids, which have special importance in material science. At melting point and above this temperature, there have been observed drastic variations in many properties of metals than at lower temperatures. It is all due to the change of phase states. At melting temperature and above, we call metal as liquid metal. A metal in liquid phase structure differs than that in crystalline phase. The crystalline phase is ordered phase whereas liquid phase is disordered. The liquid metals possess isotropic behaviour.

The Pseudopotential theory[2], which is the revived form of free electron theory of metals, is widely successful in the investigations of a number of properties of non-transition liquid metals (simple metals). In a number of conferences, the development regarding liquid metals have been discussed. In this chapter, we will discuss the important developments of different properties of non-transition metals, which have been observed in the span of last three decades - a time in which the subject of liquid metals got special attention[3-7].

Truely speaking, in a crystalline solid the bonds and nearest neighbour atoms positions are oriented along specific directions in space whereas in a disordered solid the lines joining pairs of nearest neighbour atoms point with same probability in all directions of space. However, liquid crystal exhibit a broken oriental symmetry but follow the translational invariance of a liquid. While statistical theories of liquid state may be easily applied in all disordered solids, the computer simulation and integral equations techniques are becoming very popular in studies of disordered solids. In this article, the discussions and development regarding non-transition liquid metals are fully based on all possible fundamental interactions involved in the system. It is well known fact that the exact pair interactions between most real interactions are considered and the forces between pairs of particles are defined hypothetically. Various interactions between ion-ion, electron-ion, electron-phonon and electron- electron play important role in the theory of non- transition metals.

Among static properties, the structural properties are represented in terms of distribution functions which describe the probabilities of finding given numbers of atoms at certain locations. Pair distribution functions play an important role in this direction, which give the probability of finding an atom at a distance r from a given atom. Neutron and X-ray scattering experiments as well as computer-simulation devices have solved this problem successfully, whose data are helpful in deducing thermodynamic properties of liquid metals.

In recent years, the dynamical properties associated with molecular motion in condensed matter science have got special attention. There are several experimental techniques for the study of fluid dynamics even for single particle behaviour and collective behaviour, separately. The researches of disordered solids on nuclear magnetic resonance (NMR), electron spin resonance (ESR), infra-red and Raman spectroscopy and non-linear optical techniques are appreciable. Ultimately, a quantity known as relaxation time, τ is important in all these approaches. The properties are discussed in terms of τ. In order to study the dynamical behaviour of disordered solids, the kinetic theory is very important and successful approach. The molecular collisions play significant role in determining spectra and various transport coefficients through Boltzmann treatments. The liquid state-kinetic theory considering hard sphere systems for liquid metals is a primitive approach in this direction. The molecular dynamics (MD) or computer simulations in case of disordered solids are creating interests for the scientists in this direction. The disturbances created externally within a box containing fixed number of molecules develop the induced momentum flux as a consequences of the shear. A direct proportionality between the flux and the shear gradient describes the shear viscosity. A characteristic feature of rotational dynamics in case of disordered solids is pertinent to an oscillation in the angular velocity time dependent correlation function.

4.2 DIFFERENT APPROACHES

We outline here important approaches used in the investigations of properties of non-transition liquid metals.

(a) Structural Approach

The structure factors are very useful in describing the properties of metals. The pair distribution, g(r) and the direct correlation function, C(r) may be easily determined from the Fourier transform equation of the experimental measurable quantity, S(q):

$$[g(r)-1] = \Omega \int [S(q)-1] e^{i q \cdot r} d^3q \tag{1}$$

$$C(r) = \Omega \int \left[\frac{S(q)-1}{S(q)} \right] e^{i q \cdot r} d^3q \tag{2}$$

In recent years, other than X-ray and neutron scattering experiments, the extended X-ray absorption fine structure (EXAFS) has become a very successful technique for structural studies of condensed matter through correlation functions.

March and his Co-workers[8] tried to yield an effective pair potential, $\phi(r)$ through the following eqs. of approximate theories of the liquid state.

$$\phi_{PY}(r) = k_B T \ln \left[\frac{1-C(r)}{g(r)} \right] \tag{3}$$

[Percus - Yevick (PY)]

$$\phi_{HN}(r) = k_B T \left[\{g(r)-1\} - C(r) - \ln g(r) \right] \tag{4}$$

[Hypernetted Chain (HN)]

$$\phi_{BG}(r) = k_B T \left[-n \int E(r-r') h(r') dr' - \ln g(r) \right] \tag{5}$$

[Born - Green (BG)]

Here $C(r)$ is the direct correlation function introduced by Ornstein and Zernike[9], k_B is Boltzmann constant and $E(x)$ is given by

$$E(x) = \int_{\infty}^{x} \left[\frac{g(r)}{k_B T} \right] d \{\varphi_{BG}(r)\} \qquad (6)$$

Out of above three equations of $\phi(r)$ the Born-Green equation becomes controversial as has been pointed out several times[10-12]. March[13] reviewed the situations very well.

$g(r)$ function may be used in providing informations about total electronic band structure energy $G(q)$, an experimental measurable quantity through eqs. (3) - (5) by considering interionic potential relationship,

$$V(r) = \phi(r) = \frac{Z^2 e^2}{r} - \frac{2Z^2 e^2}{\pi} \int_{0}^{\infty} G(q) \frac{\sin(qr)}{(qr)} dq \qquad (7)$$

Enderby et al.[14] used partial radial distribution function concept $g_{\alpha\beta}(r)$ as given below in the studies of knowing the logic behind the formation and stability of complex-ions for aqueous solutions (multi-component liquid systems) and molten salts

$$g_{\alpha\beta}(r) = 1 + \frac{1}{2\pi^2 nr} \int [S_{\alpha\beta}(q) - 1] q \sin(qr) dq \qquad (8)$$

Vander Lugt and Geertsma[15] and Sabounqu et al.[16] independently, at the same time, studied the case of K- Pb system and Barnes and Enderby[17] studied the case of Cu - Se systems. These workers observed that there are formation of clusters of the forms $(Pb_4)^{4-}$ and $(Se_2)^{2-}$ in these systems, respectively. The formation of these complex species was understood on the basis of the studies of partial radial distribution functions, $g_{Pb-Pb}(r)$ and $g_{Se-Se}(r)$. It is remarkable to mention that the knowledge of the factors $g_{\alpha\beta}(r)$ and $s_{\alpha\beta}(q)$ may be utilized in the description of liquid phonon spectrum of multi-component liquid systems and their elastic behaviour similar to the case of liquid metals. The Egelstaff formula[17] may be extended to such systems in the manner

$$\omega^2_{\alpha\beta}(q) = \frac{q^2 k_B T}{M S_{\alpha\beta}(q)}, \qquad (9)$$

where $S_{\alpha\beta}(q)$ also gives information about interionic potential, $V_{\alpha\beta}(q)$ through equation

$$S_{\alpha\beta}(q) = \left[\frac{V_{\alpha\beta}(q)}{k_B T} + 1 \right]^{-1}, \qquad (10)$$

which may be helpful in understanding the studies of excitations produced in liquids due to several causes.

For the system of N interacting particles representing the liquid, one may easily understand about the entropy, $S^E_{\alpha\beta}(q)$ of the assembly through standard formula of statistical thermodynamics.

$$S^E_{\alpha\beta}(q) = \sum \frac{1}{T} \hbar\omega_{\alpha\beta}(q) \left[\frac{1}{2} + \left\{ \frac{\hbar\omega_{\alpha\beta}(q)}{k_B T} - 1 \right\}^{-1} \right]$$

$$- k_B \ln \left[2 Sh[\frac{\hbar\omega_{\alpha\beta}(q)}{2 k_B T}] \right] \qquad (11)$$

In recent years[18], the collective excitations in liquids have been studied very well with the dynamical structure factors, $S(q, \omega)$, a quantity directly obtained from elastic neutron scattering experiments. There are only evidences for propagating excitations in He - monoatomic liquid for wave vectors greater than $1(\text{Å}°)^{-1}$. The theory of collective excitations in classical liquids may be used in order to determine the dynamic structure factors of metals through equation

$$S(q,\omega) = \frac{k_B T}{\pi n \omega} \left[\frac{X''_{sc}(q,\omega)}{[1 - A(q) X'_{sc}(q,\omega)]^2 + [A(q) X_{sc}(q,\omega)]^2} \right] \qquad (12)$$

where
$$A(q) = \left[\frac{k_B T}{nS(q)} + \frac{1}{X_{sc}(q,0)} \right] \qquad (13)$$

[Zero - moment rule]

which is the Fourier transform of the bare interparticle potential. $X_{sc}(q,0)$ is the second response function. $X'_{sc}(q,\omega)$ and $X''_{sc}(q,\omega)$ are real and imaginary parts of the density response functions, respectively.

For $\omega = 0$

$$S(q, 0) = S^2(q) \left(\frac{m}{4\pi k_B T q^2} \right)^{1/2} \left[3 - [S(q)]^{-1} + nP_4(q) [k_B T]^{-1} \right]^{1/2}, \qquad (14)$$

where
$$P_4(q) = \int g(r) [1 - \frac{Cos(q.r)}{q^2}].(q.\nabla)^2 \varphi(r) dr \qquad (15)$$

Here $\varphi(r)$ is the interparticle potential. $S(q,\omega)$ may be obtained directly [19] from elastic neutron scattering experiments.

Van Hove scattering function, $I(q,p)$ is related to $S(q, \omega)$ and intermediate function $F(q,t)$ by the equation [20]

$$I(q,p) = - \int_{-\infty}^{\infty} e^{-pt} \frac{\partial F(q,t)}{\partial t} dt$$

$$= \int_{-\infty}^{\infty} \frac{\omega^2}{\omega^2 + p^2} S(q,\omega)\, d\omega , \qquad (16)$$

where
$$F(q,t) = e^{i\omega t} S(q,\omega)\, d\omega \qquad (17)$$

$S(q,\omega)$ is related to self-correlation function $S_s(q,\omega)$ by the relation[21]

$$S(q,\omega) = S(q)\, S_s(q,\omega) \qquad (18)$$

A relation between Van Hove scattering function $I_s(q,p)$ and $S_s(q,\omega)$ is given by

$$I_s(q,p) = \int_{-\infty}^{\infty} \frac{\omega^2}{\omega^2 + p^2} S_s(q,\omega)\, d\omega \qquad (19)$$

where I and I_s are such that they follow the condition

$$\left(\frac{\partial I}{\partial I_s}\right)_{p=0} = \frac{S(q,0)}{S_s(q,0)} \qquad (20)$$

Since for $p = 0$, $I(q,0) = 1$ and also $I_s(q,0)$ the quantities $S(q,0)/S_s(q,0)$, $S(q)$, $[S(q)]^2$ and $S(q,\omega)$ have been studied[22,23] for liquid Ga.

One can study moment sum rules with the help of $S(q,\omega)$ function through equation

$$<\omega^m> = \frac{\int_{-\infty}^{\infty} S(q,\omega)\, \omega^m d\omega}{\int_{-\infty}^{\infty} S(q,\omega)\, d\omega} \qquad (21)$$

(b) Model Potential Approach

(i) **Pseudopotential theory of metals**: In the theory of metals, pseudopotential has played an important role in describing properties of condensed matter. One of the basic reasons for developing the pseudopotential theory of metals lies in fact that in periodic solids due to lack of knowledge in evaluating the Bloch matrix element $\psi_k(r)$ $|V_{Bloch}(r)|\,\psi_{k'}(r)$, $[\psi_k(r) = \Omega^{-1/2} e^{i k \cdot r} U_k(r)$; $U_k(r)$ is periodic in the lattice vector R_j]. One may easily employ Austin-Heine-Sham approximation[2] and thus the above matrix element may be replaced in the form $\varphi_k(r)\,|V(r)|\,\varphi_{k'}(r)$ where the pseudowave function $\varphi_k(r)$ is also of the Bloch form $\varphi_k(r)\,|V(r)|\,\varphi_{k'}(r)$. Here $U_k(r + R_j) = U_k(r)$. In self-consistent field approximation by employing Hartree treatment, one may easily express

$$<\varphi_k(r)\,|\hat{V}(r)|\,\varphi_{k'}(r)> = v(k,k')\, S(k,k') \qquad (22)$$

where
$$\hat{v}(k,k') = \frac{1}{\Omega} \int d^3r \, e^{i\mathbf{k}\cdot\mathbf{r}} W^*_k(r) \hat{V}(r) W_{k'}(r) e^{i\mathbf{k}'\cdot\mathbf{r}} \qquad (23)$$

and
$$S(k-k') = \frac{1}{N} \sum e^{-i(k-k')\cdot R_j} \qquad (24)$$

are pseudopotential form factor and structure factor, respectively, Ω_o is atomic volume and N is number of atoms in the crystal. $V(k, k')$ depends on the arrangement of the atoms through $W_k(r)$. The pseudopotential $\hat{V}(r)$, is really an operator (marked by notation Λ) which is related to the pseudopotential due to a single atom situated at the lattice site R_J in manner

$$\hat{V}(r) = \sum_j \hat{v}(r - R_j) \qquad (25)$$

In single plane wave approximation by considering

$$\varphi_k(r) = \Omega_0^{-1/2} e^{i\mathbf{k}\cdot\mathbf{r}} \qquad (26)$$

the periodicity may be absconded and thus the revived form of free-electron theory obtains much more importance in evaluating the properties of condensed matter.

The Pseudo-wave equation

$$H_P \varphi_V = (T + V_P) \varphi_V = E_V \varphi_V, \qquad (27)$$

[$H_P = H + V_R$: Pseudo-Hamiltonian ; $V_P = V + V_R$: pseudopotential ; φ_V : pseudo-wave function] with symmetric potential energy may be written in spherical co-ordinates like

$$\left[\frac{h^2}{2m} \left[\frac{1}{r^2} \frac{\partial}{\partial r}(r^2 dr) + \frac{1}{r^2 Sin\theta} \frac{\partial}{\partial \theta}(Sin\theta \frac{\partial}{\partial \theta}) + \frac{1}{r^2 \sin^2\zeta} \frac{\partial^2}{\partial \zeta^2} \right] + V_R \right] \varphi$$

$$+ [V(r) - E] \varphi = 0 \qquad (28)$$

whose solution $\varphi(r, \theta, \zeta)$ is described by

$$\varphi(r, \theta, \zeta) = F_l(r) L(\theta, \zeta) \qquad (29)$$

With this wave function the radial part of eq.(28) is given by

$$\frac{1}{r^{2l-3}} \frac{d}{dr} [r^{2l-3} \frac{d}{dr} F_l(r)] + \frac{2m}{h^2} [E - V(r) - \frac{\lambda}{r^2}] F_l(r) = 0 \qquad (30)$$

where
$$\lambda = (2l-3) \text{ and } r = \frac{x}{a} ; \alpha = \frac{2m}{\hbar^2} [E - V(r)]^{1/2} \qquad (31)$$

To construct the exact solution of the Schrodinger equation inside a sphere of model radius $r = R_M$, where potential V (r) is having spherical symmetry, it can be considered [24] that the amplitude of radial function $F_1(r)$ adjusts by multiplying the constant $S_l(x)/F_l(r)$ so that it is equal to $S_l(qR_M)$ at $r = R_M$. By using this argument, we obtain the following differential equation

$$D^2 y + \frac{(2l-3)}{x} Dy + [1 - \frac{2l-3}{x^2}] y = 0 \; ; \; [D = \frac{d}{dx}] \qquad (32)$$

whose solution is described by

$$y = S_l(x) = (-1)^{l+1} x^{2-l} J_{l-1}(x) \qquad (33)$$

Here $S_l(x)$ is called pseudo-spherical function [5] and $J_l(x)$ is ordinary Bessel function, whose relationship with spherical Bessel function $j_l(x)$ is described by

$$s_l(x) = (-1)^{l+1} x^{2-l} (\frac{\pi}{2x})^{1/2} J_{l-1/2}(x) = i (\frac{\pi}{2})^{1/2} S_{l+1/2}(x) \qquad (34)$$

$$j_1(x) = (\frac{\pi}{2x})^{1/2} J_{l+1/2}(x) = (-1)^l x^{l-1} s_{l+1}(x) \qquad (35)$$

An important recurrence relation of $S_l(x)$ function may be described like

$$S_{l+1}(x) = (-1)^{2l} (\frac{1}{x}) \frac{d}{dx} [\frac{1}{x} \sum_{l'=0}^{l} S_{l'}(x)] \qquad (36)$$

The behaviours of pseudo-spherical functions, $S_l(x)$ and radial function $F_l(r)$ for s,p,d and f electron-states have been displayed in Figs. 1-3, which describe the characteristics of these functions inside different electronic shells. It is remarkable that $S_l(x)$ and $s_l(x)$ functions have same value for $l = 0, 1$ and 2 states, which is an interesting resonastic behaviour. This differentiate it from Bessel and other spherical functions.

The display of $S_l(x)$ functions directly convey about the fluctuations of charge density in different electronic states. $S_2(x)$ function differs with $S_0(x)$ and $S_1(x)$ functions under the condition $x \longrightarrow 0$. The former has maxima whereas latter functions have minima. $F_1(r)$ function has linear behaviour whereas $F_0(r)$ and $F_2(r)$ have non-linear behaviour.

In real space the number electron density, $\rho_R(r)$ may be described by

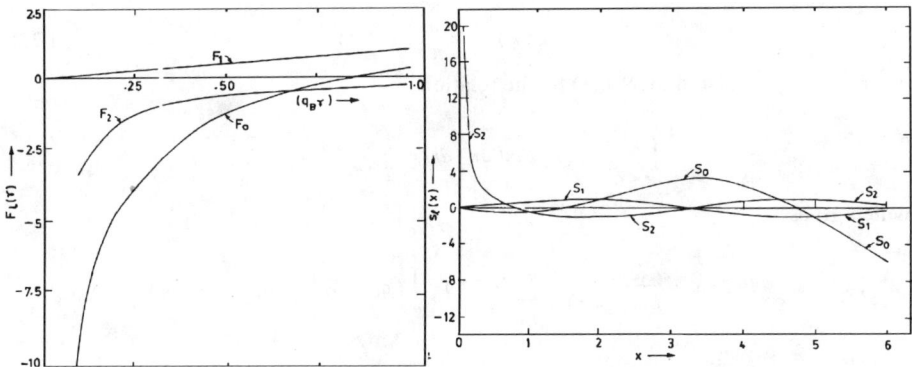

Fig. 1 - Different polynomials, $F_0(r)$, $F_1(r)$ and $F_2(r)$ of radial function $F_l(r)$

Fig. 2 - The behaviour of pseudo - spherical function, $S_l(x)$ in individual s, p, d and f electronic states.

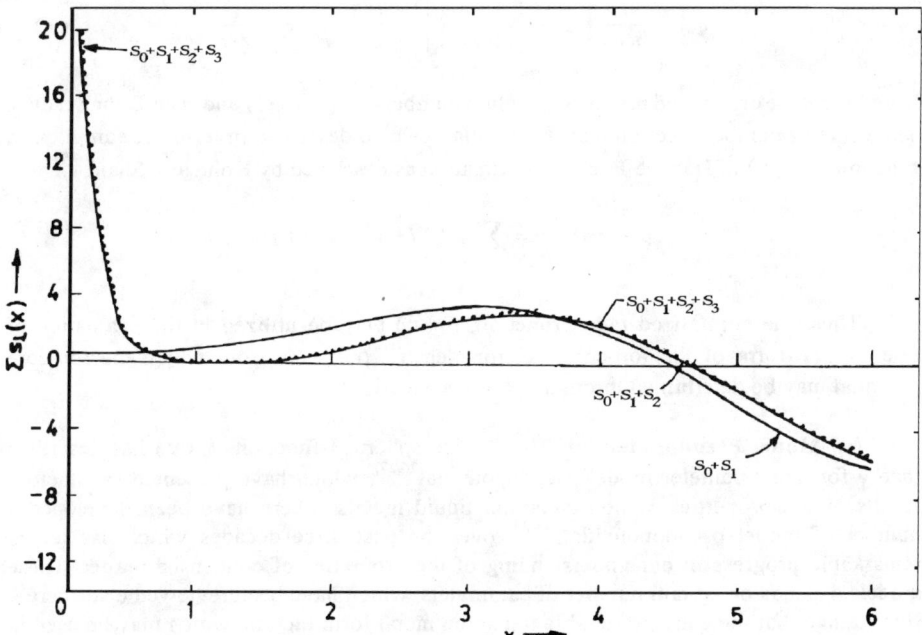

Fig. 3 - The combined behaviours of spherical function $S_l(x)$

$$\rho_R(r) = \frac{q_B^2}{4\pi r} F_l(r) \qquad (37)$$

where $F_l(r)$ is related to $S_l(x)$ by the relation

$$\int_{r<r_c} F_l(r) \sin(qr)\, dr = \frac{1}{q} \sum_{l,n} S_l(q\, r_n) \qquad (38)$$

which gives

$$F_l(r) = \frac{1}{\sin(q_B r)} \Bigg[\sum_{l,n} (-1)^{l+1} (q_B r_n)^{-l+1} J_{l-1}(q_B r_n)$$

$$- (q_B r_n)^{-l+2} J_l(q_B r_n) \Bigg] \qquad (39)$$

where q_B is Bardeen Screening parameter, which is given by

$$q_B^2 = \frac{4k_F}{\pi a_o}\left[\frac{1}{2} + \frac{1-x^2}{4x^2} \ln\left|\frac{1+x}{1-x}\right|\right] \quad ; \quad x = \frac{q}{2k_F} \qquad (40)$$

Here l and n are orbital and principal quantum numbers, respectively and a_o is Bohr's radius. The value of number electron density, ρ_R may be also described in terms of complicated function, $P_{nl}(r)$ of Hartree-Fock approximation as described by Kohn and Sham [25]

$$\rho_R(r) 4\pi r = \sum_{l,n} 2(2l+1) P^2_{n,l}(r) \qquad (41)$$

Thus, the normalized radial function, $F_l(r)$ may be utilized in the evaluation of Fourier transform of the ion-core electron density, $\rho_R(r)$, from which the ion-core potential may be determined through Poisson's equation.

(ii) **Model Pseudopotential**: The Pseudo-spherical function, $S_l(x)$ has described theory for one-parameter model pseudopotential[26,27], which have presented satisfactory results of the properties of non-transition liquid metals. There have been developed a number of model pseudopotentials[28-33] over the past three decades which has led to remarkable progress in our understanding of the properties of condensed matter. These models depends on certain number of parameters, which have been reviewed elsewhere[7]. It is strange that there are no reliable transition metal form factors, which may be used in the investigation of their properties. It is worthwhile to mention that in case of transition metals the roles of d-states and resonance are very important.

In view of the success of parametric model pseudopotential, it is remarkable that in the beginning Ziman[34], considered for small value of q the screened form factors of type

$$U(q) \sim -\frac{\frac{2}{3} E_F}{(1+q^2/k_s^2)} \qquad (42)$$

But unfortunately this emperical form could not be able to evaluate the metallic properties. On the basis of theory of associated Legendre Polynomials a suitable form of non-parametric screened pseudopotential is given by [35]

$$U(q) = -\frac{2}{3} E_F [1-(\frac{x^2}{2}) \exists (x)]^{1/2} [1-(\frac{5x^2}{2}) \exists (x)] D(x) \qquad (43)$$

where symbols have their usual meanings. A typical curve for non-transition metals is shown in Fig.4. This form has given encouraging results of electronic transport

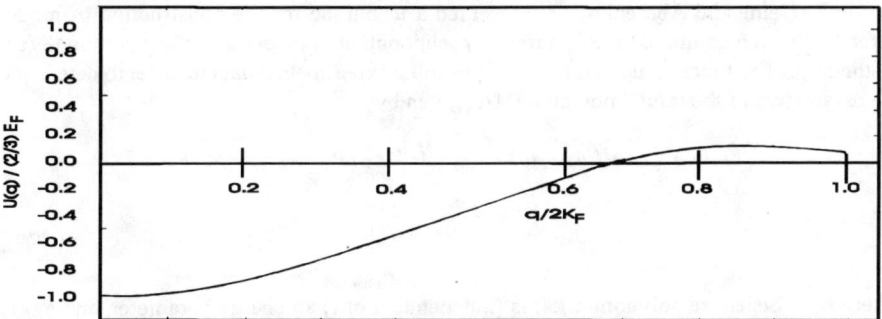

Fig.4 - General U (q)/ (2/ 3) E_F curve for simple metals of non - parametric pseudopotential

coefficients of non-transition liquid metals.

Strictly speaking in the path initiated by Cohen and Heine[36] beyond some reasonably small core radius r_c, the bare-ion potential is local and specifically Coulombic,

$$V_b(r) = -\frac{Ze^2}{r}, \; r > r_c \qquad (44)$$

Inside the core region, in addition to the Coulomb potential of the net ion charge Ze, the bare-ion potential contains a repulsive term arising from the fact that the conduction electrons are excluded from the core region by means of Pauli's principles. This repulsive term tends to cancel most of the attractive Coulomb potential in the same region [37], leaving a small potential as considered by Harrison [29] in his point-ion model, whose form factors are described by

$$V_b(q) = -\frac{4 \pi Z e^2}{\Omega q^2} + \beta', \; \text{for} \; r \le r_c \qquad (45)$$

where $\beta' = \beta/\Omega$ is strength of repulsive potential and an adjustable parameter. This model could not present satisfactory results of transport coefficients. It is remarkable that by varying form factors by .015 Ryd., there appears substantial change in transport coefficients, which reveals that the transport coefficients are sensitive to the choice of pseudopotential used.[38] Harrison modified his model later on the basis of Austin-Heine-Sham approximation [2] in the form

$$V_b(q) = -\frac{4\pi Z e^2}{\Omega q^2} + \frac{\beta}{\Omega [1+(q r_c)^2]^2} \quad (46)$$

where β and r_c are two adjustable parameters, which could be able to reproduce phonon frequencies of metals.

Heine and Abarenkov [30] presented a technique for the construction of model potentials which combined the features of pseudopotential theory and the quantum defect method [39]. Their technique requires spectroscopic experimental data in order to determine the parameters of the model potential (HA) given by

$$V_{HA}(q) = -(A_2-A_0)P_0 - (A_2-A_1)P_1 - A_2 , \quad r < R_M$$

$$= -\frac{Z e^2}{r} , \quad r \geq R_M \quad (47)$$

where P_l is Legendre polynomial, A_l is (independent of r) an energy parameter, and R_M is radius of model sphere. Thus, V_M is simply an l- and E- dependent operator. Under local approximation for the region $r < R_M$

$$<k+q|V_M|k> = \frac{4\pi}{\Omega}\sum_l (2l+1)A_l(E)P_l(\cos\theta).$$

$$\int_0^{R_M} j_l(k'r)j_l(kr) r^2 dr - \frac{4\pi Z e^2}{\Omega q^2}\cos(q.R_M) \quad (48)$$

where all $A_l = A_2$ for $l \geq 2$. The potential was made non-local and optimized in later stages. The special case of HA-model becomes for $A_l = 0$ the Ashcroft model whose bare-ion potential matrix element is given by

$$<k+q|V_b(q)|k> = -\int \frac{Z e^2}{r} e^{i|k'-k|\cdot r} d^3r$$
$$+\int \frac{Z e^2}{r} F_l(r) e^{i|k'-k|\cdot r} d^3r \quad (49)$$

where $F_l(r)$ is radial function given by eq. (39). For $l = 0$, by considering limiting values of $F_0(r)$ of empty core concept, above equation gives

$$V_b(q) = -\frac{4\pi Z e^2}{\Omega q^2} Cos(q r_c), \quad r_c : \text{ion - core radius} \qquad (50)$$

and for $l = 1$, we get

$$V_b(q) = [-\frac{4\pi Z e^2}{\Omega q^2} + \frac{\alpha}{\Omega}] \qquad (51)$$

where
$$\alpha = \frac{4\pi Z e^2}{q^2} [Sin(q r_c) - (q r_c) Cos(q r_c)] \qquad (52)$$

which is earlier one - parameter pseudopotential [40].
If we consider $A_l = Z e^2/r$ in eq. (48) the Shaw optimized model[31] is obtained.

The large numbers of model pseudopotentials available in literature show the ambiguous form regarding ion-electron-ion interaction in metals. The validity of a model depends on its utility towards a certain property for metals, which have been discussed in further sections.

(c) Phase-shift analysis approach :

The phase-shift analysis is another technique for formulating the pseudopotential in metals.

The general angular quantum number l, the solutions of radial equation, obtained in spherical coordinate systems, for a vanishing pseudopotential may be expressed by spherical Bessel function, $j_l(kr)$ and spherical Neumann function, $n_l(kr)$ at large distances as expressed respectively, by

$$j_l(kr) \sim \frac{Sin(kr - \frac{l\pi}{2})}{(kr)} \qquad (53)$$

$$n_l(kr) \sim \frac{Cos(kr - \frac{l\pi}{2})}{(kr)} \qquad (54)$$

which is singular at $r = 0$. Thus, the general solution outside the atomic cell for fixed l may be given by

$$\varphi_l = A_l [Cos\delta_l j_l(kr) - Sin\delta_l n_l(kr)] \qquad (55)$$

where A_l and δ_l are adjustable parameters, δ_l is called the phase shift.

In Born-approximation the scattering amplitude $f(\theta)$ is given by

$$f(\theta) = \frac{1}{k}\sum_{l=0}^{\infty} (2l+1) \delta_l P_l (Cos\theta) \qquad (56)$$

where P_l are Legendre polynomials, θ is scattering angle and δ_l is given by

$$\delta_l = e^{i\eta_l} Sin(\eta_l) \qquad (57)$$

Here η_l is the phase angle of the partial wave of angular momentum l. The phase shift may be obtained by an integration of the pseudo-Schrodinger equation in atomic region on the basis of perturbation theory in lowest order. δ_l is given by (Born-approximation)

$$\delta_l = -\left(\frac{2mkN}{4\pi h^2}\right) \frac{1}{N} \int_0^\infty 4\pi r^2 j_l(kr) w_l(kr) j_l(kr) dr, \quad (58)$$

where w_l is angular dependent single ionic pseudopotential called as muffin-tin potential also. The phase shifts depend on l as well as on an energy E. The Born-approximation of the perturbation theory is applicable only when phase shifts are small compared to π. A phase-shift pseudopotential may be written in terms of phase-shift. Its form factor (in opw-approximation) is defined as the pseudopotential matrix element between plane waves $k>$ and $k+q>$

$$<k+q|W|k> = N \int e^{-i(k+q)\cdot r} W(r) e^{ik\cdot r} d^3r \quad (59)$$

Now expanding the two plane waves in terms of spherical harmonics and performing angular integrations in above equation, we obtain

$$<k+q|W|k> = \sum_l N(2l+1) P_l(Cos\theta) \int 4\pi r^2 j_l(kr) W_l(r) j_l(kr) dr \quad (60)$$

Here θ is the angle between k and $k+q$, $q = 2k \, Sin \, \theta/2$. Thus in terms of δ_l, we may write

$$<k+q|W|k> = -\frac{2\pi h^2 N}{mk} \sum_{l=0}^\infty (2l+1) \delta_l P_l(Cos\theta) \quad (61)$$

On Fermi surface if k and $k+q$ occur then

$$Cos\theta = 1 - \frac{q^2}{2k_F^2} \quad (62)$$

and in terms of scattering amplitude $f(\theta)$, we may write

$$<k+q|W|k> = -\frac{2\pi h^2 N}{m} f(\theta) \quad (63)$$

i.e. $f(\theta)$ is like a matrix element of the potential or it is an element in the 't-matrix' of the scattering centre. Eq.(62) has been used by Evans and his Coworkers[41] in framing t-matrix element $t(E_F, q)$ by taking $k = k_F$ (Fermi radius). Thus

$$t(E_F, q) = -\frac{2\pi h^3 N}{m(2m E_F)^{1/2}} \sum_l (2l+1) \delta_l P_l\left(1 - \frac{q^2}{2k_F^2}\right) \quad (64)$$

δ_l or the t-matrix may also be expressed through Green function technique[42-44]. A parallel approach to t- matrix technique is reaction matrix[45] approach through which form factors may be framed out. The t-matrix method has been widely used[46-52] in the studies of pseudopotential investigations of properties of metals.

(d) Computer simulation approach :

In recent years, the simulation (molecular dynamics technique) in condensed matter studies has been an important approach in which large capacity electronic computers are used as mechanical simulations. The equations of motion of several hundred atoms by step-wise numerical integration are solved through fast computers. This technique got attention of the scientists around the year 1965. There have been fresh thoughts, activity growth and many improvements in this device for one decade, which have been reviewed by Berne and Forster[53].

First of all Rahman[54] studied this problem of spatial correlation functions using a realistic potential Lennard-Jones (LJ) in case of liquid Argon. Actually for directly stimulating many - particle systems there are basically two methods. The first is known as Monte Carlo method[55,56], which is useful for the study of equilibrium properties of condensed matter. The second is known as molecular dynamics, which is useful to study both time averaged and time dependent quantities. The molecular dynamics technique has a counterpart the inelastic neutron scattering technique. These methods have been reviewed elsewhere[57].

In this computer experiments, Rahman observed the motion of particles within velocity distribution at a particular new positions and velocities of particles with a particular time increment. For each time increment the temperature was calculated. Rahman simulated the LJ fluid at a particular reduced density and temperature. The system was allowed to evolve in time using the different equations, to generate new positions and velocities for each particle after each time step. The observations were used to derive time dependent function, $g(r)$. Now using these data the structure factors, $S(q,\omega)$ and $S(q)$ were determined. The pioneer functions[58] of Rahman on molecular dynamics of Na and later on of using the pseudopotential description of Price and Coworkers[59] is excellent achievement.

(e)   Other approaches :

There are many experimental and theoretical technique through which studies of non-transition metals may be performed successfully.

Synchrotron radiation is widely useful experimental technique for structural and electronic properties of condensed matter. We may determine energy- momentum relation $E(k)$ for electrons in metals like Cu and Ni and semiconductors like CdS and GaAs. The studies of structure of glassy amorphous materials and their phase transition problems may easily be performed with this technique. The structural studies of surfaces and catalytic activity at metals surfaces are also possible with this device as well as with electron microscope. The problems of semiconductor-metal interfaces and lithography of condensed matter may be also easily studied with this technique.

It is possible to develop pressures in excess of 1M bar in condensed matter with the help of dimond lanvil cell. Such pressure variation may change the average volume available to an atom or molecule in a solid or liquid which affect the electronic properties of condensed matter. The transformation of insulators into metals is also possible with this cell. At 2M pressure the gases like Xenon and Hydrogen become metallic. Therefore, this cell is very useful in the studies of properties of such materials in metallic- phase.

Among theoretical technique, the renormalization group methods are becoming very popular now a days for the studies of phase transition problems in melting of solids and freezing of liquids and also in magnetic properties of condensed matter. These techniques have provided a theoretical understanding of emperical relations among different properties near the phase transition of a given system. The studies of disordered magnetic materials like spin glasses are also possible with these techniques.

The density functional approach is very successful theoretical method for including electronic, exchange and correlation effects in first principles calculations of structural and vibrational properties of non-transition metallic substances. This is a parallel approach to the perturbation theory. The phonon dispersion studies are also encouraged with this theoretical approach.

In recent years new experimental technique have been used for the study of the structural and dynamical properties of the crystal surface and of atoms or molecules absorbed on to it. Other than Low energy electron diffraction (LEED) and neutron scattering experimental devices for surface properties, the theoretical approach now a days is becoming popular, which is called atom/surface scattering device. Raman spectroscopy and infra-red spectroscopy are also very useful experimental technique for surface properties. These spectroscopic devices are not useful for the structural studies of non-transition materials. The field-ion microscopy technique is extensively useful in surface studies of condensed matter. Surface structure, chemical groupings and atomic diffusion on surfaces are the surface properties which can be studied with the help of field-ion microscopy technique. By applying field pulses it is possible to strip away successive surface atoms and its effect on structural study. Laser beams are also very useful for the studies of various properties of atoms and molecules on surface studies of condensed matter. In recent years non-linear optical technique have proved very successful for the studies of surfaces and interfaces. This also gives information about the dynamics of molecular adsorption.

The model potential calculations may be performed for the understanding of the physical origin of the scattering resonances occurring in surface problems. The surface problems may be solved through the atom/surface scattering calculations which relate the interaction potential with the actual nuclear positions in the crystal. It is observed that the attractive part of the potential is long ranged and of Vander Waals types while repulsive part density in the surface layers of the crystal. The self consistent calculations of one- electron energy states may be carried out for the surface structures. The description of electronic structure of surfaces becomes complicated due to loss of translational invariance symmetry - an additional complexity.

4.3 ELECTRONIC STATES (ELECTRONIC STRUCTURE) AND ELECTRONIC PROPERTIES

In crystalline phase the periodicity of the lattice describes very well the electronic ban structure but in disordered structure this becomes irrelevant. It is sure that in simple metals the electron energies are very much the same as in the absence of the periodic potential arising form the ion. The electronic states may be framed due to orthogonalization of each states on each ion. We may proceed in same way for disordered state also. The pseudopotential perturbation theory [3] plays an important role in this direction. The techniques such as the first-principles pseudopotential, linearized augmented-plane wave, density functional and linearized muffin-tin orbital have played an important role in the description of electronic structure of complicated . The excited states of complicated systems may be understood on the basis of angle resolved techniques pertinent to photo-emission and radiation impinging on the surfaces and interfaces of disordered structure materials.

Some of the aspects of electronic structure of disordered solids are becoming very interesting particularly the many-body problems of electronic excited states. The band-theory density-functional methods may be extended for such excited problems. There are some of the neglected problems which require solutions like the evaluation of the fundamental band gaps in semiconductors in solid and liquid phases. It would be interesting to include the exchange and correlation effects in our understanding of the electronic structure of large systems with different types of defects for crystalline, liquid and amorphous states of matter. Important aspects of electronic structures have been reviewed by Casack[60].

(a) Dispersion of Electrons and Electronic Band Structure

The motion of electrons in condensed matter gives content to the idea of energy band structure. Hartree and Hartree-Fock methods provide free electron energy bands in metals. In recent years, the pseudopotential technique has got special attention in the studies of energy-band calculation problems. In real sense the pseudopotential form factors represent a most important aspect of the electronic structure.

The self-consistent field approximation is suitable in describing the electronic states in a metal due to the reason that in this approximation we may consider the effect of the interaction of a given electron with all other by a potential, which is the function of only the specific electronic coordinate. In pseudopotential approach[3] the electronic band structure of a crystalline solid may be easily determined through secular equation.

$$dt \left[\left[\frac{h^2}{2m} (k-h) E(k,h) \right] \delta_{hh'} + S_g(h-h') U(h-h') \right] = 0 \quad (65)$$

On the basis of pseudopotential perturbation theory the electronic structure of a liquid metal may be described by

$$E(k) = \frac{h^2 k^2}{2m} + N<k|U|k> + \frac{2m}{h^2} \sum_q \frac{S_g(q) S^*_g |U(q)|^2}{k^2 - (|k+q|^2)} \quad (66)$$

The quantities of right hand side are zero, first order and second order electronic energies. Thus the frequency becomes

$$E(k) = \frac{h^2 k^2}{2m} + N<k|U|k> + \frac{2m}{h^2} \sum_q \frac{S_g(q) S^*_g |U(q)|^2}{k^2 - (|k+q|^2)}$$

$$= \omega'(k) + \Delta \omega'(k) \quad (67)$$

where $\Delta \omega'(k)$ describes the perturbation frequency. The behavior of frequencies $\omega(k)$ and $\Delta \omega'(k)$ are shown in Fig. 5 (a) - (b) for author's pseudopotential[40] using experimental structure factor data[61]. $\Delta \omega'(k)$ curve shows negative values upto about k = 1.5 ($A^°$)$^{-1}$ position after which positive behavior is obtained. There is a deviation of the electronic frequencies from the free electron value. It is also evident from the figure that maximum deviation occurs in the vicinity of the first spherical Brillouin zone region. This region lies nearly at half of the distance of the first peak in the structure factor. The oscillatory behavior of $\Delta \omega'(k)$ curve within the first Brilouin zone indicates that electron dispersion may have an important effect on the electronic properties of liquid metals due to band structure effects. Ziman[34, 62)] discussed that if we look at resistivity in view of such effects, a large variation may be found in cases like Hg.

Actually in crystalline phase there exists a high density of states prominent mostly in a d-band, which corresponds to electrons of low velocity. On heating a solid the Bloch states are almost finished and in disordered phase the sharp lines of the frequencies spectrum becomes broadened, so that all the ions make large uncorrelated excursions. It does not mean that frequency spectrum of all the basis states collapses into the perturbed nearly free electron form and possess only single peak. It is not advisable to consider the equivalence of current and frequency or momentum. Truly speaking, a Bloch state differs from the sort of wave function considerably in the analysis of expectation value of Fermi-current. Also according to Falicov and Heine[63)] there appears inconsistency between the expectation value of the Fermi-current operator and the group velocity of electrons. The contribution of electron-electron interaction is appreciable in the use of current operator for Fermi surface studies. Also the effects of "back flow" currents should be taken into account when one considers the perturbations on the electron states. The tight-binding form of Bloch function may contribute to the differences between the considerations of the similar quantities ($\hbar/m$) k and group velocity for the Fermi-current calculations. But in nearly free electron situations it may be ignored. Actually the Fermi-current $j(k_F)$ becomes $e V_F$ (V_F: Fermi velocity) as it represents the charge carried by an electron in a state on the Fermi surface as considered by Edwards[64)]. The equivalence of current with group velocity is true for one particle wave function situations[7)] but not for disordered structural cases. The density of states and Fermi energy study[5)] based on electronic states described by eq. (65) indicates that there is insignificant change in the density of states on the Fermi surface at the melting point which supports the observation of Enderby et al[65)]. There is abnormal increase in volume on melting in case of Li due to which it shows abnormal behavior.

(b)   Electrical-Transport Properties of metals in liquid phase

The case of disordered materials like liquid metals is quite different than crystalline materials in respect to electrical-transport properties. This property was first of all described by Ziman[34)] on the basis of nearly free electron approximation. The electrical resistivity ρ_L formula of liquid metals is given by

$$\rho_L = \frac{3}{e^2 V_F^2 \tau N(E_F)} \quad (68)$$

where V_F and $N(E_F)$ are the velocity and density of states, respectively on the Fermi surface. τ is the electronic relaxation time which is given by

$$\frac{1}{\tau} = \int (1 - Cos\theta) f(\theta) d\Omega \quad (69)$$

where $f(\theta)$ is the probability of scattering through the angle θ into the solid angle $d\Omega$ which in terms of Golden rule is given by

$$f(\theta) = \frac{1}{4h} N(E_F) \left| < k \left| U_s \right| k' > \right|^2 \quad (70)$$

which is similar to the analysis described by Faber[4)] for inelastic scattering of neutrons of conduction electrons. Under self-consistent field approximation, we may write

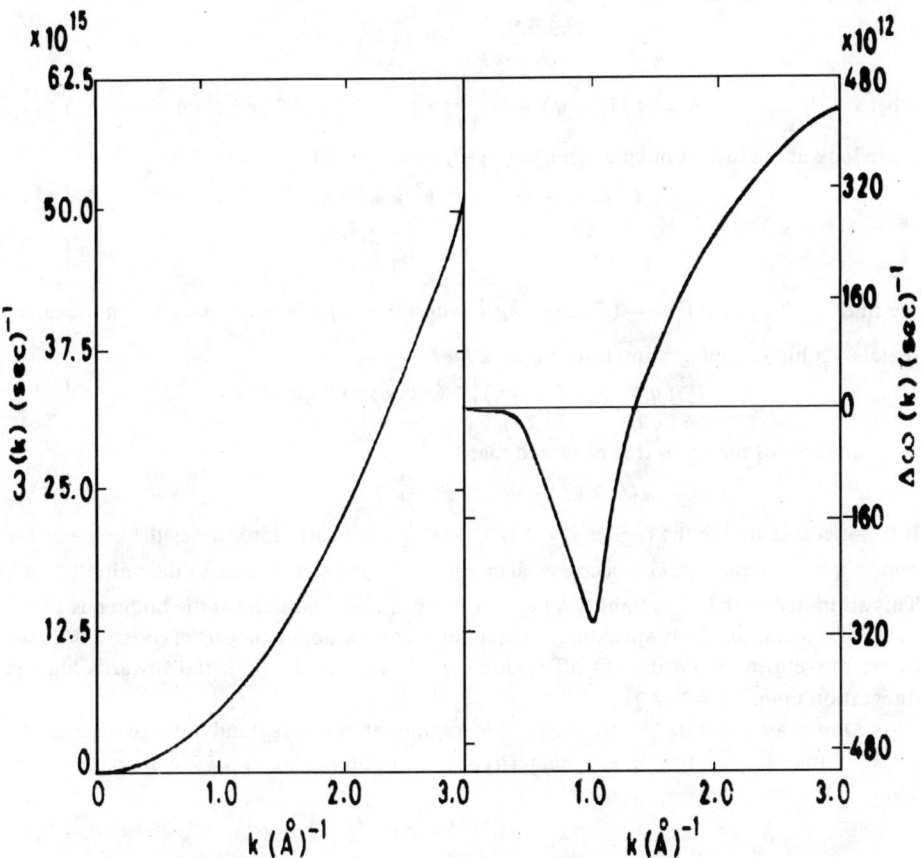

Fig. 5(a) - Electron frequency, ω (q) variations with momentum transfer q.

Fig. 5(b) - Electron perturbation frequency Δω (q) variations with momentum transfer q.

$$\left|<k|U_s|k'>\right|^2 = \frac{1}{N} S(q) |U(q)|^2 \ ; \ (q = k' - k) \tag{71}$$

where $S(q)$ and $U(q)$ are the liquid structure factors and form factors, respectively. Under Born-approximation condition

$$q = 2 k_F \sin \theta/2$$

the eq. (68) finally takes the form

$$\rho_L = \frac{3 \pi m \Omega}{2 h e^2 E_F} <S(q) U^2(q)> \tag{72}$$

where

$$<S(q) U^2(q)> = \frac{1}{4 k_F^4} \int_0^{2 k_F} S(q) U^2(q) q^3 \, dq \tag{73}$$

If we look at the lowest integration limit (q=0) in resistivity formula where

$$U^2(q) \longrightarrow |U(0)|^2 = (\frac{2}{3} Z E_F)^2 \tag{74}$$

$$S(q) \longrightarrow S(0) = (\frac{2}{3} \frac{Z E_F}{k_B T})^{-1} \tag{75}$$

we find $S(0) |U(0)|^2 = (\frac{2}{3} Z E_F) k_B T$ where $S(0)$ is nearly 0.025 for monovalent metals. At highest integration limit ($q = 2 k_B T$)

$$U^2(q) \longrightarrow |U(2 k_F)|^2 \quad (\text{very small quantity}) \tag{76}$$
$$S(q) \longrightarrow |S(2 k_F)| \quad \sim 1 \tag{77}$$

In most of the cases it is observed that

$$U(2 k_F) \sim \pm [(\frac{2}{3} Z E_F) k_B T]^{1/2} \tag{78}$$

It is noticeable that in the region $q = 2 k_F$, if $S(q)$ is nearly forty times the value of the region $q = 0$ then $U^2(2 k_F)$ becomes also smaller in the same order to the value $U^2(0)$. This all indicates that the quantity $S(q) U^2(q)$ is almost the same at the both ends of the range of integration. Truly speaking, in real integration the dominant part of resistivity arises as we move away from the cut off region ($q = 1.4 k_F$) of the potential towards highest integration point ($q = 2 k_F$).

One may separate[62] arbitrarily the region of the integrand into two parts, (i) $\rho_L^{(pl)}$ (Plasma part, $0 < q \leq 1.4 k_F$) (ii) ρ_L^{st} (Structure part, $1.4 k_F < q \leq 2 k_F$) Thus

$$\rho_L = \rho_L^{(pl)} + \rho_L^{(st)} \tag{79}$$

where

$$\rho_L^{pl} = \frac{3 \pi m \Omega}{2 h^2 e^2 E_F} \frac{1}{4 k_F^4} \int_0^{1.4 k_F} S(q) U^2(q) q^3 \, dq \tag{80}$$

and

$$\rho_L^{st} = \frac{3 \pi m \Omega}{2 h^2 e^2 E_F} \frac{1}{4 k_F^4} \int_{1.4 k_F}^{q \geq 2 k_F} S(q) U^2(q) q^3 \, dq \tag{81}$$

Generally, maxima in eq(81) is at $q = 2 k_F$. It may be said that the region ($q = 1.4 k_F$) should be also an important point in structural analysis as it is the cut off region of the potential where generally almost the value is negligible in all model pseudopotentials. In case of monovalent metals eq. (80) contributes larger as compared to eq.(81) while the situation is reversed in the case of polyvalent metals.

Paskin et al[66] developed a crude analytic procedure for resistivity of monovalent liquid metals through a relationship (AHP) ; $S(q) = [v(q)/k_B T + 1]^{-1}$ between structure factor, S(q), Fourier Component of the interatomic potential, V(q) and the screened pseudopotential U(q). According to them

$$\int_0^1 S(q) U^2(q) (\frac{q}{2k_F})^3 d(\frac{q}{2k_F}) = [\frac{e^2 k_F k_B T}{3\pi}]$$

$$\int_0^1 [\frac{1-S(q)}{\varepsilon(q)}] (\frac{q}{2k_F})^3 d(\frac{q}{2k_F}) \tag{82}$$

It is remarkable that the right hand side integral provides dominant contribution in the cut off region, which is a critical point in between plasma and structure region. Here, dielectric function $\varepsilon(q) \longrightarrow 1$ for monovalent metals. Also S(q) plays similar role in monovalent metals. According to these contributors the integral of right hand side of above equation for alkali metals follows

$$\int_0^1 [\frac{1-S(q)}{\varepsilon(q)}] (\frac{q}{2k_F}) d(\frac{q}{2k_F}) \longrightarrow \sim 0.57 \tag{83}$$

which shows that at melting point ($T = T_m$)

$$\rho_L = f(k_F, T_m) \sim 0.0028 \frac{T_m}{k_F^4} \;[\mu\Omega\,unit] \tag{84}$$

Here k_F is in atomic unit. This formulation provides about one fifth of the actual resistivity in case of Li metal. The results show that additional structural term of the resistivity would raise the value and down the temperature dependence of the resistivity as S(q) in this region does not vary as temperature, T.

The first quantitative calculation of ρ_L was performed little bit successfully by Sundstrom[67] for Heine-Abarenkov model pseudopotential using available experimental S(q) data of that time. At the same time Ashcroft and Lekner[68] achieved very satisfactory results alongwith theoretical derived structure factors for the same model potential which is accidental event as predicted by Ziman[34]. Their used S(q) data were not reliable due to instability. Young et al.[47] obtained similar satisfactory results with phase-shift analysis technique. The resistivity formula based on nearly free electron approximation is successful not only in cases of monovalent metals but in polyvalent metals also.[69-71]

The Born-approximation formula, eq.(72) of ρ_L became a controversial matter and became a matter of discussion especially from first international conference-1967 of Brookhaven on the following grounds:

(i) The formula is based upon the semi-classical Boltzmann approach and not satisfying the situation for short mean free path.

(ii) Due to Blurrings of the Fermi surfaces a correction is needed, which should be suitable to metals of short and long both mean free path situations.

(iii) The formula is not suitable[72,73] due to incorrect results of temperature coefficients of resistivity.

(iv) The formula does not take account of effective mass.

(v) There is need of using[74, 75] effective potential in the resistivity formula rather, than model pseudopotential matrix element.

Ziman[34] discussed about the validity of the formula in the first as well as in the second international conference of Brookhavan-1967 and Tokyo-1972, respectively. According to him, no doubt such types formulae were proposed earlier[76, 77] but it fulfills all the conditions for its validity and doubts raised are baseless due to one or other awareness and wrong handlings. In Tokyo conference, he said "within the usual errors and uncertainties of theoretical Physics the nearly free electron formula had been shown to agree fairly well with the experiment although minor variations of the pseudopotential could produce large variation in the calculated transport properties. But there was still a need to stabilize these calculations."

Ferraz and March[78] proposed an iteration procedure for the calculation of ρ_L the regime of short mean free path but not very good success is achieved.[79]

Thermo-electrical power Q_L is another electronic transport coefficient of liquid metals on which researches concentrated due to curiosity of its link with resistivity as described by [34,80]

$$Q_L = \frac{-\pi^2 k_B^2 T}{3e} \left[\frac{\partial \ln p(E)}{\partial E} \right]_{E=E_F}$$

$$= \frac{-\pi^2 k_B^2 T \zeta_L}{3 e E_F} \qquad (85)$$

where

$$\zeta_L = 3 - \frac{2 S(2k_F) U^2(2k_F)}{\langle S(q) U^2(q) \rangle} \qquad (86)$$

The above formula presented satisfactory results [33, 67, 81-85], of many non-transition metals but in cases like Li and Cs the refinement on the basis of energy dependent effects due to d- state were considered. Then only improvement in sign as well as magnitude were observed [26, 81,83] through relation

$$Q'_L = Q_L + \frac{\pi^2 k_F^2 T r}{6 e E_F} \qquad (87)$$

where

$$r = \frac{k_F \langle \frac{\partial}{\partial k} |U(q,k)|^2 S(q) \rangle}{\langle S(q) |U(q,k)|^2 \rangle} \qquad (88)$$

Evans[69 84] observed that the refined formula (87) is important for Hg as ζ_L contributes only value -1 where required value should be greater than 3. At the same time it is also observed that Ferraz and March[78] refinement applied to thermoelectric studies has not improved the result very much[86]. It has been observed that above melting temperature the thermoelectric power remains constant except for Zn, Hg and Ti which appears due to their thermal expansion. Ziman theory has given satisfactory results of pressure derivative of thermoelectric power of Hg. This theory has also given satisfactory results of electronic transport coefficients ρ and Q of liquid non-transition metal alloys[67 - 91]. It has been also found that ρ and Q are sensitive to the pseudopotential used for liquid metals and their alloys.

(c) Electrical-transport properties of metals in solid phase

There have been remarkable progress in past three decades in the theoretical investigations[92,93] of electrical resistivity of solid metals including first principle calculations of Ziman [94]. At the same time very little progress is towards potential studies in thermoelectrics[95]. After the appearance of the pseudopotential theory, model pseudopotentials alongwith the lattice dynamical models have given satisfactory results of low temperature electrical resistivity through following Ziman-Baym formula

$$\rho s = \frac{3}{16ne^2 h V_F^2 k_F^4} \int_0^{2k_F} S(q) U^2(q) q\, d^3q \tag{89}$$

where structure factors of solid metals were determined through lattice dynamical formula

$$S(q) = \int_{-\infty}^{\infty} S(q,\omega) \frac{\beta \omega}{(1 - e^{-\beta \omega})} d\omega \tag{90}$$

Here $\beta = h/k_B T$. ω is phonon frequency. The dynamical structure factor, $S(q,\omega)$ is the Fourier transform of time dependent pair distribution function $G(r,t)$ which is given by

$$S(q,\omega) = \frac{N}{2\pi} \iint \exp[i(q.r - \omega t)] G(r,t)\, dr.dt \tag{91}$$

On the basis of Van Hove [19] treatment finally $S(q)$ may be described by

$$S(q) = (\frac{h^2}{Mk_B T}) \sum_q (e_{q,p}.q)^2 \frac{1}{(e^{\beta \omega_{q,p}} - 1)(1 - e^{-\beta \omega_{q,p}})} \tag{92}$$

In pseudopotential approach eq.(89) has given satisfactory results [27, 96, 97] of ρ_s for monovalent as well as polyvalent metals, which reveals that the Born-approximation is also valid in the evaluation of electrical resistivity of solid metals. It is observed [96,98] that there is not much difference in the temperature dependence of resistivity at constant pressure and constant volume. Anharmonic effects are very essential in such calculations.

The pseudopotential calculations[99, 100] have revealed that following Bloch-Gruneisen resistivity formula [94] (Bloch - T^5 law) of low temperature is failure in the region $T \leq \theta_D$ (θ_D: Debye characteristic temperature)

$$\rho_{Bloch}(T) \sim A(\frac{T}{\theta_D})^5 [1 - 0.008(\frac{\theta_D}{T})^5 e^{-\frac{\theta_D}{T}}] \tag{93}$$

where A is a constant. It has been observed that resistivity follows according

$$\frac{\rho}{\rho(\theta_D)} \alpha (\frac{T}{\theta_D}); \quad T > \theta_D \tag{94}$$

and

$$\frac{\rho}{\rho(\theta_D)} = (\frac{T}{\theta_D})^{[6K(T)-1]}; \quad T < \theta_D \tag{95}$$

where

$$K(T) = \log(\frac{1}{\rho T}) / \log[\frac{1}{T}(\frac{\theta_D}{T})^5] \tag{96}$$

Below Martensitic transformation temperature θ_M it follows an exponential behavior rather than above linear relation. At very low temperature it yields a semi-empirical relation [99, 100]

$$\rho = \frac{1}{T} e^{-b/T} \qquad (97)$$

where b is some modest temperature. In temperature ranges $5°K \leq T \leq 11°K$ in case of Li, $3°K \leq T \leq 4°K$ in case of Na and $T \leq 2°K$ in case of K : b corresponds to Martensitic transformation temperature $75°K$, $40°K$ and $21°K$, respectively. From this tempeature the phonons of Umklapp process are frozen from the Bose - Einstein distribution. At very low temperature where Umklapp contribution, ρ_U does indeed disappear rapidly, the resistivity follows

$$\rho = (\rho_N + \rho_U)(1 - X) \qquad (98)$$

where

$$X = \frac{\rho_g}{(\rho_N + \rho_U)} \qquad (99)$$

Here ρ_N and ρ_g are normal and phonon drag resistivity, respectively. But pseudopotential investigations have revealed that X should follow

$$X \sim \left[1 - \left(\frac{\rho_U}{\rho_N}\right) [K(T) - 1] \left[1 + \frac{\rho_U}{\rho_N} \right] \right] \qquad (100)$$

The possibility of electron-electron interaction arises,[101] which can lead to an instability of the normal state of the metal, which causes the development of Super-Conducting State.

According to Ziman[94], at temperature $< \theta_D/10$ there is disappearance of Umklapp contribution, ρ_U but we observe this behavior in the region $T < \theta_M/10$ which gives

$$\frac{\theta_M}{\theta_D} \sim \frac{K(T)}{\beta} \qquad (101)$$

where Ziman quantity $\beta = R_D/q_0$ [R_D: Debye radius, $R_D = (6\pi^2)^{1/3} a$; a is lattice constant]. For bcc structure with a spherical Fermi surface, $\beta \sim 7$ and θ_M/θ_D comes out 0.22, 0.26 and 0.23 for Li, Na and K, respectively. It is remarkable that the value comes out < 0.25 from where the exponential behavior of resistivity arises. Such behavior may be true for transition metals also but it is to be seen that either the electron scattering from s-state to s-state is dominant or from s-state to d-state are dominant.

The thermoelectric phenomena are mostly governed by the thermoelectric power. This transport coefficient may be determined in similar manner as of liquid metals except the methodology of using structure factors which is followed by electrical resistivity determinational procedure. The role of lattice dynamics in this direction is appreciable [95, 97, 102]. There is dominant contribution of energy-momentum dependence effects on transport coefficient [95, 103]. A parallel formula to liquid metals theory for the thermoelectric power parameter of solid phase is described by

$$\zeta_S = 3 - \frac{2S(2k_F) U^2(2k_F)}{\lambda_S} \qquad (102)$$

where
$$\lambda_S = \frac{1}{4k_F^4} \int_0^{2k_F} S(q) U^2(q) d^3q \qquad (103)$$

Using author's previous model pseudopotential alongwith Kreb's lattice dynamical model a value of ζ_S ($\mu V/°k$) at 300 °K for Li, Na and Cs are -6.1, 1.3 and 2.4, respectively. The corresponding experimental value[104] for these metals are -6.7, 2.7 and 0.2, respectively and theoretical values of Robinson[102] are -0.09, 2.68 and 2.67, respectively using Debye structural data. The discrepancies in the theoretical results is due to the neglect of various anharmonic effects such as multi-phonon processes and partially due to the magnitude of pseudopotential form factors. A value of V/ρ ($\partial\rho/\partial V$)$_T$ for alkali metals in solid phase are -3.1, 1.1 and 1.2 for Li, Na and K, respectively whereas experimental values are 4.5 and 1.8 for Na and Cs, respectively.

(d) Thermal Conductivity

Generally, the metals are good conductors of heat than non-metals. The observation of Berman et al.[105] reveals that diamond - a semi conductor possesses the highest thermal conductivity 2000 ($W_m K^{-1}$) which is higher than any metallic thermal conductors like Cu 400 ($W_m K^{-1}$) or Ag 430 ($W_m K^{-1}$) and semi-metals like Bi 8.5 ($W_m K^{-1}$) and even other group members as Ga 64 ($W_m K^{-1}$) and Si 145 ($W_m K^{-1}$) at room temperature. Slack[107] observed that non-metallic crystal like BN appears a good thermal conductor having value 1300 ($W_m K^{-1}$) under certain conditions. A comparison of thermal conductors has been summarized elsewhere[107].

There has been little progress on thermal conductivity as compared to electrical conductivity in Pseudopotential approach. There are two theoretical approaches : (i) relaxation time approach and (ii) variation approach to solve Boltzmann transport equation which deals thermal conduction behavior. A probability distribution function f_q is introduced which describes the phonon distribution. It is done basically by above two methods, which generates the thermal conductivity latter on. We follow here variation approach. A variational expression for ideal thermal resistivity, ρ_{ther} of metals in weak scattering approximation is given by[108]

$$\rho_{ther} = \frac{9}{16 L_o T n e^2 \hbar V_F^2 k_F^4} \int_0^{2k_F} S(q) U^2(q) \frac{1}{q}$$
$$[\frac{q^3}{3} + (k_F^2 - \frac{q^2}{6})(\frac{\beta\omega_{q,p}}{\pi})^2] d^3q \qquad (104)$$

where $\beta = \hbar/k_B T$ and $L_o = \pi^2 k_B^2/3\beta^2$ is the Lorentz number. In order to solve above value integral the knowledge of phonon dispersions and the polarization vectors over the the irreducible part of the Brillouin zone is essential for the evaluation of the dynamical structure factors. Using Kreb's lattice dynamical model and various pseudopotentials, satisfactory results of monovalent metals have been observed. An improvement in the result may be achieved by including various anharmonic effects such as multiphonon processes, Debye-Waller factor and elastic constants variation with temperature.

A popular relation Widemann-Franz law holds good for conductivities of metals which is given by

$$\frac{\lambda}{\sigma T} = L_o \qquad (105)$$

where $\sigma = 1/\rho_{el}$: $\lambda = 1/\rho_{ther}$ and L_o is Lorentz number. A free electron value of L_o is $2.443 \times 10^{-8} v^2 \, deg^{-2}$. We have observed that our results of ρ_{el} and ρ_{ther} of pseudopotential approach follow the Widemann-Franz law.

(e) **Magnetic Properties**

(i) **Susceptibility** : It has been observed through experimental measurements[109-113] of magnetic susceptibility that for the metals of odd valency the susceptibility is greater than the free electron value and for even valency, the susceptibility is smaller than the free electron value. It is also found that there is shift in susceptibility on melting. Among alkalis lithium has abnormal behavior towards susceptibility than other group members due to deviation in the change of atomic volume during melting[114]. Pb is the only metal which possesses higher susceptibility value in liquid phase as compared to solid phase. This property is strongly temperature dependent. Generally, two theoretical approaches have been followed in the susceptibility studies (i) pseudopotential perturbation theory (ii) Green function approach. Only upto second perturbation terms to the electron-ion potential have been considered in first approach[115, 116] while higher order terms have considered in second approach[117, 118].

Generally, the magnetic susceptibility X possesses four contributions

$$X = X_p + X_i + X_d + X_L \qquad (106)$$

where the respective terms on the right hand side corresponds to Pauli-Spin, diamagnetic contributions due to ion-cores and conduction electrons and orbital interaction, respectively. Above Curie temperature θ_c, X possesses

$$X = \frac{C}{T - \theta_c} \qquad (107)$$

where C is Curie's constant. It has been observed that the possibility of a liquid alloy to be ferromagnetic is only when it is supercooled to $350\,^\circ K$ below the melting point in some of the cases. Ferromagnetism is observed only in case of some amorphous solids.

Using nearly free electron approximation in pseudopotential approach the susceptibility of metals in solid phase is described by

$$X_{solid} = \frac{3}{2} X^o [1 - \Delta P + \delta_{ex+corr}]^{-1} - \frac{1}{2} X^o [1 + \Delta L] \qquad (108)$$

where

$$X_o = \frac{\mu^2 n}{\zeta_o}$$

$$\delta_{ex+corr} = -0.166\, r_s + 0.204\, r_s^2 \,(-0.0676 \ln r_s)\, [\,in\ at.\ units\,]$$

$$\Delta p = -\frac{1}{8 \zeta_o^2} \sum_h{}' U^2(h) F_1(h)$$

$$\Delta L = -\frac{1}{8\zeta_o^2} \sum_h' U^2(h)[F_1(h) + \frac{h_x^2 + h_y^2}{h^2} F_2(h)]$$

$$F_1(h) = \frac{1}{h} \ln \left|\frac{h+2}{h-2}\right| - \frac{4}{h^2 - 4}$$

$$F_2(h) = \frac{3}{2h^2} \ln \left|h + \frac{h}{2} - 2\right| + \frac{2}{h^2}[5h^2 - \frac{12}{(h^2-4)^2}] \quad (109)$$

Here h is a reciprocal lattice vector in the units of Fermi radius, r_s is electron-spacing parameter, X_L and X_p are the ion potential contributions of the orbital and spin susceptibility, respectively. In liquid phase the corresponding values of the contributions become

$$\Delta L = -\frac{3}{2} \frac{Z}{\zeta^2} \int_0^\infty dk\, S(k) U^2(k) G_L(k)$$

$$\Delta p = -\frac{3}{2} \frac{Z}{\zeta^2} \int_0^\infty dk\, S(k) U^2(k) G_p(k)$$

$$G_L(k) = \frac{2}{k}(k^2 + 2) \ln \left|\frac{k+2}{k-2}\right| - \frac{8}{3}\left[\frac{24 - 22k^2 + k^4}{(k^2-4)^2}\right]$$

$$G_p(k) = \frac{2}{k} \ln \left|\frac{k+2}{k-2}\right| - \frac{8k^2}{k^2 - 4} \quad (110)$$

where $k = q/2k_F$. The deviated results were observed by Baltensperger[119], who found X_L as negative contribution for all metals while we believe to be of positive value in some of the cases. In this direction the results of Ford and Styles[113] are much improved. It is found that the solid and liquid phases susceptibility follow

$$\frac{(\frac{dX^{-1}}{dT})_{sol}}{(\frac{dX^{-1}}{dT})_{liq}} = C_{sl} \quad (111)$$

where C_{sl} is constant which may be negative or of positive values.

According to free electron theory the paramagnetic spin susceptibility X_p of the conduction electrons in liquid metals follows

$$X_p \; \alpha \; [N(E_F)]_{liq} \; \alpha \; \frac{m^*}{m}, \quad (112)$$

where $[N(E_F)]_{liq}$ is density of state of liquid metals at the Fermi level. Also the Landau diamagnetic susceptibility, $X_d = -\frac{1}{3} X_p$. But Timbie and White[115] deviate from such generalized argument. For transition metals the susceptibility decreases on melting because of the exchange forces between adjacent ions which favour parallel (ferromagnetic) alignment of their spins. Also the exchange coupling between paramagnetic ions is weaker in the liquid phase than in the solid phase.

(ii) Knight shift : The Knight shift 'K' is the name given to the difference between the magnetic resonant frequency of a nucleus in solid or liquid metal and that of the same nucleus in a non-metallic reference compound. This shift arises from the contact interaction between the spins of the conduction electrons and the nuclear moment. This shift is always positive. The nuclei experience a field which is always larger by some magnetic field ΔH than the applied field H. Thus, we define

$$\frac{\Delta H}{H} = \frac{\Delta v}{v} = K \quad , \tag{113}$$

where $K = 8\pi X_p \, p_F \, V/3$. v is the frequency and V is the crystal volume. p_F is the average electron density at the site of the nucleus from the conduction electron with an energy equal to Fermi-energy which is given by

$$p_F = \frac{1}{N} \sum_i \left| \psi_{k_F}(R_i) \right|^2 \tag{114}$$

R_i is position vector of an ith ion, N is the total number of ions. We define a dimensionless parameter $\exists$

$$\exists = \frac{p_F}{\left| \psi_A(0) \right|^2} \tag{115}$$

where $\psi_A(0)$ is a valence electron wave function in s-state in the free atom which may be derived spectroscopically. It is observed that for monovalent metals $\exists \sim 1$ and for polyvalent metals $\exists < 1$.

Using pseudopotential perturbation theory we may express

$$p_F = p_F^0 + p_F^1 \text{ and } K = K_0 + K_1 \; ; \; \varphi_k = \varphi_k^0 + \varphi_k^1 \tag{116}$$

where

$$\frac{p_F^1}{p_F^0} = 2 Re < \frac{1}{N} \sum_i \frac{\varphi_k^1(R_i)}{\varphi_k^0(R_i)} > k = k_F$$

$$= \left[\frac{2\Omega}{(2\pi)^3} \right] P \int \frac{U(q) S(q)}{[E_k^0 - E_{k+q}]} d^3q \tag{117}$$

P denotes the Cauchy principal value. Here φ_k is pseudo-wave function which may be described by

$$\varphi_k^0 = \frac{1}{\sqrt{V}} e^{ik \cdot r} \tag{118}$$

$$\varphi_k^1 = \frac{V}{(2\pi)^3} \int \frac{S_g(q) U(q)}{E_k^0 - E_{k+q}} \varphi_{k+q}^0 \, dq \tag{119}$$

where $E_k^0 = \hbar^2 k^2 / 2m$. $S(q) = N \left| S_g(q) \right|^2$ is the liquid structure factor. On assuming S(q) and U(q) [pseudopotential form factor] to be spherically symmetric, we may write

$$\frac{K_1}{K_0} = \frac{p_F^1}{p_F^0} = - \frac{3}{4 E_F k_F^2} \int_0^\infty S(q) U(q) q \ln \left| \frac{q + 2k_F}{q - 2k_F} \right| dq \tag{120}$$

which according to Watabe et al.[120] may be described by

$$\frac{1}{K_0} \left(\frac{\partial K_1}{\partial T} \right) V \sim \left(\frac{\partial \ln K}{\partial T} \right) V = - \frac{3}{4 T_m E_F k_F^2} \int_0^{2k_F} dq S(q) U(q) q \ln \left| \frac{q + 2k_F}{q - 2k_F} \right| \tag{121}$$

Here T_m is melting temperature. These authors extrapolated the value of above coefficient as $0.34 \times 10^{-4} K^{-1}$ for Na which is about 1/3 of experimental value, $1.00 \times 10^{-4} K^{-1}$ which is due to a crude concept similar to electrical resistivity (liquid metals) separation of plasma and structure region. When one includes complete notion of evaluation of above integral in the interval q = 0 and q = $2k_F$(Fermi level) giving emphasis to the variation of S(q) and U(q) quantities a successful quantum calculation provides a value [121] $1.01 \times 10^{-4} K^{-1}$ for Na using model -I form factors[27] along with experimental S(q) data. It is also observed that Li and Cs liquid metals show anomalous behavior which appears due to their abnormal increase in volume on melting. Their values are of negative order. Cornell evaluated a value[122] $-0.1 \times 10^{-4} K^{-1}$ in case of liquid Ga, which deviates from the experimental value $-0.55 \times 10^{-4} K^{-1}$. Dekker[128] presented a similar approach for evaluating this coefficient; the only difference is that he followed a relaxation term having relaxation time τ (a life time of an electron in a plane at the Fermi surface) which contributes about 10 to 20% extra in case of alkali metals.

The values of $(\partial \ln K/\partial T)$ in solid phase[120] also agree very well with the experimental observation in case of alkali metals. It is remarkable[124,125] that P_F and k_F increase by 8% to 27% respectively, when melting occurs in case of Cd.

A thermodynamic relationship at constant pressure (P) and constant volume (V) is described by

$$(\frac{\partial \ln K}{\partial T})_p = (\frac{\partial \ln K}{\partial T})_T (\frac{\partial \ln V}{\partial T})_p + (\frac{\partial \ln K}{\partial T}) \qquad (122)$$

where $K = K(V,T)$. The first term on the right hand side of above equation represents the contribution from the thermal expansion. It is found that above equality holds good in case of Na, Rb and Cs and values at constant pressure agree very well the observation of Mc Garvey and Gutowsky[126]. It is noticeable that among alkali metals, Cs is the only metal which possesses negative value at constant pressure.

It is strange that Faber[4] expect a deviation from above concept through the formula

$$[\frac{1}{K}(\frac{\partial K}{\partial T})_V]_{liq} \sim -\frac{2m*}{(2\pi)^2 h^2 k_F} \int_0^\infty dq.q.U(q)[\frac{\partial}{\partial T}S(q)].$$

$$\frac{1}{2}\log\left|\frac{(2K_F+q)^2 + (\frac{K_F}{2 lq})^2}{(2K_F-q)^2 + (\frac{K_F}{2 lq})^2}\right| \qquad (123)$$

alongwith the separation

$$[\frac{1}{K}(\frac{\partial K}{\partial T})]_{liq} = [\frac{1}{K}(\frac{\partial K}{\partial T})_V]_{liq} + [\frac{\alpha V}{K}(\frac{\partial K}{\partial V})]_{liq} \qquad (124)$$

where α is the volume expansion coefficient. He expressed liquid and solid phases relationship through the equation

$$\frac{(K)_{liq} - (K)_{sol}}{(K)_{liq}} = \frac{V}{K_{liq}}[(\frac{\partial K}{\partial V})_T]_{liq}\frac{dV}{V} \qquad (125)$$

which again deviates from our consideration. It requires a numerical checking for its justification.

From eq.(113) we may write for small fractional changes of Spin susceptibility, X_p and nuclear contact density ($\Omega\ P_F$) for electrons on the Fermi surface

$$\frac{\Delta K}{K} = \frac{\Delta X_p}{X_p} + \frac{\Delta(\Omega P_F)}{(\Omega P_F)} \qquad (126)$$

The nuclear contact density may be calculated [113, 127)] by explicitly orthogonalizing the conduction electron pseudo-wave functions to the core states. This procedure when adopted to evaluate the magnitudes of Knight shift, K in terms of K_0 and K_1 a departure from experimental values was observed. In an alternate procedure we may utilize the knowledge of probability density of electrons ratio P_k^1/P_k^0 in direct calculation of Knight shift K, we observe satisfactory results. The general expression of P_k^1/P_k^0 may be described by

$$\frac{P_k^1}{P_k^0} = 2Re < \frac{1}{N} \sum_i \frac{\varphi_k^1(R_i)}{\varphi_k^0(R_i)} > \qquad (127)$$

The zero order perturbation term P_k^0 of pseudo-spin density may be written as

$$P_k^0 = [P_k^0]_{K=k_F} = \frac{1}{N} <\sum_i \left| \varphi_{k_F}(R_i) \right|^2 | \qquad (128)$$

By considering the zero order, first order and mean value of the pseudo-spin density P_F the expression for the ratio P_k^1/P_k^0 may be expressed by

$$P_k^1/P_k^0 = 1 + [\frac{P_K 1}{P_K 1}]_{K=k_F} + [\frac{P_k^1}{P_k^0}]_{K=0} \qquad (129)$$

Using pseudopotential perturbation theory in the light of spherical symmetry of S(q) and U(q) we may write

$$[\frac{P_k^1}{P_k^0}]_{K=k_F} = 2Re < \frac{1}{N} \sum_i \frac{\varphi_k^1(R_i)}{\varphi_k^0(R_i)} >_{K=k_F}$$

$$= \frac{2\Omega}{(2\pi)^3} P(\frac{2m}{h^2}) \sum_q \frac{U(q) S(q)}{\kappa^2 - |k+q|^2} \qquad (130)$$

and the mean value term is given by]

$$[\frac{P_k^1}{P_k^0}]_{K=k_F} = 2Re < \frac{1}{N} \sum_i \frac{\varphi_k^1(R_i)}{\varphi_k^0(R_i)} >_{K=0}$$

$$= \frac{2\Omega}{(2\pi)^3} P(\frac{2m}{h^2}) \sum_q \frac{U(q) S(q)}{q^2} \qquad (131)$$

where P is the Cauchy principal value. U(q) and S(q) are the pseudopotential form factor and liquid structure factor, respectively. The energy momentum dependence contribution $[\kappa \partial/\partial_K (|P_k^1/P_k^0|)]_{K=k_F}$ in such calculations improves the results. For Mg and Sn the value of K_1 (%) comes out as .105 and 0.11, respectively while the value of K_0(%) comes out as 0.015 and 0.62, respectively for author's model-I. When the results of K_1 and K_0 are combined together we observe that the results of Knight shift K approaches closer to the experimental value of K (%) as 0.112 ± 0.004 and 0.747, respectively for these metals. These results are much better than earlier results, $K_1 = 0.07$ of Jena and Halder[124] in case of Mg and results $K_1 = 0.54$ of Ford and Styles[113] in case of Sn.

(iii) Hall effect: It is a transport phenomenon developed in a magnetic field. This galvenomagnetic properties has become a old method[128] for studying the electronic structure of metals.

On the basis of nearly free electron model the Hall coefficient R_H in solid phase is given by

$$R_H^0 = \frac{1}{nec} \quad (133)$$

where n is carrier concentration. This coefficient may be upset by Fermi surface distortion and it is temperature dependent quantity. In liquid phase the value deviates from the free electron value

$$(R_H)_{liq} / R_H^0 = \left(\frac{m}{m*}\right)^n = \frac{N^0(E_F)}{[N(E_F)]_{liq}} \quad (134)$$

According to Fukuyama et al.[129] n = 2 and due to Ziman[34, 62] n = 1. But actually both do not satisfy experimental observation. In both solid and liquid phases the value of this coefficient for a metal is of the same sign whereas in semiconductors the situation is different [130,131]. For p-type semiconductors due to excess of holes value is positive whereas for n-type semi- conductors due to excess of charge carriers (electrons) value is negative. For intrinsic semiconductors the value of R_H is small. In case of amorphous semiconductors the sign of is always opposite to that of thermoelectric power. For n- type doped samples and p-type doped samples the sign of R_H is positive and negative, respectively.

In general, the conductivity σ of a material is given by

$$\sigma = ne\mu \quad (135)$$

where μ is mobility. Eqs. (133) and (135) convey that both properties are associated to each other. In terms of ordinary resistivity ρ, R_H may be given as

$$R_H = \rho \frac{e\tau}{mc} \quad (136)$$

where τ is relaxation time. The measurement of R_H for a metal is not easy because of low value of R_H but for a semiconductor it can be measured very small. For most of the metals $R_H \sim 1$ except for Te which possesses a value 3.3, which decreases on heating. Eq.(127) shows the dependence of R_H on the density of states. Even and Jortner[132] observed that when Hg is heated the value of R_H increases sharply due to decrease in density which appears due to the development of pseudo-gap at low densities. Generally, it is observed that R_H is always negative for non-transition liquid metals and positive for transition metals. The change of R_H at melting points appears similar to that of of resistivity behavior.

In terms of Fermi radius, k_F we may express for metals

$$R_H = \frac{3\pi^2}{k_F^3 ec} \quad (137)$$

which shows that Fermi surface distortion may effect R_H.

For an homogenous alloy system[133)]

$$R_H = R_H^0 \left[1 - \frac{f}{3}(1-f) \beta^2 \left(\frac{\Delta \rho}{\rho^0}\right)^2 - \beta^2 f \Delta R_H \right] \quad (138)$$

where

$$\beta^{-1} = 1 + 2(1-f)\frac{\Delta \rho}{3\rho^0} \quad (139)$$

and

$$\rho = \rho^0 + \beta f \Delta \rho \quad (140)$$

Here ρ is resistivity and f is fractional volume given by

$$f = \frac{N C_1 \omega}{\Omega} = \frac{4\pi N C_1 L_0^3}{3\Omega} \quad (141)$$

For a binary alloy system

$$R_H = R_H^0 = \frac{\Omega}{(C_h Z_h + C_g Z_g) N e c} \quad (142)$$

where h : host and g : guest elements. Thus, Hall gradient becomes

$$\left(\frac{dR_H}{dC_g}\right) = \left(\frac{dR_H^0}{dC_g}\right)\left(1 - \frac{2}{3}r\right)^2 - \frac{R_H N \omega}{3\Omega} r^2 \quad (143)$$

where

$$r = \frac{\omega}{N\omega}\left(\frac{1}{\rho} - \frac{d\rho^0}{dC_g}\right) \quad (144)$$

The value of R_H / R_H^0 for most of Hg-X mercury binary alloy[134-136)] (X : any guest element) is less than unity except Hg-Sn alloy[137)] for which value is greater than unity. Te is the only element whose alloy[138)] generally possess higher values than unity of R_H / R_H^0 which is an anomalous behavior- a tendency of semiconductors which possess an incipient band gap. In metal ammonia solutions[130)] the ratio R_H / R_H^0 increases at low concentration which is due to presence of localized states. Ziman confirms that such deviations are possible and this observation is supported also by Greenfield[140)]. But for their full explanation we require a generalized acceptable theory. One may think that the failure of nearly free electron approximation in the light of short and long mean free path approximation[78)] as in case of resistivity of liquid metals.

Ziman suggested an explanation for such deviation on the basis of principle of Fermi current. According to him for liquid metals

$$R_H = \frac{eV_F}{jF} \frac{1}{nec} \quad (145)$$

At Fermi level the group velocity V_g follows

$$V_g \longrightarrow \frac{dE(k)}{dk} \longrightarrow V_F \quad (146)$$

Such explanation suggests that R_H / R_H^0 measures the ratio of the group velocity of an electron to the current that it carries. Ultimately the deviation of R_H from free electron value R_H^0 may be explained on the basis of density of states behavior.

The change of sign in R_H for the cases of transition metals has been very well explained by Zabo[141)] and Fletcher142). The phase-shift formula of Zoba for R_H is given by

$$R_H = \frac{1}{nec}\left[1 + \frac{20}{3\pi Z}\frac{\partial}{\partial \mu}[\mu\, Sin^2\delta_l\, S(2\mu^{1/2})]\right.$$
$$\left. + \frac{20}{3\pi Z}[\frac{\mu}{k_F} Sin^2\delta_l \frac{\partial}{\partial q} S(q)]_{q=2k_F}\right] \quad (147)$$

where $l = 2$. μ is chemical potential. According to Fletcher

$$Q_i = L_0 e T(\frac{\partial R_H \sigma_{el}}{\partial E})_{E=E_F}$$
$$= R_H \sigma_{el}[Q_d + L_0 e T\frac{\partial \ln R_H}{\partial E}]_{E=E_F} \quad (148)$$

where Q_i is Nernst Sttinghausen Coefficient, σ_{el} is electrical conductivity and the diffusion thermopower Q_d is given by

$$Q_d = L_0 e T[\frac{\partial \ln \sigma_{el}}{\partial E}]_{E=E_F} \quad (149)$$

$$L_0 = \frac{\sigma_{th}}{\sigma_{el}} = \frac{\pi^2 k_B^2}{3 e^2} \quad (150)$$

where σ_{th} is thermal conductivity. The value of $[\partial \ln R_H^0/\partial E]_{E=E_F}$ is given by

$$[\frac{\partial \ln R_H^0}{\partial E}] = -(\frac{2}{3}E_F)^{-1} = -N^0(E_F)(Zm^*/m)^{-1} \quad (151)$$

The deviation of R_H from the free electron value is due to the variation in the orientation or temperature.

In single site approximation we consider

$$(\delta_l)_{l=2} = \frac{m\, k_F \Omega}{(16\pi/h^2)}[\frac{\delta}{\delta q} U(q)]_{q=2k_F} \quad (152)$$

which transforms eq. (140) as

$$\frac{R_H}{R_H^0} = 1 + \frac{20}{3\pi Z}[Sin^2\delta_2 + 2E_F \delta_2 Sin\, \delta_2\, Cos\delta_2]$$
$$+ \frac{E_F}{K_F} Sin^2\delta_2 [\frac{\partial}{\partial q} S(q)]_{q=2k_F} \quad (153)$$

This approximate formula is useful in the estimation of ratio R_H/R_H^0 which has strong relationship with the ratio m/m^* and the ratio $N(E_F)/[N(E_F)]_{liq}$.

4.4 INTERATOMIC OR INTERMOLECULAR PROPERTIES

Interatomic forces in solids or intermolecular forces in liquids are responsible for the properties which are discussed in this section. Generally, these properties are evaluated through, the total energy of a system depending on ion Configurations. The metals are famous for long range forces in solid phase and short range forces in liquid phase. Both have different orientational symmetry. The bonds between nearest neighbor atoms in crystals are oriented along specific directions in space and distance vary in terms of lattice parameter 'a' of the order of quantity $2n/a$ whereas in liquids the lines joining pairs of nearest neighbour molecules or atoms will possess equal probability in all directions of space. The long range and short range forces arise due to ionic interactions modulated by the conduction electrons. The Friedel oscillations[147] in the screening field generate these forces. In metals these forces are developed due to quadrupole moments induced on the atoms during distortion. An account of these forces considerations have been described in further section.

The total energy of a crystal is described by

$$E = E_{fe} + E_{e-i} + E_{bs} + E_{es} \tag{154}$$

where E_{fe} is the energy of free electron gas depends on total volume but independent of ionic coordinates, E_{e-i} is the average energy of electrons, E_{bs} is the band structure energy, which depends on ionic configuration and E_{es} is electrostatic energy. Cochran[144] and Harrison[3] have separately given directly and indirectly methods, respectively for it. We may express E in the form [5,145]

$$E = \frac{1}{2N} \sum_{i \neq j} \frac{Z^2 e^2}{R_i - R_j} + \frac{1}{N} \sum_{k < k_F} <\psi_k|T|\psi_k> - \frac{1}{N} \int \frac{\rho_e(r)\rho_e(r')}{r-r'} d^3r \, d^3r'$$

$$+ \frac{1}{N} \int \rho(r) [E_{exch} \rho_e(r) - \mu_{exch} \rho_e(r)]$$

$$= \frac{1}{2N} \sum_{i \neq j} \frac{Z^2 e^2}{R_i - R_j} - \frac{1}{N} \sum_{q}{}' [S*_g(q) S_g(q)] - \frac{1}{N} F(q) \tag{155}$$

where

$$F(q) = \frac{2\Omega}{(2\pi)^3} \left[\int_{k<k_F} d^3k \frac{<k|U|k+q><k+q|U|k>}{(\frac{\hbar^2}{2m})(k^2 - |k+q|^2)} \right.$$

$$\left. - \frac{\Omega q^2}{8\pi e^2} W^2_e(q) \right.$$

$$= \frac{\Omega q^2}{8\pi e^2} |V_b(q)|^2 [1 - \frac{1}{\varepsilon(q)}] \tag{156}$$

Here $F(q)$ is energy wave number characteristic function (Harrison function). U is screened pseudopotential, $V_b(q)$ is bare-ion potential, $\varepsilon(q)$ is dielectric screening and electron-electron potential. $W_e(q)$ is given by

$$W_e(q) = \frac{4\pi e^2}{q^2} \rho_e(q) = V_b(q)[\frac{1}{\varepsilon(q)} - 1] = U(q) - V_b(q) \tag{157}$$

$W_e(q)$ is related to electron-electron interaction energy E_{ee} which is given by

$$E_{ee} = \frac{\Omega}{2} \sum_q{}' \rho_e(q) W_e(q) = -\frac{1}{N} \int \rho_e(r) W_e(r) d^3(r)$$

$$= \frac{\Omega q^2}{8\pi e^2} \sum_q{}' S*_g(q) S_g(q) W^2_e(q) \tag{158}$$

Within framework of pseudopotential perturbation theory the electron energy E_K is given by

$$E_k = \frac{\hbar^2 k^2}{2m} + N <k|U|k'> + N^2 \sum_q{}' S^*_g(q) S_g(q) \frac{<k+q|U|k><k|U|k+q>}{\frac{\hbar^2}{2m}(k^2 - |k+q|^2)}$$

(159)

Therefore, average electron energy becomes

$$\frac{1}{N} \sum_{K<k_f} E_k = \frac{2\Omega}{(2\pi)^3} \int E_k \, d^3k$$

$$= \frac{3}{5} Z\hbar^2 \frac{k_F^2}{2m} + Z<k|U|k'> + \sum_q{}' S^*_g(q) S_g(q) \frac{2\Omega}{(2\pi)^3} \int \sum d^3k. N^2$$

$$\frac{<k|U|k+q><k|U|k+q>}{\frac{\hbar^2}{2m}(k^2 - |k+q|^2)}$$

$$= E_0 + E_1 + E_2 \qquad (160)$$

The band structure energy E_{bs} is given by

$$E_{bs} = E_2 - E_{ee}$$

$$\left[\frac{2\Omega}{(2\pi)^2} \int d^3k N^3 \cdot \frac{<k|U|k+q><k|U|k+q>}{\frac{\hbar^2}{2m}(k^2 - |k+q|^2)} \right.$$

$$\left. - (\frac{\Omega q^2}{8\pi e^2}) W^2(q) \right] \sum_q{}' S_g(q) S^*_g(q)$$

$$= \sum_q{}' S_g(q) F(q) \qquad (161)$$

For a crystal of volume V and ion N and valency Z, we have[7] free electron density

$$\rho_o = \frac{ZN}{V} = (\frac{4\pi}{3} r_s^3)^{-1}$$

Wigner - Seitz Cell Volume $= \frac{V}{N} = \Omega = (\frac{4\pi}{3} R_a^3)$

and $E_{fe} = E_0 + E_{ex} + E_{corr}$

$$= [\frac{2.217}{r_s^2} - \frac{0.9162}{r_s} + Z(-0.115 + 0.031 \ln r_s) \, ryd] \qquad (162)$$

where E_0 is mean value of kinetic energy, E_{ex} and E_{corr} are the mean values of exchange and correlation energy of electron. Actually $E_0 + E_{ex} = E_{HF}$ is the average Hartree - Fock energy. Also

$$\bar{E}_{e-i} = \lim_{q \to 0} \frac{Z}{\Omega} [\frac{4\pi Z e^2}{\Omega q^2} + V_b(q)] \qquad (163)$$

which is the mean value of electron-ion interaction. The electrostatic energy E_{es} is given by

$$E_{es} = -\frac{\alpha Z^2}{2R_a} \qquad (164)$$

where α is Madelung's constant. According to Fuchs[146)]

$$E_{es} = -1.792 \frac{Z^{2/3}}{r_s} \qquad (165)$$

(a) Effective Ion-Ion Interaction Or Pair Potential

The configuration-dependent energy of a crystal in terms of a two body central force interactions (depending on ionic separation) is given by

$$E = \frac{1}{2N} \sum_{i,j} V(|R_i - R_j|) \qquad (166)$$

On equating above eq. with eq. (155) and inserting the structure factor $S_g(q)$ explicitly the total energy may be described by a central two-body effective interaction between ions in the form

$$V(r) = \frac{Z^2 e^2}{r} - \frac{4\pi Z^2 e^2}{N\Omega} \sum_q \frac{G(q)}{q^2} e^{iq.r} \qquad (167)$$

whose Fourier transformation v(q)

$$V(q) = \frac{4\pi Z^2 e^2}{\Omega q^2} [1 - G(q)] \qquad (168)$$

where G(q) is called total electronic band structure energy function (Cochran function) which has been normalized by F(q) function [$\lim q \longrightarrow 0\ G(q) = 1$] and related by

$$G(q) = -[\frac{\Omega q^2}{2\pi Z^2 e^2}] F(q) \qquad (169)$$

If one uses screened Hartree dielectric constant $\varepsilon^*(q)$ such that

$$\varepsilon^*(q) = 1 + [\varepsilon(q) - 1][1 - f(q)] \qquad (170)$$

then
$$G(q) = (\frac{4\pi Z e^2}{\Omega q^2})^{-2} \frac{|V_b(q)|^2}{1 - f(q)} [1 - \frac{1}{\varepsilon(q)}] \qquad (171)$$

where f(q) is associated to exchange and correlation term. The use of G(q) function in atomic properties evaluation is better due to reason that it minimizes the required summation problem. This quantity is experimentally measurable quantity[144,146,147], whose value is different in solid and liquid phases.

In theories of liquid state the effective pair potential, V(r) may be described through pair distribution function, g(r), which is related indirectly to liquid structure S(q) and bare-ion pseudopotential, $V_b(q)$. In r- space the pair potential, V(r) and bare-ion potential, $V_b(r)$ are described by

$$V(r) = \frac{Z^2 e^2}{r} - \frac{Z^2 e^2}{r} (\frac{\pi}{2})^{-1} \int_0^\infty G(q) \frac{\sin(qr)}{(qr)} d(qr)$$

$$= \frac{Z^2 e^2}{r} + V_{bs}(r) \ , \quad r < r_m$$

$$
\begin{aligned}
&= V_{bs}(r) &&, \quad r \geq r_m \\
&= 0 &&, \quad r = \sigma
\end{aligned}
\tag{172}
$$

$$
\begin{aligned}
V_b(r) &= -\frac{Ze^2}{r} + \frac{Ze^2}{r} F_l(r) \quad, \quad r < r_c \\
&= -\frac{Ze^2}{r} \quad, \quad r \geq r_c \\
&= 0 \quad \text{for} \quad F_l(r) = 1
\end{aligned}
\tag{173}
$$

where the band structure part of the potential, V_{bs} is given by

$$ V_{bs}(r) = -\frac{Z^2 e^2}{r} f_{bs}(r) $$

Here

$$ f_{bs}(r) = \left(\frac{\pi}{2}\right)^{-1} \int_0^\infty G(q) \frac{\sin(qr)}{(qr)} d(qr) \tag{174}$$

We have $\quad V(r) = V_{\min} \quad \text{and} \quad \frac{d}{dr} V(r) = 0 \quad \text{at} \quad r = r_m$

which indicate the change from attraction to repulsion. r_m is an equilibrium distance between a pair of isolated undisturbed molecules. $F_l(r)$ is normalized radial function. For solids, r_m corresponds to bond length. The depths of the minimum, $V_{\min}$ is the energy, which is required to separate such a pair and it is a measure of the strength of the binding forces. This first minimum in potential energy curve corresponds to the nearest neighbor distance in crystal and for liquids $r_m = (2)^{1/6} \sigma$ (σ: hard sphere model radius). It is very interesting to characterize a temperature $V_{\min}/k_B$ which corresponds to mean kinetic energy.

In q space the Fourier transformations of $V_r(r)$ [$eq(172)$] and $V_b(r)$ [$eq(173)$] are described by

$$
\begin{aligned}
V(q) &= \frac{4\pi Z^2 e^2}{\Omega q^2} [1 - G(q)] \quad ; (\text{For } r < r_m \text{ region}) \\
&= \frac{4\pi Z^2 e^2}{\Omega q^2} \quad ; [\text{For } r \geq r_m \text{ region where } V_{bs}(r) = 0]
\end{aligned}
\tag{175}
$$

$$
\begin{aligned}
V_b(q) &= -\frac{4\pi Z e^2}{\Omega q^2} [1 - \sum_l S_l(qr_c)] \quad ; r < r_c \\
&= -\frac{4\pi Z e^2}{\Omega q^2} \quad ; r \leq r_c \\
&= 0 \quad \text{for} \quad \sum_l S_l(qr_c) = 1
\end{aligned}
\tag{176}
$$

It is obvious that

$$ G(q) \propto \frac{V_b(q) W_e(q)}{[V_c(q)]^2} \quad ; \quad V_c(q) = -\frac{4\pi Z e^2}{\Omega q^2} \tag{177}$$

which shows that electron-potential energy plays an important role in making pair potential to be of screened nature. In solid phase the inter atomic force is of long range nature whereas in liquid phase of short range nature. At distances greater than $2r_c$ the direct Coulomb interaction between any pair of ions nearly canceled and we observe screened nature of inter-ionic potential. The cutt-off region of inter-atomic potential or pair potential may be determined through the condition

$$\int_0^\infty G(q) \frac{\sin(qr)}{(qr)} d(qr) = \frac{\pi}{2} \qquad (178)$$

In case of diatomic molecule [148] the vibrational potential energy becomes zero at equilibrium value $r = r_o$ (r_o: bond length). For $r = \infty$, $V(r) \longrightarrow D$ where $V(r) = D[1 - e^{-a(r-r_o)}]$
here D and a are constants. For $r \longrightarrow 0$, $V(r) \longrightarrow$ finite value. Since atoms are vibrating hence $V(r)$ will oscillate about position around the point at which first minimum of $V(r)$ lies. At large distance the potential oscillates with time which are Friedel oscillations and have the same origin as the Friedel oscillations in the electron density. The oscillations in the electron density will give oscillations in the electrostatic potential which will favour the neighboring ions sitting at positions of maximum electron density. Also the repulsive nature of the pseudopotential itself tends to favour the ions sitting at the positions of low electron density. It is noticeable that the long range oscillations arise due to the singularity in the electronic screening [3,5]. The same cause is the origin of Kohn effect [149] It is evident that the oscillations in the pair distribution function gives

$$\lambda_0/(k_F/2) \; \alpha \; Z^{1/3} \qquad (179)$$

whereas in case of screening charge density fluctuation or the pair potential

$$\lambda_0/(k_F/2) = \text{constant} \qquad (180)$$

where λ is wave length of oscillation.

The screening charge density, $\delta\rho(r)$ in high density region gives an oscillatory behavior whereas in low density region, a damping nature [150]

$$\delta\rho(r) = \frac{1}{2\pi^2 r} \int_0^\infty [1 - \frac{1}{\varepsilon(q)}] \qquad (181)$$

In eq.(177) a term $[1 - 1/\varepsilon(q)]$ exists similar to above eq. through G(q) function. Therefore, we observe an oscillatory nature in the potential energy curve.

On the basis of Anderson et al. theory [151] a relation between liquid structure factor S(q) and band structure part of pair potential, $V_{bs}(r)$ is described by

$$S(q) = S_\sigma(q) + \rho \int y_\rho(r) [e^{-\beta V_{bs}(r)} - e^{-\beta V_\sigma(r)}] e^{-iq\cdot r} dr \qquad (182)$$

where $\beta = 1/k_BT$. $V_\sigma(r)$ is the hard sphere potential and $S_\sigma(q)$ is the structure factor of the system of hard sphere. $y_\sigma(r) = g_\sigma(r) e^{\beta V_\sigma(r)}$ is the radial distribution function of hard sphere of the system which follows the condition

$$\int y_\rho(r) [e^{-\beta V_{bs}(r)} - e^{-\beta V_\sigma(r)}] dr \qquad (183)$$

$S_\sigma(r)$ and $y_\sigma(r)$ may be described analytically[6,8]. Thus, it is very much clear now that there appears a direct relationship between S(q) and G(q) functions as also clarifies through AHP relationship[5] which may be expressed in the region of smaller q by the eq

$$S(q) = \left[\left(\frac{4\pi Z e^2}{\Omega\, q^2} \right) \cdot \frac{G(q)}{k_B T [\varepsilon(q) - 1]} \right]^{-1} \tag{184}$$

which shows also indirectly that G(q) function of pseudopotential theory is related to pair distribution function, g(r). This equation has been utilized and discussed in further section also. Once a pair potential is established, one can easily apply Monte Carlo (MC) or molecular dynamics (MD) techniques for determining pair distribution function through the liquid structure and many structural and dynamical behaviors of liquids ; for example, mobility, charge cancellation effect, first coordination sphere penetration by species of the same charge and features of short range orders. March and Coworkers and Enderby have chosen separately the suitable forms of the model inter-atomic potentials in case of multi-component liquids(complex liquids) and for aqua-ions and then applied statistical theories of liquid states in order to produce their structures. This procedure may be extended for describing the structure of other complicated disordered - systems (multi-component systems).

The effective pair interaction theory of liquid state has been extended for the case of non-metallic atoms and to x- x,x-y and y-y types of atoms and fruitful results of phonon dynamics have been observed[152]. Beyond the first minimum the pair potentials have been found of attractive nature. The potential at short range has been found quite steep. It has been observed in cases of multi-component systems other than Friedel type of oscillations there exist some extra cosine terms in inter-atomic potential - a behavior different than metallic atoms.

The effective pair potential are very useful in describing properties such as band structure energy, inter-atomic force constant, elastic constant and dynamical properties of materials in crystalline and liquid phases. The reliability of the results depends on the approximations and methodology followed in the calculations.

(b) Crystal Structure Energy

The pseudopotential theory has played an important role in describing crystal structure on the basis of crystal energy and cohesive energy of condensed matter. The cohesive theories have stimulated the problems of structure and dynamics of materials in bulk as well as in surface states. The basic idea is to formulate the total energy of solids which has been already described. The picture becomes complicated in case of alloys.

A crystal may be considered as a system of N periodically arranged positive ions immersed in an electron gas. In an alloy system ions of two species g and h are in mixed form. For an alloy the density ρ_{el} of electron gas is

$$\rho_{el} = \frac{Z^*}{\Omega_e} \tag{185}$$

where effective valence charge Z^* is

$$Z^* = (1 - C) Z_g + C Z_h \tag{186}$$

and effective volume $\Omega_e = 3\pi^2 Z^*/k_F^3$. ($k_F$: Fermi radius). C is concentration. From above concept of alloy we can write for pure crystal say of h (host atoms) C = 0 and

$$(\rho_{el})_h = \frac{Z_h}{\Omega_h} \tag{187}$$

The crystal energy (total energy) E in real sense may be separated into two parts E_V and E_s :

$E = E_V$ (volume dependent) $+ E_s$ (Structure dependent)

where
$$E_V = E_{fe} + \overline{E}_{e-i} \tag{188}$$
and
$$E_s = E_{bs} + E_{es} \tag{189}$$

as discussed above [Sec.2.4 (a)]. E_{bs} for liquid alloys (disordered alloys) and super-structure (su.st.) like of crystals having a Cu_3Au type cubic cell are given by

$$(E_{bs})_{alloy} = \frac{3(Z_g Z_h)}{4k_F^3} \int_0^\infty V_{alloy}(q) q^2 dq \tag{190}$$

where

$$V_{alloy}(q) = \frac{\Omega_e^2 q^2}{4\pi e^2} \left[\frac{1}{\varepsilon(q)} - 1\right] \left[C_g U_g^2(q).\right.$$
$$[1 - C_g + C_g S_{gg}(q)] + C_h^2 U_h^2(q) +$$
$$[1 - C_h + C_h S_{hh}(q)] + 2U_g(q) U_h(q).$$
$$\left. C_g C_h [S_{gh}(q) - 1] \right] \tag{191}$$

$$(E_{bs})_{su.st} = \sum_{hr} S^*_g(h_r) S_g(h_r) [F(h_r)]_{su.st}$$
$$= \frac{\Omega}{8\pi e^2} \sum_{h_r} h_r^2 \overline{|U_{gh}(h_r)|^2} [\varepsilon_{h_r}(\varepsilon_{h_r} - 1)]$$
$$+ \frac{\Omega}{8\pi e^2} \sum_{h_{r_n}} h_{r_n}^2 \overline{|\Delta U_{gh}(h_{r_n})|^2} [\varepsilon_{h_{r_n}}(\varepsilon_{h_{r_n}} - 1)]$$
$$(\varepsilon(h_{r_n}) - 1) [F_{h_r}]_{hyb} \tag{192}$$

where
$$U_{gh}(h_r) = C_g U_g(h_r) + C_h U_h(h_r) \tag{193}$$
and
$$\Delta U_{gh}(h_r) = U_g(h_r) - U_h(h_r) \tag{194}$$

Here h_r and h_{r_n} are reciprocal lattice vectors of equal and unequal parity group, respectively.

For a completely disordered alloy in chaotic condition (Ch)

$$(E_{es})_{alloy-Ch} = \frac{\Omega_e}{8\pi} \sum_q' \left[\frac{1}{\varepsilon(q)} - 1\right]^2 q^2 \left[C_h [V_h^b(q)]^2 [S(q) + \frac{C_g}{N}]\right.$$
$$+ C_g [V_g^b(q)]^2 [C_g S(q) + \frac{C_h}{N}]$$
$$\left. + 2 C_g C_h V_g^b(q) V_h^b(q) [1 - \frac{1}{N}] \right] \tag{195}$$

In the presence of short range order (Sh) it becomes

$$(E_{bs})_{al.oy-Sh} = \frac{\Omega_e}{8\pi} \sum_q{}' [\frac{1}{\varepsilon(q)} - 1]^2 q^2 \Big[C_h [V_h^b(q)]^2 [S(q) + \frac{C_g}{N} \cdot$$

$$\alpha(\rho_i) e^{iq\rho_i}] + C_g [V_g^b(q)]^2 [C_g S(q) + \frac{C_h}{N} \alpha(\rho_i)$$

$$e^{-iq\rho_i}] + 2 C_g C_h V_g^b(q) V_h^b(q) [1 - \frac{1}{N}(\rho_i).e^{iq\rho_i}] \Big] \qquad (196)$$

where $\alpha_i = [1 - \rho_i^{gh}/C_g]$ is a short range order parameter, ρ_i is displacement of lattice and P_i is a probability of finding i th site occupied by an alloy molecule. The case of pure short range order has been described elsewhere. Generally, electrostatic energy E_{es} may be evaluated by Ewald method. Krasko[153] extended it for the case of alloy. For a pure metal case the contributions of crystal energy E may easily be obtained by putting $C_h = 0$. Actually energy E_S determines the structure's stability at low temperature in the absence of zero-point energy. Stround and Ashcroft[154] observed through the magnitudes of energy differences between different structures in case of disordered alloys that a fall in the crystal structure energy occurs when the Fermi sphere touches the Brillouin zone's face. The discontinuity causes a drop in the energy of the corresponding phase and therefore, it changes the sign of the energy difference between the phases. It is the reason due to which phase transition takes place when the Fermi sphere comes in contact with the first Brillouin zone boundary. The position of Fermi diameter and cut-off region of screened pseudopotential have special importance in describing crystal structure stability. Their observation also explain very well the Home-Rothery rules, as supported by Kogachi and Matsuo.[155]

The problem of structure stability in binary alloys has been discussed by Ashcroft[156] on the basis of hard sphere model approach. It is observed that near the regions $C_h = 0$ or $C_h = 1$ the effect of dilute quantities of one metal in contact with the other simply decrease the melting point of the alloy system which is associated to phase separation regions. Thus, it is clear now that freezing and phases separation are two possible instabilities for the liquid phase. The alloying effect in case of alloys may be understood in terms of the Fourier transform of characteristic alloying function $V_\alpha(q)$ [Pairwise alloying potential] which is given by

$$V_\alpha(r) = \frac{(Z_g - Z_h)^2}{4r} + \frac{2\Omega_e}{(2\pi)^3} \int [U^g(q) - U_h(q)]^2.$$

$$.\varepsilon(q) X(q) e^{iq \cdot r} dq. \qquad (197)$$

which is also called as configuration energy of an alloy. Here r is the interatomic separation. Thus, change in band structure energy, E_{bs} before and after alloying gives the alloying effect, which is described by

$$\Delta = [E_{bs}(q)]_{alloy} - [E_{bs}(q)]_{av} \qquad (198)$$

where average energy of pure components $[E_{bs}(q)]_{av}$ is given by

$$[E_{bs}(q)]_{av} = C_h [E_{bs}(q)]_h + C_g [E_{bs}(q)]_g \qquad (199)$$

It is generally observed that maximum of pairwise alloying potential, $V_\alpha(r)$ falls to the region of r for which the number of unlike neighbors is increased on alloying, i.e. total energy decreases with ordering. If for a particular alloy the ordering energy is greater than the thermal energy ($k_B T$) at the melting temperature it means an ordered state remains up to the melting point. The ordering temperature, T_c is given by

$$T_c = - C_g C_h U(\bar{g}_n) \qquad (200)$$

where $\bar{g}_n$ is that reciprocal lattice vector (i.e., cut-off region) at which screened pseudopotential U is minimum. Actually this temperature is that at which there is order - disorder phase transition.

(c) Dispersion Of Phonons

Phonon dispersion is an old lattice dynamical problem. After the first efforts of phonon frequencies of Na crystal by Toya[157] there have been a number of calculations of metals including the introduction of pseudopotential concept which has brought an improvement over the lattice dynamics of metals. Schneider and Stoll[32, 158] have utilized the measured phonon frequencies by the method of inelastic scattering of neutrons in order to determine the pseudopotential over the whole range in momentum space. Truly speaking, the pseudopotential concept has brought us more closer in understanding the whole picture of dynamics of solids.

Since last three decades after the inelastic neutron scattering studies of monoatomic liquids by Brockhouse and Pope[159], the field of phonon dispersion in liquids got special attention. This field has been stimulated largely by the advent of fast computer simulation (molecular dynamics) devices. Among others the efforts in simulation techniques of Alder and Wainwright[160] for discontinuous potentials(e.g. hard spheres) and of Rahman [54] for real potential are appreciable in this direction. After the success to the case of Lennard-Jones liquid, Schommers[161] and Rahman did efforts for cases of liquid Na and Rb.

In this section, we would describe the methodology of phonon dispersion studies first of crystalline phase then of liquid phase. This study is very important from the point of view that many properties of metals may be evaluated by the use of phonon frequencies.

(i) **Dispersion of solid phonons** : For the studies of solid phonons, generally force constant technique is used, i.e., the total energy in a solid is expressed by the phonon frequency $\omega(q)$ and wave vector q for a Bravais lattice as described by[162]

$$[M\omega^2_s(q) - \sum_{j=1}^{3} D_{ij}] e_i(q) \; ; \quad i = 1,2,3 \qquad (201)$$

with
$$D_{ij}(q) = R_{ij}(q) + C_{ij}(q) + E_{ij}(q) \qquad (202)$$

where e_i is the polarization vector, $D_{ij}(q)$ the dynamical matrix and M is the ionic mass. The Born-Mayer type of exchange interaction contribution part, $E_{ij}(q)$ is negligible for alkali metals and Coulombic contribution, $C_{ij}(q)$ may very well be developed by Ewald's method. The electronic contribution $E_{ij}(q)$ in terms of electronic band structure energy function G(q) is given by[163]

$$E_{ij}(q) = \frac{4\pi Z^2 e^2}{\Omega} \left[\sum_{h=0} \frac{(q+h)_i (q+h)_j}{q^2} G(q) \right.$$

$$- \frac{h_i h_j}{h^2} G(h) + \frac{q_i q_j}{q^2} G(q) \Big] \quad (203)$$

where h is reciprocal lattice vector. On solving eq.(210) in the symmetry directions (100), (110) and (111) the longitudinal frequency and transverse frequency $\Lambda = T$ obtained from a diagonal matrix are given by

$$\omega_s^2(q,\Lambda) = \omega_c^2(q,\Lambda) - \omega_{el}^2(q,\Lambda) \quad (204)$$

where ω_c (Coulombic contribution) may be obtained from Kellerman's method and ω_{el} is described by

$$\omega_{el}^2(q) = \frac{(e_{q,L}q)^2}{|q|^2} G(q) + \sum_{h=0} \Big[\frac{e_{q,L}\cdot(q+h)^2}{|q+h|^2} \Big] G(q+h) - \frac{(e_{q,l}\cdot h)^2 G(h)}{|h|^2}, \quad \text{for } (\Lambda = L) \quad (205)$$

$$\omega_{el}^2(q,T) = \sum_{h=0} (e_{q,T}\cdot h)^2 \frac{G(q+h)}{|q+h|^2} - \frac{G(h)}{|h|^2}; \text{ for } (\Lambda = T) \quad (206)$$

The calculation of phonon frequencies of solid phonon is more sensitive to used dielectric screening rather than electron-ion model pseudopotential.

In long wave length limit the longitudinal frequency, $\omega_{el}(q,L)$ is related to ion-plasma frequency ω_{ip} by the relation

$$\Big[\frac{\omega_{ip}(q)}{\omega_s(q)} \Big]^2 = \frac{4\pi e^2}{q^2} N(E_F) \quad (207)$$

where $N(E_F)$ is density of states at Fermi energy E_F. $\omega_{ip}(q)$ is given by

$$\omega_{ip}^2(q) = \frac{4\pi Z^2 e^2}{M\Omega} \quad (208)$$

Thus, we have

$$\omega_s^2(q) = \Big(\frac{2ZE_F}{3M}\Big) q^2 \quad (209)$$

which defines for the phonons the Bohm-Staver[164] sound velocity V_s as

$$\Big[\frac{\omega_s(q)}{q}\Big]_{q \to 0} = V_s = \Big[\frac{2ZE_F}{3M}\Big]^{1/2} \quad (210)$$

Similarly for electrons

$$V_s = \Big(\frac{Z_m V_F}{3M}\Big)^{1/2} \quad (211)$$

where V_F is Fermi velocity of electrons. For shorter wave lengths $q \to 2k_F$, $\omega_s(q) \to \infty$ which is associated to Kohn anomalies. It has been observed that Kohn anomalies and long range forces are inter-related to each other.

Besides the use of frequencies of solids phonons in case of electronic transport coefficients like electrical and thermal resistivities as discussed in sec.2.3 there are some more coefficients which are deducible and described in sec 2.5.

(ii) **Dispersion of liquid phonons:** The evaluation of liquid phonon frequencies $\omega_L(q)$ is a problem related to molecular motion in liquids, which has been reviewed elsewhere. In recent years, there has been remarkable progress in the understanding of collective excitations in liquids which is helpful in calculating dynamic structure factor, $S(q,\omega)$ from which both the phonon frequencies and life times are simultaneously forthcoming.

The liquid phonon problem may be carried through a direct analogue of the solid state program by first formulating a simple life time independent theory of the phonon spectrum of a liquid when the interatomic potential, V(q) and radial distribution function, g(r) are well known.

The first attempt in the evaluation of liquid phonon frequencies $\omega_L(q)$ of Rb in pseudopotential formalism was made by the author[165] to test the number of theories of Percus - Yevick[166], Egelstaff[167], Random phase approximation (RPA) and Hubbard-Beeby[168]. There was observed a departure in the RPA result of Rahman and experiments of Copley and Rowe[169] whereas other considered theories have given some better results. Further, recenty an independent analytic method has been described by introducing electronic band structure energy function G(q).

The liquid phonon frequencies, $\omega_L(q)$ in terms of structure factor, S(q) and interatomic potential, V(q) in low region of q are described by

$$\frac{M\omega_L^2(q)}{q^2} = \frac{k_B T}{S(q)} \qquad \text{(Egelstaff formula)} \quad (212)$$

$$\frac{M\omega_L^2(q)}{q^2} = [V(q) + k_B T] \qquad \text{(RPA formula)} \quad (213)$$

where M is ionic mass and k_B is Boltzmann constant. Actually, RPA formula[170] is the combined effect of eq.(212) with AHP relationship (valid for small q region)

$$S(q) = [\frac{V(q)}{k_B T} + 1]^{-1} \qquad (214)$$

Actually this relationship holds good only within plasma region where the Coulomb forces are dominant and it becomes poorer approximation as we move towards the position $q \sim 2 k_F$ which strongly depends on the short range part of the potential. It is noticeable that eq.(214) for $q \longrightarrow 0$ does not obey the melting formulae of Guinier-Founet[171] and Lindmann as well as Bohm-Staver sound velocity relation

$$S(0) = (k_B T X_T \rho_0) = \frac{k_B T_m}{[\frac{M\omega^2(q)}{q^2}]} = \frac{k_B T_m}{M V_s^2} \qquad (215)$$

where V_s is sound velocity. X_T is adiabatic elasticity and ρ_0 is mean density. With the help of following relation

$$[V(q)]_{AHP} = [\frac{4\pi Z^2 e^2}{\Omega q^2}]^{-1} \frac{|V_b(q)|^2}{\varepsilon(q)} \qquad (216)$$

the RPA formula in terms of G(q) function may be described by

$$\frac{M\omega^2(q)}{q^2} = \frac{4\pi Z^2 e^2}{\Omega q^2}\left[\frac{G(q)}{\varepsilon(q)-1}\right] + k_B T \qquad (217)$$

In case of Jellium model $V(q) \longrightarrow \frac{2}{3} E_F Z$ whereas for electron-ion pseudopotentials it does not happen so. Therefore, we rewrite RPA formula as[172)]

$$\frac{M\omega^2(q)}{q^2} = [\frac{2}{3}E_F Z + \Delta']G(q) + k_B T \qquad (218)$$

where $\Delta'(k_B T)$ is small correction as obtained for screened Ashcroft model potential (using Singwi et al. dielectric screening constant[18)])

$$\Delta' = \frac{4\pi Z^2 e^2 R_c^2}{\Omega} - \frac{4\pi Z^2 e^2 A B}{\Omega k_F^2}$$

$$= 2Z V_R(q=0) + Z^2[X(q=0)] - V_c(q=0) \qquad (219)$$

whereas the first term on the right hand side arises due to the repulsive part of the bare-ion pseudopotential, second term arises due to the polarization potential $X(q)[X(q) = [1-\varepsilon*(q))] q^2 / 4\pi e^2]$ and third term arises due to screened pure Coulomb potential, $V_c(q)$ (using static Hartree dielectric screening constant). On introducing a correction term $\Delta' G(q)$ we may write

$$\frac{M\omega^2(q)}{q^2} = V_{AHP}(q) + \Delta' G(q) + k_B T$$

$$= \frac{4\pi Z^2 e^2}{\Omega q^2} \frac{G(q)}{[\varepsilon(q)-1]} + \frac{4\pi Z^2 e^2}{\Omega}[R_c^2 - A B k_F^2] G(q) + k_B T \qquad (220)$$

where A and B are the parameters of Singwi et al. dielectric screening and R_c is ion-core radius of Ashcroft model potential.

It has been observed that eq.(218) has given more satisfactory results of liquid phonon frequencies $\omega_L(q)$ as compared to eqs.(217) and (220). Therefore, the RPA formula may be rewritten as

$$\frac{M\omega^2(q)}{q^2} = [V(q \to 0) G(q) + k_B T] = [\frac{2}{3} E_F Z + \Delta'] G(q) + k_B T \qquad (221)$$

From the studies of dispersion of solid and liquid phonons we observe that an experimentally measurable quantity, G(q) is useful in the description of dynamics of phonon in both phases. It makes the interatomic potential, V(q) more real and more suitable for the dynamical studies. The additional term of above eq.(221) is the effect of ion-cores and thereby called structure term important for polyvalent metals whereas the first term $2E_F Z/3$ is plasma term (important for monovalent metals). The quantity Δ' becomes dominant when one moves towards the region of wave numbers $q \sim 2 k_F$ [i.e. towards the position of first peak in the structure factor, S(q)]. The first peak lies at $q = 1.23 A°$ and $q = 1.75 A°$ in case of Rb and Na metals, respectively, which concludes the possibility of a simple law of corresponding states lying near the melting point. This formalism has improved the Bohm-Staver theory of sound velocity, V_s [$V_s = [\omega(q)/q]_{q \to 0}$]. The results of V_s by using eq.(218) are summarized in Table - 1. There appears possibility of developing electron-liquid phonon interaction similar to solid phonon.

Table - 1. Velocity of sound, V_S (m/sec) of liquid metals and non-metals

* Metallic ion	Present Calculated Values	Bohm - Staver	Exp.
Li	3189	5765	2526
Na	1709	2847	1890
K	2961	1993	1330
Rb	1591	1291	
Cs	1438	999	
**Non - Metallic ions			
H	2135	4230	
O	2687	1503	4459
N	4314	2182	
B	3194	1792	
P	2670	1188	

*Using pseudopotential model-I [S.K.. Srivastava, J. Phys.. Chem. Solids, **36** (1975) 993] and Dielectric Screening of Singwi et al. [K.S. Singwi, M.P. Tosi, L.H.. Land and A. Sjolander, Phys. Rev., **176** (1968) 589; ibid **B1** (1970) 1044].

** Using pseudopotential model-I and new dielectric screening of author [S.K. Srivastava, J. Phys. Chem. Solids, **36** (1975) 993].

The obtained values of liquid structure factors, $S(q)$ obtained through pseudopotential formalism [eq.(218)] as discussed above in case of alkali metals have been summarized in Table-2, which show good agreement with experiment and theoretical results of Ashcroft and Lenker[68]. The discrepancies in results lie due to approximations used in the description.

4.5 Thermodynamic Properties

The phonon frequencies may be directly used in the evaluation of thermodynamic coefficient such as heat capacity, Gruneisen parameter and entropy etc.

(a) Heat Capacity

The heat capacity of a material is associated to structural behavior mostly. First of all Animalu et al. described a pseudopotential method for evaluating heat capacity of metals in solid phase which is based on the harmonic approximations theory. In their method the

Table - 2. * Liquid Structure Factors, S (q)

q/2k$_F$	Li$^+$	Na$^+$	K$^+$	Rb$^+$	Cs$^+$
0.1	.01345	.01546	.02158	.02196	.02508
0.2	.01450	.01622	.02268	.02313	.02638
0.3	.01553	.01809	.02541	.02604	.02962
0.4	.01842	.02173	.03083	.03186	.03611
0.5	.02381	.02859	.04137	.04331	.04884
0.6	.03444	.04236	.06373	.06823	.07646
0.7	.05833	.07419	.12069	.13471	.14945
0.8	.12542	.16736	.31428	.37717	.40777
0.9	.38740	.53044	.93443	.99734	.99892
1.0	.99998	.96075	.62565	.50636	.52092
1.1	.54809	.49293	.31558	.24606	.27786
1.2	.29044	.28224	.20531	.17811	.18909
1.3	.19501	.19939	.15946	.14215	.15158
1.4	.15295	.16172	.13903	.12676	.13550
1.5	.13297	.14416	.13175	.12262	.13130
1.6	.12460	.13788	.13344	.12678	.13589
1.7	.12393	.13972	.14320	.13906	.14915
1.8	.12980	.14898	.16222	.16142	.17313
1.9	.14256	.16660	.19385	.19837	.21256
2.0	.16400	.19526	.24460	.25858	.27638
2.1	.19772	.23996	.32605	.35757	.38005
2.2	.25019	.30931	.45656	.51868	.54520
2.3	.33244	.41689	.65372	.75248	.77579
2.4	.46121	.57857	.88625	.96794	.97487
2.5	.65170	.79031	.57857	.99994	.95861

* Using Pseudopotential model -I of author [*J. Phys. Chem. Solids*, **36** (1975) 993] and dielectric screening of Singwi et al. [K.S. Singwi, M.P. Tosi, L.H. Land and A. Sjolander, *Phys. Rev.* **176** (1968) 589 ; *ibid* **B1** (1970) 1044].

$q/2k_F$	H^-	O^{--}	N^{---}	B^{+++}	P^{++++}
0.1	.00140	.00124	.00115	.00217	.00286
0.2	.00160	.00147	.00133	.00243	.00312
0.3	.00202	.00182	.00170	.00298	.00365
0.4	.00281	.00255	.00241	.00401	.00460
0.5	.00428	.00391	.00371	.00592	.00621
0.6	.00719	.00662	.00630	.00973	.00900
0.7	.01395	.01590	.01232	.01848	.01409
0.8	.03454	.03212	.03075	.04479	.02454
0.9	.14351	.13495	.12999	.17778	.05024
1.0	.89995	.89996	.89997	.89980	.14061
1.1	.25194	.23911	.23166	.30255	.51435
1.2	.10173	.09571	.09226	.12637	.98975
1.3	.06227	.05849	.05632	.07789	.53777
1.4	.04668	.08382	.04218	.05848	.28854
1.5	.03951	.03708	.03570	.04950	.18932
1.6	.03640	.03418	.03290	.04559	.14294
1.7	.03586	.03368	.03242	.04486	.11868
1.8	.03743	.03516	.03386	.04675	.10557
1.9	.04124	.03876	.03733	.05141	.09909
2.0	.04801	.04514	.04349	.05966	.09727
2.1	.05924	.05517	.05375	.07333	.09932
2.2	.07817	.07368	.07108	.09621	.10518
2.3	.11187	.10570	.10212	.13634	.11538
2.4	.11740	.16835	.16310	.21241	.13114
2.5	.31943	.30614	.29806	.36907	.15470

These values have been obtained by using pseudopotential model - I of author and new dielectric screening of author [*J. Phys. Chem. Solids*, **36** (1975) 993] phonon frequencies in solid phase were determined through well known lattice dynamical procedure.

Bratby et al.[173] evaluated the heat capacity of liquid metals by using the inequality of the long wave limit in the behavior at melting point that both the specific heats C_P and C_V are almost equal. Therefore, they evaluated C_V through a pair potential procedure as described by

$$C_V = \frac{\partial E(\rho)}{\partial T} \tag{222}$$

where $\quad E(\rho) = \frac{3}{2} k_B T_m + E_0(\rho) + \frac{\rho}{2} \int V(r,\rho) g(r,\rho) d^3r \tag{223}$

here $V(r,\rho)$ and $g(r,\rho)$ are effective pair potential between ions separated at a distance r and equilibrium pair correlation function, respectively. Their method is quite different than the procedure of Elsenschitz - Wilford[174] and Tooms[175], which is basically based on the collective movement approach in which the model Hamiltonian of the system consists of harmonic part as well as perturbation part, which arises due to different coupling.

(b) Gruneisen parameter

A crude analysis about Gruneisen parameter, γ_G may be obtained through relation

$$\frac{\partial \ln K_i}{\partial \ln V} = \frac{\partial \ln \rho}{\partial \ln V} - 2\gamma_G = -\frac{2}{3}\zeta_Q = -2 + \frac{4}{3}q \tag{224}$$

where

$$q = \frac{S(2k_F) U^2(2k_F)}{\frac{1}{4k_F^4} \int_0^{2k_F} S(q) U^2(q) q^3 dq} \tag{225}$$

here ζ_Q is thermoelectric power parameter of liquid metals, K_i is electron-lattice interaction parameter. In case of monovalent metals it has been observed[176] that

$$\frac{\partial \ln K_i}{\partial \ln V}\bigg/\zeta_A = \frac{\partial \ln E_F}{\partial \ln V} = -\frac{2}{3} \tag{226}$$

By finding ζ_Q in pseudopotential approach for one parameter models using Kerb's lattice dynamical model the Gruneisen parameter γ_G of alkali metals have been computed. For $\partial \ln \rho / \partial \ln V$ the reported experimental values[178] have been used. The obtained values of γ_G have been depicted in Table 3 together with measured values. A good agreement is found.

The Gruneisen parameter, γ_G is defined by

$$\gamma_G = -\frac{d \ln \omega_{q,p}}{d \ln \Omega} \tag{227}$$

Table - 3. Gruneisen Parameter, γ_G.

Metal	Temp (°K)	γ_G	
		Author Model	Ashcroft Model
Li	50	-30.40	-17.20
	75	-45.90	-26.90
	100	-36.30	-21.70
	300	-23.40	-15.00
Na	50	-11.70	-28.70
	75	+00.75	-02.00
	100	+02.50	+01.80
	300	+03.00	+01.65
K	50	+03.05	+02.70
	75	+03.15	+02.60
	100	+03.00	+02.75
	300	+03.10	+02.90
Rb	50	+03.10	+02.70
	75	+03.05	+02.65
	100	+03.15	+02.90
	300	+03.10	+02.75
Cs	50	+02.25	+02.20
	75	+02.20	+02.10
	100	+02.20	+02.15
	300	+02.35	+02.30

which is related to other thermodynamic coefficients such as thermal expansion coefficient, α_p and isothermal compressibility, β_L by the relation

$$\frac{\alpha_P \Omega}{3 \beta_L k_B} = \frac{1}{3} V \sum \gamma_{q,p} E(x) ; \quad x = \frac{h \omega_{q,p}}{k_B T} \quad (228)$$

where
$$E(x) = y(y+1) ; \quad y = [e^x - 1]^{-1} \quad (229)$$

At high temperature, an idea about macroscopic Gruneisen constant may be achieved by employing solid phonon frequencies $\omega_{q,p}$ in above equations.

Wallace[179] procedure of determining $\gamma_{q,p}$ along symmetry directions is based on the average over k of the trace of the dynamical matrix

$$3M<\omega^2> = \alpha\gamma \sum_n \exp(-\gamma.r_n)[\gamma - \frac{2}{r_n}] + \sum_q q^2.F(q).$$
$$\left[1 - \sum_n \exp(-iq.r_n)\right] \quad (230)$$

where α and β are the parameters of the Born-Mayer repulsive energy $(U_k = \frac{1}{2}\sum_{n=n'} nn'\ \alpha\ \exp(\gamma|r - r_n|))$ and Hellmann-Feymann theorem.[180)^{n=n'}]

(c) Thermal Expansion

First of all thermal expansion coefficient α_p was determined by Wallace using pseudopotential perturbation theory and lattice dynamics. It is related to Helmholtz free energy function F by the equation

$$\beta_L\gamma_V = -\beta_L(\frac{\partial^2 F}{\partial T \partial \Omega})_{\Omega,T} = \beta_L(\frac{\partial P}{\partial T})_V \quad (231)$$

Here β_L is compressibility, γ_V is the thermal pressure coefficient and function F is given by
$$F = (G - PV) = U - TS \quad (232)$$
where G is Gibb's free energy function, S is entropy and U is internal energy which is given by
$$U = E + E_M \quad (233)$$

The crystal energy E is determined by eq.(155). The Modelung energy, E_M (ion-contribution) is given by

$$E_M = \frac{1}{2}\sum_{q=0} \frac{8\pi Z^2}{q^2} [S(q) - 1]\ ryd. \quad (234)$$

For a hard sphere model
$$E_M = [-1.705 + (\eta - 0.41)]Z^{5/3}\ r_s\ (ryd/ion) \quad (235)$$
here η is packing density function. Thus, it is clear that function, F or function, G for solid or liquid metals depend on band structure energy, E_{bs}, Madelung energy, E_M and entropy, S.

The calculated values of α_p of Wallace are somewhat high at intermediate temperature. At high temperature region the results are satisfactory. Using hard sphere model analysis Ross and Greenwood[181] and Shimoji[182] obtained not very satisfactory results of non-transition metals.

The most satisfactory effort in this direction is of Hasegawa and Watabe[183], whose theory is based on electron theory of metals. On the basis of uniform compression approximation these authors firstly derived the isothermal compressibility β_L and then used internal energy (U) and pressure (P) relationship. Their expression is described by[184]

$$\gamma_V = n_i k_B + \frac{n_i}{2(2\pi)^3}\int dq\ [U_2(q,n_e) + \frac{q}{3}\frac{\partial}{\partial q}U_2(q,n_e)]$$

$$+ n_e \frac{\partial}{\partial n_e} U_2(q, n_e)][\frac{\partial}{\partial T} S(q)] V \qquad (236)$$

where n_i and n_e are ionic and electron number densities, respectively. For the evaluation of temperature coefficient of structure factor S(q) they used the theory of Anderson et al.[185]. Thus by knowing β_L and γ_V the coefficient α_p was evaluated, satisfactory. More satisfactory result may be obtained by including the effect of the many body forces, which would be more fruitful for polyvalent metals.

It has been observed that a dimensionless parameter V_s (velocity of sound) for liquid Hg and H$_2$O increases with compression, contrary to the behavior of γ_G [$\gamma_G(V,T) = K_a \alpha_p V/C_p$] K_a is adiabatic bulk modulus for solids. On integration, we get for solids

$$P(V,T) = P_0(V,0) + \gamma_G(V)[U(V,T)/V] \qquad (237)$$

(d) Self-Diffusion Coefficient

The self-diffusion coefficient is generally measured by three techniques, (i) nuclear magnetic resonance (NMR), (ii) inelastic neutron scattering, (iii) trace analysis. These methods are very well reviewed by Nachtrieb[186]. Theoretically, important approaches are (i) model approach based on some transport coefficient, (ii) molecular dynamics, (iii) correlative - based on corresponding state principle, (iv) dynamical theories of solids. It requires the analytic solution of the equations of motion in terms of distribution functions.

The self-diffusion coefficient D_s is related to friction coefficient ζ_s by the relation[187,188]

$$D_s = \frac{k_B T}{\zeta_s} \qquad (238)$$

where

$$\zeta_s = \frac{8}{3} \rho g(\sigma) \sigma^2 (\pi m k_B T)^{1/2} \qquad (239)$$

Here $g(\sigma)$ is pair distribution function. For soft collision[189]

$$\zeta_s = -\frac{\rho}{12\pi^2} (\pi m k_B T)^{1/2} \int_0^\infty q^3 V_{bs}(q) S(q) dq \qquad (240)$$

where $V_{bs}(q)$ [$V_{bs}(q) = \{4\pi Z^2 e^2/\Omega\ q^2\} G(q)$] is Fourier transform of the long range part of the pair potential. It is observed that ζ_s derived for polyvalent metals are most satisfactory as compared to alkali metals[190]. As the hard sphere model approach is concerned the finding of Swalin[191] is better than that of Cohen and Turnbull[192] although former method is only applicable to systems of constant volume and does not take into account of correlation. The results of Ascarelli and Paskin[193] derived through combination of hard sphere model, molecular dynamics approach and of Helfand and Rice[194] related to correlative approach are also encourageable.

A very good interesting approach has been followed by Hicter and his coworkers[195] based on a thermodynamic model of liquid metals. Their expression is given by

$$D_s = (D_s)_v + (D_s)_{tr}$$

$$= \frac{8\pi^2 m k_B \partial^2 [T - (J)_{kr} + Th^2(J)_{kr}]}{8\pi^2 mh [(J)_{kr} + (J)_{tr}]} \qquad (241)$$

where the vibrational J_{kr} and translational J_{tr} Partition functions are given by

$$J_{kr} = \frac{T}{\theta}[1 - e^{\frac{-E}{k_B T}}] \qquad (242)$$

$$J_{tr} = \frac{(2\pi m k_B T)^{1/2}}{h} \delta e^{-E/k_B T} \qquad (243)$$

Here δ ($\delta = V^{1/3}$) is translational amplitude, θ is Einstein temperature and volume dependence as this coefficient may be easily studied.

(e) Compressibility

The well known Ornstein and Zernike relation for the compressibility of liquid metals at melting point is given by[9]

$$\beta_L = \frac{\Omega S_L(0)}{k_B T_m} = \frac{3\Omega}{2 Z E_F} \qquad (244)$$

where $\quad S_L(0) = [1 - C(q)]^{-1}{}_{q \to 0} = 1 + \rho \int [g(0) - 1] d^3 r \qquad (245)$

Here β_L is isothermal compressibility and $\Omega = V/N$ ρ^{-1}. Also for isotropic solid(s) the adiabatic compressibility, β_s is described by

$$\beta_s = \frac{3\Omega}{(k_B T_m)(1 + \sigma)(1 - \sigma)} S_{is}(0) \qquad (246)$$

where σ is Poisson's ratio and $S_{is}(0)$ is for q lying in the first Brillouin zone(for q=0). In terms of liquid phonon frequency $\omega(q)$ an information about β_L may be obtained from

$$\beta_L = \frac{\Omega}{[V(0) + k_B T_m]} = [\frac{\Omega q^2}{n\omega^2(q)}]_{q \to 0} \qquad (247)$$

where $V(0)$ is long wave length limit of pair potential, V(q). From hard sphere model theory

$$\beta_L = \frac{\Omega}{k_B T_m} [\frac{(1-\eta)^4}{(1+2\eta)^2}] \qquad (248)$$

By including structure independent contribution β_L in above eq. as considered by Ascarelli also, Shimoji[182] obtained satisfactory results of for polyvalent metals. The very satisfactory result of β_L in case of alkali metals were obtained by Watabe and coworkers[183].

The temperature coefficient of compressibility, $\partial \beta_s / \partial T$ may be studied by the following relation

$$\frac{\partial \beta_s}{\partial T} = [1 - (\frac{\beta_s}{\beta_s(int)})] (\frac{\beta_s}{T}) \qquad (249)$$

where $\quad \beta_s(int) = (\frac{\Omega}{N})^2 [W(0) + \frac{\Omega}{4\pi^2 N} \int_0^\infty [\frac{2}{15} q \frac{d}{dq} S(q)$

$$+ \frac{1}{5} q^2 \frac{\partial^2}{\partial q^2} S(q)] W(q) q^2 dq] \qquad (250)$$

and
$$W(q) = \frac{4\pi Z^2 e^2}{\Omega q^2} - \frac{k_B T \gamma}{\Omega}[\varepsilon(q)-1][\frac{1}{S(q)} - 1] \qquad (251)$$

In terms of thermal expansion coefficient α_p, Egelstaff described the compressibility ratio[196]

$$\frac{\beta_L}{\beta_s} = \frac{S_L(0)}{S_{is}(0)} \frac{(1+\sigma)}{3(1-\sigma)} \sim \frac{1+(\frac{T_m \alpha_p V}{E})}{1+(\frac{PV}{E})} \qquad (252)$$

which may be used in the study of melting phenomenon.

It is remarkable that at melting point there is change of volume in case of metals. Although metals expand less than 5% whereas it is more than 15% in case of Ar. We observe at melting point

$$\frac{1}{\beta} - \frac{1}{\beta_0} = \Omega (\frac{\partial^2 E}{\partial \Omega^2}) T \qquad (253)$$

$$(\frac{\beta}{\beta_0})_L \cdot (\frac{\beta}{\beta_0})_s = (\frac{\Delta\Omega}{\Delta\Omega_0})^2 \qquad (254)$$

This thermodynamic coefficient is useful in testing the validity of Lindmann's melting law and in other melting properties.

The adiabatic compressibility, β_s ($\beta_s = \frac{1}{\rho V_s^2}$) of a binary alloy may be described by[197]

$$(\beta_s)_{ideal} = \frac{C_g V_g (\beta_s)_g + C_h V_h (\beta_s)_h}{C_g V_g + C_h V_h} \qquad (255)$$

where V_g and V_h are molar volumes of guest (g) and host(h) elements, respectively. By finding measured values of sound velocity, V_s of host and guest elements at different densities, above formula may be used for compressibility evaluation of an alloy system. McAlister et. al. applied above analysis in case of liquid alloy Mg-Bi; they found that their measured values deviate from the ideal curve which varies linearly with concentration. They observed a distinct maximum at the 60 at. %Mg - composition which appears due to formation of intermediate compound $Mg_3 Bi_2$ and there exists some local order in the system. In another study McAlister[198] from his electromagnetic detection technique measured V_s in case of several Na- based systems and on Na-Mg in particular. When his data were fitted in a theory suggested by Faber a large deviation appears due to formation of intermediate compound $Na Mg_2$ in Na-Mg alloy system. A similar deviation was also observed in case of Na-K system reported by Abowitz and Gordon[199]. Localized states much affect in such studies. A good effort is needed in this direction.

(f) Entropy

Some of secrets about structure of a material lie inside its entropy, S. The entropy of a perfect gas (g) is given by[200]

$$S_g = \int \frac{dQ}{T} = \int \frac{dU + PdV}{T} = Nk_B \log[\frac{\Omega}{N}(\frac{2\pi e m_A k_B T}{h^2})^{3/2}] \qquad (256)$$

where Q is quantity of heat developed and U is internal energy. During melting
$$dQ = L = \Delta U + P \Delta V = T_m \Delta S \tag{257}$$
Here Δ denotes the difference between liquid and solid quantities. The evaluation of U and P has been already discussed in earlier section. For calculating the change in volume, ΔV during melting one may use eq.(254). Hartman[201] found poor results in many metals especially in Li and polyvalent metals. It may be because of neglecting the quantity $P \Delta V$.

A quantity ($S - S_g$) may be termed as the excess entropy which is of course, a negative quantity since molecules are more ordered in the liquid phase than for a perfect gas. For metals generally, ($S - S_g$) is of value in between -3 and -5 except for semiconductors Si and Ge for which value is -2.5 and -1.0, respectively. Also for bcc metals is 0.85 k_B and for close-packed structural metals is around 1.15 k_B. For semi-metals and semiconductor the value is quite large in between 2.5 k_B to 3.5 k_B. Thus it is clear that there is no close relationship between ($S - S_g$) and S. Actually when these substances melt they loose their structural behavior and follow a common pattern. On the basis of Mott criteria ΔS is described by[202]

$$\Delta S = \frac{3}{2} N k_B \log \left(\frac{\rho_L}{\rho_S} \right) \tag{258}$$

which holds good in most of the metals. Here ρ_L and ρ_S are the electrical resistivities in liquid and solid phases, respectively.

The entropy of an assembly of particles is described by[94]

$$S = -k_B \int [f_k \ln f_k + (1 - f_k) \ln(1 - f_k) dk] \tag{259}$$

where f_k is distribution function. One may consider Fermi distribution function, f_k

$$f_k = \frac{1}{e^{\beta(E - U_{ch})} + 1} \tag{260}$$

where the chemical potential, U_{ch} for electron is given by
$$U_{ch} = (-0.125 + 0.031 \ln r_s) \, ryd. \tag{261}$$
Total energy E may be determined as earlier. Thus, S may be evaluated.

From Partition function, J [eqs. (242) and (243)] also entropy, S may be developed through thermodynamical potential, φ

$$\varphi = \int_0^\infty J \left(\frac{\partial f_k}{\partial E} \right) dE \tag{262}$$

which may be described by

$$S = 3 k_B \ln[(J)_{vr} + (J)_{tr}] + \frac{3E}{\theta(J_{vr} + J_{tr})}$$
$$[1 + \left(\frac{J_{vr}}{J_{vr} + J_{tr}} \right)] + \frac{3E}{T} - \frac{3E}{\theta(J_{vr} + J_{tr})} \tag{263}$$

By calculating J function the entropy, S may be determined.

A standard formula of entropy in liquid phase on the basis of statistical thermodynamics is described by

$$S_L = (\frac{L}{2\pi})^3 4\pi \int_0^\infty q^2 \, dq \, [\frac{1}{T_m} \hbar\omega(q) \, [\frac{1}{2} + [e^{\frac{\hbar\omega}{k_B T_m}} - 1]^{-1}]$$
$$- k_B \ln [2Sh [\frac{\hbar\omega(q)}{2k_B T_m}]]] \tag{264}$$

where
$$(\frac{L}{2\pi})^3 \frac{4}{3} \pi Q^3 = 3N \tag{265}$$

or
$$Q = (18\pi^2 \frac{\rho}{m})^{1/3}; [\rho = \frac{mN}{L^3}] \tag{266}$$

Here $(\frac{L}{2\pi})^3$ is number density. Thus, we may write[203)]

$$S_L = 9Nk_B (\frac{1}{2} I_0 + I_1 + I_2) \tag{267}$$

where
$$I_0 = \int_0^1 y^2 \frac{\hbar\omega(q)}{k_B T} dy$$
$$I_1 = \int_0^1 y^2 [\frac{\hbar\omega(q)}{k_B T_m}] \, [e^{\hbar\omega(y)/k_B T_m} - 1]^{-1} dy$$
$$I_2 = -\int_0^1 y^2 \ln [2Sh \frac{\hbar\omega(y)}{2k_B T_m}] dy \tag{268}$$

The molar entropy change ΔS at melting is given by
$$\frac{\Delta S}{R} = \frac{S_L - S_s}{R} = 1.5 \ln (\frac{\rho_L}{\rho_S}) \tag{269}$$

where R ($R = N/k_B$) is a gas constant and S_s is the molar entropy of solid at melting temperature. By using Ziman values of ρ_L/ρ_S, $\Delta S/R$ for most of the metals has been found satisfactory. Except in case of Na, Rb and Ga in which about 25% to 30% discrepancies are there, other metals show good results. Mott criteria show discrepancies in between 10% to 20% in case of alkali metals. Especially for Na and Pb it lies in between 10% to 12%. The values of ΔS for alkali metals have been evaluated by using liquid phonon frequencies $\omega(q)$ from eqs.(267) - (269). The obtained values are depicted in Table - 4 together with experimental and theoretical values as reported by Omni[203)]. In general present values are largely satisfactory. The discrepancies appear due to approximations involved in theory like that of Percus-Yevick structure factor and correlation between S(q) and pair potentials.

The above described theories are extendible to the case of binary alloys. The entropy formula for a binary alloy system may be described by

$$\frac{S}{R} = -(C_g \log C_g + C_h \log C_h) + C_g S_g (\Omega_g) \, C_h S_h (\Omega_h)$$
$$- C_g C_h (\Omega_h - \Omega_g) (\frac{\alpha_h}{\beta_1} - \frac{\alpha_g}{\beta_0}) \tag{270}$$

where α_h and α_g are the volume expansion coefficients of host and guest elements at constant pressure.

Table - 4. $\Delta S/R$, Change in Entropy at Melting Point (per atom, in units k_B) for alkali metals

Metals	$[\frac{\Delta S}{R}]\text{exp}$	Using eq. (268)	
		For exp. S (q) Present value	For theo. S (q) Omini values [203]
Li	0.80	0.87	0.82
Na	0.84	0.73	0.62
K	0.82	0.85	0.80
Rb	0.83	0.78	0.63
Cs	0.80	0.74	0.70

The last term in above equation represents a non-ideal entropy of mixing which could have either sign. It is remarkable that one should always consider the influence of entropy on the equilibrium configuration in such calculations. Always total energy should be minimized rather than free energy.

(g) Viscosity

The viscosity is a structure sensitive property similar to compressibility and widely useful in melting phenomenon.

With use of a pair distribution function, g(r) the coefficient of viscosity, η_v is given by

$$\eta_v \sim \frac{2}{15} (m_a/k_B T)^{1/2} \pi \rho (0) \int_0^\infty g(r) [\frac{d}{dr} V(r)] r^4 dr \qquad (271)$$

where m_a is the mass of the atom, $\rho(o)$ is average number density and V (r) is interionic potential. But poor results are observed. Therefore, Shimoji[182] modified into

$$\eta_v \sim \frac{16}{15} (\frac{m_a}{k_B T})^{1/2} F_T \qquad (272)$$

where surface tension F_T may be determined through Born- Green and Percus-Yevick equations. The satisfactory results were obtained. It has been also observed[204] that Lenord-Jones potential through simulation devices (molecular dynamics) presents also satisfactory results of η_v. In high frequency modes region the spectral density of particle velocity also evaluates good results of viscosity.

A relation between η_v and diffusion coefficient D may described by[205]

$$\frac{\eta^* V^*}{D^*} = \frac{V}{D} \qquad (273)$$

where $\eta^* = (\frac{\eta}{mE})^{1/2}$, $D^* = (\frac{m}{E})^{1/2}$ and $V^* = \frac{V}{Nd^3}$ (274)

Here η^* and D^* are the reduced (dimensionless) coefficient for shear viscosity and self-diffusion, respectively. V^* is reduced volume. η^* is kinematic viscosity and d is a distance parameter.

The collective excitations in liquids may also be utilized for understanding of viscosity through auto correlation functions and frequency moments of the spectral functions[206,207]. Alder et.al.[204] described relationship between η_v and D as

$$\eta_v D \sigma k_B T_m = 0.168 \quad (275)$$

which after using the value of D becomes

$$\eta_v = C(MRT)^{1/2} V^{-2/3} \quad (276)$$

where

$$C = 0.92 \left(\frac{\mu}{\mu m}\right) \frac{1}{N\sigma^2} (\pi)^{1/2} V^{2/3} [9.385 (\frac{\rho}{T_m}) - 1] \quad (277)$$

Here N is Avogadro's number, M is molecular weight. ρ is density, σ is hard sphere diameter and μ is packing density.

(h) Surface Tension

This is also structure sensitive phenomena like viscosity. It mainly depends on surface energy. A number of authors have developed statistical analysis of this coefficient. F^T in terms of pair potential and pair distribution functions and observed satisfactory results of liquid metals. The energy band structure at the surface may be useful in evaluating this coefficient as it has been found that excess energy which exist on the surface may be considered as sum of (i) electronic energy (associated to electron density variation), (ii) electrostatic energy (associated to dipole layer), (iii) interaction energy (associated to electron-ion interaction), (iv) cleavage energy (associated to ion-ion interaction). The coefficient of viscosity, may also give an idea about this coefficient through eq.(271) and we may write

$$F^T \sim \frac{\pi}{8} \rho(0) \int_0^\infty g(r) [\frac{d}{dr} v(r)] r^4 dr \quad (278)$$

where symbols have their usual meaning. Using statistical procedure of Kirkwood and Buff[208] from following thermodynamical relations

$$F^T = T \left(\frac{\partial F_T}{\partial T}\right) - U^S$$

$$F^T = \frac{\partial F}{\partial A} = \left(\frac{k_B T}{J}\right) [\frac{\partial J}{\partial A}] \quad (279)$$

where J is Partition function and U^S is surface energy, Evans[209] calculated the surface-tension F^T of a number of liquid metals. His observation shows that $\partial F_T/\partial T$ is a negative quantity above melting point, which indicates that the surface region is in some sense more disordered than bulk state due to which there lies excess entropy on this surface which is against the observation of White[210]. Such controversy requires justification.

REFERENCES

1. A.J. Walton, *Three Phase of Matter* (ELBS/Oxford University press, Oxford, 1976).
2. B.J. Austin, V. Heine and L.J. Sham, *Phys. Rev.* **127**, (1962) 276.
3. W.A. Harrison, *Pseudopotential theory of metals*(Benjamin Inc., New York, 1966); *Solid State Theory*(Tata McGraw Hill Pub.Co.Ltd.,New Delhi, 1970).
4. T.E. Faber, *Introduction to the theory of liquid metals*-(Cambridge University Press, 1972).
5. S.K. Srivastava, *Pseudopotential in Physics and Chemistry of Solids* (T.P.I. Printers, Allahabad, India, 1977); D.Sc. Thesis, *The Investigation of Properties of Solids*,(University of Allahabad, Allahabad, India, 1975).
6. (a) Proceeding of First International Conference on properties of Liquid Metals, Brookhaven, USA; *Adv.Phys.* **16**, (1966) 147 - 744.
   (b) Proceedings of Second International Conference *Theory of Liquid Metals*(Taylor & Francis, London,1972)ed.,S. Takeuchi.
   (c) Proceedings of Third International Conference (Inst. Physics, Bristol, 1973), ed., R. Evans and D.A.Greenwood.
   (d) Proceedings of Fourth International Conference, *Liquid and Amorphous Metals*,(Greenoble, France, 1980).
7. L.I. Yastriebov and A.A. Katsnelson, *Foudations of one electron theory of solids* (Mir. Publishers, Moscow, 1987).
8. M.D. Johnson and N.H. March, *Phys. Lett.*, **3** (1963) 313; M.D. Johnson, A.P. Hutchenson and N.H. March, *Proc. Roy. Soc.*, **A282** (1964) 283.
9. L.S. Ornstein and F. Zernike, *Proc. Sect. Sci. K. Ned. Akad. Wet.*, **17** (1914)793 (Reprinted in Ftisch and Lebowitz, Opt. (it.).
10. P.C. Gehlen and J.E. Enderby, *J. Chem. Phys.*, **51** (1969) 547.
11. R. Kumaravadivel, R. Evans,and D.A..Greenwood, *J. Phys., F.***4** (1974)1839
12. M. Tanaka, Ref. No. 6 c p. 164.
13. N.H. March, *Can. J. Phys.*, **65** (1987) 219.
14. J.E. Enderby, S. Cuminmgs, G.J. Herdman, G.W.Neilson, P.S. Salmon and N. Skipper, *J. Phys. Chemistry* **91** (1987) 5851, J.E.Enderby , *Phil. Mag.*,**A58** (1988) 5.
15. W. Vader Lugt and G. Geertsnam, *Can. J.Phy.*,**65** (1987) 326.
16. M.L. Saboungi, R. Blomguist, K.J. Volin and D.L. Price, *J.Chem. Phys.*, **87** (1987) 2273.
17. A.C. Barnes and J.E. Enderby, *Z. Phys. Chem. Neue Folge,***156** (1988) 529; *VI International Conf. Liquid and Amorphous Metals - Lam6* (Garmisch, Partenkirchem, West Germany,1985).
18. K.S. Singwi, K. Sjolander and M.P. Tosi, *Phys. Rev.*, **A1** (1970) 2427; P. Vashistha and K.S.Singwi *Phys. Rev.*,**B6** (1972) 249.
19. L.Van Hove, *Phys. Rev.*,**95** (1954) 249.
20. G.L. Gyorffy and N.H. March, *Phys. and Chem. Liquid*, **2** (1971)197.
21. G.H. Vineyard, *Phys. Rev.*, **110** (1958) 999.

22. M.I. Barker, M.W. Johnson, N.H. March and D.I.Page, Ref. No. 6 b. (1972) 99.
23. W. Glasser, S. Hagen, U. Loffer, J.B. Suck, and W. Schommers, Ref. No.6 b (1972) 111.
24. V. Heine, *Solid State Physics*, **24,** ed. Ehendreich. Seitz and urnbull (Academic press, New York, 1970) p.1. M.H. Cohen, and V. Heine, *ibid.* **24** (1970) 37, 247.
25. W. Kohn and L.J. Sham, *Phys. Rev.,* **A140** (1965)1133.
26. N.W. Ashcroft, *Phys. Letters* **23** (1966) 480; *J.Phys.C.* **1** (1968) 238 ; N.W.Ashcroft and D.C.Langreth, *Phys. Rev.,* **155** (1967) 682.
27. S.K. Srivastava, *Phys. Stat. Sol.* 61 b 1974 (731); *ibid* 64 b (1964) 679, *J. Phys. and Chem. Solids* **36** (1975) 993.
28. M.H. Cohen, *J. Phys. Radium* **23** (1962) 643.
29. W.A. Harrison, *Phys. Rev.* **131** (1963) 2453; *ibid* 136 (1964) A1167; *ibid* **139** (1965) A1167.
30. V. Heine and I. Abarenkov, *Phil.Mag.* **9** (1964) 451; I. Abarenkov and V. Heine, *Phil.Mag.* **12** (1965) 529 ; A.O.E. Animalu and V.Heine, *Phil. Mag* **12** (1965) 1249; A.O.E. Animalu, *Phil.Mag.* **11** (1965) 379.
31. R.W. Shaw, and W.A. Harrison, *Phys. Rev.,* **163** (1967) 6041; R..W. Shaw, *Phys. Rev.* **174** (1968) 769.
32. T.Schneider and E.Stoll, *Phys. Kondens Mat.* **5** (1966) 331,364.
33. A.J. Greenfield and N. Wiser, *J. Phys. F.* **5** (1975) 289.
34. J.M. Ziman, *Phil. Mag.* **6** (1961) 1013, *Adv Phys* **16** (1967) 551.
35. S.K. Srivastava, *Phys. Stat. Sol.* 66 b (1974) 93.
36. M.H. Cohen and V. Heine, *Phys. Rev.* **122** (1961)1821.
37. L.J. Sham and J.M. Ziman, *Solid State Phys.,***15**,ed F. Seitz and D. Turnbull (Academic press, Inc. New York,1963) p.221.
38. S.K. Srivastava and P.k. Sharma, *Physica* **54** (1971) 29.
39. H. Brooks and F.S. Ham, *Phys. Rev.* **122** (1958) 3443. F.S. Ham, *Solid State Phys.,***1**,ed. F.Seitz and D.Turnbull (Academic press, Inc.New York,1964) p.127.
40. S.K. Srivastava and P.K. Sharma, *Sol. State. Comm.* **8** (1970) 703.
41. R. Evans, G.D. Gaspari and B.L. Gyorffy,*J.Phys. F.***3** (1973) 39.
42. R. Evans,V.K. Ratti and B.L. Gyorffy, *J. Phys. F.* **3L** (1973) 199.
43. J.L. Beeby, *Proc. Roy. Soc.* **B 279** (1964) 82.
44. J.S. Rousseau,J.C. Stoddart and N.H. March, Ref.No. 6 b p.249.
45. De Dycker and P. Phariseau, *Adv. Phys.* **16** (1962) 401.
46. M. Lax ,*Phys. Rev.* **85** (1952) 621.
47. J.M. Dickey, A. Mayer and W.H. Young, *Phys. Rev. Lett.* **16** (1966) 727 ; W.H. Young, A. Mayer,and G.L. Kilby, *Phys. Rev.* **160** (1967) 482 ; A. Mayer, W.H. Young and M. Dickey,*J. Phys.*C.**1** (1968) 486 ; A.Mayor, W.H. Young and Deivllers,*J.Phys.* **F4** (1974) 394.
48. M.J.G. Lec, *Phys. Rev* **178** (1969) 953, *ibid* , **B4** (1971) 673.
49. N.W. Ashcroft and W. Schaich, *Phys. Rev.* **B1**(1971) 370.
50. J.M. Ziman, *Solid State Phys.***26,** ed. H. Ehrenreich, F. Seitz and D.Turnbull (Academic Press, Inc., New York, 1971),p1.

51. L. Schwartz and H. Ehrenrich, *Ann. Phys.* **64** (1971) 100.
52. S.N. Khanna and A. Jain, *J. Phys. F.***4** (1974) 1982.
53. B.J. Berne and D. Forster, *Ann. Phys. Chem.* **22** (1971) 563.
54. A. Rahman, *Phys. Rev.* **136** (1964) A405; *Phys. Rev. Lett.* **19**(1971) 420.
55. N.Metropolis, A.W. Rosenbluth, M.N. Rosenbluth, A.H. Teller and F. Teller, *J. Chem. Phys.* **21** (1953)1087.
56. W.W. Wood, *Physics of Simple Liquids*, ed. H.N.V. Temperley, J.S. Rowlinson and G.S. Rousbrooks, (Amesterdam, North-Holland, 1968) p.115.
57. F.H. Ree, *Physical Chemistry*, An Adv. Treat. Vol. VIIIA (Academic press, Inc., New York, 1971) p.157.
58. A. Paskin and Rahman, *Phys. Rev.* **16** (1966) 30; A. Rahman, *Phys. Lett.* **32** (1974) 53, *Phys. Rev.* **A1**(1974)1667; Statistical Mechanics(University of Chicago press, Chicago, 1972).
59. D.L. Price, K.S. Singwi and M.P. Tosi, *Phys. Rev.***B2**(1970) 2983.
60. N.E. Cusack, Ref. No. 6 b (1972) 157.
61. A.J. Greenfield, J. Wallendorf, and N. Wiser, *Phys. Rev.* **A4** (1971)1607.
62. J.M. Ziman, *Adv. Phys.* **13** (1965) 89, *Proc. Phys. Soc.***86** (1965) 337.
63. L.M. Falicov and V. Heine, *Adv. Phys.* **10** (1961) 57.
64. S.F. Edwards, *Phil.Mag.* **3** (1958)1020; *Proc. Roy. Soc.* **A267** (1962) 518.
65. J.E. Enderby, J.M. Titman and G.D. Wignall, *Phill. Mag.* **10** (1964) 63.
66. A. Paskin, A., Ref. No. 6 a p. 223 ; A. Paskin, R.J. Harrison, . and P. Ascarelli, p.263; P. Ascarelli, R.J. Harrison, and A. Paskin, p.717.
67. L.J. Sundstorm, *Phil.Mag.* **11** (1965) 657.
68. N.W. Ashcroft and J. Lekner, *Phys. Rev.* **145** (1966) 83.
69. R. Evans, D.A. Greenwood, P.L. Loyd and J.M. Ziman, *Phys. Lett.* **304** (1969) 313, R. Evans, *J. Phys. C.***2** (1970) 137.
70. Y. Waseda and K. Suzuki, *Phys. Stat. Sol.* **40** (1970) 183.
71. J.B. Venzytfeld, J.E. Enderby and E.W. Collings, *J. Phys.F.***2** (1972) 73.
72. P.D. Adams and N.W. Ashcroft, Ref. No. 6 a p.597.
73. A.J. Greenfield and N. Wiser, Ref. No. 6a p.591,601.
74. T.E. Faber, *Adv. Phys.* **15** (1966) 547.
75. N.W. Ashcroft and W. Schaich, *Phys. Rev.* **B1** (1970) 1370.
76. A.B. Bhatia and K.S. Krishnan, *Proc. Roy. Soc.* **A194** (1948) 185.
77. H. Gertenkorn, *Ann. Phys.* **10** (1952) 49.
78. A. Ferraz and N.H. March, *Phys. Chem. Liq.* **8** (1979) 271.
79. S.K. Srivastava and D.R. Singh, *Ind. J. Pure & Appl. Phys.* **25** (1987) 292.
80. C.C. Bradley, T.E. Faber, E.G. Wilson and J.M. Ziman, *Phil.Mag.***7** (1962) 865.
81. V. Bortolani and C. Calandra, *Nouv. Cim* **58 B** (1968) 393.
82. S.K. Srivastava and P.K. Sharma, *Lett. al. Nouv Cim.* **2** (1969) 1960.
83. S.K. Srivastava and P.k.Sharma, *Phys,* **45** (1970) 225.
84. R.Evans, *Phys. Chem. Liq.* **2** (1971) 279.
85. S.Wang and C.B.So, *J. Phys.F.***7** (1977) 1439.

86. S.K.Srivastava and D.R.Singh, *Proceedings of Ist International Conf. on Thermoelectrics,* 1984, Cardiff,U.K.
87. T.Takeuchi and S.Noguchi, Ref. No. 6 b p. 219.
88. W.Vander Lugdt, J.F. Devlin, J. Hennephot and M.R. Leenstra Ref. No. 6 b, p.345.
89. S. Takeuchi, K. Suzuki, F. Itoh, K. Kai, M. Misawa and K. Murakami, Ref. No. 6 b p.69.
90. R. Oshima, Y. Kota, H. Endo, S. Minomura and Y. Onoda, Ref. No. 6 b p.213.
91. J.C. Joshi and S.K. Srivastava, *N.P. and SSP Symposium,VEC Project,* BARC, Calcutta, 1975.
92. J. Bardeen, *Phys. Rev.* **52** (1937) 688.
93. M. Bailyn, *Phys. Rev.* **120** (1960) 381.
94. J.M. Ziman, *Electrons and Phonons* (Clarendon, Oxford, 1960).
95. S.K. Srivastava, D.R. Singh and C.L. Rastogi, *Ind. J. Pure and Appl. Phys.,***28** (1990) 321; S.K. Srivastava, Maduri Khare and Ram Dawar, *VIII International Conference on Thermoelectric energy conversion,* Nancy-France (1989) 205.
96. R.C. Dynes and J.P. Carbotte, *Can. J. Phys.* **49** (1971) 1952.
97. P.K. Sharma and S.K. Srivastava, *Can. J. Phys.* **50** (1972) 1907.
98. B. Hayman and J.P. Carbotte, *Can. J. Phys.* **49** (1971) 1952.
99. M. Kaveh and N. Wiser, *Phys. Rev.* **B6** (1972) 36480; M. Kaveh and N. Wiser, *Phys. Rev. Lett.* **26** (1971) 635; *ibid* **29** (1971) 1374.
100. S.K. Srivastava, *Acta. Phys. Pol.* **A47** (1975) 3.
101. J.E. Ekin, *Phys. Rev. Lett.,* **26** (1971)1550.
102. J.E. Robinson, *Phys. Rev.* **161** (1967) 533.
103. A. Meyer and W.H. Young, *Phys. Rev.* **184** (1969)1003.
104. J.S. Dugdale, *Science,* **134** (1961) 77.
105. R. Berman, F.E. Simon and J. Wilks, *Nature,* **168** (1951) 277.
106. G.A. Slack, *J.Phys. Chem. Solids* **34** (1973) 321.
107. C.M. Bhandari, and D.M. Rowe, *Thermal Conduction in Semi-conductors* (Wiley Eastern ltd., New Delhi, 1988).
108. S.K. Srivastava, *Austr. J. Phys.,* **28** (1975) 403.
109. P.W. Selddod, *Magnetochemistry* (Wiley Ineterscience, Inc., New Y0rk, 1956).
110. G. Bush and S. Yuan, *Phys. Kondens Materie* **1** (1968) 37.
111. E.W.Collings, *Phys. Konders Materie,* **3** (1965) 336.
112. W.W. Warren, *Phys Rev.* **B3**, (1975) 3708.
113. C.J. Ford and G.A. Stylus, Ref. No. 6 b p.189.
114. J.E. Enderby, J.M. Titman and G.D. Wignall, *Phil Mag.* **10** (1964) 633.
115. J.P. Timbie and R.M. White, *Phys. Rev.* **B1** (1970) 2409.
116. R. Dupree and E.F.W. Seymour, *Phys. Kond. Materie,***12** (1970) 97.
117. Y. Takahashi and M. Shimizu, *J. Phys. Soc., Japan* **34** (1973) 942; *ibid,* **35** (1973) 1046.
118. A.G. Samoilovich and Ya, E. Rabin Ovich, *Fiz. Tverd.Tela,* **5** (1963) 778 (English Transl.: *Soviet Phys. Solid State* **5** (1963) 778.
119. W. Baltensperger, *Phys. Kond. Materie,* **5** (1966) 115.

120. M. Watabe, M. Tanaka, H. Endo and B.K. Jones, *Phil.Mag.* **12** (1965) 347.
121. S.K. Srivastava and P.K. Sharma, *Il Nuovo Cim.* **70B** (1970) 90.
122. Dekker, *Liquid Metals Chemistry and Physics*, ed, S.Z. Beer, (Convert Enterprises, Inc. Syracuse, New York, 1972).
123. D.A. Cornell, *Phys. Rev.* **153** (1967) 208.
124. R.V. Kawowski, *Phys. Rev.* **187** (1969) 891.
125. P. Jena and N.C. Halder, *Phys. Rev. Lett.* **26** (1971) 1024; *Phys. Rev.* **B4** (1971) 11.
126. B.R. Mc Garvey and H.S. Gustowsky, *J. Chem. Phys.* **21** (1953) 2114.
127. J. Heighway and E.F.M. Seymour, *Phys. Kondens Mater* **13** (1971) 1.
128. E. Fawcett, *Adv. Phys.* **13** (1964) 139.
129. H. Fukuyama, H. Ebisawa and Y. Nada, *Prog. Theor. Phys.* **42** (1939) 494.
130. L. Bayai and A. Aldea, *Phys Rev.* **143** (1986) 652.
131. T.V. Ramakrishnan, *Ph.D. Thesis* (Columbia University,1971).
132. U. Even and J. Jortner, *Phys. Rev. Lett.* **28** (1972) 31; *Phil.Mag.* **25** (1972) 715.
133. C. Herring, *J. Appl. Phys,* **31** (1960) 1939.
134. N.E. Cusack and P. Kendall, *Phil. Mag.* **8** (1963) 157.
135. H.J. Guntherodt, A. Menth, and Y. Tieche, *Phys. Kond.Mat.* **5** (1966) 392.
136. H.A. Davies, J.S.L. Leach and P.H. Draper, *Phil.Mag.***23** (1971) 116.
137. A.A.Andre'ev, and A.R., Regel, *Sov. Phys. Sol. State* **8** (1967) 295.
138. J.E.Enderby and C.J. Simmons, *Phil.Mag.***20** (1969) 125; J.E.Enderby and L., Walsh, *Phys. Lett.***14** (1965) 9.
139. D.S. Kyser, and J.C., Thompson, *J. Chem. Phys.* **42** (1965) 3910.
140. A.J. Greenfield, *Phys. Rev.* **A135** (1964) 1589.
141. S. Zabo, *J. Phys. C.*, **5** (1972) L241.
142. R. Fleteher, *Phil.Mag.* **29** (1974) 1331.
143. J.Fridel, *Phil.Mag.* **43** (1952) 153 *Adv. Phys.* **3** (1954) 446 ; *Nuovo cim.* **2** (1958) 287.
144. W. Cochran, *Proc. Roy. Soc.* **A276** (1963) 308. *In Inelastic Scattering of Neutrons in Solids and Liquids,***1** (I.A.E.A., Vienna, 1965) p.3; In Lattice Dynamics, ed. p. 75.
145. W.M.Shyu,and G.D. Gaspari, *Phys. Rev.* **163** (1967) 667 *ibid,* **170** (1968) 687.
146. W. Fuchs, *Proc. Roy. Soc.* **A151** (1935) 585.
147. R.A.Cowley,A.D.B. Woods, J. Dolling *Phys. Rev.***150** (1966) 487.
148. S. Glastone, *Physical Chemistry*, 2nd Ed., D. Van (Nostrand Co. Inc. U.S.A., 1974); reprinted (MacMillan India Ltd. Madras, 1984).
149. W. Kohn, *Phys. Rev. Lett* **2** (1959) 393.
150. S.K. Srivastava, *J. Phys. Chem. Sol.* **38** (1977) 451.
151. J.D.Anderson, Weeks and D. Chandler, *Phys. Rev.* **A4** (1971) 1597; D. Weeks, D. Chandler, and H.C. Anderson, *J. Chem. Phys.* **54** (1971) 5237; D. Chandler and J.D. Weeks, *Phys Rev. Lett.* **25** (1970) 149.
152. Sudha Srivastava, S.K. Srivastava and S.K. Dey, *Symp. Ind. Agr. Chem. Soc.*, Allahabad, 1988; S.K. Srivastava and Madhuri Khare, *J. Non Cryst. Solids,* **150,** (1992) 216.
153. G.L. Krasko, *Pisma Zh Exper. Teor. Fix.* **13** (1971) 218.

154. D. Stroud and N.W. Ashcroft, *J. Phys,* **F.1** (1971) 113.
155. M.Kogachi and Y. Matsuo, *J. Phys. Chem. Sol.* **32** (1971) 2393.
156. N.W. Achcroft, Ref. No. 69c (1976) 39.
157. T. Toya, *J. Res. Inst. Catalysis,* Hokkaido University, **6** (1958) 161; *ibid* **6** (1958) 183.
158. T.Schneider and E. Stoll, *Sol. State Comm.* **4** (1966) 79.
159. B.N. Brockhouse and N.k. Pope, *Phys. Rev. Lett.* **3** (1959) 259; B.N. Brockhousel,*"Phonons and Phonon Interaction,* ed. Back, T.A. (Benjamin, Inc., New York, 1964).
160. B.J. Alder and T.E. Wainwright, *J. Chem. Phys.* **27** (1957) 1208; *ibid,* **31** (1959) 459.
161. W. Schommer, *Ph.D. Thesis* (University of Karisruhe,1972, unpublished).
162. S.H. Vosko, K. Teylor, and G.H. Keech, *Canad. J. Phys.* **43** (1965) 1187; D.J.W. Gelder, and S.H. Vosko, *Canad. J.Phys.* **44** (1966) 2137; J.S.L. Langer, and S.H. Vosko, *J.Phys. Chem. Sol.* **12** (1969) 196.
163. A.O.E. Animalu, V. Bortolani and P. Ottaviani, *Nuovo Cim.* **44** (1966) 159.
164. D. Bohm and Staver, *Phys. Rev.* **84** (1951) 836.
165. S.K. Srivastava, *J. Phys.* **F.12** (1982) 649.
166. J.K. Percus and G.J. Yevick, *Phys. Rev.* **110** (1958) 1.
167. P.A. Egelstaff, *Rep. Prog.Phys.* **29** (1966) 333.
168. J. Hubbard and J. Beedy, *J. Phys. C.* **2** (1969) 556.
169. J.R.D.Copley and J.M. Rowe, *Phys. Rev. Lett.* **32** (1974) 49; *Phys. Rev.* **A9** (1974) 1656.
170. N.H.March and M.P. Tosi, *Atomic Dynamic in Liquids* (London, Macmillan, 1976).
171. A. Guidnier and G. Founet, *Small Angle Scattering of X-rays* (John Wiley and Sons, Inc., New York, 1965).
172. S.K. Srivastava, and Madhuri Khare, *Ind. Jour. Pure & Appl. Phys.* **28** (1990) 397.
173. P. Bartby, T. Gaskell and N.H. March, *Phys. Chem. Liq.* **2** (1970) 53.
174. R.Eisenschitz and M.J. Wilford, *Proc. Phys. Soc.* **80** (1972) 1078.
175. G.A. Tooma, *Proc. Phys.Soc* **86** (1965) 277.
176. S.K. Srivastava and P.K. Sharma, *Sol. St. Com.***7** (1969) 601.
177. K. Krebs *Phys. Lett.* **10,** (1964) 12; *Phys. Rev.* **A143** (1965) 138.
178. M. Born and K. Huang *Dynamical Theory of Crystal Lattices,* (New York, 1956).
179. D.C. Wallace *Phys. Rev.* **162** (1967) 776; *ibid* **176** (1968) 827,832; *ibid,* **182** (1969) 778; *Rev. Mod.Phys.* **37** (1965) 57.
180. H. Hellmann *Einfuhrung in dei Quanten-chemie* (Franz Deuticke, Leipzig, 1937).
181. R.G. Ross and D. A. Greenwood *Prog. Mater. Sci.* **14** (1969) 173.
182. M. Shimoji Ref. No. 6a p. 705, Ref. No. 6b p. 421.
183. M.Hasegawa and M. Wetabe, *J. Phys. Soc., Japan* **32** (1972) 14; Ref. No. 6b p. 133, 439.
184. L.Kanopoff and J.N. Shapiro, *Phys. Rev.* **B1** (1970) 3893.
185. H.C. Anderson, J.D. Weeks and D. Chandler, *Phys. Rev.***A4** (1971) 1597.
186. N.H. Nachtrieb, Ref. No. 6 b, p.521.
187. J.C. Kirkwood, *J. Chem. Phys.* **14** (1946) 180.

188. M. Born, and M.S. Green, *Proc. Roy. Soc.* **A188** (1946) 10.
189. E. Helfand, *Phys. Fluids* **4** (1961) 681.
190. Y. Waseda and K. Suzuki, *Phys. Stat. Soc.* **47 b** (1972) 203; *ibid* **49 b** (1972) 643.
191. R.A. Swalin, *Acta Metall* **7** (1959) 736; Z. Naturf. **A23** (1968) 805.
192. M.H. Cohen and D.Turnbull, *J. Chem. Phys.* **31** (1969) 1164.
193. P. Ascarelli and A. Paskin, *Phys. Rev.* **165** (1968) 222.
194. E. Helfand and S.A. Rice, *J. Chem. Phys.* **32** (1969) 1642.
195. P. Hicter, F. Dur and E. Bannier, H. Rupersberg and P. Hicter, Ref. No. 6 b, p.57.
196. P.A. Ascarelli, *Phys. Rev.* **173** (1968) 271; P.A. Ascarelli and R.J. Harrison, *Phys. Rev. Lett.* **22** (1969) 385.
197. T.E. Faber, Rev. No. 6 b p. 433, 607.
198. S.P. Mcalister and R. Turner *J. Phys. F.* **2** (1972) L51; S.P. Mcalister, E.D. Crozier and J.F. Cochran, Ref. No. 6 b , p.445.
199. G. Abowitz and R.B. Gorden *J. Chem. Phys.* **37** (1962) 125.
200. R.R. Hultgren, R.L. Orr, P.O. Anderson and K.K. Kelley, *Selected Values of Thermodynamic Properties and Alloys* (Willey Eastern, New York, 1963).
201. W.M. Hartmann, *Phys. Rev. Lett.* **26** (1971) 1640.
202. N.F. Mott, Proc. Roy. Soc. A146, 465 (1934); N.F. Mott and F. Jones, *The Theory of Properties of Metals and Alloys* (Clarendon Press, Oxford, 1936)..
203. M. Omini, (1972) (Private communication).
204. J.H. Dymand and J. Alder, *Chem. Phys.* **45** (1966) 2061; B.J. Alder, D.M. Gass and T.E. Wain Wright, *J. Chem. Phys.Chem.* **53** (1970) 3813.
205. A.D. Pasternak, *Phys. Chem. Liquids,* **3** (1972) 41.
206. R. Kishore and K.N. Pathak, *Phys. Rev.* **183** (1969) 672.
207. R. Bansal, *J. Phys. C.* **6** (1973) 1204.
208. J.G. Kirkwood and F.P. Buff, *J. Chem. Phys.* **17** (1949) 338.
209. R. Evans, *J. Phys. C.* **7** (1974) 2808.
210. D.W.G. White *Trans. Metall. Soc. AIME* **236** (1966) 796.

Chapter 5

TRANSPORT PROPERTIES : MAINLY IN LIQUID METALS AND AMORPHOUS SILICON

by N.H. March
Theoretical Chemistry Department
University of Oxford
5 South Parks Road
Oxford, OX1 3UB
U.K.

5.1. Introduction

In this Chapter, a brief survey will be given of the theory of ionic and electronic transport in liquid metals. A final section treats thermal conductivity in amorphous silicon. The following article will then deal in some detail with observations on both electrical conductivity and thermoelectricity of another type of disordered system : a metallic binary continuous solid solution.

The outline of the present work is as follows. In section 5.2 below, the self function $S_s(k,\omega)$ and the dynamical structure factor $S(k,\omega)$, accessible to neutron scattering experiments, will be connected via the so-called Kubo–Green formulae with ionic transport coefficients. This will then lead into a simple discussion of regularities among diffusion coefficients and shear viscosity just above the melting point for a variety of liquid metals. Section 5.3 then takes up a phenomenological discussion of the relation between longitudinal and transverse effects in ionic and electronic transport. In section 5.4, a force–force correlation approach to electronic transport in liquid metals is surveyed, the nearly–free electron theory emerging as the simplest limiting case. The role of electron correlation is illustrated here by reference to properties of the expanded fluid alkali metals along the liquid–vapour coexistence curve. Finally, in section 5.5 the thermal conductivity of amorphous silicon will be considered.

5.2. Self–motion $S_s(k,\omega)$ and dynamical structure factor $S(k,\omega)$ in liquids

Let us start by discussing the motion of a single ion, which meanders in time in the liquid after starting out from the origin $r = 0$ at time $t = 0$. Then the probability that the ion is, at a later time t, at position r is written $G_s(r,t)$, the subscript s denoting self motion (the ensuing discussion is restricted to motion in classical liquids). In the hydrodynamic regime (actually large r and long time t), $G_s(r,t)$ will satisfy the diffusion equation

$$\nabla^2 G_s(r,t) = \frac{1}{D}\frac{\partial G_s(r,t)}{\partial t} \tag{5.2.1}$$

where D is the self–diffusion coefficient. The Fourier transform on $r(\to k)$ and on $t(\to \omega)$ of $G_s(r,t)$, namely $S_s(k,\omega)$, is readily obtained from the solution of eqn(5.2.1) satisfying the boundary condition already referred to: namely

$$G_s(r,t)\big|_{t=0} = \delta(r), \tag{5.2.2}$$

this solution being readily verified as

$$G_s(r,t) = \frac{1}{(4\pi Dt)^{3/2}} \exp\left[-\frac{r^2}{4Dt}\right]. \tag{5.2.3}$$

The double Fourier transform of eqn (5.2.3) readily yields

$$S_s(k,\omega) = \frac{Dk^2}{\pi[\omega^2 + (Dk^2)^2]} \tag{5.2.4}$$

which, because of the range of validity (large r and t) of eqn (5.2.3) applies for small k and ω. It follows straightforwardly from eqn(5.2.4) that one can write for the transport coefficient D:

$$\frac{D}{\pi} = \lim_{\omega \to 0} \omega^2 \lim_{k \to 0} \frac{S_s(k,\omega)}{k^2}, \tag{5.2.5}$$

which is the first of the so–called Kubo–Green formulae.

5.2.1 Van Hove Dynamical Structure Factor $S(k,\omega)$

Having introduced a formula for diffusion, let us turn to the dynamic generalization $S(k,\omega)$ of the static liquid structure factor $S(k)$, which has already been discussed extensively. The first point to be stressed is that $S(k,\omega)$ has the physical interpretation that it is the probability that a neutron incident on the liquid transfers momentum $\hbar k$ and energy $\hbar \omega$ to the liquid. The second point is that the integral of the dynamical structure factor $S(k,\omega)$ over all energy transfers $\hbar \omega$

leads back to $S(k)$:

$$\int_{-\infty}^{\infty} S(k,\omega)d\omega = S(k) \qquad (5.2.6)$$

In addition to this so-called zero moment theorem, it can also be shown for a classical liquid consisting of ions of mass M that

$$\int_{-\infty}^{\infty} \omega^2 S(k,\omega)\, d\omega = \frac{k^2 k_B T}{M}, \qquad (5.2.7)$$

the second-moment theorem.

There is a corresponding Kubo-Green formula for $S(k,\omega)$ to that of eqn(5.2.5), namely

$$\lim_{\omega \to 0} \omega^4 \lim_{k \to 0} \frac{S(k,\omega)}{k^4} = \frac{k_B T}{\pi} \frac{(4/3\eta + \zeta)}{\rho M^2} \qquad (5.2.8)$$

where ρ is the ionic number density of the liquid while η and ζ represent shear and bulk viscosities respectively.

5.2.2 Approximate Formulae Relating D and η for Liquid Metals Just Above the Freezing Point

The above Kubo-Green formulae were used by Brown and March[1]. For liquid metals just above the melting temperature T_m, these workers exploited the fact that the self-correlation function $S_s(k,\omega)$ and the dynamical structure factor $S(k,\omega)$ entering the Kubo-Green formulae have a rather well defined frequency range $0 < \omega < \omega_D$, the Debye frequency ω_D being analogous to that in a crystalline solid. Relating ω_D to the melting temperature T_m using Lindemann's law of melting, Brown and March obtained the approximate relations for the ionic transport coefficients at T_m:

$$\frac{DM^{1/2}\rho^{1/3}}{T_m^{1/2}} = \text{constant} \qquad (5.2.9)$$

and

$$\frac{\eta}{T_m^{1/2} M^{1/2} \rho^{2/3}} = \text{constant}. \qquad (5.2.10)$$

The formula (5.2.10), which was obtained earlier from kinetic theory arguments by Andrade[2] which would no longer be regarded as convincing[3], leads to quantitative results if the constant is chosen empirically, as illustrated in Table 5.1 below.

Table 5.1

Shear viscosities (cp) of liquid metals at freezing

	Li	Na	K	Rb	Cs	Cu	Ag	Au	In	Sn
Experiment	0.60	0.69	0.54	0.67	0.69	4.1	3.9	5.4	1.9	2.1
Equation (2.10)	0.56	0.62	0.50	0.62	0.66	4.2	4.1	5.8	2.0	2.1

The result (5.2.9) is less impressive (for refinements, see Appendix 5.1). One reason for this is that whereas eqns (5.2.9) and (5.2.10) yield a relation between D and η at melting of the form

$$(D\eta/k_B T)/\rho^{1/3} = \text{constant} \qquad (5.2.11)$$

subsequent work of Zwanzig[4], discussed in Appendix 5.1 below, yields more generally

$$\left[\frac{D\eta}{k_B T}\right] \Omega^{1/3} = 0.0658 \left[2 + \frac{\eta}{\eta_\ell}\right] \qquad (5.2.12)$$

where Ω is the ionic volume while η_ℓ is the longitudinal viscosity. When results for η/η_ℓ eventually become available, it will be of obvious interest to test eqn (5.2.12) for liquid metals at T_m in more detail than proves possible at the time of writing.

5.3. Relation between longitudinal and transverse effects in ionic and electronic transport

One source of motivation for the present section is the so-called Haeffner effect[5]. When electrodes are placed across liquid Hg, Haeffner observed that the light isotope of the naturally occurring isotopic mixture moves towards the positively charged electrode. Subsequent experimental work has revealed no exceptions to this for a variety of liquid metals studied to date; see, for example, the review by Ginoza and March[6].

In the present section, following March and Paranjape[7], we shall begin by forging a close analogy between the treatment of a longitudinal thermoelectric effect[8]; see also the article by Burkov and Vedernikov[9]; characterized by the Thomson coefficient, σ_T say, and the Haeffner effect described above.

Immediately below, therefore, we shall apply the results of Moreau[10], of Bridgman[11], plus the Wiedemann–Franz law[12]; see also Appendix 5.3, to relate the Thomson coefficient to two transverse phenomena involving magnetic fields, the Ettingshausen[13] and Hall effects, via the Lorenz number.

5.3.1 Relation of Thomson Coefficient to Ettingshausen and Hall Coefficients in a Liquid Metal

In this section, we shall apply the considerations of Moreau[10] and of Bridgman[11]. These workers, by arguments of a very general character, established relations between various transverse effects. We start out from the result of Moreau[10] which relates the Nernst coefficient Q to the Hall coefficient R, via the electrical conductivity $\sigma = \rho^{-1}$ and the Thomson coefficient σ_T:

$$Q = \frac{\sigma_T R}{\rho} . \qquad (5.3.1)$$

Next, we combine the Moreau relation (5.3.1) with a further result, established by Bridgman[11]: namely

$$Q = \frac{\kappa}{T} P . \qquad (5.3.2)$$

Here κ is the thermal conductivity, while P denotes the Ettingshausen coefficient. It follows immediately from eqns (5.3.1) and (5.3.2) that

$$\frac{\sigma_T}{(\rho\kappa/T)} = \frac{P}{R} \qquad (5.3.3)$$

where in the denominator of the LHS we have grouped the combination of electrical and thermal conductivities and absolute temperature which enter the Wiedemann–Franz law. Specifically, this law reads (see also Appendix 5.3):

$$\frac{\kappa \rho}{T} = L \qquad (5.3.4)$$

where L is the Lorenz number. Numerical values of the LHS of eqn (5.3.4) are collected for a number of liquid metals in Appendix 5.3.

Eqn (5.3.4) evidently represents a widely useful approximation in such materials and hence, combining eqns (5.3.3) and (5.3.4):

$$\sigma_T = L \frac{P}{R} . \qquad (5.3.5)$$

Thus, in a liquid metal, the Thomson coefficient σ_T is given by the ratio: Ettingshausen coefficient to Hall coefficient involving a transverse magnetic field, the scale being determined by the Lorenz number L.

5.3.2 Ionic Transport: Longitudinal and Transverse Effects

Eqn (5.3.5) above is dominated by the itinerant electrons in a liquid metal. We turn now to discuss, by close analogy, a relation between longitudinal and transverse effects in ionic transport. Though we consider the case of an isotopically pure liquid metal, we note from the Haeffner effect that the light isotope in a binary isotopic mixture moves preferentially to the positively charged electrode (taken at x = 0, where the x axis is the direction of current flow: i.e. longitudinal). Thus, in a pure isotope such as liquid Li^7, we define a longitudinal coefficient for ions σ_h, by analogy with the Thomson coefficient σ_T for thermoelectricity, through

$$E_x = \sigma_h \frac{\partial n_i}{\partial x} , \qquad (5.3.6)$$

wher E_x is the electric potential gradient. It is the ionic density gradient $\frac{\partial n_i}{\partial x}$ we wish to estimate as the first step towards explaining the Haeffner effect in an isotopic mixture (see March and Paranjape[7]).

Next we write, again by analogy with eqn (5.3.5), σ_h as the ratio of two transverse effects. Corresponding to the Ettingshausen coefficient P entering eqn (5.3.2), we define an ionic counterpart, now depending on a transverse ionic gradient $\partial n_i/\partial y$, by

$$P_i = \frac{\partial n_i}{\partial y} / jH \qquad (5.3.7)$$

where j is the longitudinal current density while H is the transverse magnetic field in the Hall configuration.

Via a 'coupling constant' L_i, analogous to the Lorenz number entering eqn (5.3.5) we write

$$\sigma_h = L_i \frac{P_i}{R} . \qquad (5.3.8)$$

Using eqns (5.3.6) and (5.3.8), it follows that the longitudinal ionic density gradient can be written as

$$\frac{\partial n_i}{\partial x} = \frac{E_x}{\sigma_h} = \frac{E_x R}{L_i P_i} . \qquad (5.3.9)$$

Employing the definition (5.3.7), longitudinal and transverse density gradients are evidently related inversely by

$$\frac{\partial n_i}{\partial x} = \frac{E_x R j H}{L_i} \bigg/ \frac{\partial n_i}{\partial y} . \qquad (5.3.10)$$

5.3.3 Calculation of Transverse Ionic Density Profile

To use eqn (5.3.10) to relate to the longitudinal Haeffner effect, we must next calculate the ionic density gradient $\partial n_i/\partial y$.

In the transverse magnetic field characterizing the Hall configuration, an electric field E_H is created which is dominated by the electrons. This implies that the force on electrons due to the Hall field is almost entirely balanced by the Lorenz force. The result is that the same electrical force on the ions, namely eE_H, drives them in the direction of the Hall field, the magnetic force being negligible since the ions carry very little current[14].

The ions in a liquid metal being classical, the transverse ionic density profile then takes the Boltzmann form

$$n_i(y) = n_i(o) \exp\left[-\frac{eE_H y}{k_B T}\right] , \qquad (5.3.11)$$

the resultant potential energy $-eE_H y$ corresponding to the electrical force on the ions leading to a transverse ionic density gradient. Thus, in eqn (5.3.10) we can replace $\partial n_i/\partial y$ by

$$\left.\frac{\partial n_i}{\partial y}\right|_{y=0} = n_i(o)\left[-\frac{eE_H}{k_B T}\right] . \qquad (5.3.12)$$

Hence eqn (5.3.10) now reads

$$\left|\frac{\partial n_i}{\partial x}\right| = \frac{k_B T (eE_x)}{L_i \, n_i(o) \, e^2} . \qquad (5.3.13)$$

Introducing the classical Debye length ℓ_D of the ions through

$$\ell_D^2 = \frac{k_B T}{4\pi \, n_i(o) \, e^2} , \qquad (5.3.14)$$

one obtains the desired result for the longitudinal ionic density gradient created by the electric field E_x as

$$\left|\frac{\partial n_i}{\partial x}\right| = \frac{4\pi \ell_D^2 (eE_x)}{L_i}. \qquad (5.3.15)$$

While a full microscopic theory will eventually be required to calculate the coupling constant L_i, we note from eqn (5.3.15) that it has dimensions $ML^7 t^{-2}$.

5.3.4 The Haeffner Effect : Qualitative Explanation

Into the first isotope, we now introduce a (small) concentration of lighter isotopes, say Li^6 into liquid metal Li^7. The Haeffner effect is then explained if the inequality

$$\frac{dn_i}{dx} > \frac{dn_{i2}}{dx}, \qquad (5.3.16)$$

is satisfied as depicted in Figure 5.1, with the subscript 2 denoting the lighter isotope.

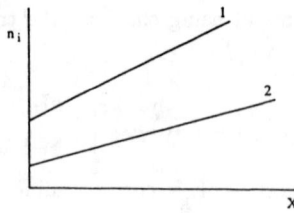

Fig.5.1. The reduction of longitudinal ionic density gradient for the lighter isotope 2 in a binary isotopic fluid metal mixture.

The origin of the inequality (5.3.16) arises in the (dilute) isotopic mixture as follows. The electrons bombard the ions and transfer momentum. With their preferential drift to the positively charged electrode they oppose the tendency of the light isotopes to drift to the negative electrode, more than that of the heavier isotope. The observed effect then is enrichment of the lighter isotope at the positively charged electrode.

In summary, the main results of the present section are:

(i) The recognition for thermoelectricity in liquid metals that the longitudinal Thomson coefficient σ_T can be expressed as the ratio of two transverse effects, namely Ettingshausen and Hall coefficients, the scale being fixed by the Lorenz number L.

(ii) The corresponding result for the coupling of ionic mass transport to a longitudinal electrical potential gradient. This leads to an inverse relation between longitudinal and transverse ionic density gradients. This has as its analogue such a relation between temperature gradients in thermoelectricty.

(iii) Equation (5.3.12) for the transverse ionic density gradient, demonstrating that it is determined by the potential energy associated with the Hall electric field in units of thermal energy $k_B T$, via a Boltzmann factor.

(iv) Use of the inverse relation in (ii) above between longitudinal and transverse ionic density gradients as the first step in explaining the Haeffner effect. The second step is simply the preferential momentum transfer from electrons drifting to the positively charged electrode to the lighter isotope. (see also the subsequent study by Robson et al[14], of the Haeffner effect in plasma).

It is only in (i) and (iv) above, to date, that experimental observations exist. However, it may be of interest for the future to study carefully the practicability of observing the transverse ionic density gradient referred to in (iii) above.

Following the above thermodynamic plus phenomenological discussion of ionic and electronic transport, we shall now take up a rather full discussion of the microscopic theory of electrical resistivity in liquid metals.

5.4. Force–force correlation function and electrical resistivity of liquid metals

In this section, the microscopic theory of electronic transport will be presented. As a starting point, let us consider the excess resistivity of a dilute metallic alloy following Huang[15].

5.4.1 Exact Resistivity Formula for Finite–Range Spherical Potential

Huang[15] derived an exact expression in terms of phase shifts $\eta_l = \eta_l(k_f)$ for the excess resistivity of a dilute metallic alloy in which independent free electrons are scattered by a spherical potential $V(r)$. Apart from constants, which are omitted here, the nub of the Huang formula is the sum

$$S = \sum_{l=0}^{\infty} [(2l+1)\sin^2\eta_l - 2l\sin\eta_l \sin\eta_{l-1} \cos(\eta_l - \eta_{l-1})], \qquad (5.4.1)$$

which is readily shown to be equivalent to

$$S = \sum_{l=1}^{\infty} l\sin^2(\eta_{l-1} - \eta_l). \qquad (5.4.2)$$

Following the writer[16], a result of Gerjuoy[17] that was rediscovered by Gaspari and Gyorffy[18] can now be inserted into eqn (5.4.2). These workers show that the radial wave functions R_l for such scattering from a potential $V(r)$ of finite range but arbitrary strength satisfy the exact relation

$$\int_0^\infty dr\, r^2 R_{l-1}(r) \frac{\partial V(r)}{\partial r} R_l(r) = \sin(\eta_l - \eta_{l-1}) \qquad (5.4.3)$$

where, outside the range of V, R_l has the form

$$R_l(r) = j_l \cos\eta_l - n_l \sin\eta_l, \qquad (5.4.4)$$

j_l and n_l being spherical Bessel and Neumann functions, respectively. Thus, from eqns (5.4.2) and (5.4.3)

$$S = \sum_{l=1}^{\infty} \int_0^\infty dr_1\, r_1^2 R_{l-1}(r_1) \frac{\partial V(r_1)}{\partial r_1} R_l(r_1)$$

$$\times \int_0^\infty dr_2\, r_2^2 R_{l-1}(r_2) \frac{\partial V(r_2)}{\partial r_2} R_l(r_2) \qquad (5.4.5)$$

At this point, it is to be noted that an inverse transport theory proposed by Rousseau, Stoddart, and March[19] (RSM: see also Appendix 5.2) has as its nub the force–force correlation function, F say, again stripped of unimportant multiplying factors:

$$F = \int d\mathbf{r}_1\, d\mathbf{r}_2\, \frac{\partial V(\mathbf{r}_1)}{\partial \mathbf{r}_1} \cdot \frac{\partial V(\mathbf{r}_2)}{\partial \mathbf{r}_2} |\sigma(\mathbf{r}_1,\mathbf{r}_2)|^2, \qquad (5.4.6)$$

where $\sigma(r_1 r_2)$ is the energy derivative of the Dirac density matrix for potential V evaluated at the Fermi energy $E_f = k_f^2/2$. Using the result of March and Murray[20]

$$\sigma(r_1 r_2) = \sum_l (2l + 1)\sigma_l(r_1 r_2) P_l(\cos \gamma) \qquad (5.4.7)$$

where γ is the angle between r_1 and r_2, the integrand, f, say, in eqn (5.4.6) can be written as

$$f = \frac{\partial V(r_1)}{\partial r_1} \cdot \frac{\partial V(r_2)}{\partial r_2} |\sigma(r_1 r_2)|^2,$$

$$= \sum_l \sum_j (2l + 1)\sigma_l(r_1 r_2) P_l(\cos \gamma)$$

$$\times (2j + 1)\sigma_j(r_1 r_2) P_j(\cos \gamma) \frac{\partial V}{\partial r_1} \frac{\partial V}{\partial r_2} \cos\gamma, \qquad (5.4.8)$$

the last step in eqn. (5.4.8) following because the potential is spherical. According to eqn. (5.4.6), on multiplying by $\sin\gamma \, d\gamma$ and integrating over γ from 0 to π:

$$\int_0^\pi f \sin\gamma \, d\gamma = \sum_l \sum_j (2l + 1)\sigma(r_1 r_2)(2j + 1)\sigma_j(r_1 r_2)$$

$$\times \frac{\partial V(r_1)}{\partial r_1} \cdot \frac{\partial V(r_2)}{\partial r_2} \int_0^\pi \cos\gamma P_l(\cos\gamma) P_j(\cos\gamma) \sin\gamma \, d\gamma. \qquad (5.4.9)$$

The integral in eqn. (5.4.9) is related to Clebsch–Gordon coefficients, but the answer can be written in elementary fashion by using the identity

$$(2l + 1)\cos\gamma P_l = (l + 1)P_{l+1} + l P_{l-1} \qquad (5.4.10)$$

On substituting eqn. (5.4.10) into eqn (5.4.9) and performing the integration, a short calculation gives

$$\int_0^\pi f \sin\gamma \, d\gamma = 4 \sum_{l=1}^\infty l \sigma_{l-1}(r_1 r_2) \sigma_l(r_1 r_2) \frac{\partial V(r_1)}{\partial r_1} \frac{\partial V(r_2)}{\partial r_2}. \qquad (5.4.11)$$

But on writing (see March and Murray[20], Rousseau[21]; Harris[22])

$$\sigma_l(r_1 r_2) \propto R_l(r_1) R_l(r_2), \qquad (5.4.12)$$

it is readily seen from eqns.(5.4.12), (5.4.11) and (5.4.6) that, apart from constants, the Huang[15] formula, rewritten in the form (5.4.5), has exactly the same structure as the RSM formula. They are both force–force correlation functions of the type (5.4.6), as shown by March[23].

5.4.2 Weak Scattering Theory of Electrical Resistivity

The idea behind weak scattering theory is to represent the total potential energy $V(r)$ scattering the conduction electrons by a sum of screened potentials $v(r)$ at the ionic sites $\mathbf{R}_i$, where one has taken a "snapshot" of the ions at a particular time:

$$V(\mathbf{r}) = \sum_i v(|\mathbf{r} - \mathbf{R}_i|). \tag{5.4.13}$$

Since the inverse transport theory expression for F in eqn. (5.4.6) already contains the potential energy V to second order explicitly, one can evaluate this force–force correlation function to second order in V by replacing the energy derivative σ of the density matrix by its free electron value. The density matrix ρ for free electrons is readily calculated for plane waves $\mathscr{V}^{-1/2}\exp(i\mathbf{k}\cdot\mathbf{r})$ normalized in a volume $\mathscr{V}$ as

$$\rho_0(\mathbf{r}_1\mathbf{r}_2) = \sum_{|\mathbf{k}|<k_f} \mathscr{V}^{-1}\exp(i\mathbf{k}\cdot\mathbf{r}_1 - \mathbf{r}_2). \tag{5.4.14}$$

Replacing the summation of $\mathbf{k}$ by an integration with the usual constant density of states in $\mathbf{k}$ space as $(8\pi^3)^{-1}\mathscr{V}$, one finds

$$\rho_0(\mathbf{r}_1\mathbf{r}_2) = \frac{k_f^2}{\pi^2}\frac{j_1(k_f|\mathbf{r}_1 - \mathbf{r}_2|)}{|\mathbf{r}_1 - \mathbf{r}_2|}. \tag{5.4.15}$$

The energy derivative follows by using $k_f = (2E)^{1/2}$ and then forming $(\partial\rho_0/\partial E)_{E=E_f=k_f^2/2}$, say $\sigma_0(\mathbf{r}_1\mathbf{r}_2 E_f)$. The resistivity ρ is then found by using this result for σ_0 in eqn.(5.4.6); clearly, as only pairs of sites $\mathbf{R}_i$ are now correlated, taking the liquid average one obtains a result in terms of the structure factor $S(k)$ and the Fourier transform of the localized potential, $\tilde{v}(k)$ say, in eqn. (5.4.13). The result, when one puts back all the numerical factors (see section 5.4.1), is for weak scattering with a sharp Fermi surface of diameter $2k_f$:

$$\rho = \frac{3\pi}{\hbar e^2 v_f^2 \rho_i} \frac{1}{(2k_f)^4}\int_0^{2k_f} S(k)|\tilde{v}(k)|^2 4k^3\, dk. \tag{5.4.16}$$

This is the basic formula for the electrical resistivity of simple (s–p) nearly free electron metals such as Na and K. In eqn. (5.4.16), ρ_i is the ionic number density, and since S(k) is measurable by diffraction experiments, the only quantity needed to determine ρ is the Fourier transform of the localized atomic–like screened potential energy $\tilde{v}(k)$. Some discussion of the way approximations may be set up for this quantity has already been given in Chapter 3. It is also relevant to note that real liquid metals have blurred Fermi surfaces, in accord with the Heisenberg uncertainty principle

$$l\Delta k_f \sim 1, \qquad (5.4.17)$$

where Δk_f is the blurring of k_f and l is the electronic mean free path.

5.4.3 Strong Correlations in Electrical Resistivity : Relation to Magnetic Susceptibility

The purpose of this section is to:

(i) Propose an approximate generalization of the force–force correlation function theory to treat resistivity in a strongly correlated electronic system

and

(ii) Relate resistivity thus obtained to magnetic susceptibility.

(a) <u>Inverse treatment and strong electronic correlations</u>. It has to be said that there is presently no fully satisfactory inverse transport theory for strongly correlated electrons. Therefore to gain orientation, and to provide also motivation for a particular presentation of experimental data on electrical resistivity R and magnetic susceptibility χ below for expanded Cs, let us return to eqn (5.4.6). The one–body character of this equation is reflected in:

(i) The appearance of the one–body scattering potential V(r)

and

(ii) The use of the Dirac density matrix, which satisfies the idempotentcy condition $\rho^2 = \rho$.

As to (i) above, one has the equation of motion for the Dirac density matrix as

$$\nabla^2_{\mathbf{r}_1}\rho - \nabla^2_{\mathbf{r}_2}\rho = \frac{2m}{\hbar^2}[V(\mathbf{r}_1) - V(\mathbf{r}_2)]\rho. \tag{5.4.18}$$

Dividing both sides of ρ, one can then form both $\frac{\partial V(\mathbf{r}_1)}{\partial \mathbf{r}_1}$ and $\frac{\partial V(\mathbf{r}_2)}{\partial \mathbf{r}_2}$ solely in terms of the Dirac density matrix ρ, and hence these forces can be eliminated from eqn. (5.4.6). So far, this procedure is equivalent to eqn. (5.4.6), except that now R is solely a functional of the Dirac density matrix.

However, for orientation, let us now assume that R can be characterized in the many–electron case by replacing the Dirac density matrix ρ by the first–order density matrix γ, which satisfies $\gamma^2 < \gamma$ because, essentially, of the Pauli exclusion principle. With jellium as an example, one has

$$\rho_0 \, \alpha \sum_{|\mathbf{k}| \leq k_f} \exp(i\,\mathbf{k}.\mathbf{r}_1 - \mathbf{r}_2) \tag{5.4.19}$$

whereas

$$\gamma_0 \, \alpha \sum_{\text{all } \mathbf{k}} n(\mathbf{k}) \exp(i\,\mathbf{k}.\mathbf{r}_1 - \mathbf{r}_2). \tag{5.4.20}$$

The final step then is to expand $R \equiv R[\gamma(\mathbf{r}_1,\mathbf{r}_2)]$ when q is small as

$$R = R_0 + q\,R_1 + \ldots \tag{5.4.21}$$

Eqn. (5.4.21) focusses on the discontinuity q at the Fermi surface in the electronic momentum distribution. This same quantity q also enters the theory of the magnetic susceptibility χ, as will now be outlined.

(b) <u>Theory of magnetic susceptibility</u>. Below a brief outline will be given of an extension of the phenomenological theory of March, Suzuki and Parrinello[24], which was posed for metal–insulator transitions occurring at absolute zero, to elevated temperatures (see also Chapman and March[25], where a microscopic treatment based on ideas borrowed from heavy Fermion theory at $T \neq 0$ was developed). One writes, phenomenologically, the free energy per atom as an expansion in the magnetization per atom m and the discontinuity (quasiparticle renormalization factor) q in the electronic momentum distribution n(k). Thus

$$F(m,q,T) = E_0 + a(T)m^2 + \ldots + b(T)q$$
$$+ c(T) q^2 + \ldots + e(T) qm^2 + \ldots \quad (5.4.22)$$

Here, the interpretation of q in terms of the average number of doubly occupied sites (see March, Suzuki and Parrinello[24]) at $T = 0$) is more appropriate, since the discontinuity in the single–particle occupation number will not be such a well–defined quantity when $T \neq 0$. Furthermore, as $T \to \infty$, one expects (Warren[26]) a reversion to Curie-like behaviour in the (paramagnetic contribution to) magnetic susceptibility, as the degeneracy temperature of the Fermi liquid is exceeded. This is achieved if $a(T)$ in eqn (5.4.21) is proportional to T, with the coefficient $e(T)$ much less strongly dependent on T, and remaining finite as $T \to 0$. Thus, if $a = \alpha T$, then as $T \to \infty$, $\chi \to n_0 \mu_0 \mu_B^2 / 2 \alpha T$, and for lower temperatures χ is always less than this limiting value, for all densities. If α is chosen as $(1/2)k_B$ this is simply the Curie law for the electrons. In the opposite $T \to 0$ limit, one has $\chi \to n_0 \mu_0 \mu_B^2 / 2eq$ and one requires $1/2e \to N(E_f)$, the density of states at the Fermi level in the limit of high density ($q \to 1$), to regain the Pauli susceptibility. Thus, one expects a gradual transition between these two limiting results, when $a \sim eq$: ie $2eq \simeq k_B T$.

Chapman and March have used heavy Fermion theory to estimate q at the maximum in χ shown in Figure 5.2 (see Warren[26], Freyland[27], Bottyan et al[28]. They find $q \sim 0.18$, a result which has subsequently been confirmed by Warren[29] using NMR data.

It is important to contrast the above with the result of the jellium model at the same density: here, from the work of Lantto[30], q is 0.53. The conclusion is clearly that at the density of 0.8 gcm^{-3} corresponding to the maximum in χ in Figure 5.2, both electron–electron interactions (from reduction of q in jellium from 1 in the non–interacting case to 0.53 at 0.8 gcm^{-3}) and electron–ion interactions ($q \sim 0.18$ at same density) are important in determining the properties of expanded fluid Cs.

(c) <u>Relation between R and χ for expanded Cs</u>. Having discussed theories of R and χ in sections 5.4.3(a) and 5.4.3(b) above respectively, the purpose of this section is to expose an intimate link between the two properties which is forged via the discontinuity q. In Figure 5.3(a), a schematic representation is shown of the electronic momentum distribution n(k) as a function of k/k_f in the jellium model. Figure 5.3(b) depicts schematically the reduction in q, at the same density as in Figure 5.3(a), due to the introduction of granular ions. While in (a) it is evident that electron–electron correlation is already strong, (b) shows that this is also the case for the electron–ion interaction (at a density $\sim$ 0.8 gcm^{-3} in expanded Cs: see Figure 5.2).

Fig.5.2

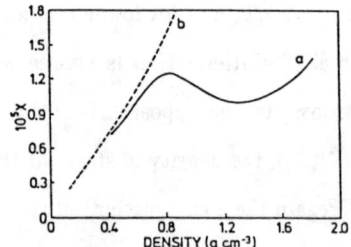

Measured magnetic susceptibility of liquid metal Cs as a function of density, along liquid–vapour coexistence curve.

Fig.5.3

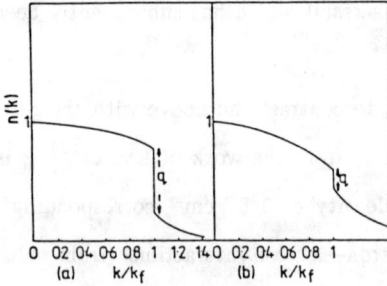

Schematic representation of momentum distribution. (a) Result of jellium model. (b) For liquid Cs, at same density as in (a).

To relate χ and R, let us note first from the discussion of section 5.4.3(b) that if the Curie limiting susceptibility is denoted by χ_c then

$$\frac{1}{\chi} - \frac{1}{\chi_c} \alpha \; q, \qquad (5.4.23)$$

leading to χ tending to the dashed curve in Figure 5.2 as q is reduced on approaching the critical point.

Since as noted in eqn (5.4.23) $(1/\chi - 1/\chi_c)$ provides an empirical measure of the discontinuity q, then substitution in the (assumed) expansion (5.4.21) motivates the final plot, using experimental data, of R vs $(1/\chi - 1/\chi_c)$. There is a striking correlation[31], as shown in Figure 5.4.

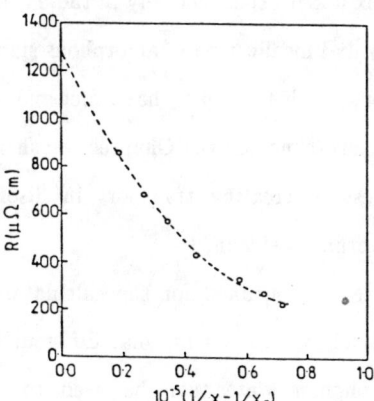

Fig.5.4

Electrical resistivity R versus $(1/\chi - 1/\chi_c)$ for expanded liquid metal Cs along coexistence curve.

Via the discontinuity q in the electronic momentum distribution n(k), it has been demonstrated that at the density $0 \cdot 8$ gcm^{-3} corresponding to the maximum in the paramagnetic susceptibility χ vs density, both electron–electron and electron–ion interactions are strong. Therefore the Ziman nearly–free electron theory of electrical resistivity R, while it works well at the higher densities along the liquid–vapour co–existence curve, breaks down completely at lower densities, as must be expected from (a) its independent electron nature and (b) its basic assumption of weak electron–ion interaction.

Fortunately, the availability of the discontinuity q as an order parameter in the theory of magnetic susceptibility χ, and its intimate relation to $(1/\chi - 1/\chi_{\text{curie}})$ in the strong coupling regime near the critical point, allows a link to be forged with the electrical resistivity R. This intimate relation is borne out in Figure 5.4. It is to be expected that the theory presented here for R and χ for expanded Cs should be valid for the other alkali metals, even though the maximum in χ for Cs has not yet been exposed for the lighter alkalis, presumably because of the more limited range of χ which has been explored experimentally in these other cases.

5.5 Thermal conductivity and localization in amorphous silicon

The preceding sections of this Chapter have focussed exclusively on normal and expanded fluid metals. To conclude the Chapter, we shall take one example to give the flavour of progress in treating transport in disordered solids; namely thermal conductivity in amorphous silicon.

In Chapter 2, a route was outlined for the calculation of phonon states in disordered systems. What follows, in a somewhat different framework, illustrates how knowledge of such phonon states can be used to estimate the thermal conductivity of amorphous silicon. The account below is based closely on the work of Feldman et al[32], supplemented mainly by references to important earlier work. Essentially, the focus below is on thermal conductivity and localization in glasses, as exemplified by a numerical study of a model of amorphous silicon.

In the work of Feldman et al[32], explicit numerical calculations of the thermal conductivity $\kappa(T)$ are reported for realistic atomic structure models of amorphous silicon with 1000 atoms and periodic boundary conditions. Using Stillinger–Weber forces[32], the vibrational eigenstates were computed by exact diagonalization in the harmonic approximation. Only the uppermost 3% of the states were found to be localized. The finite size of the system prevents fully quantitative information being gained about low energy vibrations, but the 98% of the modes with energies above 10 meV are densely enough represented to permit a good deal of information to be extracted. Each harmonic mode has an intrinsic (harmonic) diffusivity (defined by the Kubo formula : see[32]), which one can accurately calculate for $\omega > 10\text{meV}$. If the mode could be described by a wavevector $\vec{k}$ and a velocity $\vec{v} = \partial\omega/\partial\vec{k}$, then Boltzmann theory would assign a diffusivity $D_{\vec{k}} = \frac{1}{3}v\ell$ where ℓ is the mean free path. Feldman et al[32] find that they cannot define a wavevector for the majority of the states, but the intrinsic harmonic diffusivity is nevertheless still well-defined and has a numerical value similar to that obtained by using the Boltzmann result: replacing v by a sound velocity and ℓ by an interatomic distance a. In order to fit the experimental thermal conductivity $\kappa(T)$ it proves necessary to add a Debye-like continuation from 10meV down to 0. The harmonic diffusivity becomes a Rayleigh ω^{-4} law and gives a divergent $\kappa(T)$ as $T \to 0$. To eliminate this Feldman et al[32] make the standard assumption of resonant plus relaxational absorption from two-level systems (this is an anharmonic effect which would lie outside their model even if it did contain two-level systems implicitly.) A reasonable fit and explanation then results for the behaviour of $\kappa(T)$ in all temperature regimes. Feldman et al[32] also study the effect of increasing the harmonic disorder by substitutional mass defects (modelling amorphous Si/Ge alloys.) The additional disorder increases the fraction of localized states, but delocalized states still dominate. However, the diffusivity of the delocalized states is diminished.

5.5.1 Some background and experimental findings

The thermal conductivity κ of amorphous silicon (a–Si) has been measured by Pompe and Hegenbarth[33], and by Cahill et al.[34]. Two factors make this an attractive case for detailed theoretical study. First, a–Si has technological interest, and especially for potential thermoelectric applications a deeper understanding of thermal conductivity seems important. Second, from a purely theoretical point of view, a–Si can be used as a model system for studying generic effects seen in κ of glasses. Fig. 5.5 shows the experimental data, and also the theoretical fits of Feldman et al[32] discussed below. The behaviour of $\kappa(T)$ is customarily classified into three regimes: (1) low temperatures, where only low-energy vibrations are excited; here $\kappa(T)$ is approximately[32] a quadratic function of T. The model of scattering from 2-level systems[35] seems to give a satisfactory fit. (2) At somewhat higher temperatures, typically 10–30K as seen in Fig.5.5, there is a "plateau" region. (3) At T > 30K, $\kappa(T)$ rises smoothly to a T–independent or "saturated"

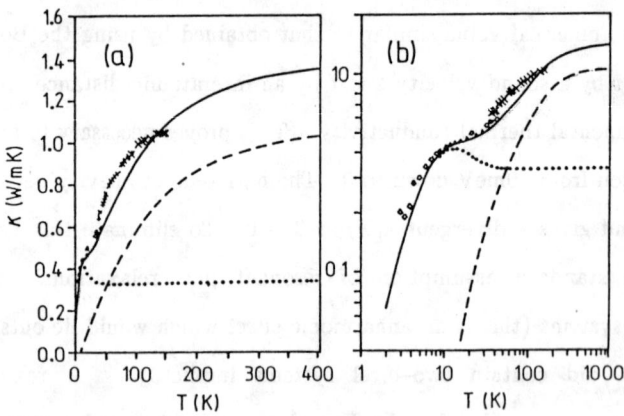

Fig.5.5

Thermal conductivity *versus* temperature in amorphous Si. Diamonds are data from ref.[33] X's are from ref.[34]. Lines are the theoretical fits of Feldman et al.[32] In addition to the total $\kappa(T)$ (solid line), the theoretical contributions are shown separately from vibrations with energies above (long dashes) and below (short dashes) 5meV.

value. In this last regime, Feldman et al suppose that the dominant mechanism is the intrinsic harmonic diffusion of higher energy delocalized vibrations. These modes have not been well described by most previous theories. Feldman et al[32] suggest also that most of the effects seen in $\kappa(T)$ of glasses have close analogues in electrical transport of disordered metals, discussed earlier in the present Chapter.

The vibrations which dominate the high–T heat transport in the treatment of Feldman et al[32] lie near the borderline of Anderson localization. However, except for a small minority ($\approx$ 3%), the vibrational eigenstates are not localized. Neither are they "propagating" in the sense that there is no way to assign them a wavevector or velocity. Nevertheless, they carry heat, contributing to $\kappa(T)$ an amount $C_i(T)D_i/V$ per mode i, where the specific heat C_i is k_B at high T and D_i is a temperature independent "mode diffusivity" (see Allen and Feldman[36]). Boltzmann transport theory (which is not applicable) would have given $D_i = v_i \ell_i / 3$. As observed by Kittel[37], if the sound velocity is used instead of v_i and the interatomic spacing is employed in place of the mean free path ℓ_i, then $\kappa(T)$ is semiquantitatively fitted at temperatures above the plateau region. Slack[38] has discussed how such a model is also useful for crystalline insulators with strong scattering. In their work, Feldman et al[32] report numerical calculations on realistic finite size models which agree reasonably well with experiment.

This same picture is also supported by the simulation results of Sheng and Zhou[39], and by a phenomenological analysis of Love and Anderson[40]. The work of Feldman et al[32] does not ignore localization. Their system sizes are large enough so that the location of the mobility edge is relatively precise and unambiguous. Both the standard participation ratio, and the D_i calculation show a sharp breakpoint at the onset of localized states, where D_i becomes zero modulo exponentially small finite–size errors. Their theory remains within harmonic approximation, and therefore there are no inelastic processes which could allow localized states to carry heat at any temperature. The accuracy of the harmonic restriction depends on the material under consideration. In an earlier simulation, Payton et al[41] presented evidence that turning on anharmonicity increased the value of κ at higher T. However, silicon in crystalline form is very harmonic (its thermal conductivity rivals Cu at room temperature, and exceeds it at nitrogen temperature.) The classical simulations of Feldman et al[32] on a–Si with the fully anharmonic Stillinger–Weber[32](SW) potential confirm that anharmonic effects are weak in this model.

There is a somewhat common view that eigenstates are either propagating or localized, and that states near the margin of localization cannot give very large transport currents, perhaps because the regime near the margin is presumed to be narrow. Further, it is sometimes assumed that most vibrational states in glasses are localized. As an example, it is often supposed that the Ioffe–Regel condition signals localization. The "Ioffe-Regel condition" means $\ell \approx a$, i.e. states not propagating, and is ubiquitous in glasses. Feldman et al[32] assert that their work demonstrates that the identification of the Ioffe-Regel condition with localization is oversimplified. They find only a narrow region of localized states at high frequencies, but a wide region where the Ioffe–Regel condition is obeyed, with sufficient heat transport in these non–propagating delocalized states to explain the experimental $\kappa(T)$ at temperatures above the plateau. A similar situation applies in the case of electron eigenstates in metallic glasses or disordered alloys, and it is worth presenting the arguments because electrons are in one sense simpler (Feldman et al[32]). In principle, vibrations can never completely localize (the long wavelength hydrodynamic modes always can propagate) whereas the relevant electrons within $\approx 10 k_B T$ of the Fermi level of a metal can in principle all localize. An "Anderson heat insulator" is forbidden, but an "Anderson electrical insulator" is not. Highly disordered metals often have resistivities within a factor of two of $150 \mu\Omega$cm and show only weak T–dependence. This is just the value that one obtains when each one–electron state has a diffusivity $v_F \ell/3$ and ℓ is set equal to the interatomic spacing a, i.e. it is the Ioffe–Regel condition. These one–electron states are not localized, as can be deduced from three experimental observations. (1) The resistivity does not diverge as $T \to 0$ but instead remains almost independent of T. (2) At higher T the resistivity is still T–independent, whereas conduction in localized states should increase with T because of thermally activated hopping and also possible thermal activation of carriers into delocalized states. (3) The majority of these metals cannot be made into Anderson insulators by any form of treatment (alloying, substitutions or radiation damage) other than diluting them severely with nonmetallic elements like Ar. Appropriating somewhat loosely the language of attractors and fixed points, under a wide range of disordering conditions, both electrical conductors and electrically insulating heat conductors collect at the "*diffusive fixed point*" where they are neither Anderson insulators nor quasiparticle conductors. This is probably not a mathematical fixed point of a well–defined class of Hamiltonians, but instead a phenomenological fixed point (much as the notion of a Fermi liquid as a physical fixed point seems more general than can be warranted at the time of writing by the scaling properties of real Hamiltonians : see Feldman et al[32].)

The usefulness of Si as a model system derives from the fact that there are microscopic models for interatomic forces and for atomic coordinates, both of which are needed for microscopic theory. In many ways it seems that a–Si can be considered a typical glass. Sometimes the word "glass" is reserved for systems which can be produced in bulk rather than only thin film form but Feldman et al[32] argue that this distinction refers simply to relative heights of energy barriers to recrystallization, and is not of fundamental importance in the present context.

The work of Allen and Feldman (AF)[36] gives the formalism needed, i.e., the use of exact harmonic eigenstates of specific disordered atomistic models containing N atoms with periodic boundary conditions. The resulting 3N eigenfrequencies ω_i and corresponding eigenvectors ϵ_i were then used to evaluate the Kubo formula for κ. Their fit to the thermal conductivity of amorphous Si has already been depicted in Fig.5.5. Three fitting parameters were used to describe inelastic effects necessary for the low frequency vibrations. Higher frequency vibrations are only weakly influenced by these inelastic processes, and their intrinsic harmonic diffusivity dominates. This part of their calculation has no adjustable parameters. For the computation of the intrinsic harmonic effects, and finite size influences, the work of Feldman et al[32] should be consulted. Specifically, they examine the dependence of calculated values of $\kappa(T)$ on N, on the choice of boundary conditions at the edges of the unit cell, and on the direction of heat flow. The method of extrapolation to zero frequency is discussed by Feldman et al[32]. The problematic contribution of low frequency modes is cured by a phenomenological treatment of inelastic processes.

5.5.2 <u>Results for pure amorphous silicon</u>

A measure of the realism of the model used by Feldman et al[32] is shown in Fig.5.6, which depicts the vibrational density of states compared with neutron

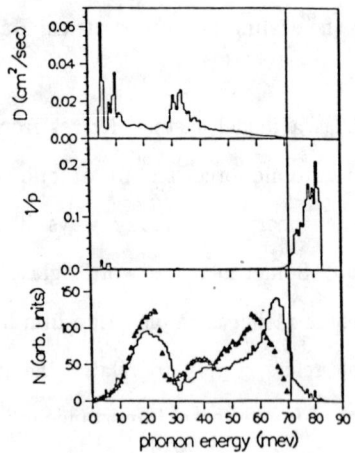

Fig.5.6

Vibrational density of states $N(\omega)$ of amorphous Si. Also shown are mode diffusivities and inverse participation ratios following Feldman et al[32]. The vertical line at ~ 70meV shows the location of the mobility edge.

scattering experiments by Kamitakahara et al[42]. The mathematical definition

$$N(\omega) = \sum_i \bar{\delta}\,(\omega - \omega_i) \qquad (5.5.1)$$

was used, where $\bar{\delta}$ is a rectangular approximation to a δ-function centred at discrete frequencies ω with width equal to 0.715meV. The energy of the upper end of the spectrum is overestimated by about 20%. This seems to be a limitation of the SW potential. The SW model makes the same overestimate when applied to crystalline Si[32]. When compared with the crystalline case, a–Si has a fairly similar phonon density of states, the most important difference being the softening of the lowest energy peak in the amorphous state.

The central result of the treatment of Feldman et al[32] is then the intrinsic harmonic diffusivity D_i, shown in Fig.5.6. Actually, Fig.5.6 shows D_i averaged over modes in a frequency interval : ie what is plotted is

$$D(\omega) = \sum_i D_i \delta(\omega-\omega_i)/N(\omega) , \qquad (5.5.2)$$

a broadened δ–function with a width greater than the level spacing being employed. In principle, the width of the δ–function goes to zero after the system size goes to infinity. Feldman et al[32] used a Lorentzian of width $\sim$ 0.04meV. The resulting values of D_i showed large fluctuations (rms fluctuations of $\approx$ 25% within the bins used in eqn. (5.5.2).). A larger width of the Lorentzian would probably have reduced these fluctuations, but the possibility exists that, in the infinite size limit, the value of D_i has significant random fluctuations from state to state, reflecting the idiosyncratic nature of these delocalized but non–propagating states. In any event, they average out smoothly in eqn (5.5.2) and Fig.5.2.

It can be seen that there are two regions of large diffusivity; the lowest frequencies, and the region around 35meV. In crystalline Si, this second region is a portion of the spectrum lying above the transverse acoustic vibrations, where longitudinal modes have large velocities and are probably highly effective carriers of heat. The reason for the high diffusivity in the amorphous state is that there is a local minimum in the density of states, so that the structural disorder mixes these pseudo–longitudinal vibrations with relatively few other modes, preserving some aspects of their propagating character. It may be possible to assign these modes a velocity and mean free path. As a crude estimate, one can set the diffusivity $D \approx 2 \times 10^{-6} m^2/s$ equal to $\frac{1}{3}v\ell$ and choose for v the longitudinal sound velocity found for this same model of a–Si, $v_L = 7.6 \times 10^3 \text{m/s}$[32]. This yields an estimated mean free path of 8Å (safely smaller than the 28Å cell size used by Feldman et al) for this especially diffusive region of the spectrum. With such a short mean free path, the notions of wavevector, velocity, and mean free path can only be marginally meaningful.

Above 40meV there is a smooth decrease of diffusivity, approximately linear in energy ($D_i \alpha (\omega_c - \omega_i)^p$ with a critical exponent $p \approx 1$) and a critical energy $\omega_c \doteq 70$ meV where D_i vanishes to within a very small noise level. The critical exponent of 1 agrees with scaling and other theories of Anderson localization[43], and this seems to locate the mobility edge above which the diffusivity would be strictly zero in the infinite size limit. As a further test, Feldman et al calculated the "inverse participation ratio" $1/p_i$, defined as

$$1/p_i = \sum_\ell [\sum_\alpha \epsilon_i(\ell, \alpha)^2]^2 \qquad (5.5.3)$$

where $\epsilon_i(\ell,\alpha)$ is the α–th Cartesian component of the normalized polarization vector of the i–th mode on the ℓ–th atom. The square of this polarization vector when summed over all atoms gives unity. In the work of Feldman et al[32], it is squared a second time before summing. If the vibration were localized on a single atom, the result would still be 1, whereas if the vibration were equally distributed on all atoms, the answer would be 1/N. Thus p_i measures n_i, the number of atoms on which with i–th vibrational mode has significant amplitide. For a delocalized mode, $1/p_i \to 0$ as the system size $\to \infty$, whereas for a localized mode, $1/p_i$ remains non–zero. Fig. 5.5 shows a sharp increase in $1/p_i$ at $\omega_i \doteq 70$ meV; just where the diffusivity approaches zero. The 3% of eigenstates above this energy seem clearly to be localized. On the scale of the cell size (28Å) 97% of the phonons are delocalized. Feldman et al[32] believe that this estimate is not sensitive to system size, or to fine details of the structural model. The same result holds for both their smaller 216–atom models, and has been confirmed by Biswas et al[44]. Feldman et al suppose that this is the normal situation for vibrations in glasses, provided that they are reasonably dense, i.e. not too porous.

There are possibly a very few localized modes at $\omega_i \doteq 30\text{meV}$, where there is a sharp minimum in the density of states separating the low energy transverse peak from higher energy modes. The lowest non–zero mode also appears localized, an effect which was also seen by Biswas et al[44]. It seems likely that finite size effects enter here. In a larger system, there would be more modes in this energy region. It seems probably that these vibrations would be pseudo–acoustic propagating modes. At the lowest energy, finite size effects, even for large model systems, will cause an unphysical gap at the bottom of the spectrum, and the states closest to this gap might be, or appear to be, localized in the finite size theory, although not localized but propagating in a macroscopic sample. It is well documented by specific heat and neutron scattering[45,46] that glasses have "extra" low energy modes. These are probably related to defects or "soft" parts of the sample, and are unlikely to be correctly incorporated in a model like that of Feldman et al[32].

The workers then calculate the thermal conductivity $\kappa(T)$, using the standard result

$$\kappa(T) = \int_0^{\omega_{max}} d\omega N(\omega) C(\omega/T) D(\omega)/V \qquad (5.5.4)$$

$$C(\omega/T) = k_B[(\hbar\omega/2k_B T)/\sinh(\hbar\omega/2k_B T)]^2 \qquad (5.5.5)$$

where the latter formula is the specific heat of an oscillator of energy ω. The results are shown as the middle curve in Fig.5.3. There is a significant finite–size problem in this treatment which is discussed fully by Feldman et al[32]. By inspection of Fig.5.6 it is obvious that statistics are quite poor for properties of modes with $\omega < 10\text{meV}$. Although the density of states $N(\omega)$ is small, the diffusivity $D(\omega)$ is large, and the heat capacity for all but low T is saturated at its mamimum value, k_B. The product $N(\omega)D(\omega)$ is not small and is fluctuating widely because of poor statistics arising from the finite size of the system. The nature of the problem is exacerbated when one realizes that in principle the harmonic diffusivity at low ω is a property of propagating modes which must obey a Rayleigh law $D(\omega) \propto \omega^{-4}$ at low ω, and this behaviour is sufficiently singular for the integral (5.5.4) for $\kappa(T)$ to diverge at all T. (For a fictitious harmonic material, the divergence is cut off by the system size, and κ will scale as $L^{\frac{1}{4}}$.) This property of harmonic systems is well known[32], although it is perhaps surprising to realize that it applies not just to weakly disordered crystalline matter, but also to strongly disordered amorphous material.

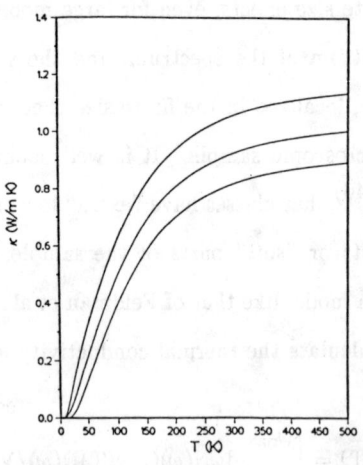

Fig.5.7

Thermal conductivity *versus* temperature for amorphous Si. The middle curve was obtained by integrating the numerical $N(\omega)$ and $D(\omega)$ curves of Fig.5.6; the lower curve is the same with the integrals cut off at a lower frequency of 10meV; the upper curve has $N(\omega)$ and $D(\omega)$ smoothly continued to low ω but the integrals are cut off below 5meV.

One way to estimate the importance of the low-ω region is to replace the low frequency parts of $N(\omega)$ and $D(\omega)$ in Fig.5.6 by appropriately chosen smooth extrapolations. For the density of states, Feldman et al[32] used a Debye extrapolation, $N(\omega) = C_D \omega^2$ with the coefficient C_D determined by the sound velocities $v_L \doteq 7.6 \times 10^3$m/s and $v_T \doteq 3.77 \times 10^3$m/s previously calculated for this model[45]. This not only matches well with the numerical results at $\omega \approx 10$meV, but also fits fairly well with the measured T^3 specific heat[46] (the corresponding theoretical Debye temperature is $\sim$ 450K, while the specific heat gives 530 ± 20K). To extrapolate $D(\omega)$ is more hazardous because there are no firm guidelines, but Feldman et al[32] argue that the method is not very important. Therefore they somewhat arbitrarily chose the Rayleigh ω^{-4} law and used that to continue smoothly on to their calculation at $\omega \doteq$ 10meV where $D(\omega)$ first becomes smooth. Fig.5.7 shows three curves, the middle one being just the raw calculation which is uncontrolled at low ω. The upper one depicts the calculation using the smooth continuation just described for $D(\omega)$ and $N(\omega)$, but with a lower frequency cutoff of 5meV to avoid the divergence. The lowest curve represents the integral of the raw calculations, but with a lower cutoff of 10meV which avoids the region where the calculation is uncontrolled. These results show that the low-ω region in harmonic theory is making a significant contribution to $\kappa(T)$ at all T.

Real materials always have anharmonic interactions, and even the mild anharmonicity of crystalline matter suffices to cure the divergence. However, glassy materials have anomalously small low-T conductivity, which requires an exceptionally effective inelastic process. The accepted description involves scattering from "two-level systems" (TLS). At ω low enough that modes propagate and a Boltzmann description applies, the usual lowest order treatment gives Mattheissen's rule, that is a summation of the various resistive mechanisms:

$$1/D(\omega) = 1/D_h(\omega) + 1/D_{res}(\omega,T) + 1/D_{rel}(\omega,T). \qquad (5.5.6)$$

The first term is the intrinsic harmonic term which Feldman et al calculate numerically at higher ω and continue smoothly at lower ω. Since the last two terms dominate at low ω, the answers ought not to be sensitive to the extrapolation procedure used for D_h. The second term denotes resonant absorption where a propagating vibration of energy ω is absorbed by a TLS of energy splitting ω which is initially in its lower energy state:

$$1/D_{res}(\omega,T) = d_{res}^{-1}(\hbar\omega/k_B T_0)\tanh(\hbar\omega/2k_B T). \qquad (5.5.7)$$

The third term refers to absorption by structural relaxation processes. Although logically independent of the second term, TLS gives one mechanism of relaxational absorption. In the regime where the phonon–induced TLS flipping rate is greater than its energy splitting, the environmental damping has destroyed the simple two–level problem. Vibrational modes perceive this as a classical viscous damping. Hunklinger and Arnold[47] summarize a unified theory of these effects, which yields a somewhat complicated formula for the additional absorption due to relaxational effects. A simplified version of this formula is

$$1/D_{rel}(\omega) = d_{rel}^{-1} \frac{(T/T_0)^3}{1 + (k_B T_0/\hbar\omega)(T/T_0)^3} \qquad (5.5.8)$$

This form was used by Sheng and Zhou[39], with the additional constraint that the coefficients d_{rel} and d_{res} had the ratio 2. Feldman et al relax this constraint, and use three independent adjustable parameters, $T_0 = 20K$, $d_{rel} = 2.5 \times 10^{-4} m^2/s$, and $d_{res} = 1.7 \times 10^{-4} m^2/s$. The ratio is quite close to 2; fixing the ratio at 2 would have little effect on the quality of the fit to experiment.

Although the Mattheissen's rule form (5.5.6) is only justified for the low–frequency propagating modes, the extra inelastic terms in eqn.(5.5.6) are fairly unimportant at higher ω where the harmonic resistivity is large (D_h is small.) Therefore one can in fact use the form of eqn. (5.5.6) for all frequencies. The results are shown in Fig.5.5. Also depicted there are the separated contributions of the frequencies above and below 5meV. The contribution from above 5meV should be compared with the upper curve of Fig.5.7, which is the same thing without the additional inelastic terms, and therefore slightly larger.

An interpretation of the plateau behaviour can now be offered, following Feldman et al. As seen in Figs.5.5 and 5.7 the total thermal conductivity can be viewed as the sum of two different conduction channels. The first, which dominates at low T, is heat carried in a conventional way by propagating long wavelength acoustic modes, scattering strongly from the special inelastic processes available in a glass. These processes essentially annull the low–frequency contribution at higher temperatures, leaving a peak at ≈ 20K which becomes the plateau. The second channel, although less familiar, is quite elementary[32]. It consists of the heat carried by non–propagating modes which are strongly influenced by the glassy disorder but mostly not localized and therefore able to conduct the intrinsic harmonic diffusivity. This is a very smooth term which closely resembles the specific heat and saturates like the specific heat at high T. The sum of these terms will produce a plateau–like feature for a fairly broad range of parameters. This picture is a sort of "shunt resistor model", reminiscent of the model which has often been used to describe the resistivity of metals with strong electron–phonon scattering[48]. When the propagating modes become quite damped and are no longer able to carry much heat, the great reservoir of delocalized and poorly conducting vibrations (the "shunt") takes over, giving a net result in accord with Kittel's picture.

Precisely the same explanation of thermal conductivity was given by Shen and Zhou[39], and by Allen and Feldman[36]. The work of Sheng and Zhou used a somewhat less realistic model, but calculated diffusivities by direct simulation which enabled larger systems to be explored.

APPENDIX 5.1

MODIFICATION OF STOKES–EINSTEIN RELATION FOR LIQUID METALS

When all the particles are identical, the random-walk formula $<r^2> = 6Dt$ for D, with $t \to \infty$, is equivalent to.

$$D = \left[\frac{1}{3N}\right] \int_0^\infty dt \sum_j <v_j(t) \cdot v_j(0)>. \tag{A5.1.1}$$

If the displacements in a cell obey harmonic oscillator dynamics, then a normal mode transformation (Zwanzig[4]) diagonalizes the force constant matrix, leading to the introduction of normal mode frequencies ω. The sum over coordinates in (A5.1.1) can then be viewed as a sum over normal modes. Most of the normal modes may be treated as localized in some subvolumes, denoted as V^*; long wavelength modes cannot be localized in this way, but these make up only a small fraction of the entire spectrum. The time dependence of a single normal mode contribution in (A5.1.1) varies as $\cos \omega t$, until a cell jump interrupts this motion. Normal modes in different subvolumes are interrupted at different times. This is accounted for in the work of Zwanzig[4] by introducing a factor $\exp(-t/\tau)$, the waiting time distribution for cell jumps that destroy coherence of the oscillations in any V^*. The long wavelength modes are not substantially affected by cell jumps, but, as already stressed, these are but a small fraction of the entire spectrum. Then one obtains the following expression for the self-diffusion coefficient D:

$$D = \left[\frac{k_B T}{3MN}\right] \int_0^\infty dt \sum_\omega \cos \omega t \, \exp\left[\frac{-t}{\tau}\right], \tag{A5.1.2}$$

the sum being taken over all the 3N normal mode frequencies. The time integration is easily carried out to yield

$$D = \left[\frac{k_B T}{3MN}\right] \sum_\omega \frac{\tau}{1 + \omega^2 \tau^2}. \tag{A5.1.3}$$

In the absence of much detailed information about the actual frequency distribution, Zwanzig now utilizes a Debye spectrum. Treating longitudinal and transverse oscillations separately, each with its own Debye cutoff q_0:

$$\omega_l(q) = v_l q, \quad \omega_t = v_t q \tag{A5.1.4}$$

for $0 < q < q_0$, v_l and v_t being longitudinal and transverse velocities of sound, respectively. The cutoff q_0 is chosen such that there are N modes in each branch of the spectrum, so that

$$\left[\frac{4\pi}{3}\right]\left[\frac{q_0}{2\pi}\right]^3 = \frac{N}{V} \quad (A5.1.5)$$

where N/V is the number density of ions in the liquid metal. When the sum over frequencies in a single branch of the spectrum is replaced by an integral over q, the result is

$$\left[\frac{1}{N}\right]\sum_q \tau(1 + v^2 q^2 \tau^2) = \left[\frac{V}{N}\right]\left[\frac{1}{2\pi}\right]^3 (4\pi q_0/v^2\tau)\left[1 - \left[\frac{1}{q_0 v\tau}\right]\arctan(q_0 v\tau)\right]. \quad (A5.1.6)$$

The preceding dynamical picture is physically meaningful only if the waiting time τ is much longer than a Debye period, or $q_0 v\tau \gg 1$. Then one can neglect the arctan term in eqn. (A5.1.6). Introducing the mass density $d = MN/V$, one then arrives at the Zwanzig expression

$$D = \left[\frac{k_B T}{3\pi}\right]\left[\frac{3N}{4\pi V}\right]^{1/3}\left[\frac{1}{dv_l^2 \tau} + \frac{2}{dv_t^2 \tau}\right]. \quad (A5.1.7)$$

One next notes that in eqn. (A5.1.7), dv_l^2 is an elastic bulk modulus, whereas dv_t^2 is a corresponding shear modulus. As Zwanzig notes, one expects that $dv_l^2 \tau$ and $dv_t^2 \tau$ are actually the longitudinal and shear viscosities η_ℓ and η, respectively, which he interprets as follows. The viscosity coefficients are determined by time integrals of stress correlation functions. These correlation functions are short–ranged in space, so that one may focus on the local viscosity of the region V*. As long as the system remains in the neighbourhood of a potential minimum, its motions are those of an elastic Debye continuum, and the local stress correlation function of the subvolume V* provides a time–independent local elastic modulus. But when the system leaves this cell, all correlations in V* are lost, and the stress correlation function of this region vanishes. Thus the viscosity is the time integral of the constant elastic modulus multiplied by the exponential waiting time distribution, and the expected results follow. Then the self–diffusion coefficient becomes

$$D = \left[\frac{k_B T}{3\pi}\right]\left[\frac{3N}{4\pi V}\right]^{1/3}\left[\frac{1}{\eta_\ell} + \frac{2}{\eta}\right]. \quad (A5.1.8)$$

Equation (A5.1.8) can then be rewritten in a form like the Stokes–Einstein relation by defining the ionic volume $\Omega = V/N$. The quantity $\Omega^{1/3}$ is a length per particle and takes over the role of the molecular diameter σ. Then one can write

$$\left[\frac{D\eta}{k_B T}\right]\Omega^{1/3} = 0.0658\left(2 + \frac{\eta}{\eta_l}\right) = C'. \qquad (A5.1.9)$$

Although the actual value of C', as defined by this equation, clearly depends on the ratio of the shear viscosity to the longitudinal viscosity, which is still often not known quantitatively, Zwanzig[4] notes that C' as defined can vary only between 0.13 and 0.18. In the text eqn. (A5.1.9) has been brought into contact with transport coefficients of liquid metals just above their melting points, following the study of Brown and March[1]; see also March[4].

APPENDIX 5.2

DERIVATION OF INVERSE–TRANSPORT THEORY FOR NON–INTERACTING ELECTRONS

In the body of this article rather extensive use of the inverse–transport theory of non–interacting electrons has been made. The treatment below leads back to the results of Rousseau et al[19](RSM), with a polarization–denominator contribution relating to the role of bound states in electronic transport. To date, the theory has not been derived for interacting electrons (a problem for further work).

Following McCaskill and March[49], independent electrons moving in a total scattering potential V will be considered below. The inverse–transport theory is known from the work of Leung and March[50] not to be exact at finite temperature, as a result of a distribution of drift velocities rather than a single value $\tilde{v}$. Below, the finite–temperature expressions will be retained, but the anlaysis is only exact in the limit as the Fermi function f(E) approaches the step function characteristic of Fermi statistics at T = 0.

The structure of this Appendix is as follows. Immediately below, a force balance is set up on the asymptotic timescale to be defined. It will be demonstrated that this leads to a polarization denominator in the inverse–transport theory of RSM. We then account specifically for bound states without any a priori assumptions, showing thereby how their influence is calculated within the theory. Following this, a discussion using time–dependent techniques will be given. In conclusion, the relation to the decomposition of the forces in electromigration theory will be briefly referred to.

It will first be shown that the balance of force in the direct response to an external field, as in the Greenwood[51] calculation of the current, does not describe a transport situation: an additional force is needed. The inverse–transport theory meets this requirement, if in addition, the terms resulting from the direct linear response are retained[52]. One of these then contributes a polarization contribution to the denominator, thereby transcending the work of RSM[19].

Using the gauge defined by the vector potential $\mathbf{A} = (-cFt, 0, 0)$ where F is the electric-field strength and c the velocity of light, and periodic boundary conditions, McCaskill and March[49] obtained for the linear response to the equilibrium density matrix f(H),

$$\rho = f(H) + g, \qquad (A5.2.1)$$

the off-diagonal matrix with elements

$$g_{nm} = i\hbar e\, Fv_{nm} \frac{f_n - f_m}{(E_n - E_m)^2} \left\{ 1 - \exp\left[\frac{(E_n - E_m)t}{i\hbar}\right] \right\} \qquad (A5.2.2)$$

in the adiabatic basis defined by

$$H(t)|n(t)\rangle = E_n(t)|n(t)\rangle \qquad (A5.2.3)$$

for the perturbed Hamiltonian

$$H(t) = \frac{1}{2m}\left[\left[\frac{\hbar}{i}\frac{\partial}{\partial x} + eFt\right]^2 + \left[\frac{\hbar}{i}\frac{\partial}{\partial y}\right]^2 + \left[\frac{\hbar}{i}\frac{\partial}{\partial z}\right]^2\right] + V(x,y,z). \qquad (A5.2.4)$$

Here e is the electronic charge, V the electronic potential in the condensed matter, f(E) the Fermi function as above and v the velocity operator $(\hbar/mi)\partial/\partial x$.

The force on the electrons, after perturbation by the electric field, may be directly computed as

$$F_{xdir}(t) = -2\mathrm{Re}\left\{ \mathrm{Tr}\left[\frac{\partial V}{\partial x} g(t)\right] \right\}. \qquad (A5.2.5)$$

Inserting eqn.(A5.2.2) and using the Fourier-Laplace transform to take the $t \to \infty$ limit, we obtain

$$\lim_{t \to \infty} E_{xdir} = 2\pi\hbar e F \sum_{m \neq n} \mathrm{Re}(V'_{nm} v_{mn})\left[\frac{\partial f}{\partial E}\right]_{E_n} \delta(E_n - E_m)$$
$$- 2\hbar e F \sum_{m \neq n} \mathrm{Im}(V'_{nm} v_{mn}) \frac{f_n - f_m}{E_n - E_m} \frac{P}{E_n - E_m}, \qquad (A5.2.6)$$

where P denotes the principal part in integrals. Only on the timescale $\hbar/k_B T \ll t \ll \hbar/\Delta E$ does the quasidelta function in the first term provide a time-independent result, and the evaluation of the second term requires similar care. Here ΔE is a characteristic energy spacing, as discussed at a similar stage of the current calculation by Greenwood[51].

Using the above treatment, the resistivity R is calculated by McCaskill and March[49] to have the RSM form $N/(1 + D)$, where N is the numerator of RSM. The 'correction' D in the denominator is then given by

$$D = \frac{2\pi\hbar}{N_e} \sum_{m \neq n} \text{Re}\left[\left[\frac{\partial V}{\partial x}\right]_{nm} v_{mn}\right] \left[\frac{\partial f}{\partial E}\right]_{E_n} \delta(E_n - E_m)$$
$$+ \frac{2\hbar}{N_e} \sum_{m \neq n} \text{Im}\left[\left[\frac{\partial V}{\partial x}\right]_{mn} V_{mn}\right] \frac{f_n - f_m}{E_n - E_m} \frac{P}{E_n - E_m}. \quad (A5.2.7)$$

If there exists a timescale that separates the polarization behaviour of localized states from those delocalized over the entire system –

$$\frac{\hbar}{\Delta E_b} \ll t \ll \frac{\hbar}{\Delta E_d}, \quad (A5.2.8)$$

where ΔE_b and ΔE_d are characteristic energy differences between nearest levels for bound and delocalized states – then the denominator yields a constant value, which is calculated by McCaskill and March as

$$D = -N_b/N_e, \quad (A5.2.9)$$

i.e. the ratio of the number of bound electrons to the total number of electrons. The effect of the denominator when eqn.(A5.2.9) holds is to yield a resistivity that is just that of the numerator N but with N_e modified to $N_e - N_b$. However, in the asymptotic timescale the bound electrons do not drift with respect to the potential V, and hence the density ρ_e should also be modified to $(N_e - N_b)/\mathscr{V}$, where $\mathscr{V}$ is the volume. This provides a total correction $[N_e/(N_e - N_b)]^2$ to the RSM expression for the resistivity. The bound electrons make no contribution to the numerator, which would vanish with box boundary conditions.

The clear separation of the behaviour of bound and delocalized electrons is no longer possible in the vicinity of the metal–insulator transition in particular, where the polarization time for some marginally bound state may be long. In any case, in the absence of an a priori estimate of N_b, the resistivity may still be calculated as

$$R = \frac{N}{(1 + D)^2}, \quad (A5.2.10)$$

with D given by eqn.(A5.2.7) and N by the RSM numerator, but with ρ_e the total electron density. Equation (A5.2.10) describes the structure of a transport theory independently of assumptions about the number of electrons participating. The above argument avoids the difficulties associated with the definition of such a description, at least at a formal level.

In concluding this Appendix, it is of interest to stress, following McCaskill and March[49], that the decomposition of the force that occurs in inverse–transport theory outlined above has a parallel history in the theory of electromigration[53-55]. In particular, the discussion there has revolved around the ambiguity of a division of forces into electron–wind and direct–field parts. Here the numerator plays the role of the 'ergodic' electron–wind contribution, and the polarization denominator that of the direct–field force. The degree to which polarization can take place on a physical timescale (asymptotic–time domain) is limited for nearly free electrons (and the numerator provides the dominant force), but absolute for bound electrons and significant for strong–scattering disordered condensed matter.

APPENDIX 5.3

WIEDEMANN–FRANZ LAW

In section 5.3, the Wiedemann–Franz law was utilized. The purpose of this Appendix is twofold:

(i) To present results for the Lorenz number for a variety of liquid metals at or near the melting temperature T_m

and

(ii) To summarize arguments showing the conditions under which the Wiedemann–Franz law could remain valid up to the metal–insulator transition, say along the liquid–vapour coexistence curve.

As used already, the Wiedemann–Franz law asserts the constancy of $\kappa_e/\sigma T$, where κ_e is the electronic contribution to the thermal conductivity while σ denotes the electrical conductivity. Theory relates $\kappa_e/\sigma T$ precisely to the so-called Lorenz number, through

$$\frac{\kappa_e}{\sigma T} = \frac{\pi^2 k_B^2}{3e^2} = L. \qquad (A5.3.1)$$

In the units employed in Table A5.3.1 below, where experimental data on a variety of liquid metals have been collected, L is 2.45.

The expression (A5.3.1) rests on solution of the Boltzmann equation, which is given in various places[56),57)]. In the customary arguments, it is stressed that the result (A5.3.1) does not depend on specific assumptions about Fermi surfaces, density of electron states, nature of scattering potential, and the like. It does, however, involve the general assumption that the scattering is elastic.

Table A5.3.1

$\frac{\kappa}{\sigma T}$ at melting point from experiment

Li	2.6	Aℓ	2.4
Na	2.2	Ga	2.07
K	2.1	Tℓ	3.2
Cs	2.4	Sn	2.9
Cd	2.5	Pb	2.4
Hg	2.75	Sb	2.6
Zn	3.2	Bi	2.5

It can be seen from Table A5.3.1 that while there are some 12 cases in which the Lorenz number lies in the (somewhat arbitrarily chosen) range 2.1 to 2.8, the cases of Zn and Tℓ lie outside these limits. To the writer's knowledge, no solution has been proposed to date for the spread of values around the Lorenz number.

A5.3.1 Fermi Liquid Model of Wiedemann–Franz Law up to Metal–Insulator Transition

Let us turn to the second objective of this Appendix. This is to consider the validity of the Wiedemann–Franz law in a specific model, namely that of Landau's Fermi liquid theory[58], well away from the melting point, detailed in Table A5.3.1, and indeed right up to the critical point along, say, the liquid–vapour coexistence curve. This problem has been considered by Castellani et al[59]. Specifically, these workers have studied the heat diffusion and the thermal conductivity of an interacting disordered electron liquid in the metallic regime close to the metal–insulator transition. The heat-diffusion constant provides a direct measurement of the dressed, or quasi–particle diffusion constant, which scales differently from the charge diffusion constant. Castellani et al[59] demonstrate, within the framework of the Fermi liquid model, that the thermal conductivity scales like the electrical conductivity, thereby establishing in this model the validity of the Wiedemann–Franz law (A5.3.1) up to the metal–insulator transition.

While in no way denying the considerable interest of the above theory, Chapman and March[25] have cautioned that there is evidence that, in taking the fluid alkalis up towards the critical point (see also section 5.4.3), the Fermi liquid picture does not hold up to the metal–insulator transition. It remains to be seen, therefore, whether there are other liquid–metallic systems where the above model will come into its own.

References

1). R.C. Brown and N.H. March, Phys.Chem.Liquids, **1**, 141 (1968).

2). E.N. da C. Andrade, Phil.Mag. **17**, 497 (1934).

3). See, for instance, T.E. Faber, Theory of Liquid Metals (Cambridge University Press), (1972).

4). R.W. Zwanzig, J.Chem.Phys. **79**, 4507 (1983); see also, N.H. March, J.Chem.Phys. **80**, 5345 (1984).

5). E. Haeffner, Nature, **172**, 775 (1975).

6). M. Ginoza and N.H. March, Phys.Chem.Liquids, **15**, 75 (1985).

7). N.H. March and B.V. Paranjape, J.Phys.Chem.Solids, **53**, 1055 (1992).

8). See, for example, A.H. Wilson, The Theory of Metals, 2nd Edition (Cambridge University Press) (1965).

9). Burkov and Vedernikov, this Volume.

10). M. Moreau, C.R.Acad.Sci.Paris, **130**, 412, 562 (1900).

11). P.W. Bridgman, Phys.Rev. **24**, 644 (1924).

12). See, for instance, N.H. March, Liquid Metals : Concepts and Theory (Cambridge University Press) 1990.

13). See B.V. Paranjape and J.S. Levinger, Phys.Rev. **120**, 437 (1960)

14). See, for instance, R.E. Robson, N.H. March and B.V. Paranjape, Plasma Physics, to appear 1994.

15). K. Huang, Proc.Phys.Soc. **60**, 161 (1948).

16). N.H. March, Phil.Mag. **32**, 497 (1975).

17). E. Gerguoy, J. Math.Phys. **6**, 993 (1965).

18). G.D. Gaspari and B.L. Gyorffy, Phys.Rev.Lett. **28**, 801 (1972).

19). J.S. Rousseau, J.C. Stoddart and N.H. March, J.Phys. **C4**, L59 (1971).

20). N.H. March and A.M. Murray, Phys.Rev. **120**, 830 (1960).

21). J.S. Rousseau, J.Phys. **C4**, L351 (1971).

22). R. Harris, J.Phys. **C5**, L56 (1972).

23). See N.H. March, Linear and Non–Linear Electron Transport in Solids (Plenum : New York) Eds. J.T. Devreese and V.E. van Doren, p131, 1976.

24). N.H. March, M. Suzuki and M. Parrinello, Phys.Rev. **B19**, 2027, (1979).

25). R.G. Chapman and N.H. March, Phys.Rev. **B38**, 792 (1988).

26). W.W. Warren. Phys.Rev. **B29**, 7012 (1984).

27). W. Freyland, Phys.Rev. **B20**, 5104 (1979).

28). L. Bottyan, R. Dupree and W. Freyland, J.Phys. **F13**, L173 (1983).

29). W.W. Warren, J.Phys. Condensed Matter **5**, Suppl. **34B**, B211 (1993).

30). L.J. Lantto, Phys.Rev. **B22**, 1380 (1980).

31). N.H. March, Phys.Chem. Liquids, **22**, 191 (1990).

32). J.L. Feldman, M.D. Kluge. P.B. Allen and F. Wooten, Phys.Rev. **B48**, 12589 (1993); see also F.H. Stillinger and T.A. Weber, Phys.Rev. **B31**, 5262 (1985).

33). G. Pompe and E. Hegenbarth, Phys.States Solidi **B147**, 103 (1988).

34). D.G. Cahill, H.E. Fischer, T. Klitsner, E.T. Swartz and R.O. Pohl, J.Vac.Sci.Technol. **A7**, 1259 (1989).

35). P.W. Anderson, B.I. Halperin and C.M. Varma, Phil.Mag. **25**, 1 (1972); see also W.A. Phillips, J.Low Temp.Phys. **7**, 351 (1972).

36). P.B. Allen and J.L. Feldman, Phys.Rev. **B48**, 12581 (1993).

37). C. Kittel, Phys.Rev. **75**, 972 (1948).

38). G.A. Slack, Solid State Physics, **34**, 1 (1979).

39). P. Sheng and M. Thou, Science **253**, 539 (1991).

40). M.S. Love and A.C. Anderson, Phys.Rev. **B42**, 1845 (1990).

41). D.N. Payton, M. Rich and W.M. Visscher, Phys.Rev. **160**, 706 (1967).

42). W.A. Kamitakahara. C.M. Soukoulis, H.R. Shanks, V. Buchenau and G.S. Grest, Phys.Rev. **B36**, 6539 (1987).

43). See Chapter by F. Siringo, present Volume.

44). R. Biswas, A.M. Bouchard, W.A. Kamitakahara, G.S. Grest and C.M. Soukoulis, Phys.Rev.Lett. **60**, 2280 (1988).

45). V. Buchenau, Yu.M. Galperin, V.L. Gurevich, D.A. Parshin, M.A. Ramos and H.R. Schober, Phys.Rev. **B46**, 2798 (1992).

46). M. Garcia–Hernandez, A. Burriel, F.J. Mermejo, C. Piqué and J.L. Martinez, J.Phys.Condens.Matter **4**, 9581 (1992).

47). S. Hunklinger and W. Arnold, in Physical Acoustics: Eds. W.P. Masort and R.N. Thurston (Academic: New York, 1976) Vol.**12**, p155.

48). See, for example, P.B. Allen, in Superconductivity in d– and f– Band Metals, Eds. H. Suhl and M.B. Maple, (Academic : New York, 1980), p.291.

49). J.S. McCaskill and N.H. March, J.Phys.Chem.Solids, **45**, 215 (1984).

50). C.H. Leung and N.H. March, Plasma Physics (GB) **11**, 277 (1977).

51). D.A. Greenwood, Proc.Phys.Soc. **71**, 585 (1958).

52. See, for example, ref.49.

53). R.S. Sorbello and P.R. Rimbey, Phys.Rev. **B38**, 1095 (1988).

54). W. Jones and H.N. Dunleavy, J.Phys. **F9**, 1541 (1979).

55). W. Jones and G.C. Barker, Phys.Chem.Liquids **17**, 105 (1987).

56). See, for instance, J.M. Ziman, Electrons and Phonons (Oxford University Press) 1960.

57). See also G.V. Chester and A. Thellung, Proc.Phys.Soc. **77**, 1005 (1961).

58). See, for instance, P. Fulde, J.Phys. **F18**, 601 (1988).

59). C. Castellani, C. DiCastro, G. Kotliar and P.A. Lee, Phys.Rev.Lett. **59**, 477 (1987).

CHAPTER 6
ELECTRICAL AND THERMOELECTRIC PROPERTIES OF DISORDERED METALLIC BINARY CONTINUOUS SOLID SOLUTIONS

A.T. Burkov, M.V. Vedernikov

A.F. Ioffe Physico - Technical Institute, Academy of Sciences of the USSR,
1942021 St. Petersburg, USSR

CONTENTS

6.1.	Introduction	362
6.2.	Theory	364
6.3.	Experimental Data on the Electrical Resistivity and Thermopower of BCSS	366
	(a) Isoelectronic Metal Alloys	370
	(i) Potassium - Rubidium Alloy (K - Rb)	370
	(ii) Potassium - Cesium Alloys (K - Cs)	370
	(iii) Rubidium - Cesium Alloys (Rb - Cs)	371
	(iv) Calcium - Stronitium Alloys (Ca - Sr)	371
	(v) Scandium - Yttrium Alloys (Sc - Y)	371
	(vi) Titanium - Zirconium Alloys (Ti - Zr)	371
	(vii) Titanium - Hafnium Alloys (Ti - Hf)	371
	(viii) Zirconium - Hafnium Alloys (Zr - Hf)	372
	(ix) Vanadium - Niobium Alloys (V - Nb)	372
	(x) Niobium - Tantalum Alloys (Nb - Ta)	372
	(xi) Chromium - Molybedenum Alloys (Cr - Mo)	372
	(xii) Chromium - Tungsten Alloys (Cr - W)	372
	(xiii) Molybedenum - Tungsten Alloys (Mo - W)	372
	(xiv) Rhodium - Iridium Alloys (Rh - Ir)	372
	(xv) Palladium - Platinum Alloys (Pd - Pt)	392
	(xvi) Silver - Gold Alloys (Ag - Au)	392
	(b) Nonisoelectronic Metal Alloys	392
	(i) Scandium - Zirconium Alloys (Sc - Zr)	392
	(ii) Vanadium - Chromium Alloys (V - Cr)	395
	(iii) Vanadium - Molybedenum Alloys (V - Mo)	395
	(iv) Vanadium - Tungsten Alloys (V - W)	395
	(v) Niobium - Molybedenum Alloys (Nb - Mo)	395
	(vi) Niobium - Tungsten Alloys (Nb - W)	395
	(vii) Tantalum - Molybedenum Alloys (Ta-Mo)	395
	(viii) Tantalum - Tungsten Alloys (Ta - W)	395
	(ix) Rhodium - Palladium Alloys (Rh - Pd)	413
	(x) Rhodium - Platinum Alloys (Rh - Pt)	413
	(xi) Palladium - Silver Alloys (Pd - Ag)	413
	(xii) Palladium - Gold Alloys (Pd - Au)	415
	(xiii) Rhenium - Osmium Alloys (Re - Os)	415
6.4.	Discussion of Experimental Data	419
6.5.	Conclusion	421
References		422

6.1 INTRODUCTION

The review covers experimental data on thermopower and electrical resistivity of the simplest but very important class of metallic alloys, viz. binary continuous disordered substitutional solid solutions (BCSS). The term "continuous" means here that for any concentration of the components of an $A_x B_{1-x}$ alloy, no new compounds with a structure different from that of the initial components will form, the probability for a given lattice site being occupied by an atom of species A being proportional only to the concentration x of the atoms of this species in the alloy. The alloys of this class occupy a particular place in the physics of metals since, because of their comparative simplicity, many of them serve as model objects for testing the current theoretical ideas concerning the electronic structure and properties of disordered metallic systems, which play an ever increasing role in many modern technologies. Note that there exists only limited number of alloys of this class. If we consider only systems in which complete mutual solubility prevails throughout the region of existence of the solid state, i.e. from 0°K to the alloy melting point, and exclude magnetically ordered systems then we will find only about 30 such BCSS systems. Many properties of the BCSS have been a subject of comprehensive experimental and theoretical investigation. In particular, the dependence of the electrical resistivity of BCSS on composition was found to follow an empirical rule of Kurnakov which was theoretically substantiated by Norgheim. This rule, however, was formulated basing on an analysis of a fairly limited set of experimental data, indeed, as this became clear later on, it is not capable of describing all available experimental information. The first attempt at a systematic analysis of the electrical resistivity of BCSS was made by author of [1,2]. It was established that all BCSS can be divided into two types according to the dependence of their electrical resistivity on alloy composition, namely, in the alloys of the first type the impurity resistivity depends on composition by the rule of Kurnakov-Nordheim :

$$\rho_{imp} = A.x(1-x) , \qquad (1)$$

where x is the atomic concentration of one of the alloy components. In alloys of the second type the electrical resistivity do not obey this rule, the dependence of impurity resistivity on alloy composition being depicted by a curve which is asymmetric relative to the equiatomic composition. The authors established also that the alloys of the first type can be formed only of isoelectronic (i.e. belonging to the same group of the Periodic Table) metals, whereas those of the second type are made primarily of nonisoelectronic metals. Later this classification was confirmed also by an analysis of thermopower[3,4]. Information on the electrical resistivity and thermopower of some BCSS can be found in the recent edition of Landolt-Bornstein[5]. No complete review of the available experimental resistivity of BCSS has, however, been published up to now. We have recently published a review of experimental data on the thermopower of 29 systems over the temperature range of 100 °K to 2000 °K[4]. A more comprehensive experimental investigation of the properties of BCSS posed a number of question. It was established earlier that at high (above Debye) temperatures the thermopower and electrical resistivity of many metals, rather than vary with temperature by theoretical relations of Mott and Bloch-Gruneisen [6,7], follow more complex laws. While the electrical resistivity of a number of metals at high temperatures grows with temperature slower than by the linear law and may reveal trend to saturation, there are metals whose resistivity increases with temperature faster than linear rate. The high temperature thermopower exhibits a diverse behavior which can not be described by a simple linear dependence following from Mott's expression even in the last approximation. A significant observation was made on pure metals : all metals belonging to the same group of the Periodic Table can, in the first approximation, be characterized by a common dependence of the diffusion thermopower on temperature but temperature dependences for metals from different groups are qualitatively different[8]. At the same time there are indications[9,10] that the temperature behavior, primarily of thermopower, is intimately connected with the structure of the electronic spectrum near the Fermi level. In alloys there are at least two other factors that can affect the transport properties : (i) additional conduction electron scattering due to disordered distribution of the alloy component atoms over the lattice sites ,and (ii) the change of the electronic structure of a metal induced by the dissolution of the second component. Both these factors may influence the absolute values of the thermopower and electrical resistivity, and their temperature behavior. In alloys of isoelectronic metals which are close in electronic structure, impurity scattering is a dominant factor. In alloys of nonisoelectronic metals, however, both these factors may play an important role. This should manifest itself in the way the temperature dependences of thermopower and electrical resistivity transform when the composition of alloys of isoelectronic or nonisoelectronic metals is varied. A parallel analysis of data for these both types of alloys should permit, in our opinion, separating to a certain extent the effects of these factors and following the influence of changes in the electronic structure of a conductor on the temperature dependences of the electronic transport properties.

The present paper is an attempt at making as complete a review as possible of the available experimental data on the electrical resistivity and thermopower of BCSS vs alloy composition and temperature. A considerable fraction of these data has been obtained recently by us and part of them is being published for the first time. We do not consider here alloys revealing magnetic ordering, the only exception being alloys of Chromium which is an antiferromagnet with a Neel temperature of 312 °K . Therefore the paper does not cover the vast group of rare earth metals alloys. The paper also does not discuss such problems as phonon drag, the Kondo effect, and quantum interference correction to the electrical conductivity and thermopower. These effects being observed usually at low temperatures, substantially below Debye temperature, we will place emphasis in our analysis of experimental data on the high temperature behavior of thermopower and electrical resistivity, i.e. in the range of the order and above the Debye temperature.

The paper is constructed in the following way. The first section gives the concise review of the formal theory of electronic transport properties with an analysis of the validity of the approximations employed. The second section presents the available literature and our own experimental data on the thermopower and electrical resistivity of BCSS versus alloy composition and temperature. The third section discusses the experimental data and formulate the principal relations connecting thermopower and electrical resistivity with composition and temperature. The approach to the interpretation of these relations is briefly outlined.

6.2 THEORY

The electrical current j generated in conductor to which an electric field E is applied is determined by the electrical conductivity σ of this conductor (or its inverse, the electrical resistivity $\rho = 1/\sigma$):

$$j = \sigma.E, \quad \nabla T = o \quad (2)$$

The Seebeck effect consists in the generation in a conductor of an electric field E in the presence in it of a temperature gradient ∇T under the condition of zero net current:

$$E = S.\nabla T, \quad j = 0 \quad (3)$$

Here S is thermopower coefficient. Generally speaking, the coefficients σ and S are tensors of second rank, however for isotropic media these tensors degenerate into scalars. Since we are going to consider in what follows systems which are at least macroscopically isotropic, all the expressions will be given in the scalar form. Besides thermopower, there are other thermoelectric phenomena, namely, the effects of Peltier (π) and Thomson (μ_t). However they are not independent, and the corresponding coefficients (π and μ_t) can be expressed in terms of the thermopower coefficients. The relation between the various thermoelectric effects follows from Onsager's principle of symmetry of kinetic coefficients which, in its term, is a consequence of the reversibility of the equations of motion[7,11]. The relations between the principle thermoelectric coefficients called Thomson's relations can be written [7,11]

$$\pi = T.S$$
$$\mu_t = T.dS/dT \quad (4)$$

Microscopic consideration of the electronic transport phenomena is usually carried out basing on Boltzmann equation to the description of the transport of heat and charge in a conductor is the possibility of separating carrier drift in an external field into free motion and the scattering processes. Solving this equation for the case of stationary and homogenous external fields in an approximation linear in these fields results in the following expressions for the electrical resistivity and thermopower:

$$\sigma(T) = \int_0^\infty \omega(\varepsilon, T).(-\partial f^0/\partial \varepsilon).d\varepsilon \quad (5)$$

$$S(T) = \frac{1}{e.T} \frac{\int_0^\infty \omega(\varepsilon, T).(\varepsilon, \mu).(-\partial f^0/\partial \varepsilon).d\varepsilon}{\int_0^\infty \omega(\varepsilon, T).(-\partial f^0/\partial \varepsilon).d\varepsilon} \quad (6)$$

$$\omega(\varepsilon, T) = \frac{e^2}{12\pi^3 h} \oint \frac{\tau(\varepsilon, k, T).v^2}{|\nabla \varepsilon|} .dA \quad (7)$$

where f^0 is Fermi-Dirac distribution function, $\tau(\varepsilon, k, T)$, v and μ – relaxation time, velocity and electrochemical potential of electrons respectively. The integral in eq. (7) is taken over the surface of constant energy $\varepsilon(k) =$ const. in wave vector k-space.
Equations (5)-(7) are of a general nature and can be applied to a description of the electrical conductivity and thermopower of conductors irrespective of the degeneracy of the electron gas. The area of their validity is limited by the boundaries of applicability of the linearized Boltzmann equation. For metals this approximation may be considered fully substantiated, since it leads, e.g., to Ohm law for electrical conductivity which is confirmed with a high precision experimentally. For metals one employs usually one more approximation associated with the fact that the electron gas in metals is strongly degenerated. In this case the function $-\partial f^0/\partial\varepsilon$ has a sharp peak at $\varepsilon = \mu$, the integrand in (5) and (6) being nonzero in a narrrow (compared to the Fermi energy) region near the Fermi energy. Expanding the function $\omega(\varepsilon, T)$ in a Taylor series in the vicinity of $\varepsilon = \mu$ and restricting oneself only to the first nonvanishing terms, one can readily obtain the following expressions for the electrical conductivity and thermopower[7,11].

$$\sigma(T) = \omega(\mu, T) \qquad (8)$$

$$S(T) = -\frac{\pi^2 K_B^2}{3|e|} T \left[\frac{1}{\omega}\frac{\partial\omega}{\partial\varepsilon}\right]_{\varepsilon=\mu} \qquad (9)$$

Equation (8) leads to the well known Bloch-Gruneisen law which predicts for the lattice electrical resistivity a linear dependence on temperature above the Debye temperature. Eq.(9) is called Mott's expression for thermopower. According to this expression, thermopower should also be a linear function of temperature. However, as already pointed out, experimental data suggest that the electrical resistivity and, particularly, thermopower not always follow these laws. The complex behavior of these properties at high temperature may be connected with the existence of fine structure in the electron energy spectrum in the vicinity of the Fermi energy with a characteristic energy scale $\approx 0.05 - 0.1$ eV. The existence of such a structure, in particular, in the density of electronic states, is supported by present calculations of the band structure of metals[12]. In this case the approximation of strong degeneracy used in the derivation of eqs. (8) and (9) can't be considered as substantiated particulary at high temperatures,where the carrier thermal energies are comparable with the scale of the energy spectrum features. This implies that the high temperature behavior of thermopower and electrical resistivity should be analyzed basing on the more general relations (5)-(7). It should be pointed out that this approach is presently not universly acccepted in the physics of metals. Usually,even in studies of high temperature properties and of their anomalies the considerations of the inapplicability of the strong degeneracy approximation are disregarded. In our opinion, however, this point is of a cruicial importance for the interpretation of the complex behavior of electrical resistivity and, in particular, of thermopower at high temperatures.

In case of disordered metal and alloys the situation becomes still more complicated by the existance of the radical difficulties in the description of the electronic structure of disordered metals systems, which have not yet been overcome. The theory of electronic transport in disordered metal alloys is still less developed. Therefore there are at present no sufficiently general and well substantiated theoretical models which would connect the electrical resistivity and thermopower of a disordered alloy with its compositions. We have already mentioned the empirical rule of Kurnakov-Nordheim for impurity electrical resistivity of a disordered alloy. According to this rule, the electrical resistivity ρ_{imp} associated with electron scattering from potential fluctuations at lattice sites of a disordered alloy is connected with the concentration x of the alloy components through a simple relation :

$$\rho_{imp} = A \cdot x(1-x) \qquad (10)$$

where $A = \text{const}(x)$. This simple relation, however, describes satisfactorily the impurity part of the resistivity only for some alloys formed of isoelectronic metals. Another relation which widely used in the analysis of resistivity experimental data, namely, Matthiessen rule, states that the total electrical resistivity of a dilute alloy can be represented in the form :

$$\rho = \rho_{imp} + \rho_{ph}(T) \qquad (11)$$

where $\rho_{ph}(T)$ is temperature-dependent electrical resistivity of the pure metal solvent. This rule is sufficiently well validated theoretically only for very dilute alloys and for a low scattering potential of the impurity. Nevertheless, even for such alloys experiment reveals noticeable deviations from it[13]. Quite frequently this rule is treated in a generalized way, and it is assumed that the resistivity of any disordered alloy can be represented in the form of two mutual independent parts, namely, a temperature-dependent one which is associated with conduction electron scattering from thermal exitations, and a temperature-independent impurity contribution. Generally speaking, however, this approach is theoretically not validated. A consequence of Matthiessen rule for the electrical resistivity and Mott's expression for thermopower is the Nordheim-Gorter rule which relates the thermopower and resistivity of an alloy at a fixed temperature[11] :

$$S = (S_{ph} - S_{imp}) \cdot \rho_{ph}/(\rho_{ph} + \rho_{imp}) + S_{imp} \qquad (12)$$

As follows from (11), the dependence of the total thermopower S of an alloy on total reciprocal resistivity should be linear. From the slope of this dependence and the intercept with the vertical axis one can determine the parameters S_{ph} and S_{imp} which are interpreted as the thermopowers associated with electron scattering from phonons and impurites, respectively. The significance of this treatment, however, is not very high. The derivation of the Nordheim - Gorter expression (just as the theoretical derivtion of Matthiessen rule) is based on fairly rigid assumptions, namely, (i) independence of electronic structure on alloy composition, and (ii) independence between the electron-phonon and electron - impurity scattering. Both these conditions can apparently be realized only in dilute alloys of isoelectronic metals.

6.3 EXPERIMENTAL DATA ON THE ELECTRICAL RESISTIVITY AND THERMOPOWER OF BCSS

In presenting here the experimental data, we will follow a scheme based on the only common relation aaassociaated with the dependence of the electronic transport properties on BCSS composition. This relation has already formulated in the introduction. It consists essentially in that all BCSS can be divided into two groups according to the character of the dependences of thermopower and resistiviy on the composition. The first group of alloys combines BCSS formed of isoelectronic metals. The second group encompasses BCSS formed of nonisoelectronic metals, In accordance with such a classification, we will first present here data on alloys of the first type and, subsequently, information on alloys of second type. All the considered alloys are listed in Table 1 in the same order. The content of the alloy components is specified in atomic per cent.

Prior to presenting the experimental data , some comments on errors in electrical resistivity and thermopower measurements are in order. The precision in measuring the electrical resistivity is connected primarily with determining the sample demensions (effective cross section and distance between the potential probes). Most of the data on the resistivity of BCSS were obtained by the four-probe technique on short bulk samples. The accuracy of determining the dimensions of the sample of pores, microcracks and other mechanical defects. As result of presence of such defects, the effective dimensions of sample may differ from the measured geometrical dimensions, that will lead to errors in determining the absolute value of resistivity and the slope of its temperature dependence. These errrors exceed consideraably in most cases the errors in the measurement of electric signals and those caused by disregarding the thermal expansion of the sample. Therefore we assume the accuracy in the determination of the absolute value of resistivity, unless otherwise specified, to be not better than 1%.

The accuracy of determining the thermopower is essentially limited by the present accuracy of the absolute thermoelectic scale. The absolute thermoelectric scale was established by measuring Thomson heat for a number of reference materials used as primary reference electrodes. The absolute thermopower of these materials was calculated from the measured values of Thomson heat using the relation :

$$S(T) = \int_0^T \frac{\mu_t}{t} dt \qquad (13)$$

At low temperatures (0 - 300 °K) lead is used as the reference material. The most precise data on the absolute thermopower of lead were obtained by Roberts [14]. In the high temperature domain one uses as materials for references electrodes copper , platinum, molybdenum, tungsten. There are sufficiently reliable data on the absolute thermopower of copper and platinum approximately up to 1700 °K[11]. At still higher temperatures one uses molybdenum and tungsten, however, information on the thermopower of these metals above 1700 °K is considerably less reliable. It is based on the only measurements of Thomson heat made by Lander[15]. These measurements were used to calculate the absolute thermopower for a number of metals upto 2400 °K[16]. A comparison of different data on absolute thermopower of reference electrodes and an analysis of the precision in its determination permit one to evaluate the accuarcy of the existing thermoelectric scale as $\mp 0.2\, \mu\, V/K$ for temperatures up to 1000 °K, $\mp 0.5\, \mu\, V/K$ in the range 1000 - 1700 °K, and $\mp 2\, \mu\, V/K$ above 2000 °K.

Since a considerable part of the experimental data on the thermopower of BCSS, particulary in the high - temperature domain, were obtained by the present authors, we consider it reasonable to present here information on the reference electrodes used in our thermopower measurements. In the temperature range 100 °K to 500 °K we used as electrode material copper calibrated in the low temperature region (up to 350 °K) against lead, using literature data[14] on its thermopower. In the range from 300 °K to 1000 °K we

Table 1. BINARY CONTINUOUS SOLID SOLUTIONS $A_x B_{1-x}$

	Metal A	Metal B	$T_{melt}[K]$ A	$T_{melt}[K]$ B	Lattice
1	Potassium	Rubidium	336.6	312.6	bcc
2	Potassium	Caesium	336.6	301.5	bcc
3	Rubidium	Caesium	312.6	301.5	bcc
4	Calcium	Strontium	1125	1043	fcc
5	Scandium	Yttrium	1803	1793	hcp
6	Titanium	Zirconium	1941	2125	hcp
7	Titanium	Hafnium	1941	2503	hcp
8	Zirconium	Hafnium	2125	2503	hcp
9	Vanadium	Niobium	2190	2742	bcc
10	Niobium	Tantalum	2742	3270	bcc
11	Chromium	Molybdenum	2150	2895	bcc
12	Chromium	Tungsten	2150	3653	bcc
13	Molybdenum	Tungsten	2895	3653	bcc
14	Rhodium	Iridium	2239	2720	fcc
15	Palladium	Platinum	1827	2045	fcc
16	Silver	Gold	1233.5	1337	fcc
17	Scandium	Zirconium	1803	2125	hcp
18	Vanadium	Chromium	2190	2150	bcc
19	Vanadium	Molybdenum	2190	2895	bcc
20	Vanadium	Tungsten	2190	3653	bcc
21	Niobium	Molybdenum	2742	2895	bcc
22	Niobium	Tungsten	2742	3653	bcc
23	Tantalum	Molybdenum	3270	2895	bcc
24	Tantalum	Tungsten	3270	3653	bcc
25	Rhodium	Palladium	2239	1827	fcc
26	Rhodium	Platinum	2239	2045	fcc
27	Palladium	Silver	1827	1233.5	fcc
28	Palladium	Gold	1827	1337	fcc
29	Rhenium	Osmium	3453	3320	hcp

used pure platinum, information on its absolute thermopower being also available[11]. In the 500 °K - 2000 °K range we made use of a rhenium 20% - tungsten alloy. A wire made of this alloy was calibrated at temperatures up to 1700 °K against platinum, and above 1700 °K molybdenum was demployed as the primary reference, whose thermopower up to 1700 °K was determined relative to copper and platinum, and taken from literature[16] for temperatures above 1700 °K. However in order to obtain a continuous dependence of thermopower on temperature, we had to introduce a temperature independent correction to the data given in ref.[16]. The reason for this lies possibly in that Lander[15] used in his work an insufficiently pure metal. Fig. 1 presents the temperature dependence of the thermopower of molybdenum obtained by us and used in the calibration of our reference electrode at high temperatures, as well as the data from ref.[16].

Now a last comment on the actual form of the data presentation in the graphs. The data taken from literature are presented in the form they are given in the corresponding papers. Our original data on temperature dependences of thermopower and electrical resistivity contain very many experimental points, the interval between them, as a rule, not

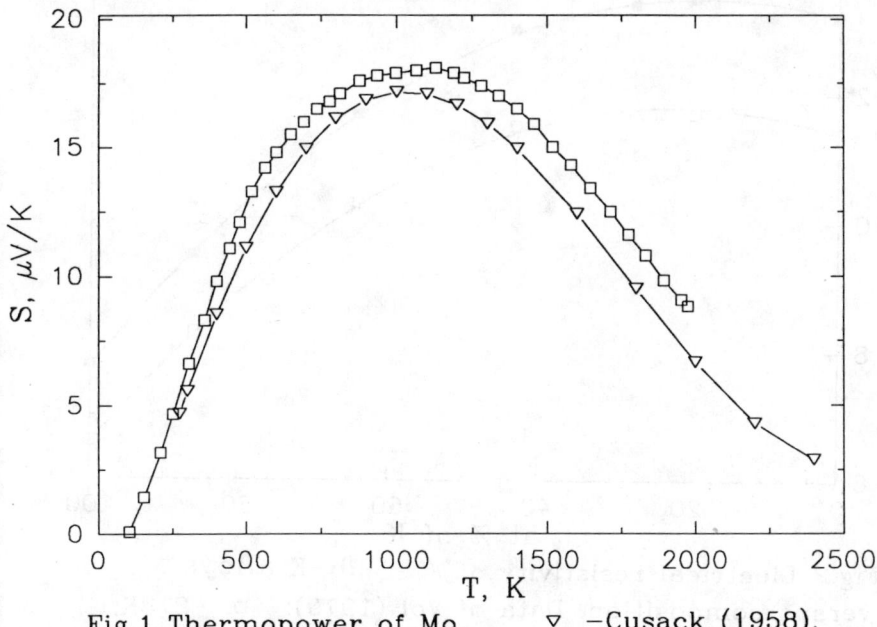

Fig.1 Thermopower of Mo.   ▽ −Cusack (1958),
   □ −present authors.

execeed 5 °K.. To make the graphs easily readable, the experimental points are given at intervals of 100 °K..The lines connecting the points were drawn by spline interpolation.

(a) Isoelectronic Metal Alloys.

In the figures depicting the dependences of the electrical resistivity on alloy composition experimental data are represented by points. The lines display the second order regrassion:

$$\rho = A \cdot (1-x) \cdot x$$

which corresponds to the rule of Kurnakov - Nordheim and reflects the accuracy with the experimental electrical resistivity vs composition dependence follow this rule.

(i) Potassium - Rubidium alloys (K-Rb)

The electrical resistivity of these alloys was measured as a function of comsipotion at 273 °K and 298 °K [17]. These data are presented in Fig. 2. Data on thermopower are available only for very low temperatures and a limited number of compositions on the K side[18-20]. The results of McDonald [18,19] at T = 20 °K are given in Fig 3.

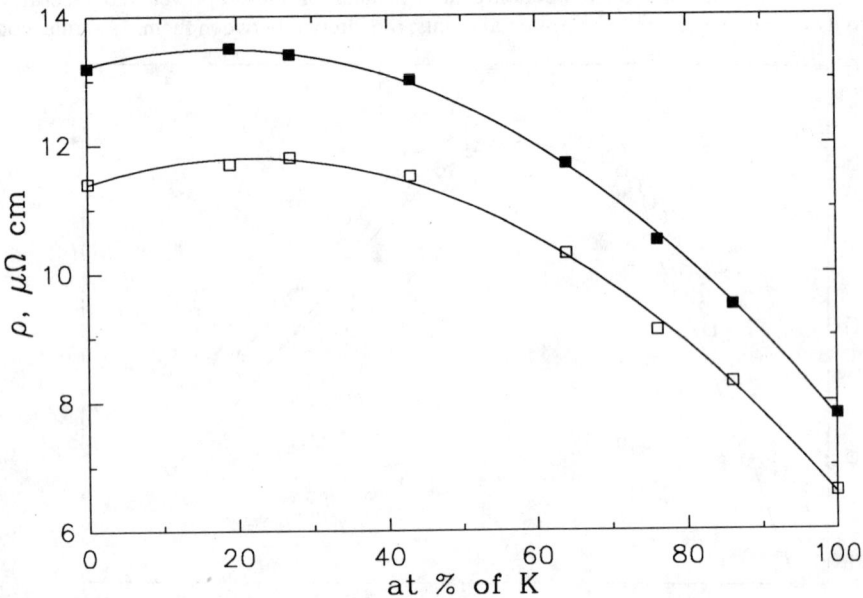

Fig.2 Electrical resistivity of Rb-K alloys versus composition. Data of Vol (1979): □ -273K; ■ -293K.

(ii) Potassium - Cesium alloys (K- Cs)

We know of no data on the electrical resistivity of concentrated alloys. The thermopower of these alloys was measured only at very low temperatures (down to 3 °K) and for a limited number of composition on the K side[20].

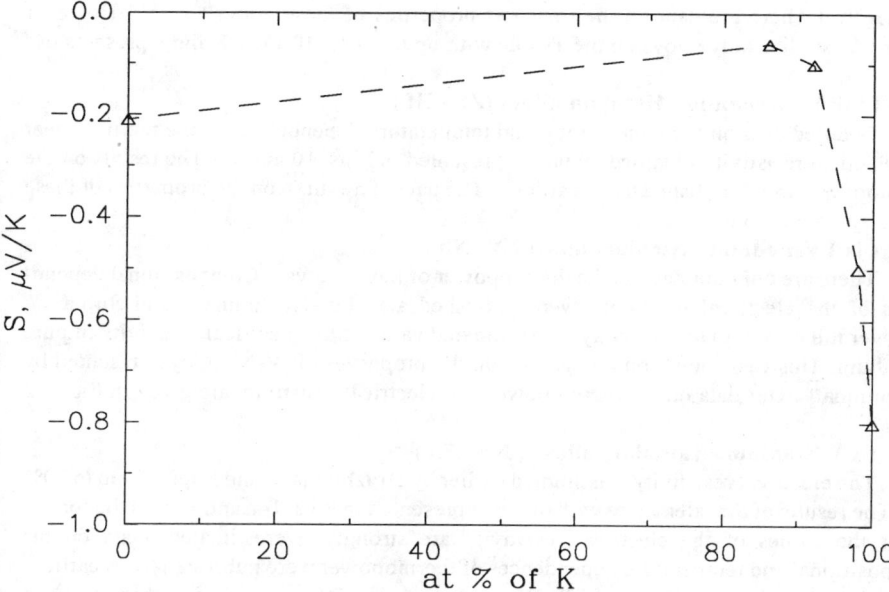

Fig.3. Thermopower of Rb−K alloys versus composition. Data of McDonald (1953) at 20 K.

(iii) Rubidium - Cesium alloys (Rb - Cs)
The situation is the same as in K - Cs case. There are data on the thermopower of several alloys obtained at very low temperature (down to 3 °K).

(iv) Calcium - Strontium alloys (Ca - Sr)
We know of no data on the electrical resistivity and thermopower of these alloys.

(v) Scandium - Yttrium alloys (Sc - Y)
We know of no literature data on electrical resistivity and thermopower of the alloys. The results of our studies of the thermopower were published earlier[4]. Our unpublished electrical resistivity data and data on thermopower from[4] are presented in Figs. 4 and 5. The thermopower on the Sc side is strongly composition - dependent, however concentrated alloys practically do not reveal any dependence of thermopower on composition.

(vi) Titanium - Zirconium alloys (Ti- Zr)
Both metals forming these alloys undergo a structural transformation at high temperatures. The transformation temperatures are T_{tr} = 1155 °K for Ti and T_{tr} = 1135 °K for Zr. At high temperatures both metals are bcc. There are data on the transport properties of these alloys[1,3,21]. The results of these studies are in a good agreement. Figs. 6 and 7 presents our more detailed data.

(vii)Titanium - Hafnium alloys (Ti - Hf)
Hafnium, just as titanium and zirconium, undergoes a polymorphic transformation at high temperatures, however the temperature at which it occurs is considerably higher

(2013 °K) There are data on the transport properties of these alloys[1,3,21,22]. However Tilkina[22] studies only alloys on the Ti side with up to 30 % Hf. Figs. 8 and 9 presents our data.

(viii) Zirconium - Hafnium alloys (Zr - Hf)

Detailed data on the composition and temperature dependences of the thermopower and electical resistivity obtained by us are presented in Figs. 10 and 11. The results on the thermopower were published by us earlier[4]. The data of Savin[21] on the properties of these alloys agree with ours.

(ix) Vanadium - Niobium alloys (V- Nb)

There are only our data on the thermopower of these alloys[4]. Compositional dependences of the electrical resistivity were published also by Druzhinina[23] and Sirota[24], however the former gives a stongly overestimated value of the electrical resistivity of pure vanadium. This casts doubt on all the data on the properties of V-Nb alloys presented by Druzhinina[23]. Our data on the thermopower and electrical resistivity are given in Figs. 12 and 13.

(x) Niobium - Tantalum alloys (Nb - Ta)

The electrical resistivity was studied earlier by Druzhinina[25] and Rapp[26] (up to 20% Ta). The results of the later agree with our data presented in Figs. 14a and 15a. In the former paper the values of the electrical resistivity are strongly overestimated. Data on the compositional and temperature dependences of thermopower were published by us earlier[4] and are presented in Figs. 14b and 15b. We are not aware of any other data on thermopower.

(xi) Chromium - Molybdenum alloys (Cr - Mo)

Chromium is an antiferromagnet with a Neel temperature of 312 °K, which decreases rapidly in alloys. Therefore all data presented in Figs. 16 and 17 relate to the paramagnetic state. The electrical resistivity of the alloys containing up to 20% Mo was studied by Mamiya[27]. The thermopower was studied in more details[3,4,28,29]. The results of these studies agree well with one another and with our data.

(xii) Chromium - Tungsten alloys (Cr - W)

We are aware only of data on the thermopower of dilute alloys of tungsten with chromium (up to 5% of W) obtained at temperatures only up to 360 °K[29,30]. The results of these studies are depicted in Fig. 18.

(xiii) Molybdenum - Tungsten alloys (Mo - W)

Kieffer[31] investigated the compositional dependence of the electrical resistivity at room temperature. The only published data on thermopower were obtained by us[4]. Figs. 19 and 20 present the dependences of both properties on composition and temperature obtained by us.

(xvi) Rhodium - Iridium alloys (Rh - Ir)

Beliaev[32] presents the electrical resistivity versus alloy composition at room temperature, as well as the integral thermopower in combination with platinum. We used these data to calculate the absolute thermopower of the alloys. The results for the electrical resistivity and thermopower are shown in Fig. 21.

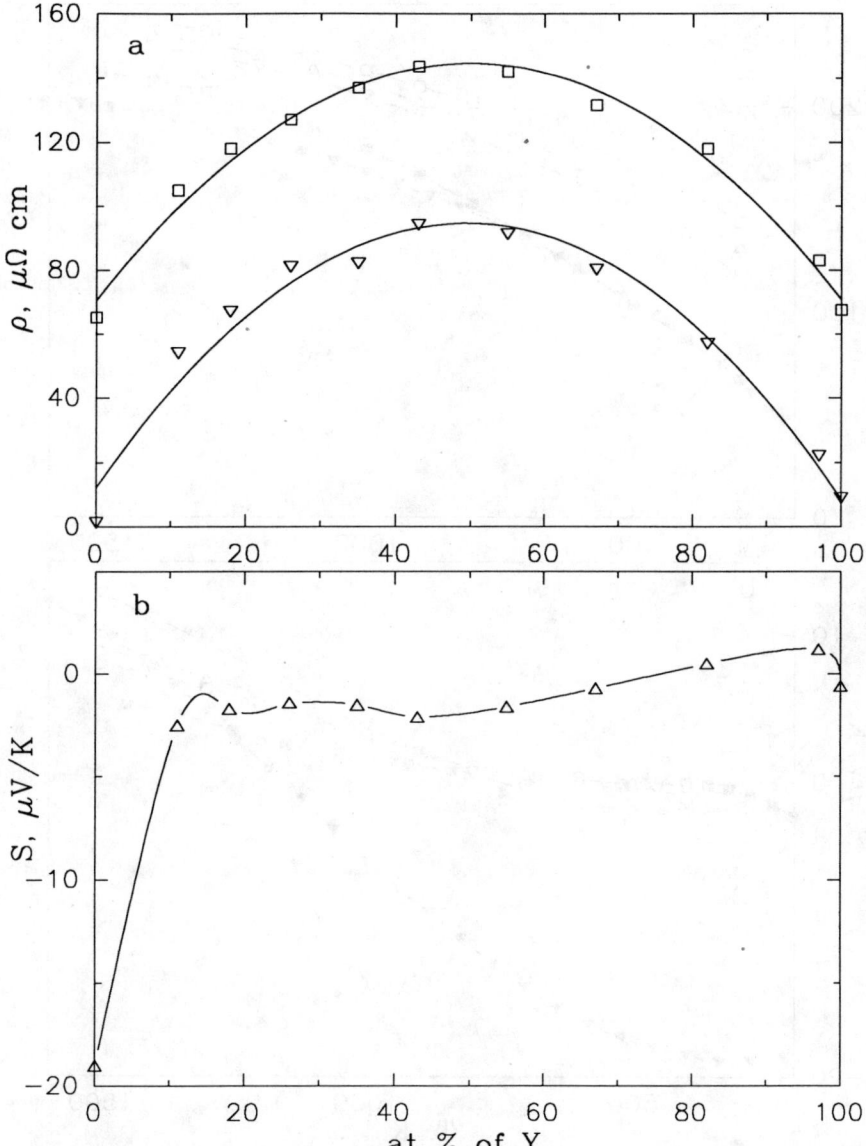

Fig.4. Electrical resistivity (a) and thermopower (b) of Sc–Y alloys versus composition. Present authors: ▽ – at 4.2K; □ – 300 K; △ – S at 300 K.

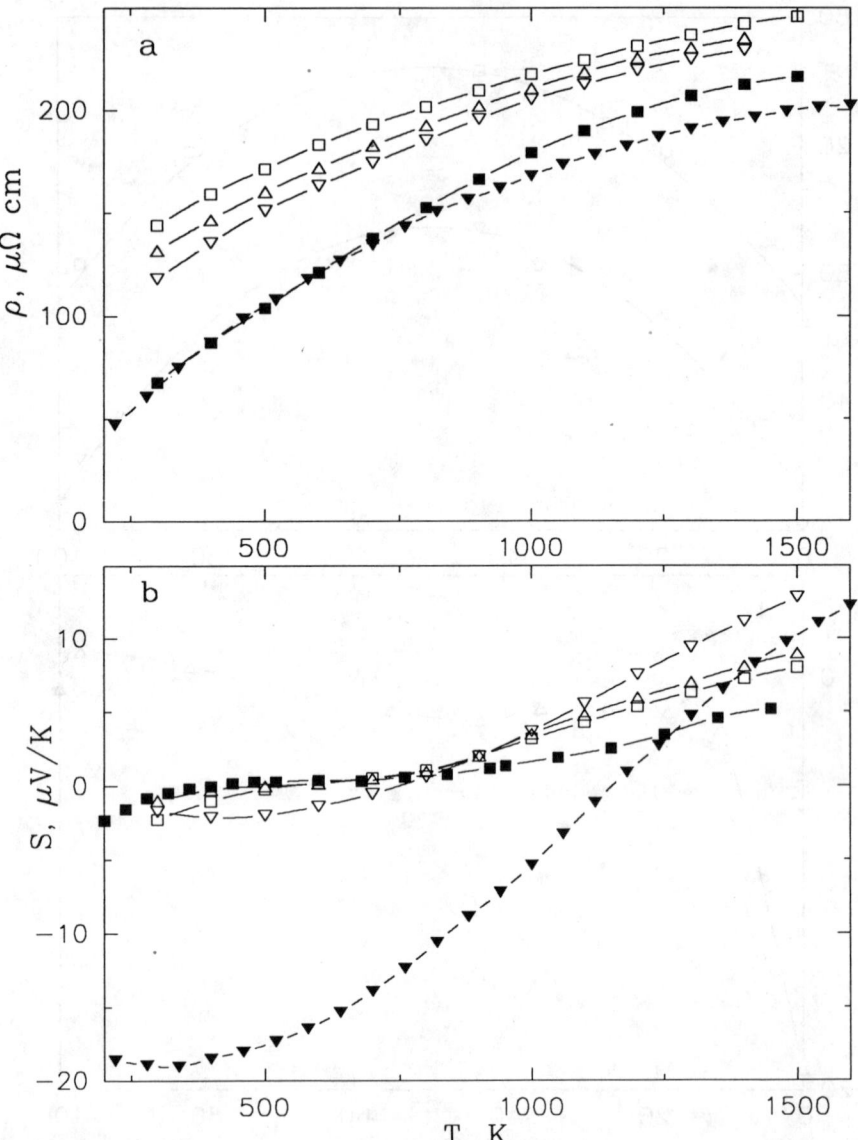

Fig.5. Electrical resistivity (a) and thermopower (b) of Sc–Y alloys. Present authors: ▼ –Sc; ▽ –18%Y; □ –43%Y; △ –67%Y; ■ –100%Y.

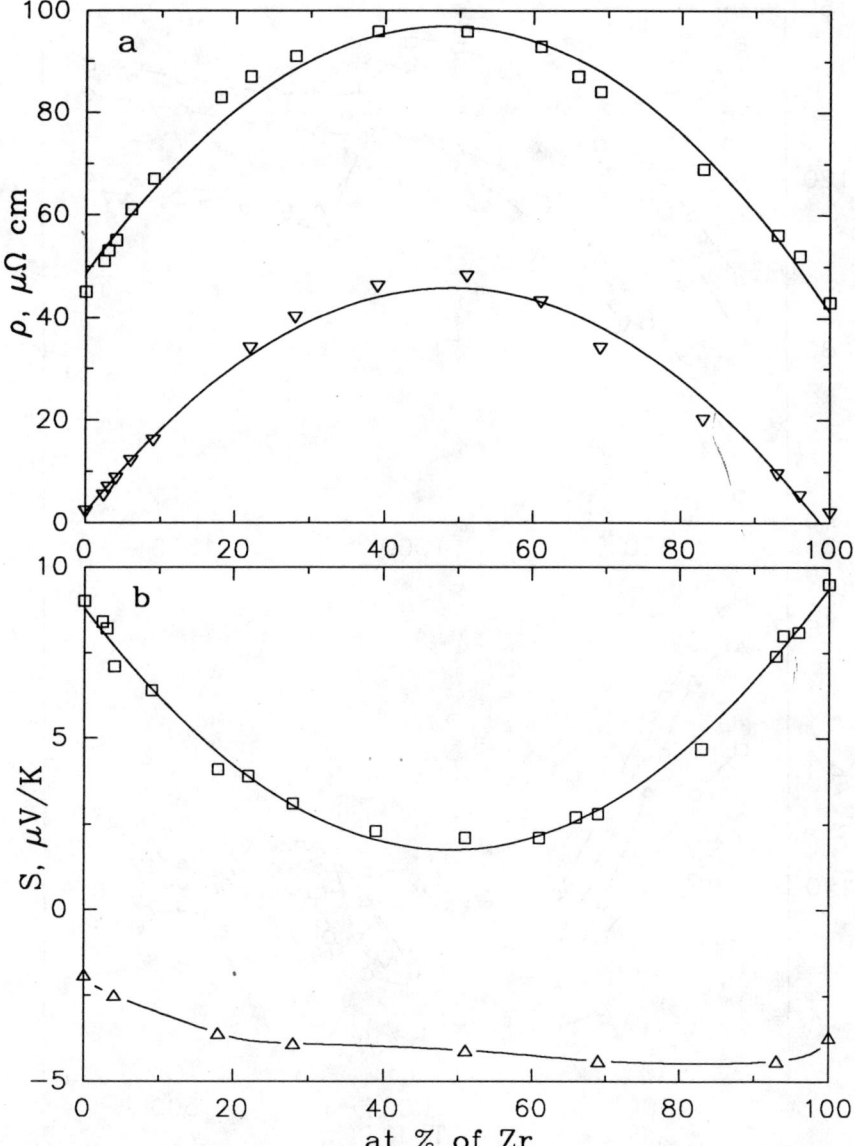

Fig.6. Electrical resistivity (a) and thermopower (b) of Ti–Zr alloys versus composition. Present authors: ▽ –at 4.2K; □ –293K; △ –800K.

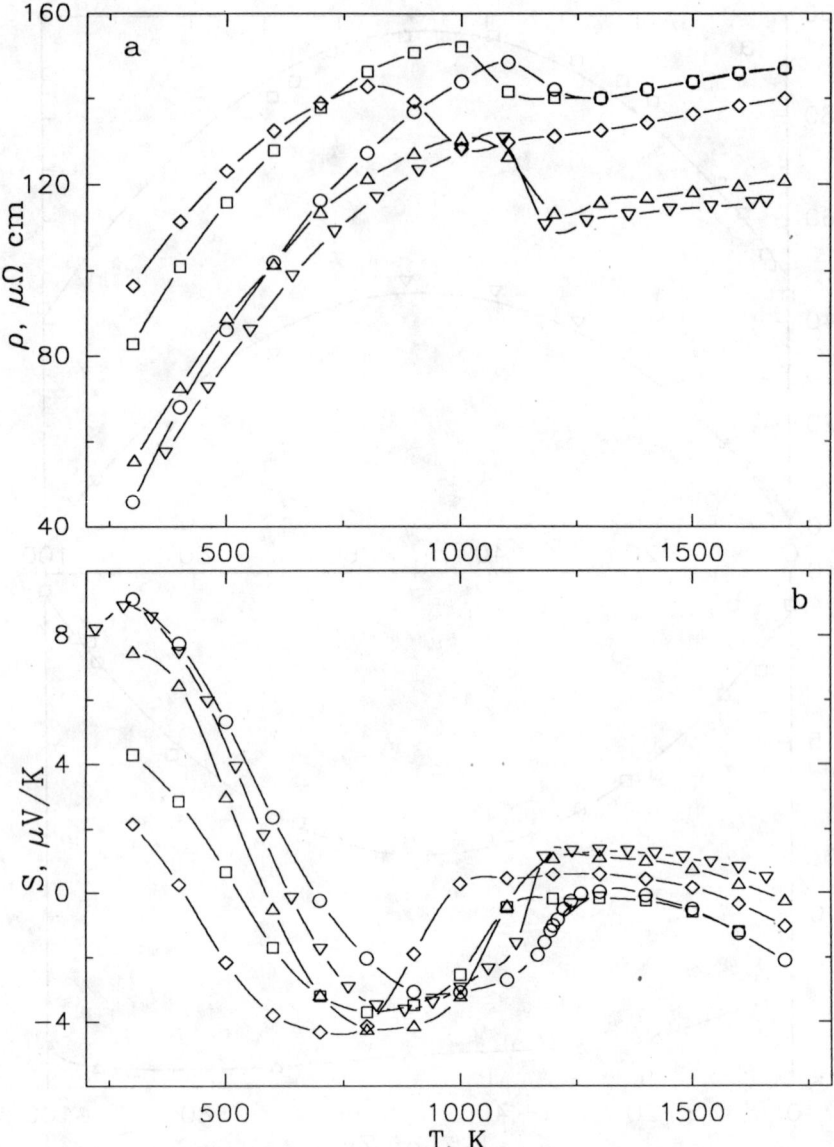

Fig.7. Electrical resistivity (a) and thermopower (b) of Ti–Zr alloys. Present authors: ○ –Ti; □ –18%Zr; ◇ –51%Zr; △ –93%Zr; ▽ –100%Zr.

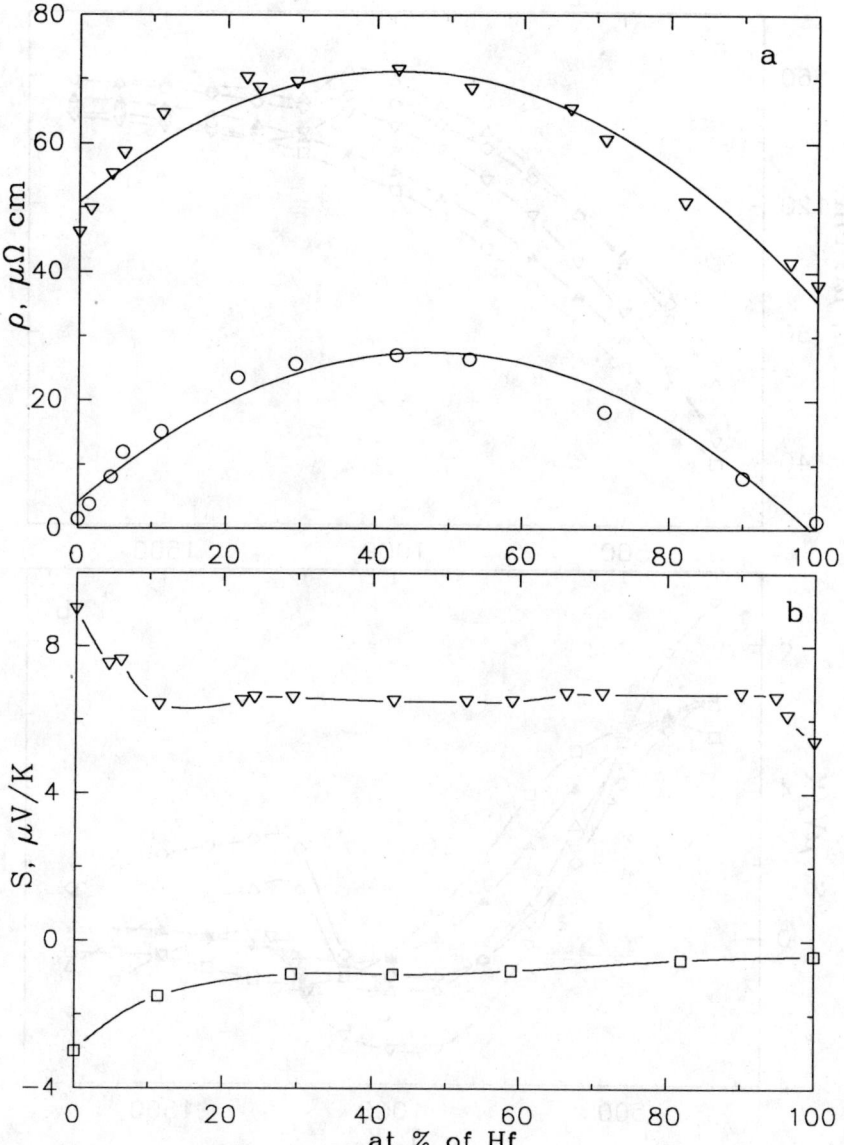

Fig.8. Electrical resistivity (a) and thermopower (b) of Ti−Hf alloys. Present authors: ○ −at 4.2K, ▽ −293K; □ −1000K.

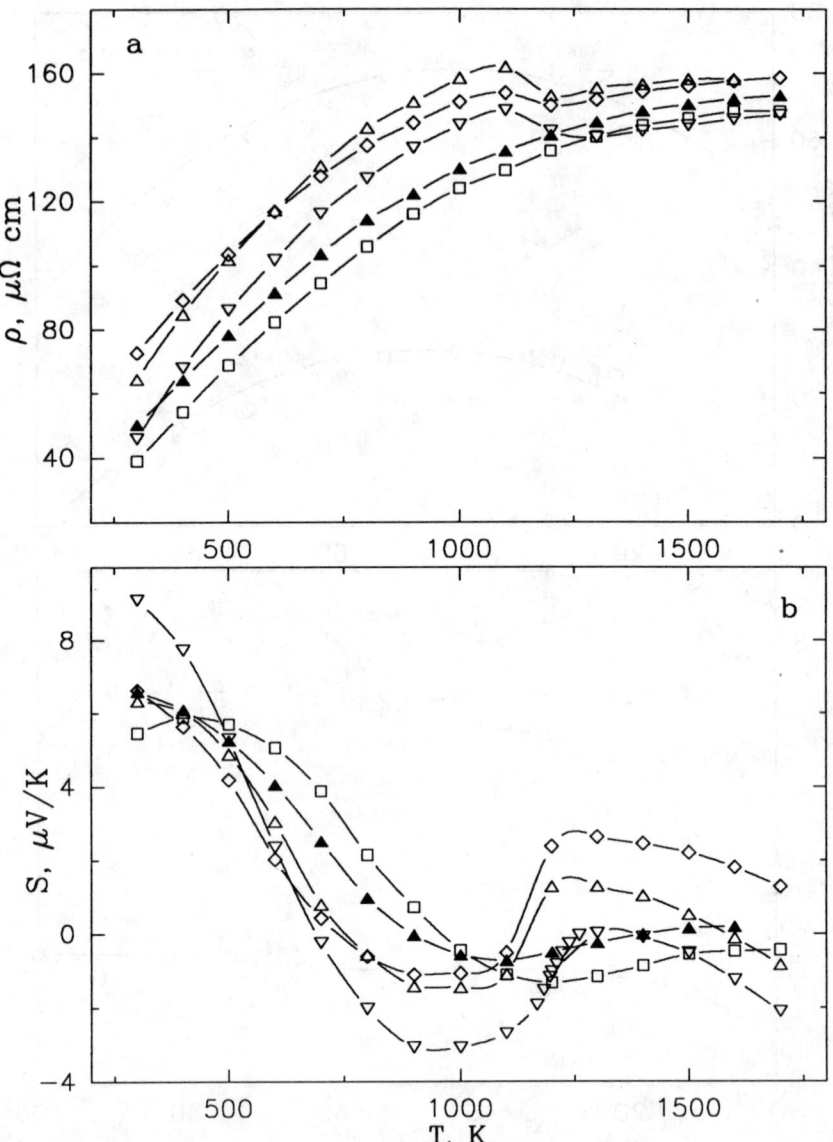

Fig.9. Electrical resistivity (a) and thermopower (b) of Ti–Hf alloys. Present authors: ▽ –Ti; △ –11%Hf ◇ –43%Hf; ▲ –82%Hf; □ –100%Hf.

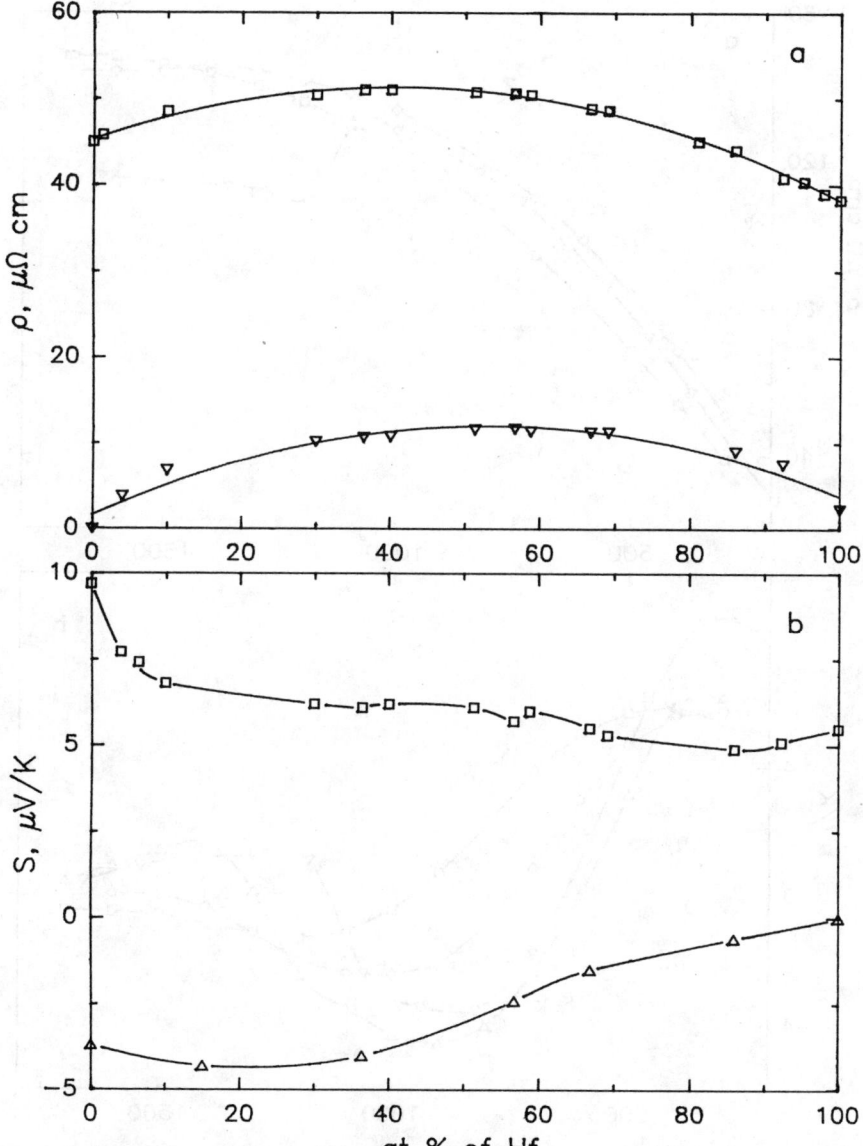

Fig.10. Electrical resistivity (a) and thermopower (b) of Zr–Hf alloys versus composition. Present authors. ▽ –at 4.2K; □ –293K; △ –1000K.

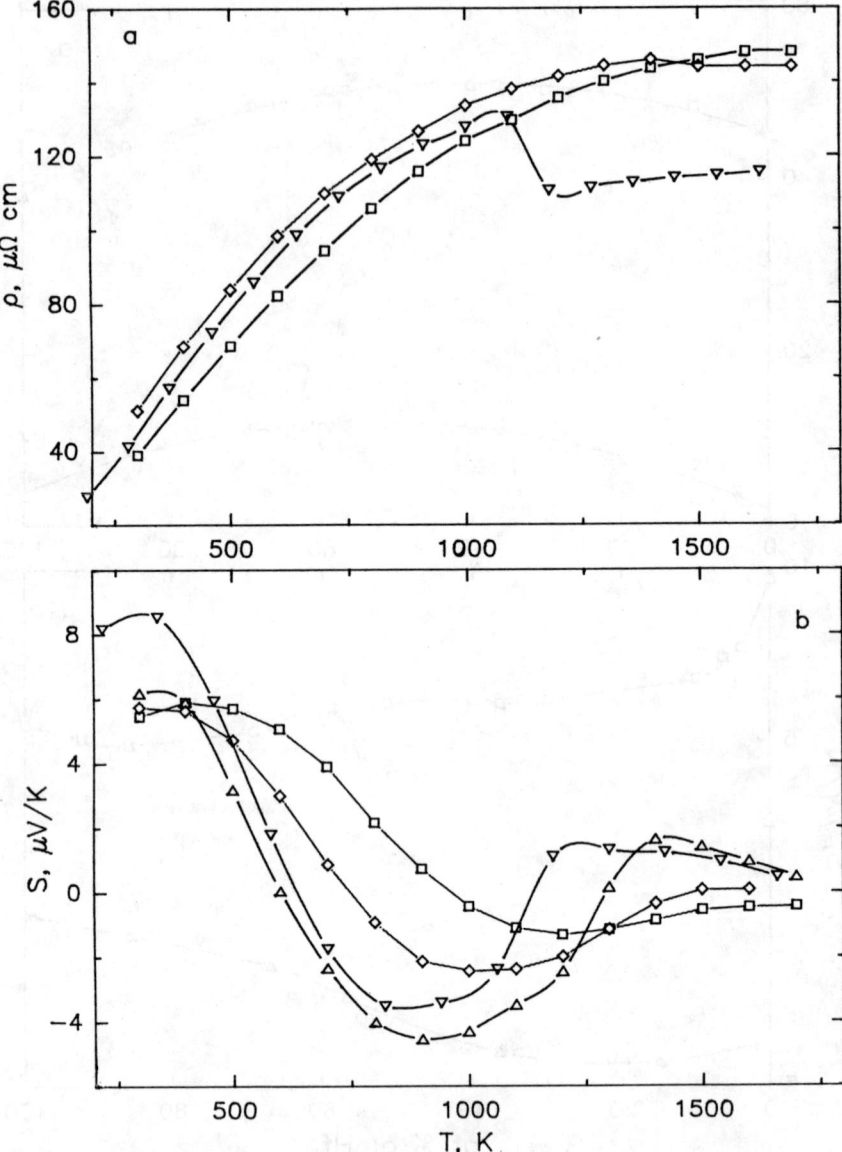

Fig.11. Electrical resistivity (a) and thermopower (b) of Zr–Hf alloys. Present authors. ▽ –Zr; △ –15%Hf; ◇ –57%Hf; □ –100%Hf.

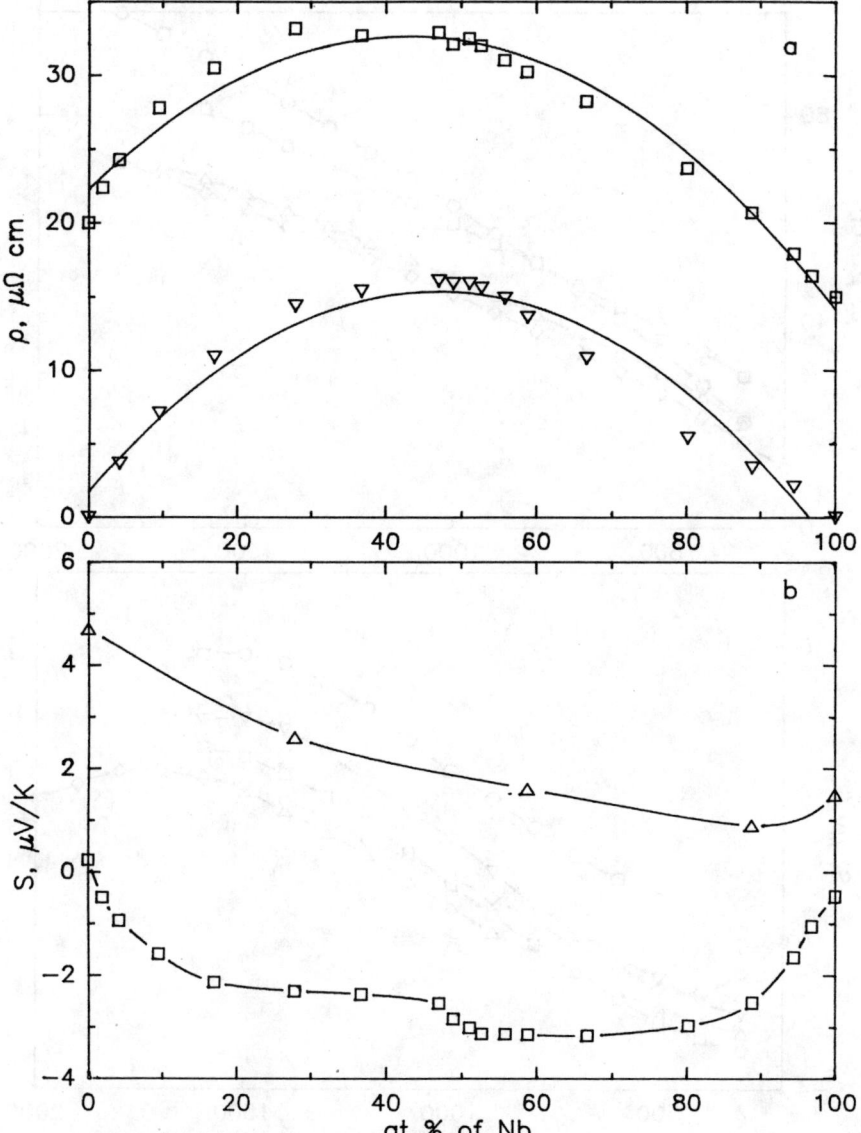

Fig.12. Electrical resistivity (a) and thermopower (b) of V–Nb alloys versus composition. Present authors.
▽ – at 4.2K; □ – 293K; △ – 1000K.

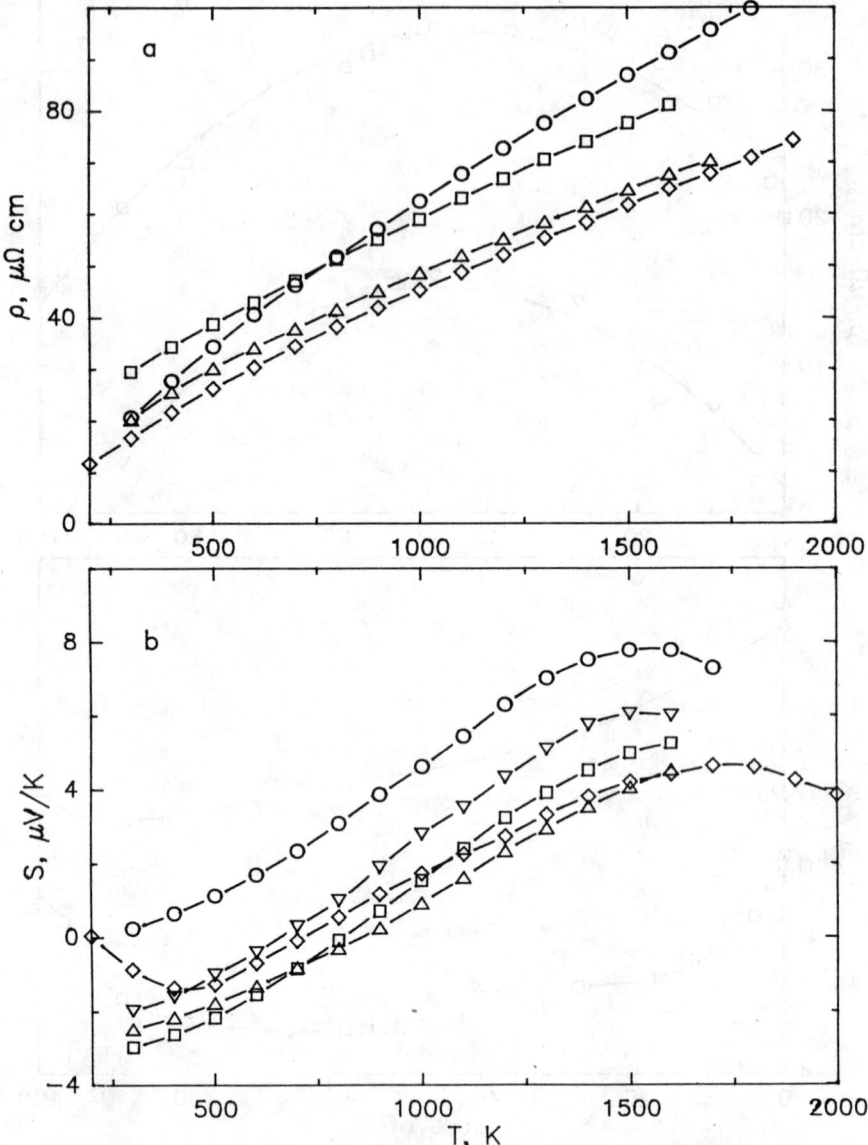

Fig.13. Electrical resistivity (a) and thermopower (b) of V—Nb alloys. Present authors. ○ —V; ▽ —28%Nb; □ —58%Nb; △ —88%Nb; ◇ —100%Nb.

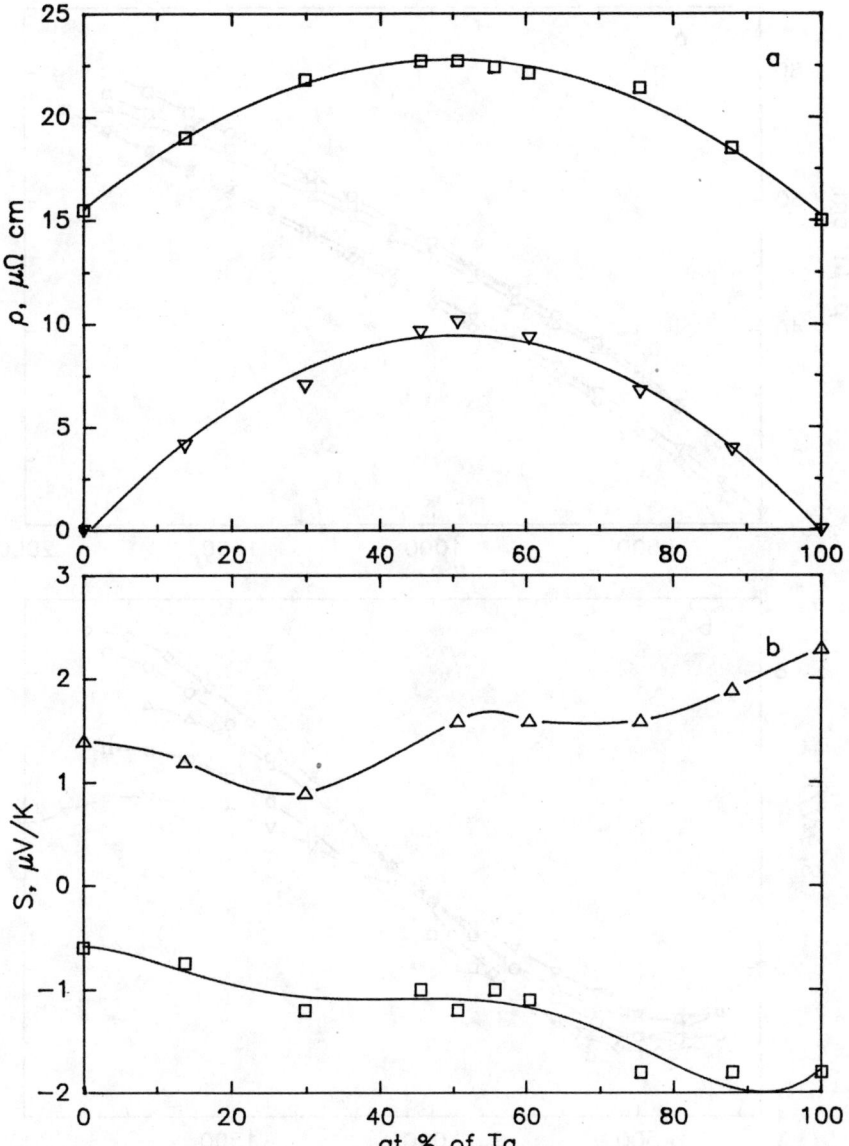

Fig.14. Electrical resistivity (a) and thermopower (b) of Nb–Ta alloys versus composition. Present authors.
▽ –at 4.2K; □ –293K; △ –1000K.

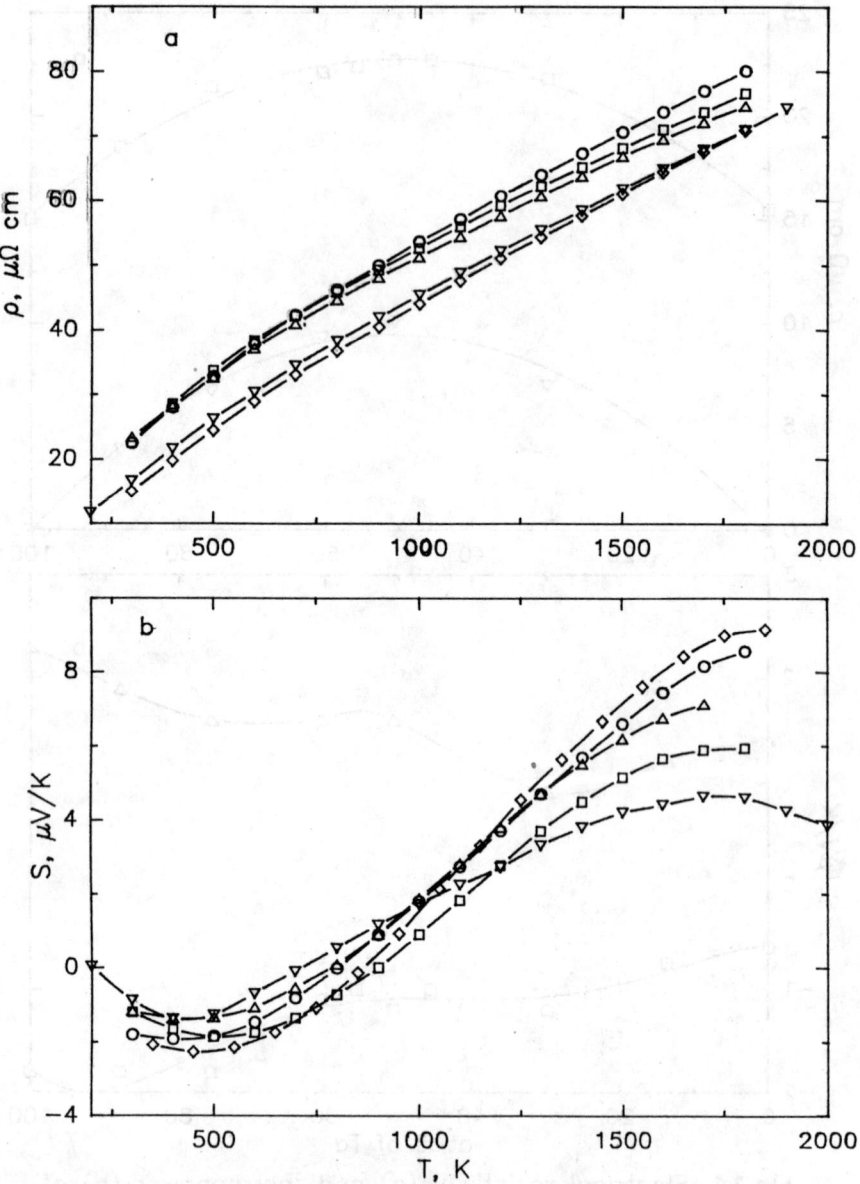

Fig.15. Electrical resistivity (a) and thermopower (b) of Nb–Ta alloys. Present authors. ▽ –Nb; □ –30%Ta; △ –50%Ta; ○ –75%Ta; ◇ –100%Ta.

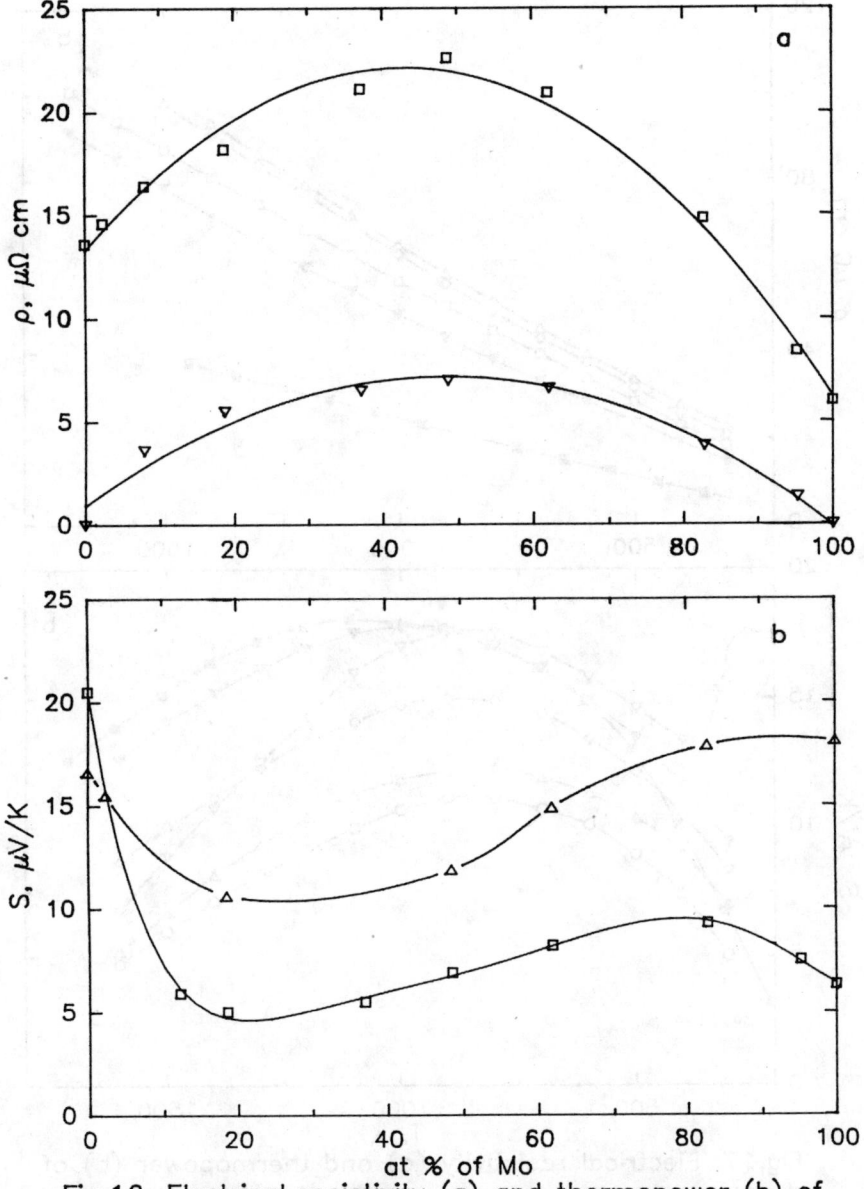

Fig.16. Electrical resistivity (a) and thermopower (b) of Cr–Mo alloys versus composition. Present authors.
▽ – at 4.2K; □ – 293K; △ – 1000K.

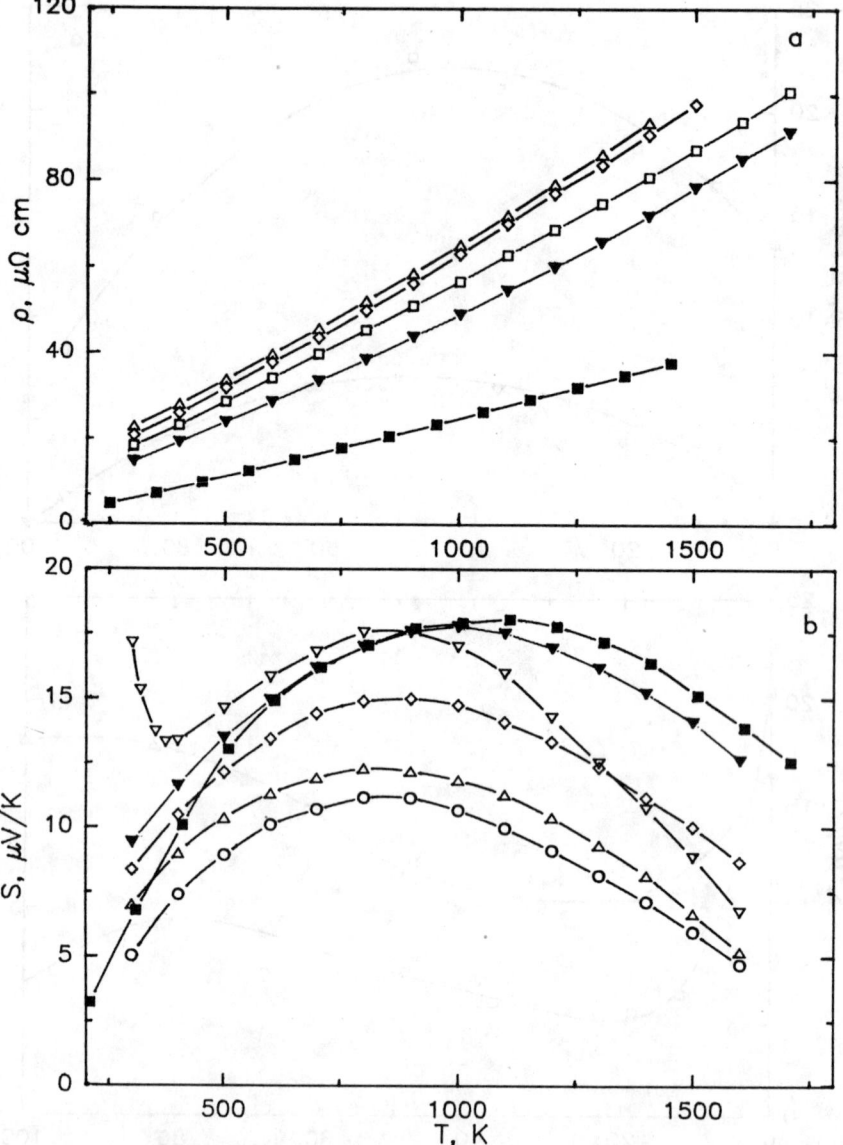

Fig.17. Electrical resistivity (a) and thermopower (b) of Cr–Mo alloys. Present authors. ▽ —Cr; □ —19%Mo; ○ —31%Mo; △ —49%Mo; ◇ —61%Mo; ▼ —83%Mo; ■ —100%Mo.

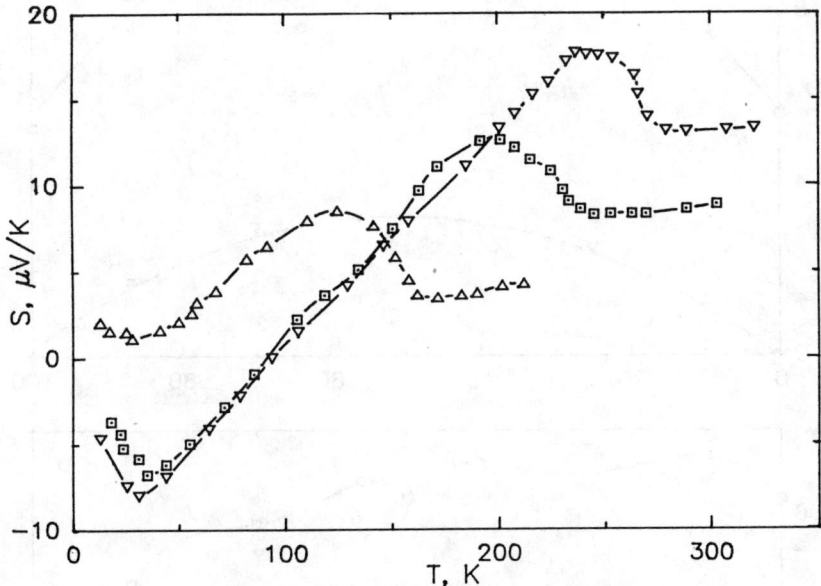

Fig.18. Thermopower of Cr–W alloys. Data of Trego (1968).
▽ —1.74%W; ◻ —3.64%W; △ —4.98%W.

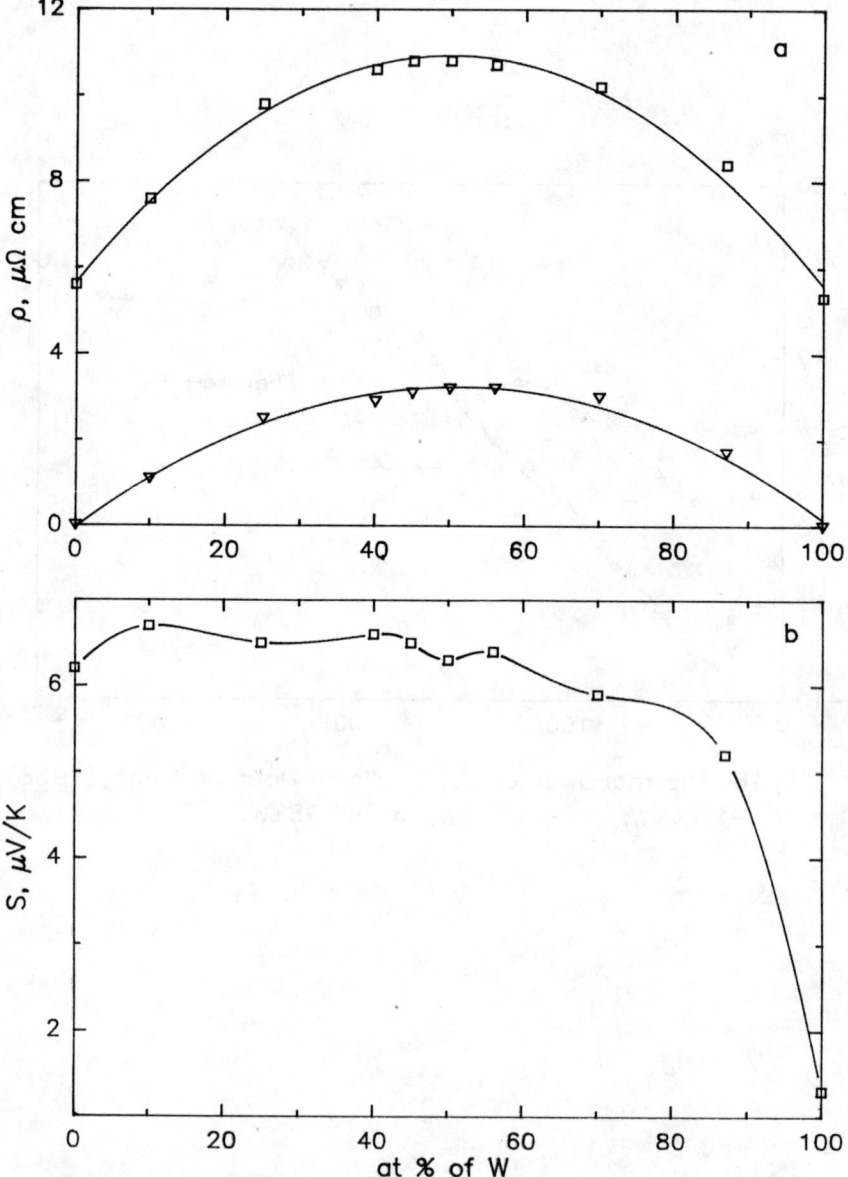

Fig.19. Electrical resistivity (a) and thermopower (b) of Mo—W alloys versus composition. Present authors.
▽ —at 4.2K; □ —293K.

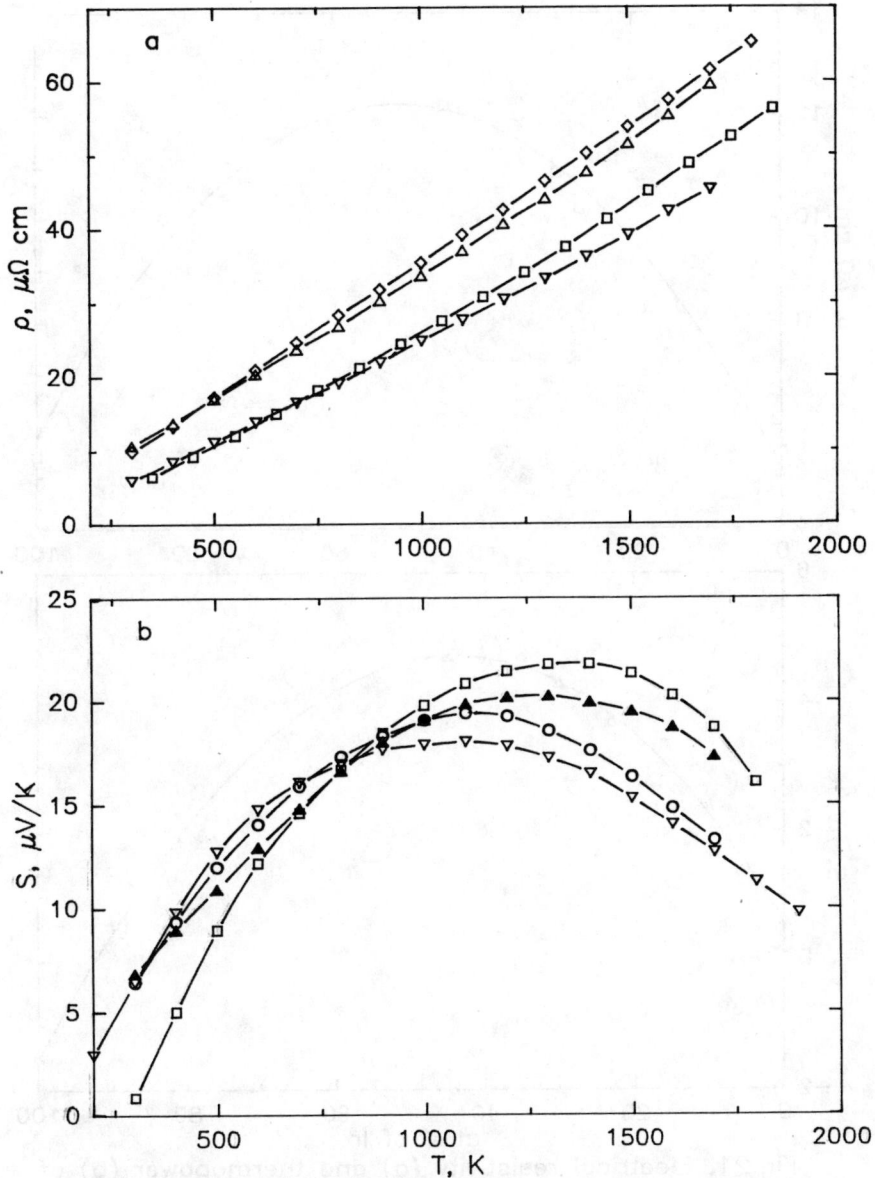

Fig.20. Electrical resistivity (a) and thermopower (b) of Mo—W alloys. Present authors. ▽ —Mo; □ —100%W; ○ —10%W; ▲ —40%W; △ —45%W; ◇ —70%W.

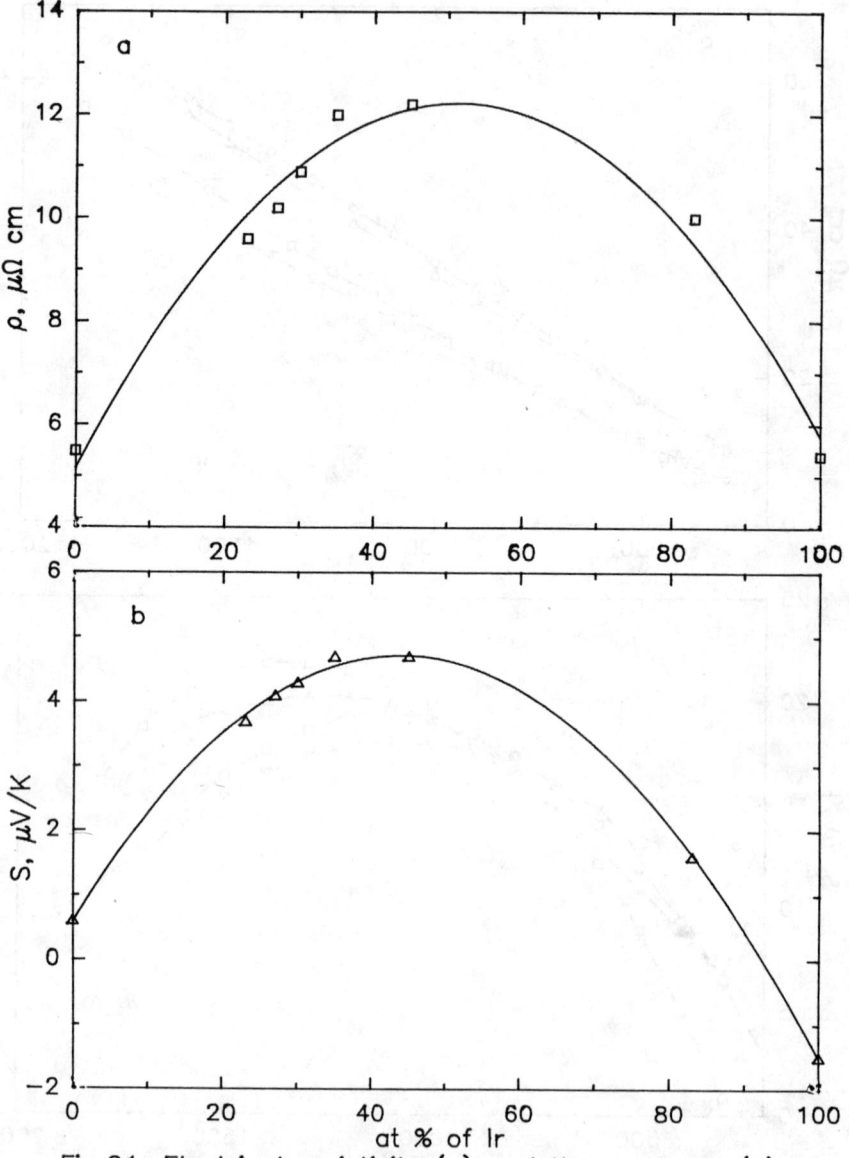

Fig.21. Electrical resistivity (a) and thermopower (b) of Rh–Ir alloys versus composition. Beliaev (1974).
□ —at 293K; △ —at 573K.

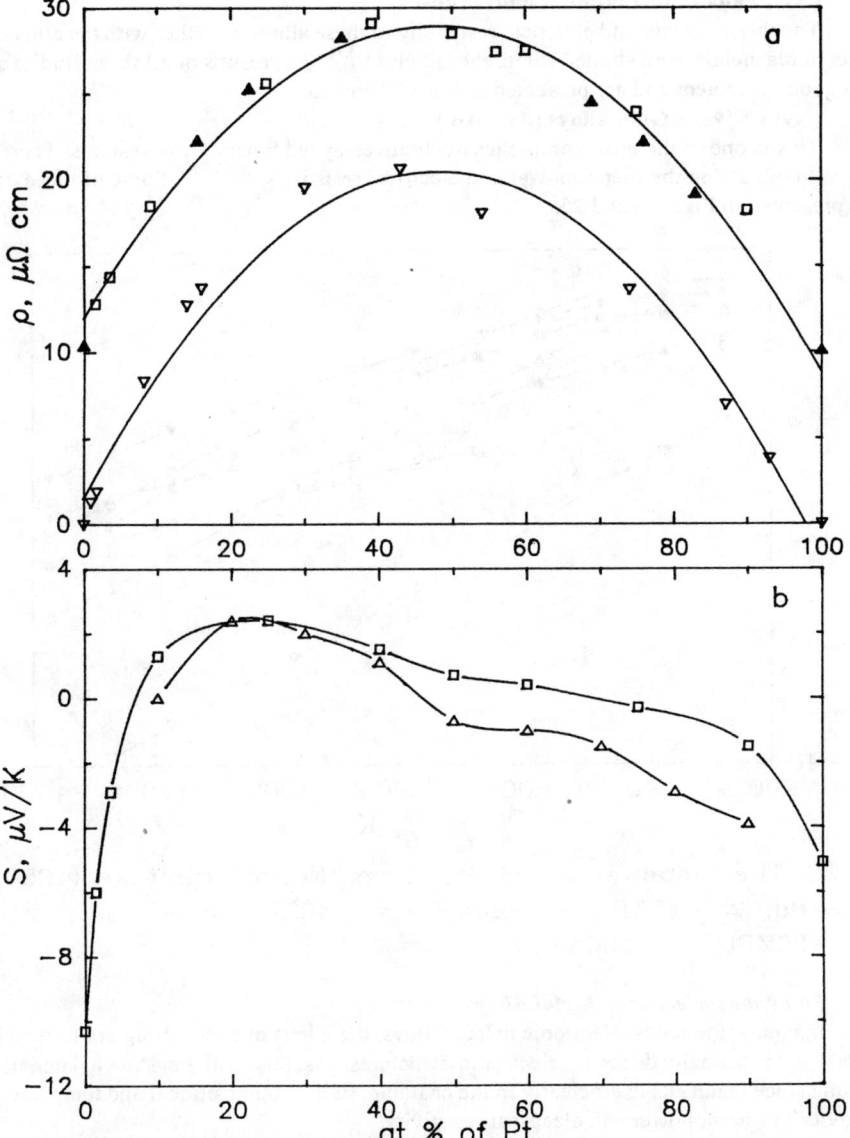

Fig.22. Electrical resistivity (a) and thermopower (b) of Pd–Pt alloys versus composition. ▽ —at 1.K, Blood (1972); ▲ —293K, Beliaev (1974), ▫ —293K, Koster (1963); △ —293K, Rudnitskii (1956).

(xv) Palladium-Platinum alloys (Pd-Pt)

The thermopower and electrical resistivity of these alloys, together with the alloys of other noble metals were studied comprehensively[33-35]. The results of all these studies are in a good agreement and are presented in Figs. 22 and 23.

(xvi) Silver - Gold alloys (Ag - Au)

This is one of the most comprehensively investigated binary alloy systems. There is a wealth of data on the thermopower and electrical resistivity[33,36-38]. Some of these data are presented in Figs. 24 and 25.

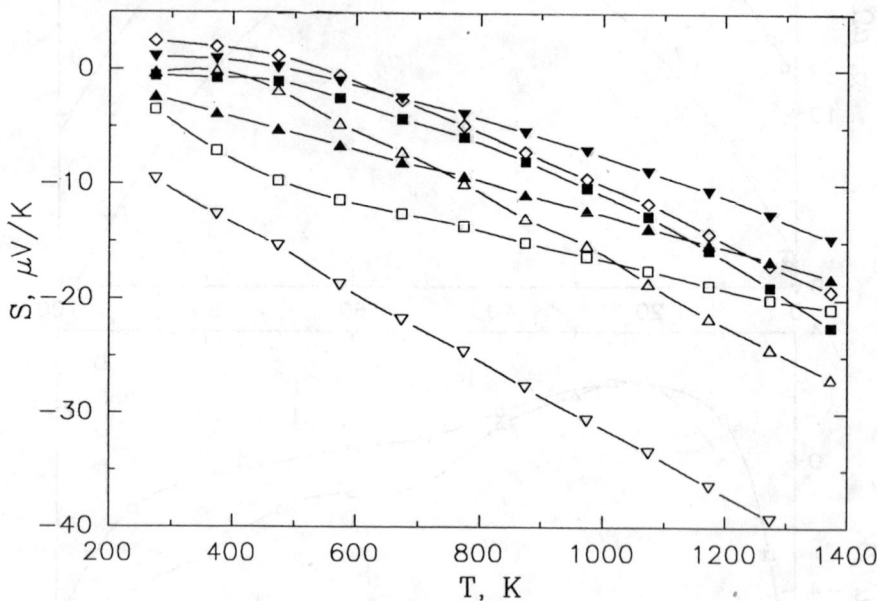

Fig.23. Thermopower of Pd−Pt alloys. From Rudnitskii (1956).
▽ −Pd; △ −10%Pt; ◇ −20%Pt; ▼ −40%Pt; ■ −70%Pt;
▲ −80%Pt; □ −100%Pt.

(b) Nonisoelectronic Metal Alloys

In contrast to the isoelectronic metals alloys, the alloys of these group are formed of metals with strongly different electronic structures. As this will be shown later, this circumstance manifests itself clearly in the character of the compositional and temperature behavior of thermopower and electrical resistivity.

(i) Scandium-Zirconium alloys (Sc-Zr)

The only available experimental data on the thermopower and electrical resistivity of these alloys were obtained by the present authors. They were partially published earlier[3,4]. More complete data are depicted in Figs. 26 and 27. The alloys were prepared using

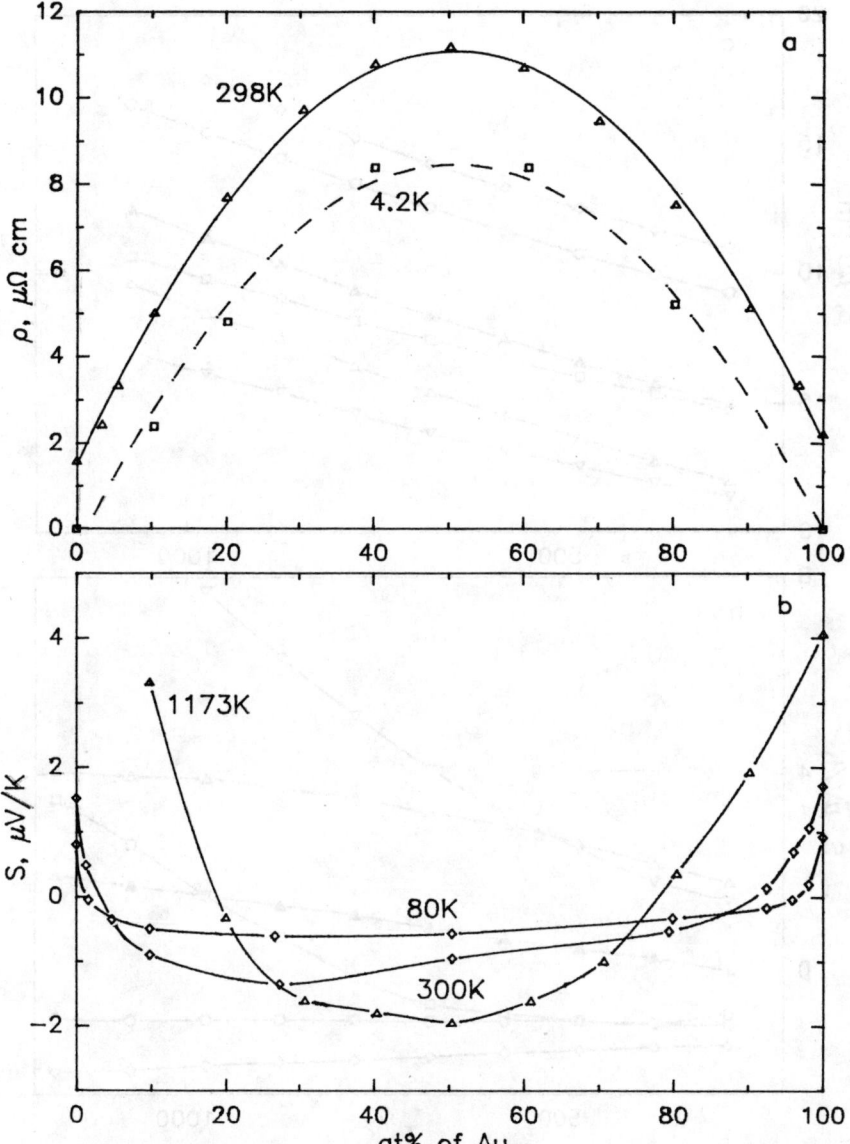

Fig.24 Electrical resistivity (a) and thermopower (b) of Ag—Au alloys △ —Rudnitskii (1956), ◇ —Crisp (1970), □ —Borelius (1937).

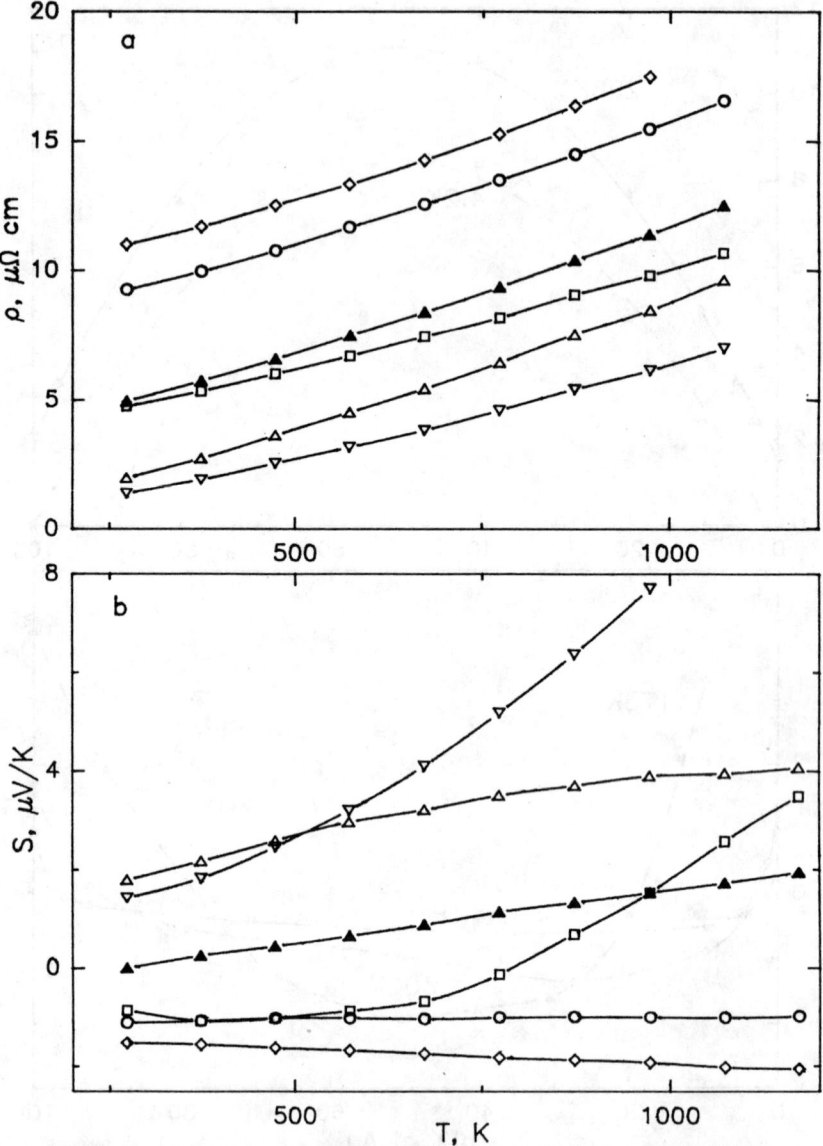

Fig.25. Electrical resistivity (a) and thermopower (b) of Ag–Au alloys. Rudnitskii (1956). ▽ —Ag; □ —10%Au; ◇ —50%Au; ○ —70%Au; ▲ —90%Au; △ —100%Au.

scandium with an integral purity characterized by the ratio RRR = $\rho(300)/\rho(4.2) = 3$. Apart from this, we measured the properties of scandium prepared by high vacuum distillation and having RRR = 50. The data on the properties of these metals are also given in Fig. 27 and reveal an extremely high sensitivity of the properties of scandium to the presence of small amounts of impurities.

(ii) **Vanadium-Chromium alloys (V-Cr)**

There several publications on the transport properties of these alloys. Kashido [39] studied the low temperature electrical resistivity of alloys containing upto 15% V. The results on the electrical resistivity of Giannuzi [40] are clearly strongly overestimated. The data of Kashido[39] and Chiu [41] agree well with our more detailed results presented in Figs. 28a and 29a. The results of thermopower measurements obtained by several authors[4,29,40,41] agree well with one another. Figs. 28b and 29b display our most detailed data.

(iii) **Vanadium-Molybdenum alloys (V-Mo)**

Data on the electrical resistivity can be found in Terekhov [42] and on thermopower in the publications of Vedernikov [1] and Zhumagulov[2]. Our data are displayed in Figs 30 and 31.

(iv) **Vanadium - Tungsten alloys (V-W)**

There are published data only on the thermopower[4]. Our data on the thermopower and electrical resistivity are presented graphically in Figs. 32 and 33.

(v) **Niobium - Molybdenum alloys (Nb-Mo)**

There are publication on the electrical resistivity and thermpower of Vedernikov[1], Cox[43] and Lin[44], respectively. The results of these studies agree well with one another and our recent data presented in Figs. 34 and 35.

(vi) **Niobium-Tungsten alloys (Nb-W)**

On the electrical resistivity there are three publications - Kieffer [31], Krimer[45] and Micheev [46]; and on the thermopower only one - Vedernikov[31]. Fig 36a displays our data on compositional dependences of electrical resistivity, and those of Kieffer [31]. While there is a noticeable difference between the results of the two publications, however both data sets reveal clearly of nonisoelectronic metals, namely, these dependences are strongly nonsymmetrical with respect to the equiatomic composition. The temperature dependences of the electrical resistivity and thermopower obtained by us are displayed in Fig. 37.

(vii) **Tantalum - Molybdenum alloys (Ta-Mo)**

There are published data on the electrical resistivity only in the low temperature domain - Torn [47] and detailed information on the thermopower obtained by us[4]. Figs. 38 and 39 depict our data on electrical resistivity and thermopower.

(viii) **Tantalum - Tungsten alloys (Ta - W)**

The electrical resistivity of these alloys was studies by Kieffer[31] and Tomas [48], the range of measurements in the latter publication being extended up to 2400 °K. There are also data on thermopower Vedernikov[4] and Lin [44]. The published data and our compositional dependences of the electrical resistivity an thermopower are depict in Fig. 40. The data are seen to disagree noticeably. A comparison or our data or the Ta-W alloys with the corresponding data for the analog Ta-Mo gives us grounds to consider our data to be more

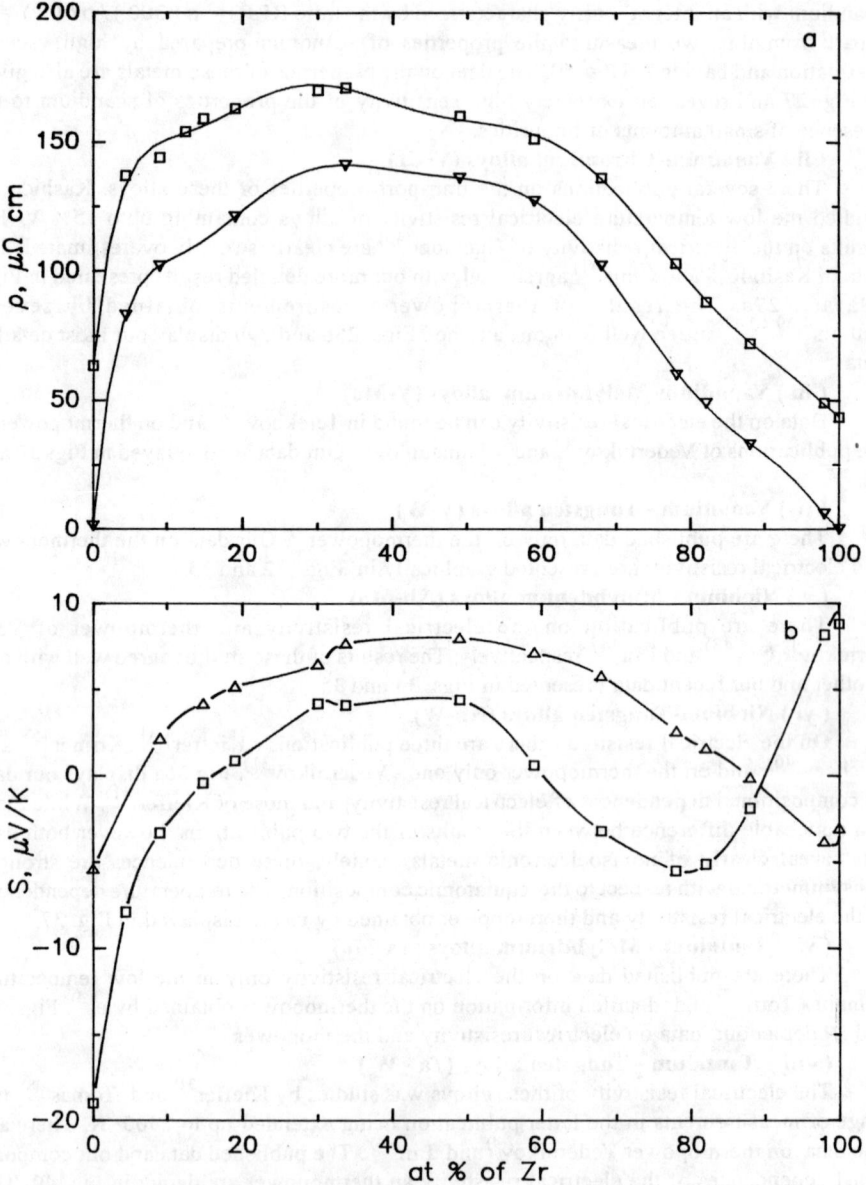

Fig.26. Electrical resistivity (a) and thermopower (b) of Sc–Zr alloys versus composition. Present authors.
▽ – at 4.2K; ▫ – 293K; △ – 1000K.

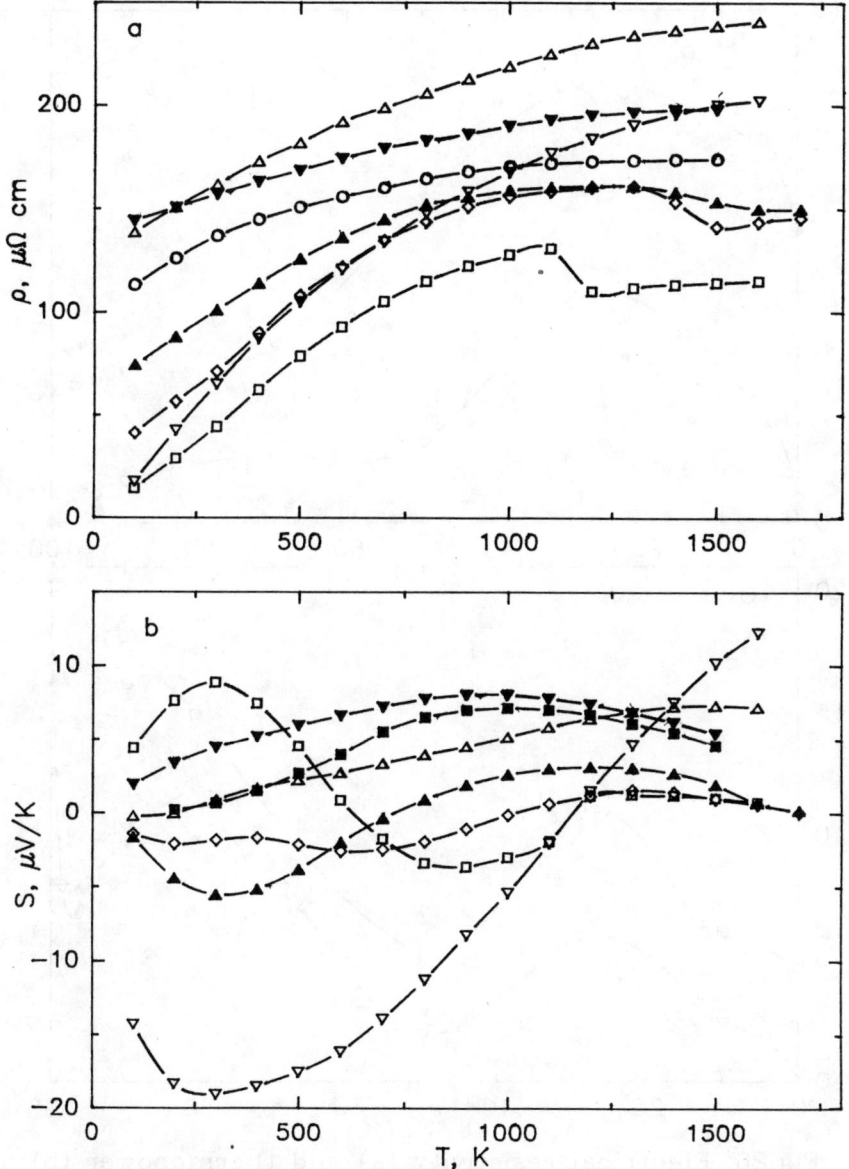

Fig.27. Electrical resistivity (a) and thermopower (b) of Sc-Zr alloys. Present authors. ▽ —Sc; □ —100%Zr; △ —19%Zr; ▼ —49%Zr; ■ —59%Zr; ○ —68%Zr; ▲ —78%Zr; ◇ —88%Zr.

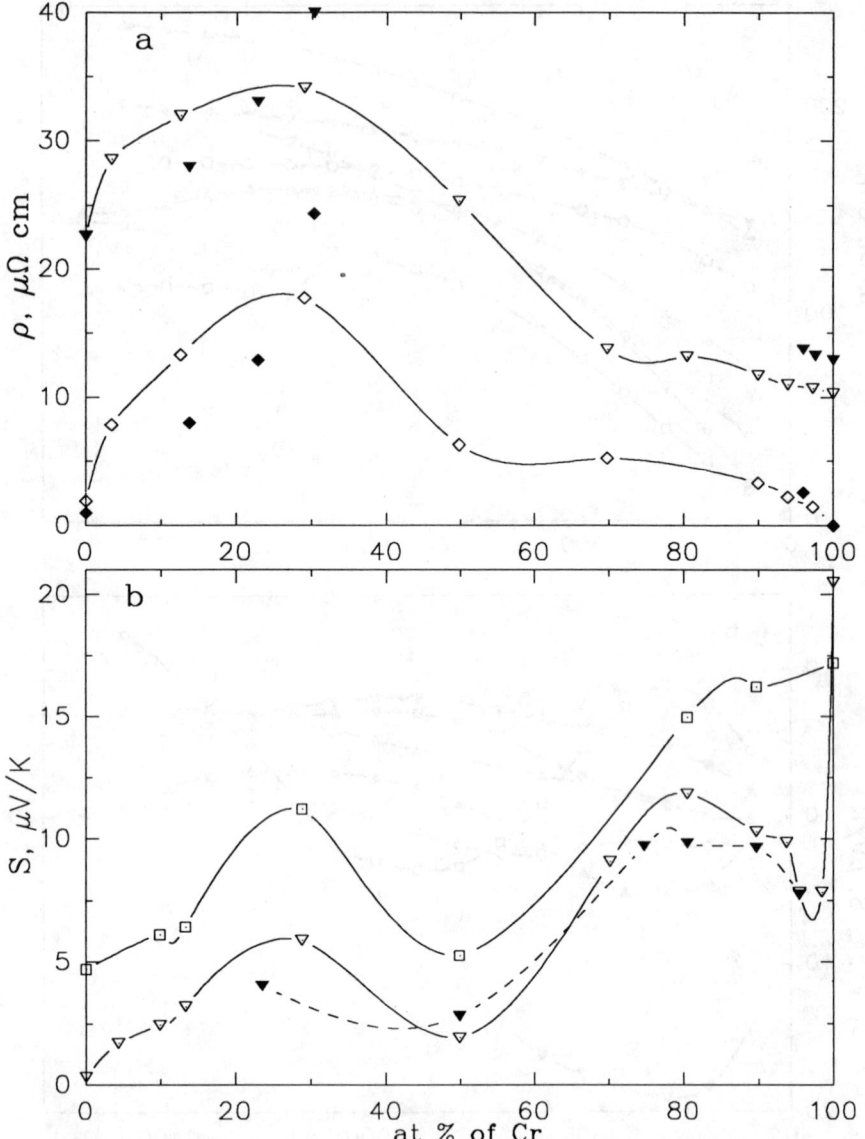

Fig.28. Electrical resistivity (a) and thermopower (b) of V–Cr alloys versus composition. ◇ —at 4.2K; ▽ —293K; □ —1000K. Hollow symbols—present authors, filled symbols —data of Chiu (1976) O(ρ) and Giannuzi (1970) (S).

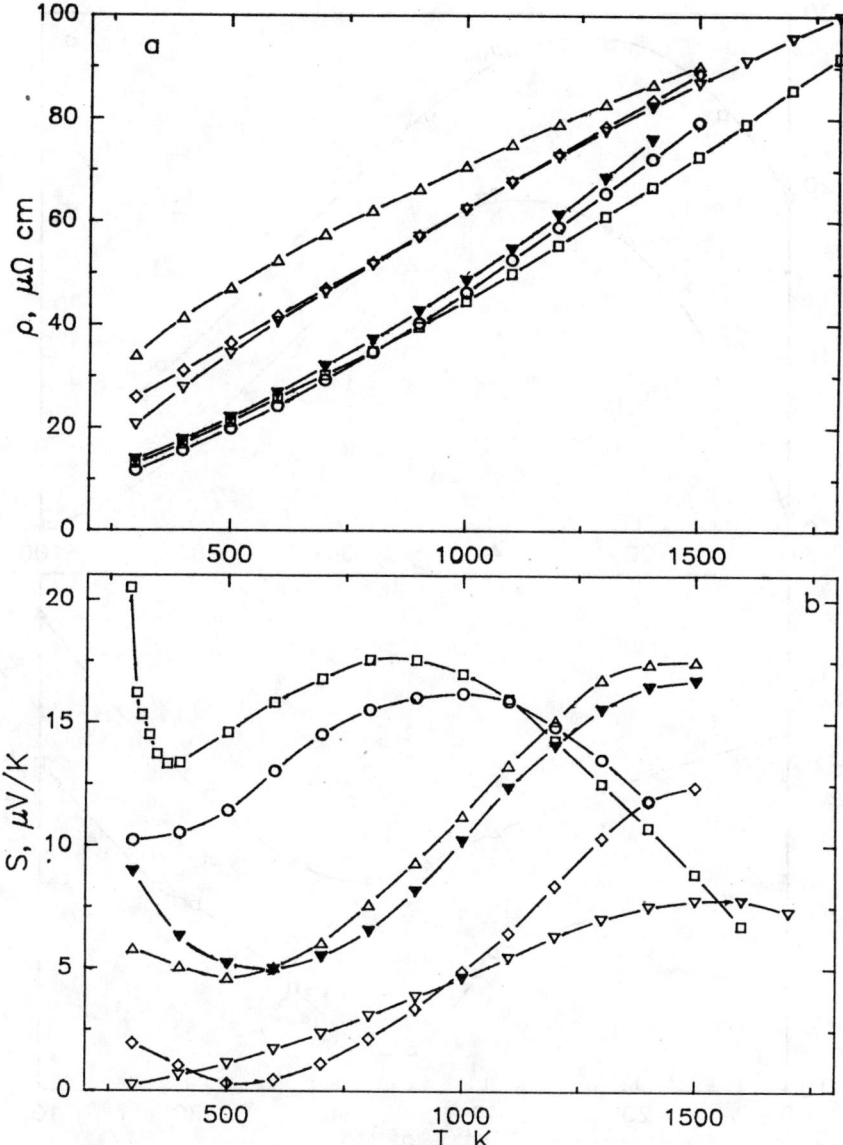

Fig.29. Electrical resistivity (a) and thermopower of V–Cr alloys. Present authors. ▽ —V; △ —30%Cr; ◇ —51%Cr; ▼ —70%Cr; ○ —90%Cr; □ —100%Cr.

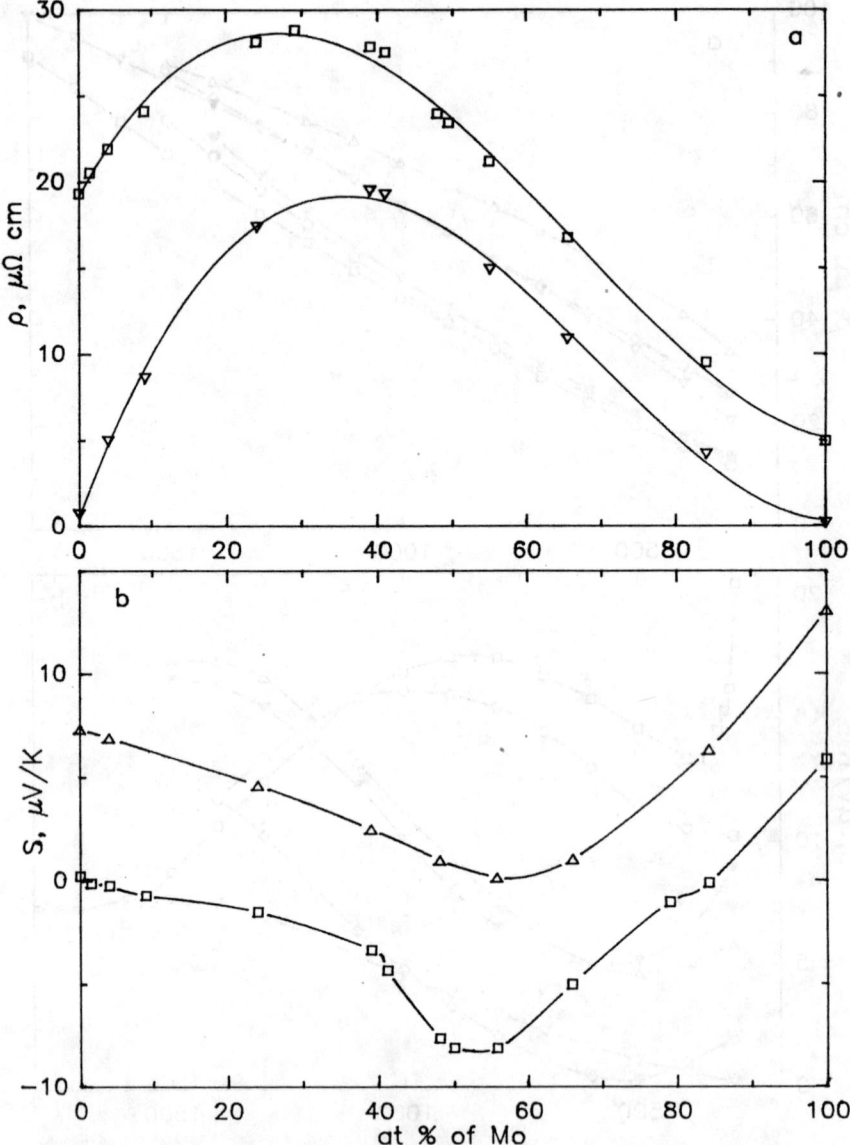

Fig.30. Electrical resistivity (a) and thermopower (b) of V–Mo alloys versus composition. Present authors.
▽ —at 4.2K; ▫ —293K; △ —1700K.

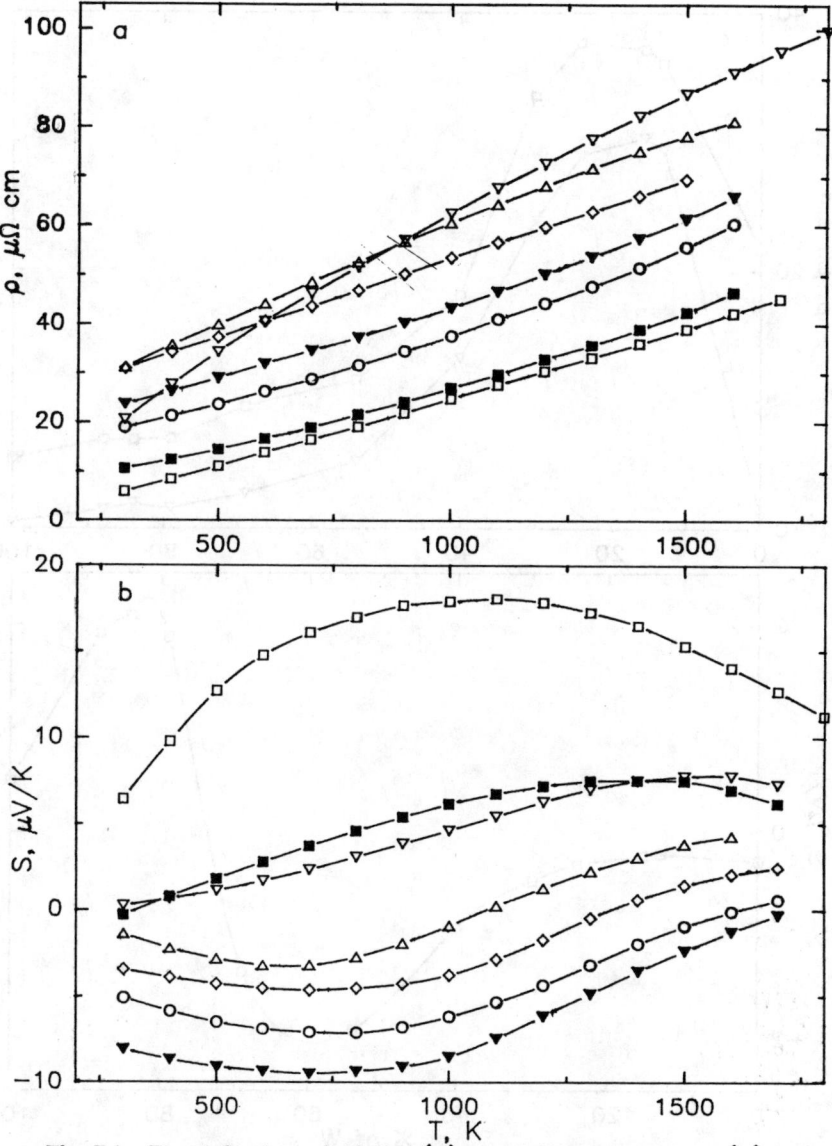

Fig.31. Electrical resistivity (a) and thermopower (b) of V–Mo alloys. Present authors. ▽ —V; △ —24%Mo; ◇ —38%Mo; ▼ —55%Mo; ○ —66%Mo; ■ —84%Mo; □ —100%Mo.

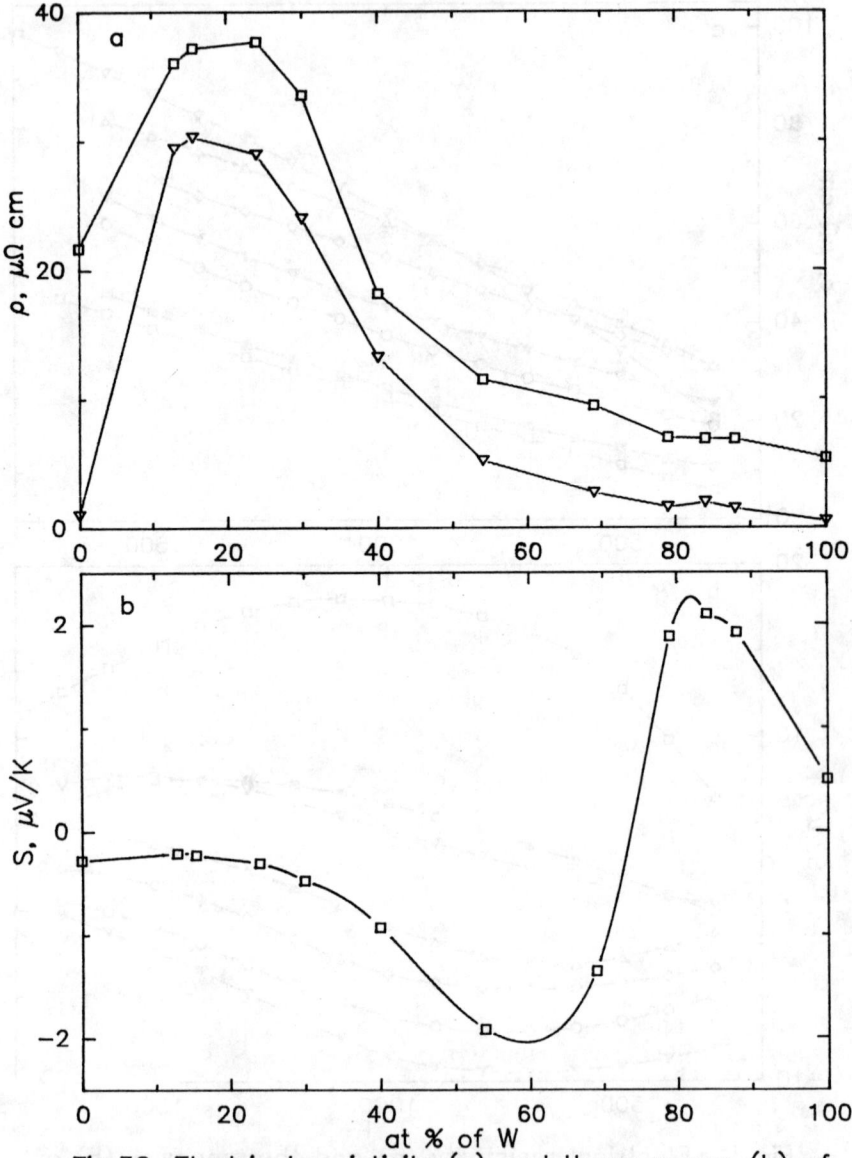

Fig.32. Electrical resistivity (a) and thermopower (b) of V–W alloys versus composition. ▽ —at 4.2%; □ —293K.

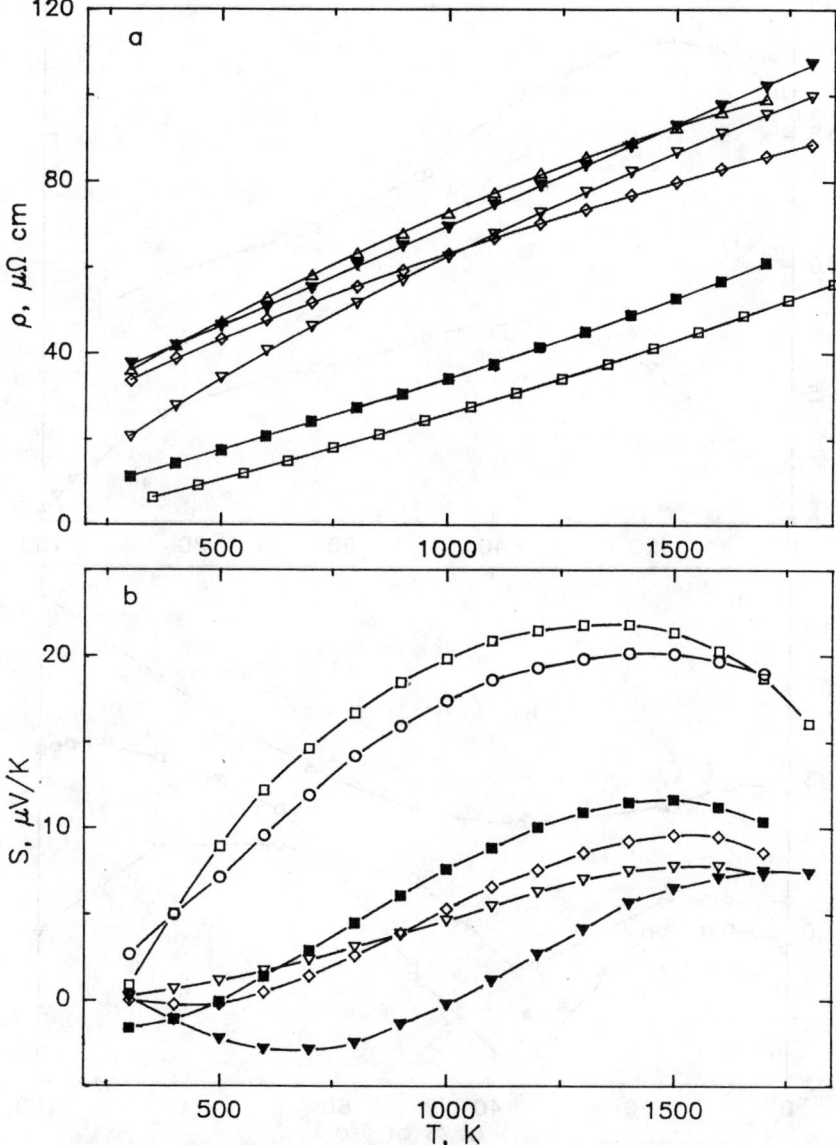

Fig.33. Electrical resistivity (a) and thermopower (b) of V–W alloys. Present authors. ▽ —V; △ —13%W; ▼ —24%W; ◇ —30%W; ■ —55%W; ○ —79%W; □ —W.

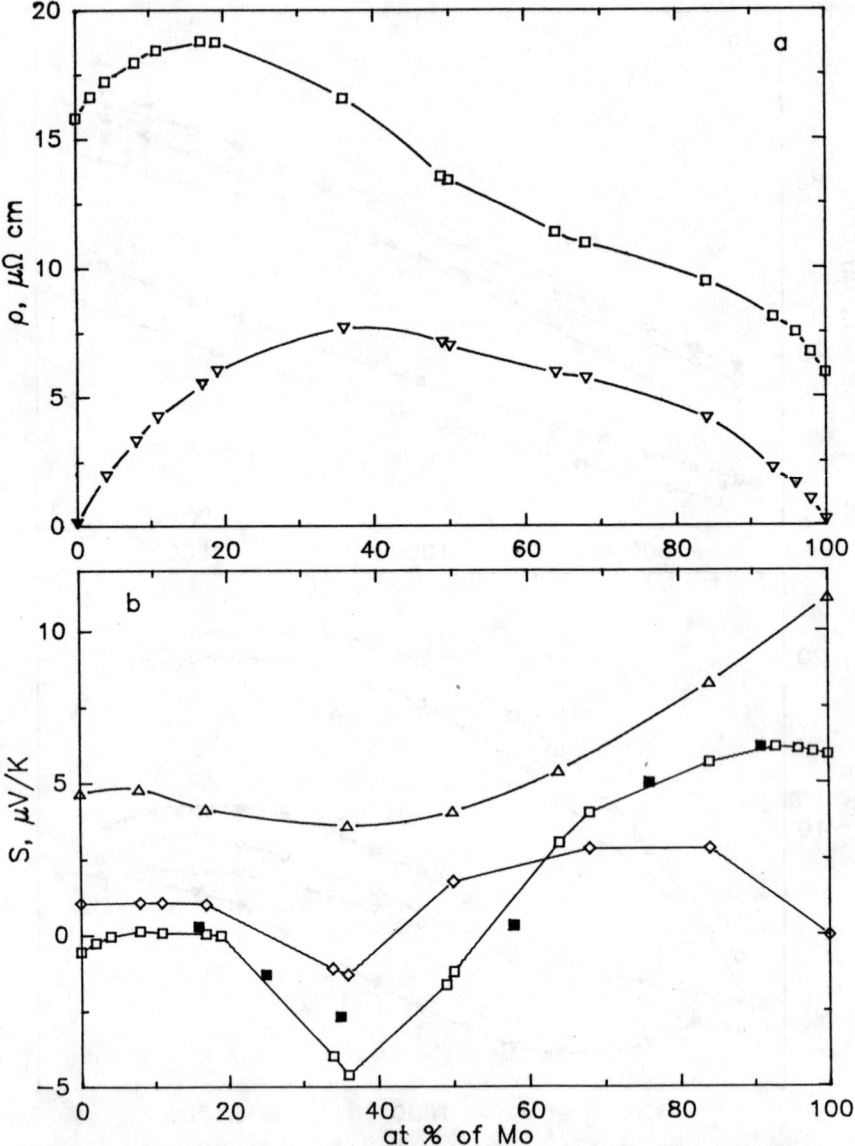

Fig.34. Electrical resistivity (a) and thermopower (b) of Nb–Mo alloys. Hollow symbols —present authors, filled symbols—Lin (1980). ▽ —at 4.2K; ◇ —100K; □ —300K; △ —1700K.

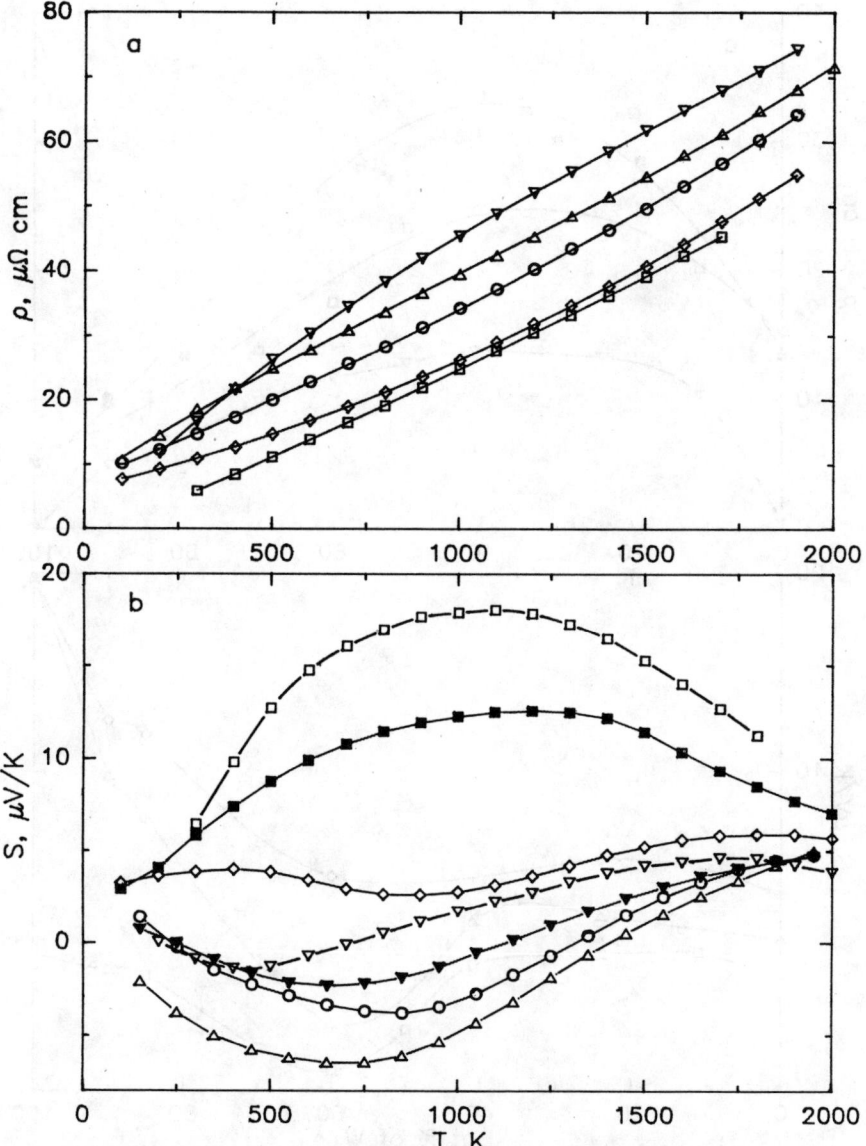

Fig.35. Electrical resistivity (a) and thermopower (b) of Nb—Mo alloys. Present authors. ▽ —Nb; ▼ —17%Mo; △ —36%Mo; ○ —50%Mo; ◇ —64%Mo; ■ —84%Mo; □ —100%Mo.

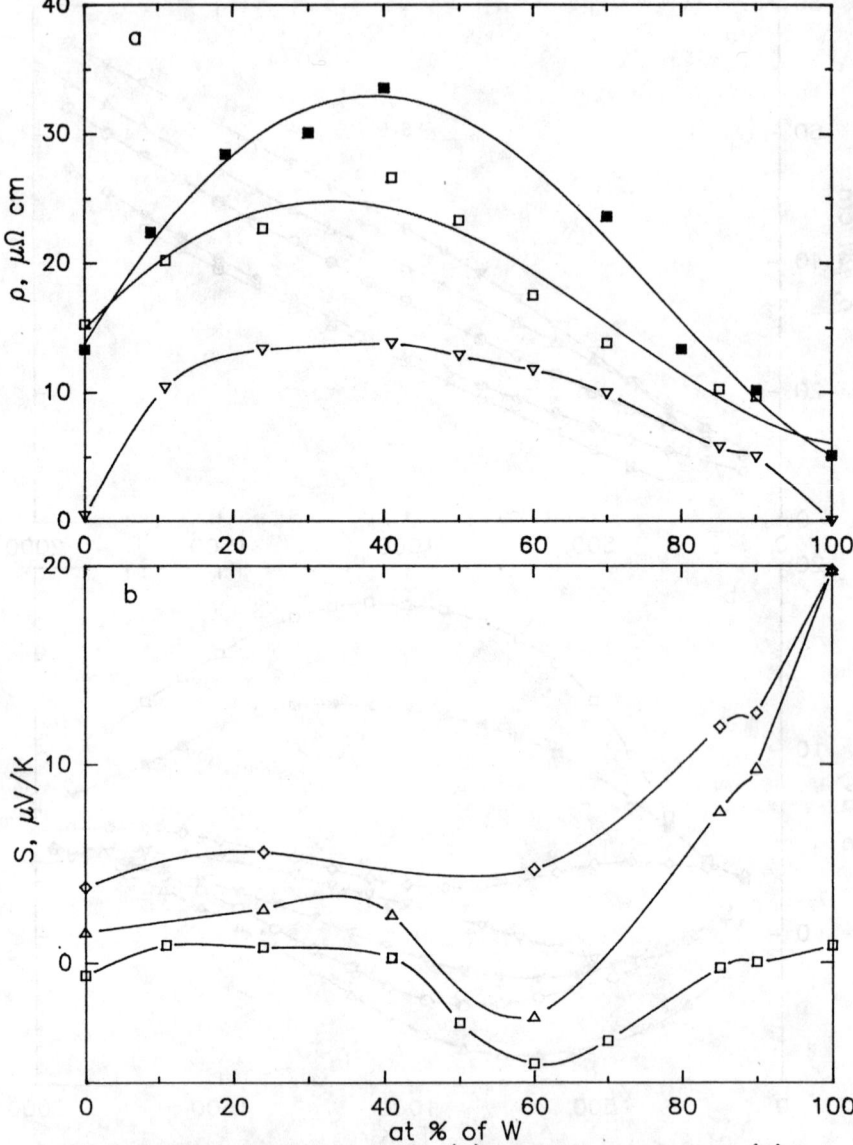

Fig.36. Electrical resistivity (a) and thermopower (b) of Nb–W alloys versus composition. Hollow symbols —present authors. Filled symbols—Kieffer(1959). ▽ —at 4.2K; □ —293K; △ —1000K; ◇ —1700K.

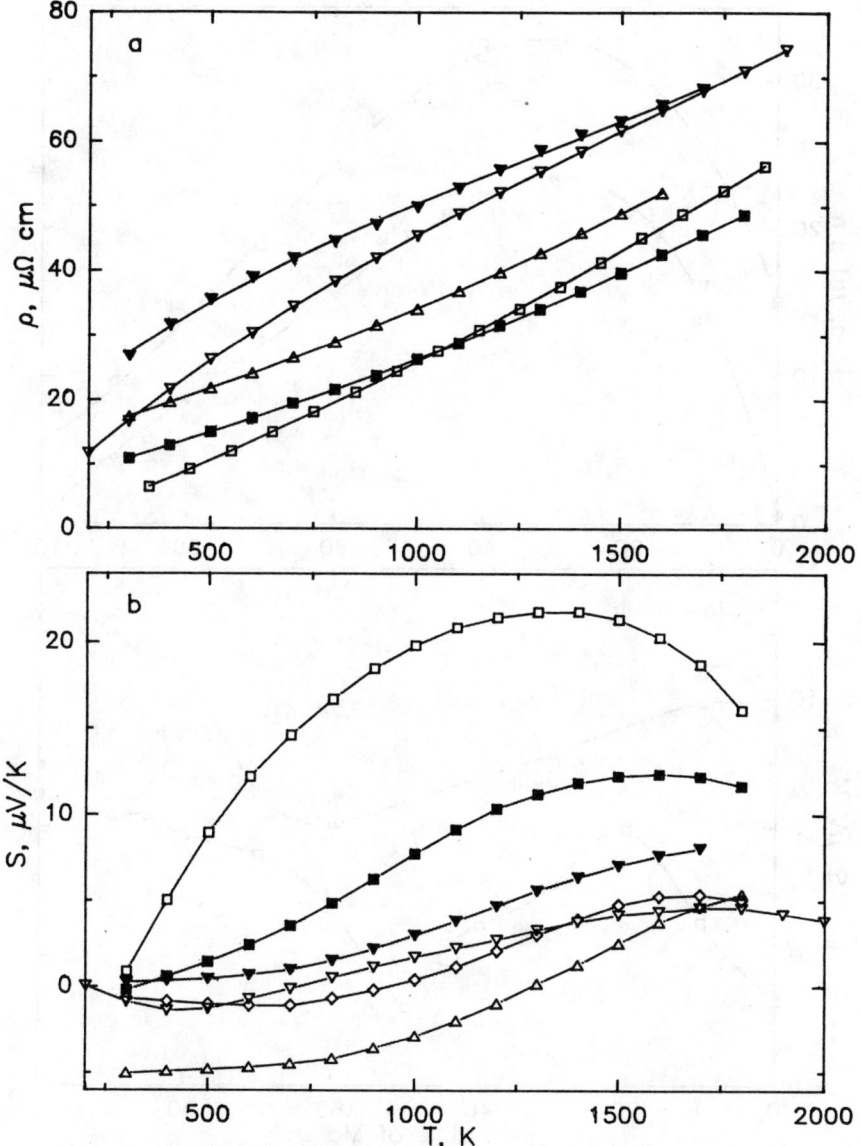

Fig.37. Electrical resistivity (a) and thermopower (b) of Nb—W alloys. Present authors. ▽ —Nb; ◇ —25%W; ▼ —41%W; △ —60%W; ■ —85%W; □ —100%W.

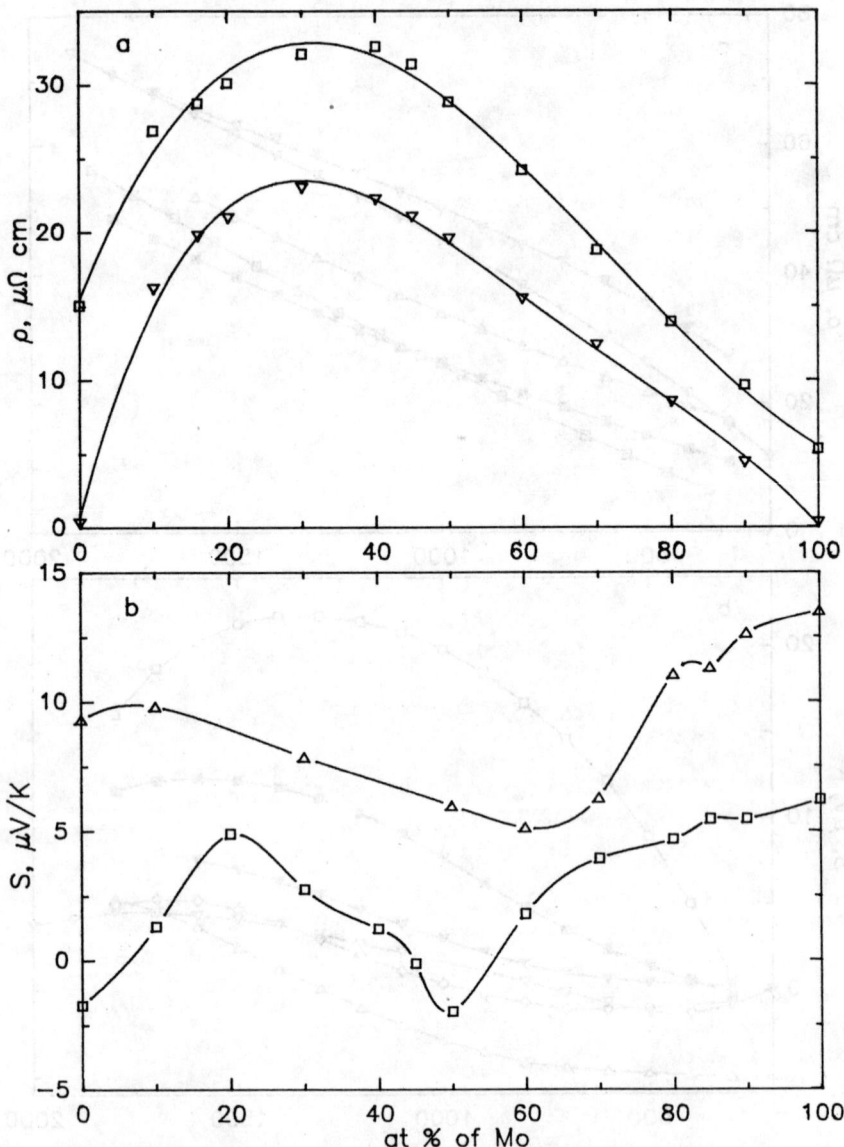

Fig.38. Electrical resistivity (a) and thermopower (b) of Ta—Mo alloys versus composition. Present authors.
▽ —at 4.2K; □ —293K; △ —1700K.

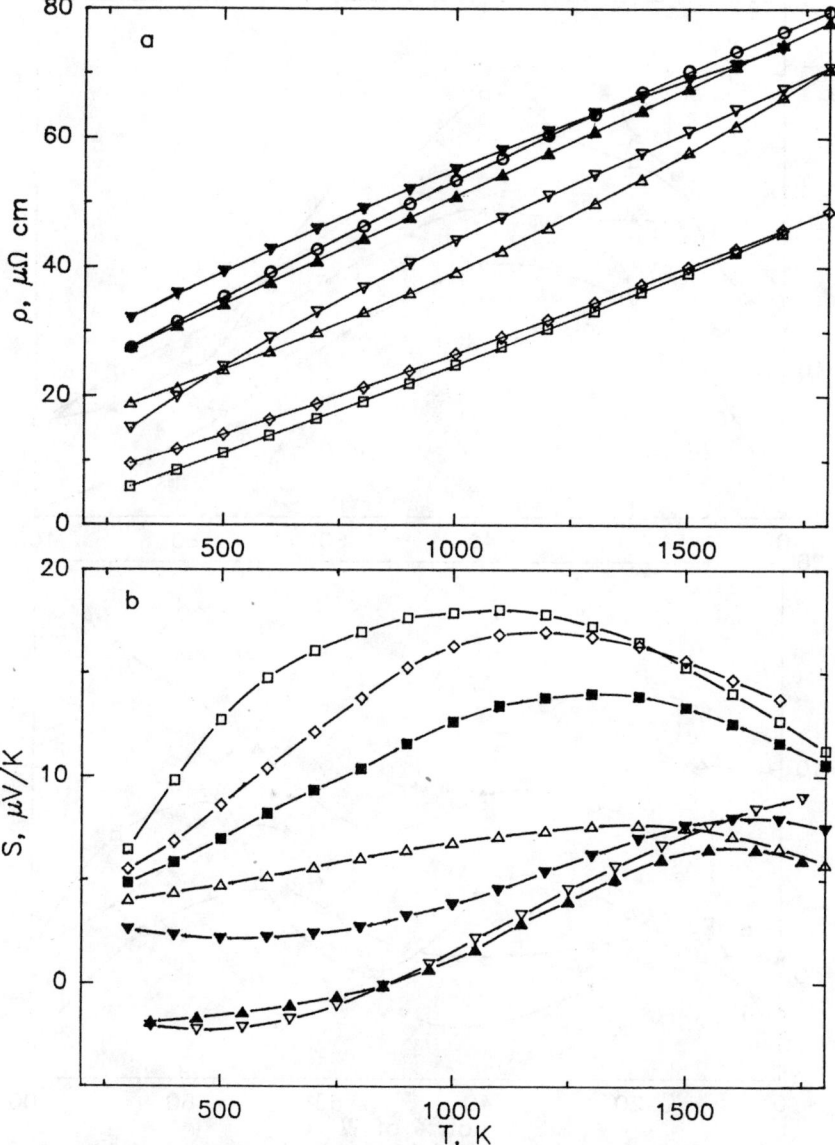

Fig.39. Electrical resistivity (a) and thermopower (b) of Ta–Mo alloys. Present authors. ▽ —Ta; ○ —10%Mo; ▼ —30%Mo; ▲ —50%Mo; △ —70%Mo; ■ —80%Mo; □ —100%Mo.

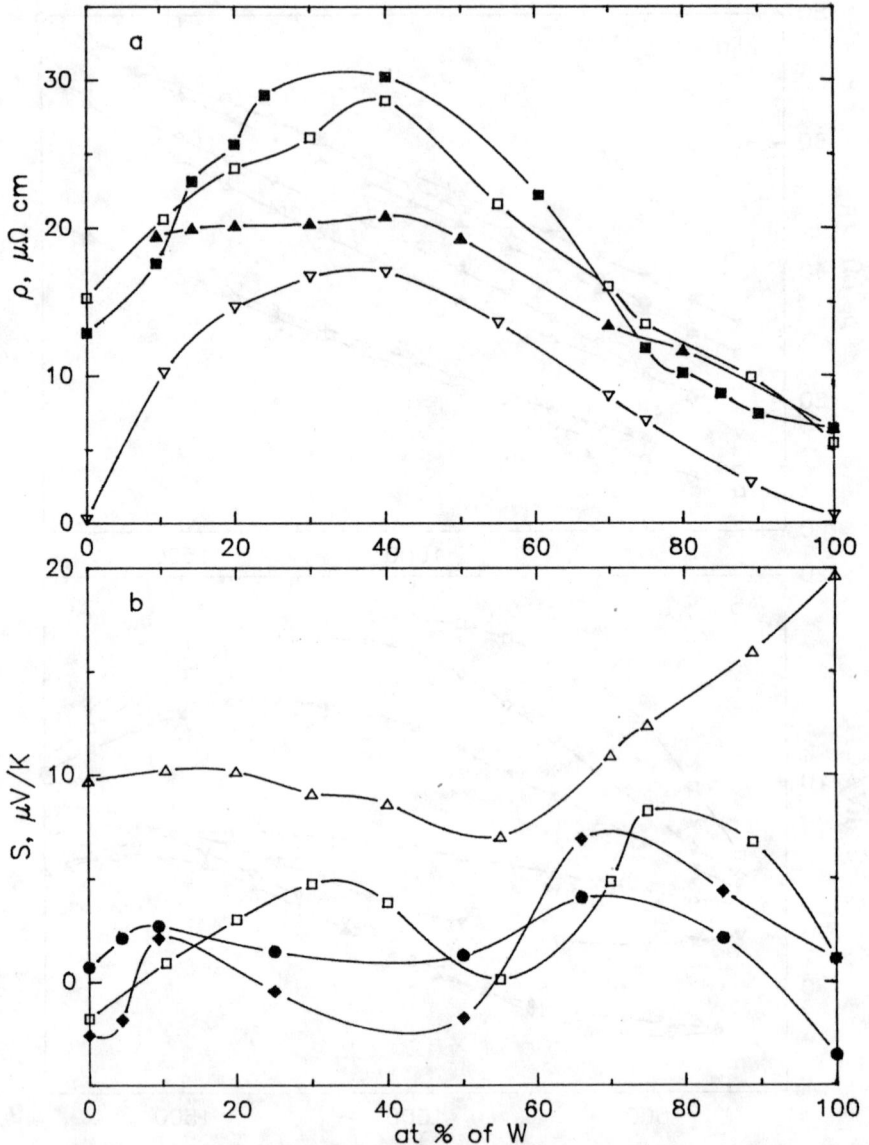

Fig.40. Electrical resistivity (a) and thermopower (b) of Ta–W alloys. Hollow symbols—present authors. ▽ —at 4.2K; □ — 293K; △ —1700K. ▲ —Tomas (1968) at 293K; ■ —Kieffer (1959) at 293K; ●, ◆ —Lin (1980) at 100K and 295K.

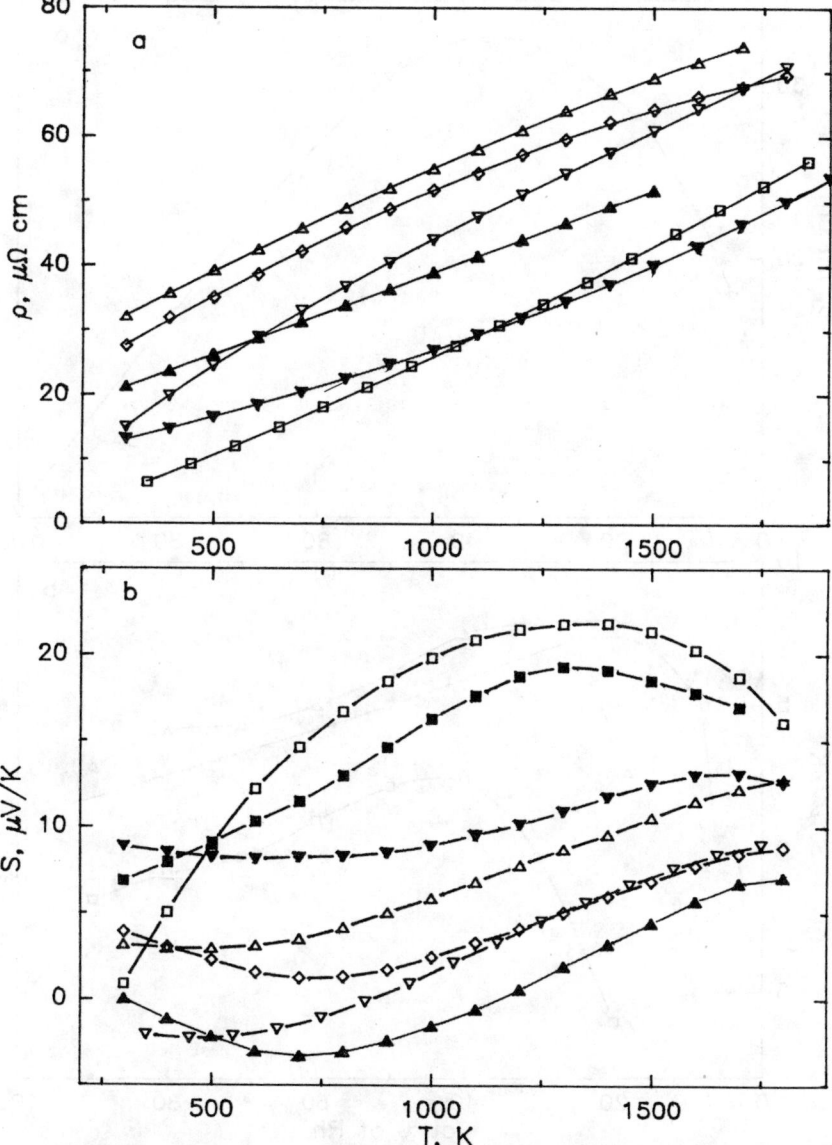

Fig.41. Electrical resistivity (a) and thermopower (b) of Ta-W alloys. Present authors. ▽ —Ta; △ —20%W; ◇ —42%W; ▲ —55%W; ▼ —75%W; ■ —89%W; □ —100%W.

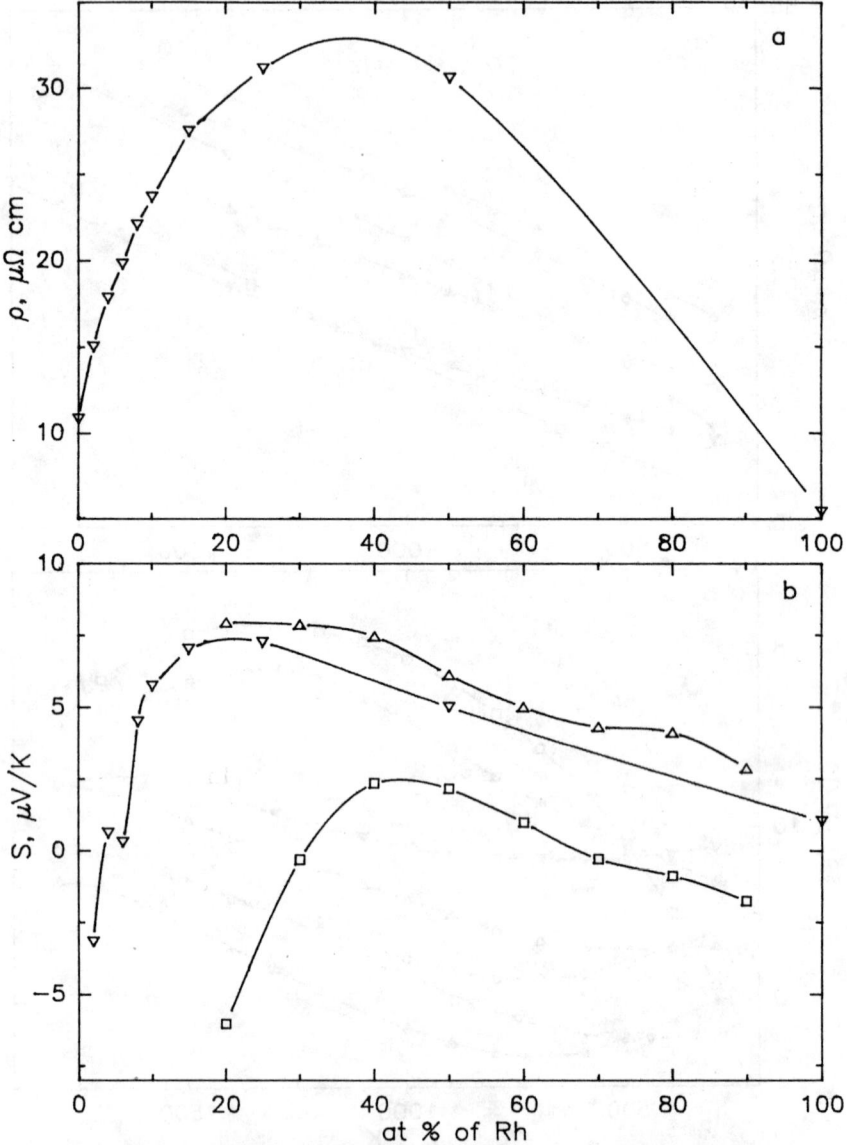

Fig.42. Electrical resistivity (a) and thermopower (b) of Pd–Rh alloys versus composition. ▽ –Koster (1963) at room temperature. △ –Rudnitskii (1956) at 373K; □ –Rudnitskii (1956) at 1473K.

reliable. Fig. 41 presents our results on the temperature dependence of the thermopower and electrical resistivity.

(ix) Rhodium - Palladium alloys (Rh - Pd)

The electrical resistivity and thermopower of these alloys were studied comprehensively by Rudnitskii[33] and Koster[34]. These data are displayed in Figs. 42 and 43.

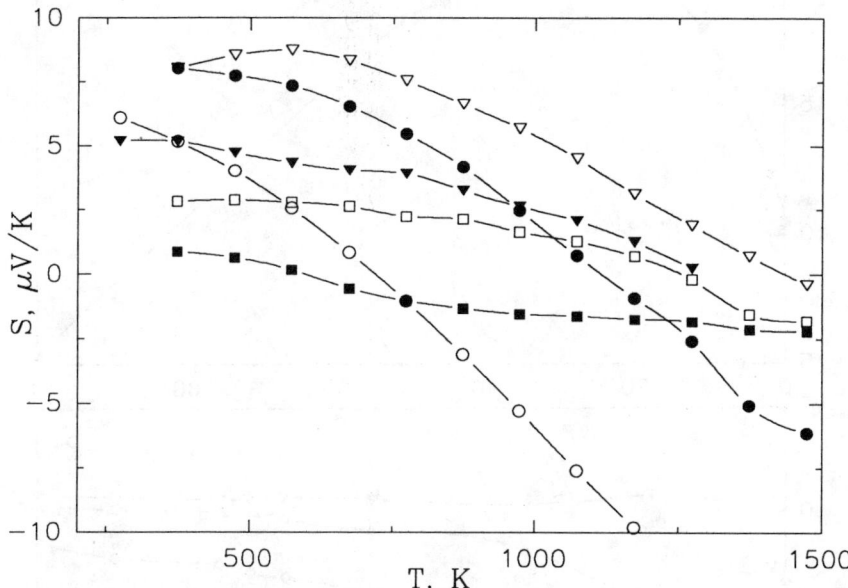

Fig.43. Thermopower of Pd−Rh alloys. Rudnitskii (1956).
○ −Pd+10%Rh; ● −20%Rh; ▽ −30%Rh; ▼ −60%Rh;
□ 90%Rh; ■ −100%Rh.

(x) Rhodium - Platinum alloys (Rh - Pt)

Some of these alloys have been used for a long time as thermocouple materials and are therefore well studied. However the electrical resistivity has been investigated only for alloys containing up to 60% of Rh - Beliaev [32]. A first experimental data on the absolute thermopower was published by Nystrom [49]. The temperature and composition dependence taken from this paper are depicted graphically in Figs. 44 and 45.

(xi) Palladium - Silver alloys (Pd - Ag)

These are apparently the most comprehensively and reliably investigated alloys. The total list of publications dealing with the properties of these alloys contains tens of references. We are presenting here only some of the data which we believe to be the most complete and reliable. The dependence of the electrical resistivity on composition at room

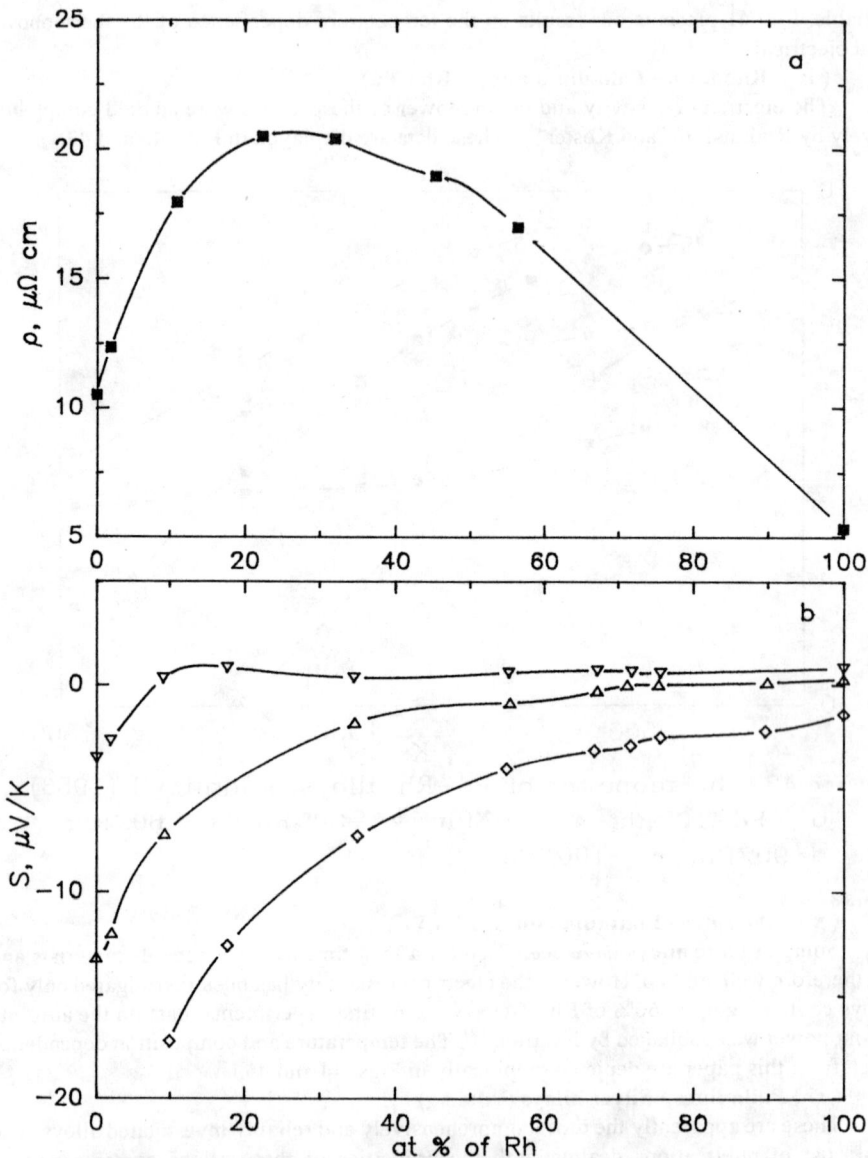

Fig.44. Electrical resistivity (a) and thermopower (b) of Pt–Rh alloys versus composition. ■ —Beliaev (1974) at 293K; Hollow symbols—Nystrom (1959): ▽ —273K; △ —773K; ◇ —1473K.

temperature was studied by Rudnitskii[33], Kemp[50], Seemann[52] and Coles[53]. The results

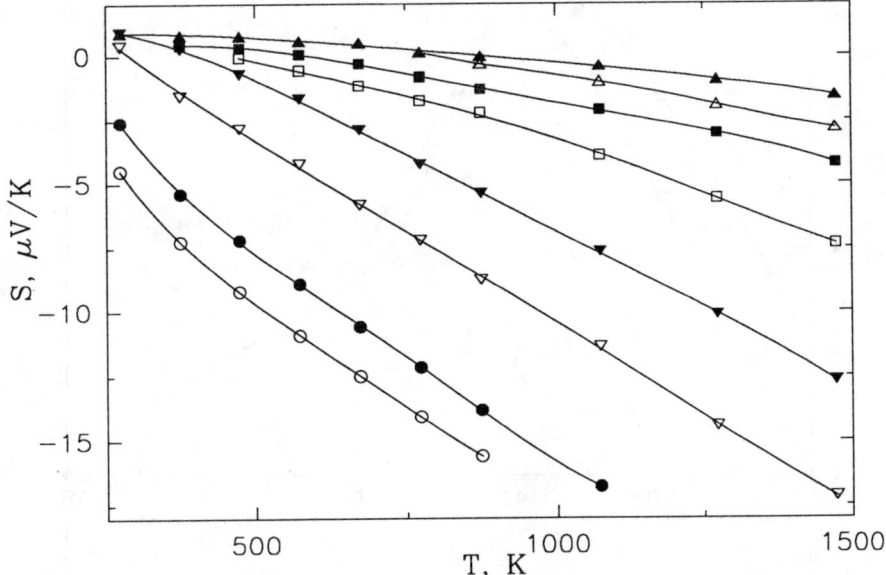

Fig.45. Thermopower of Pt–Rh alloys. Nystrom (1959).
○ –Pt; ● –1.9%Rh; ▽ –9.07%Rh; ▼ –17.4%Rh;
□ –34.3%Rh; ■ –54.8%Rh; △ –71.2%Rh; ▲ –Rh.

of these studies are in a good agreement and are displayed in Fig. 46a. The temperature dependences of the electrical resistivity obtained by Ricker[51] are given in Fig. 47a. Note the anomalous behavior of the 40% Ag alloy where the temperature dependence of electrical resistivity reveals a broad minimum in the region of 700 °K. While this feature was reproduced by Ahmad[54], later its existence was not confirmed - Arajs[55]. The latter study suggests that the minimum is observed in unannealed alloys and is associated with the existence of short-range order caused by thermal and mechanical effects- Chen[56]. This problem cannot be considered a closed. The thermopower was also investigated in many papers - Rudnitskii[33], Ricker (1966)[51], Chen[56], Taylor[57], Chupina[58], Koster[59] and Fletcher[60,61]. The results of these studies are in a good agreement with some of them presented in Figs. 46b and 47b.

(xii) **Palladium - Gold alloys (Pd-Au)**
These alloys are analogs of the Pd -Ag alloys and are also comprehensive studies. Figs. 48 and 49 present data obtained by Rudnitskii[33] and Kim[62].

(xiii) **Rhenium- Osmium alloys (Re-Os)**
We are not aware of any experimental data on the thermopower and electrical resistivity of these alloys.

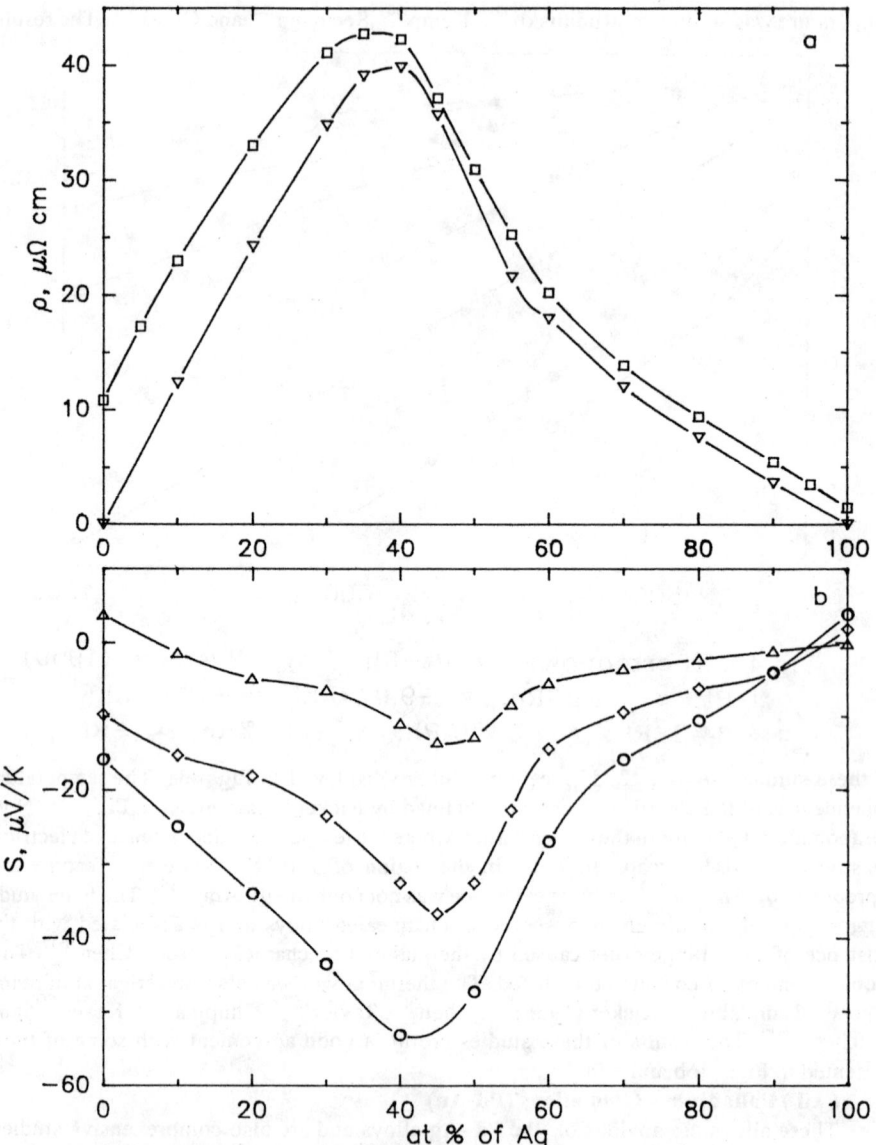

Fig.46. Electrical resistivity (a) and thermopower (b) of Pd–Ag alloys. ▽ —Coles (1962) at 4.2K; □ —Seemann (1966) and Kemp (1956) at 300K. △ —83K, ◇ —273K: Seemann (1966); ○ —Ricker (1966) at 473K.

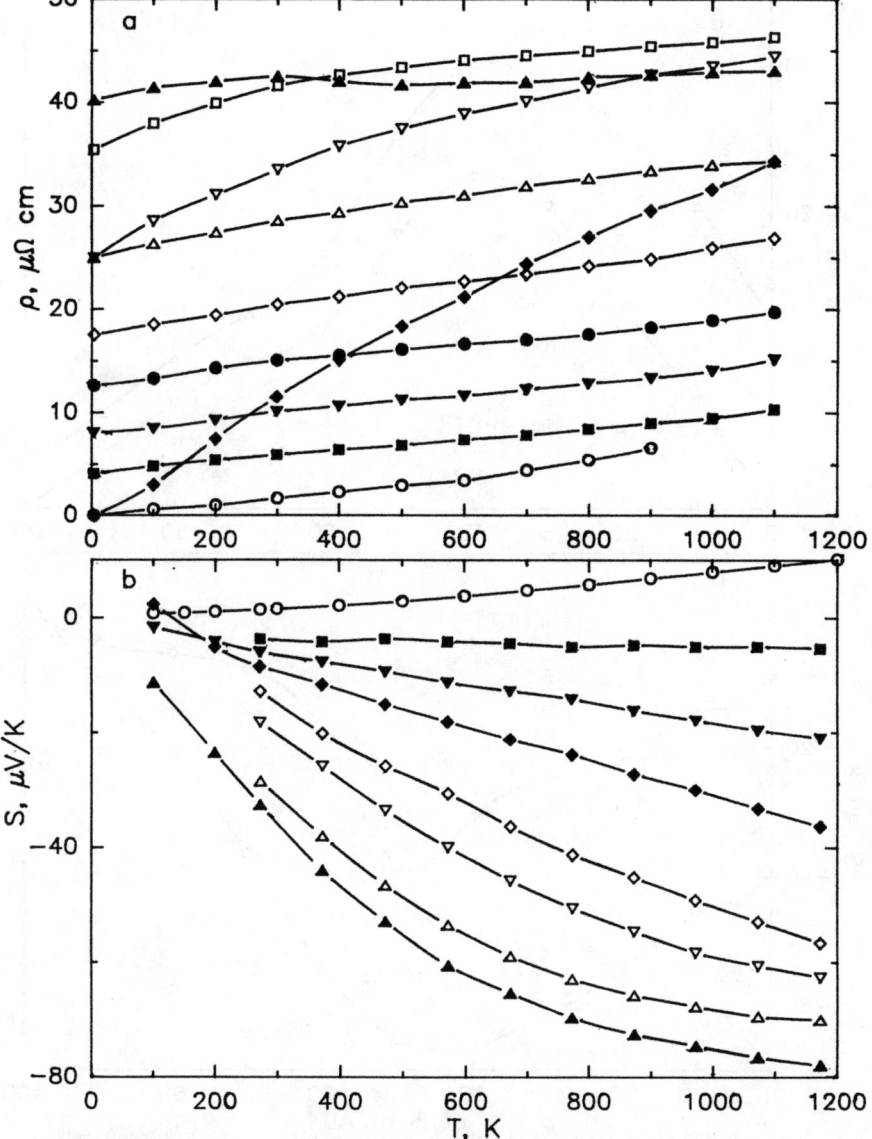

Fig.47. Resistivity (a) (Ricker (1966)) and thermopower (b) (Rudnitskii (1956) and Taylor (1956)) of Pd–Ag alloys.
♦ —Pd; ▽ —20%Ag; □ —30%Ag; ▲ —40%Ag; △ —50%Ag
◇ —60%Ag; ● —70%Ag; ▼ —80%Ag; ■ —90%Ag; ○ —Ag.

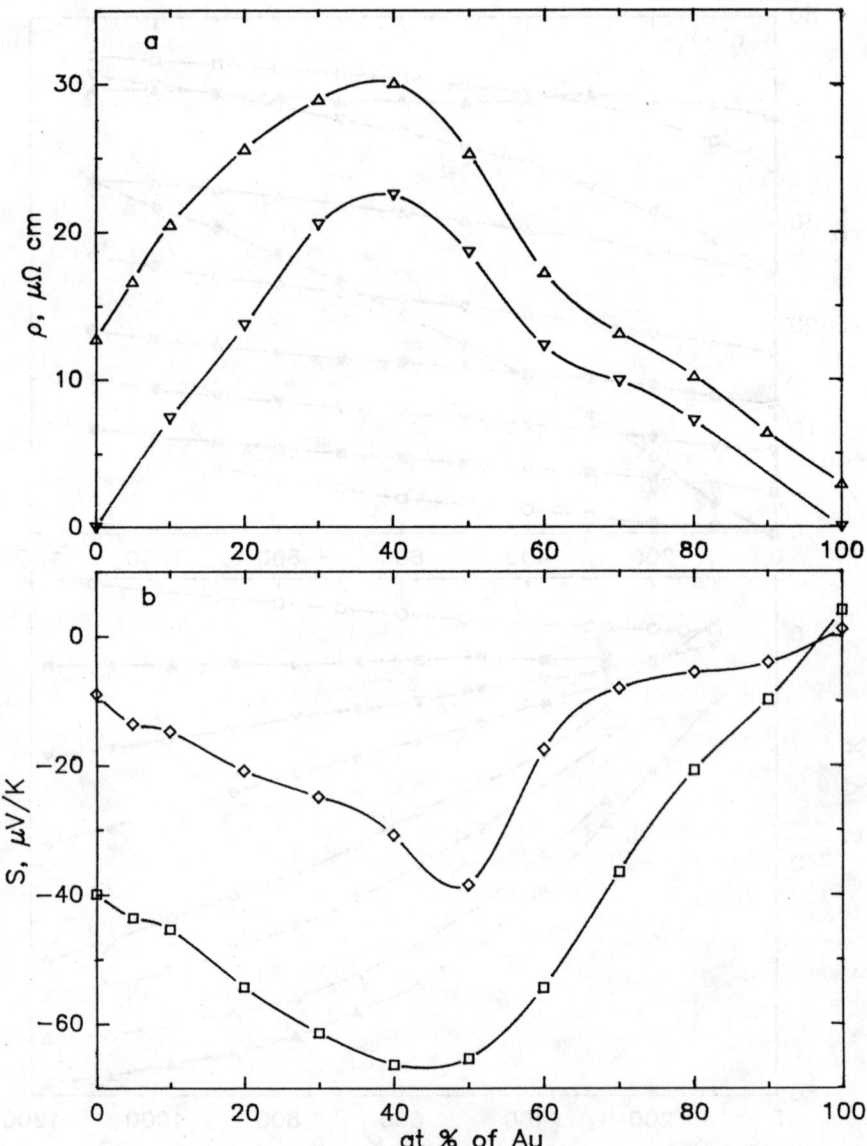

Fig.48. Electrical resistivity (a) and thermopower (b) of Pd–Au alloys versus composition. Rudnitskii (1956) and Kim (1967). ▽ – at 4.2K; △ – 293K; ◇ – 273K; □ – 1273K.

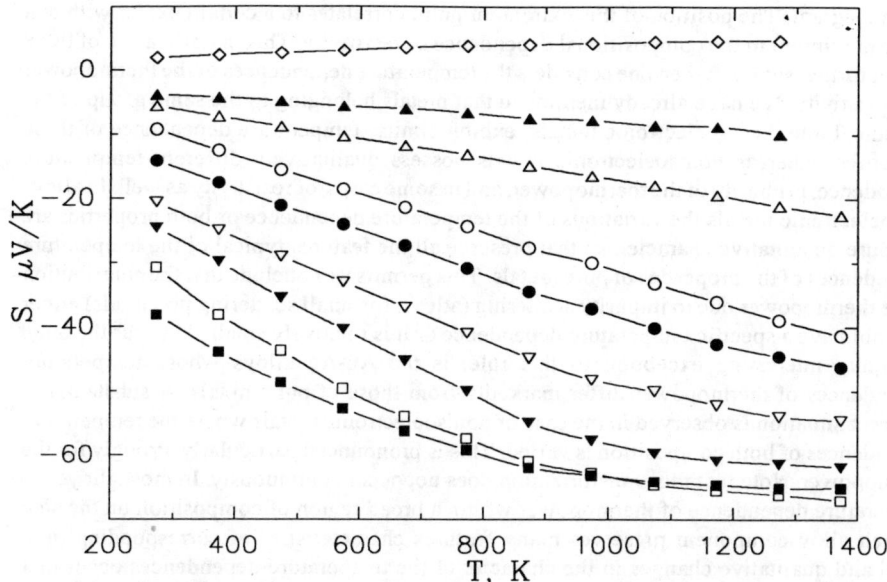

Fig.49. Thermopower of Pd—Au alloys. Rudnitskii (1956).
○ —Pd; ● —10%Au; ▽ —20%Au; ▼ —30%Au; □ 40%Au;
■ —50%Au; △ —80%Au; ▲ —90%Au; ◇ —Au.

6.4 DISCUSSION OF EXPERIMENTAL DATA

Let us first of all separate the characteristic features in the compositional and temperature dependences of the alloy thermopower and electrical resistivity permitting one to divide all BCSS into two types. As evident from the experimental data presented in the preceding section, the characteristic feature of the alloys of the first type formed of isoelectronic metals consists in that the dependence of the electrical resistivity on their composition obeys the Kurnakov-Nordheim rule. The impurity contribution to resistivity is described bt a parabolic profile symmetric relative relative to the equiatomic composition. The thermopower of large fraction of alloys of this type depends only weakly on composition in concentrated alloys, a stronger dependence being observed in dilute alloys, up to 10 - 15 % of the second component. The composition dependence of the resistivity for alloys of the second type formed of nonisoelectronic metals deviates noticeably from the Kurnakov - Nordheim rule and represents a curve which is nonsymmetric relative to the equiatomic composition and whose maximum is shifted toward the nominally lower-valence metal and the larger derivative $\partial \rho / \partial T$ at lower temperatures. The thermopower of these alloys depends strongly on composition for any component ratio, with a clearly pronounced extremum being usually observed in the middle of the thermopower -composition diagram. The position of this extremum point correlates to a certain extent with that of the maximum in the compositional dependence of resistivity. This classification of BCSS finds a further support when one considers the temperature dependences of the thermopower and resistivity. We have already mentioned that metals belonging to the same group of the Periodic Table, i.e. isoelectronic metals, exhibit similar temperature dependence of these properties, whereas nonisoelectronic metals possess qualitatively different temperature dependence, primarily of the thermopower, and in some cases, of resistivity as well. In alloys of isoelectronic metals the variations of the temperature dependence of both properties are of a pure quantitative character, so that preserve all the features typical of the temperature dependences of the properties of pure metals. This permits us conclude that the contribution to the thermopower due to impurity scattering (atleast for small scattering potentials) either does not have a specific temperature dependence or it is relatively small. Thus far the only, and quite interesting exception to this rule, is the Au-Ag alloys whose temperature dependences of thermopower differ markedly from those of pure metals. A substantially different situation is observed in the case of nonisoelectronic metals where the temperature dependences of both composition is varied. This is pronounced particularly strongly for the thermopower. Note that this transformation does not occur continuously. In most alloys the temperature dependence of thermopower within a broad region of composition on the side of each alloy component preserves many features characteristic of corresponding pure metal and qualitative changes in the character of the temperature dependences occur in a comparatively narrow region of alloys compositions.

Thus the comprehensive empirical analysis is carried out for the electrical resistivity and thermopower of metallic BCSS at high temperatures. It revealed more general rule than Kurnakov-Nordheim's one for dependence of electrical resistivity and thermopower o BCSS versus their composition. The conclusion on existence of such a more general rule presents itself as proved enough since it bases on the analysis of properties of almost all existing BCSS systems; on the parallel analysis of two independent properties; on a problem arises how to interpret this generalized empirical relations theoretically. In principle, this can be done basing on eqs. (5 - 7) for the electrical conductivity and thermopower, however, one has to know for its purpose the function $\omega(\varepsilon, T)$ which includes all the details of electronic structure and conduction electron interactions with other excitations in a metal. Obviously, one could hardly calculate this function presently from first principle in a realistic model of the electronic spectrum and scattering mechanisms.(Recently some progress was made in theoretical calculation of compositional dependence of the BCSS electrical resistivity. For survey see for example Rossiter[63]. However, the techniques used for the calculation are quite specialised and require powerful computing facilities). This is why it seems to us as sufficiently promising to use a simple phenomenological approach, by means of which, however, it is possible to get interesting qualitative results. Because the lack of space we will only try to outline briefly this approach here.

Let us assume that the function $\omega(\varepsilon, T)$ can be represented as a product :
$$\omega(\varepsilon, T) \; \theta(T) . \varphi(\varepsilon) \tag{14}$$
This approximation means that the nature of scattering processes and the electronic structure do not very substantially in the temperature range under consideration. Under this

conditions θ(T) depends only on the intensity of the conduction electron scattering and vanishes in the expression for the thermopower. At the same time function leads to a monoatomic increase of electrical resistivity with temperature (rule of Bloch - Gruneisen). The function φ(ε) reflects the structure of the electronic density of states function and the energy dependence of scattering process in the vicinity of Fermi energy. Using this form of representation of the function ω(ε, T), one can readily obtain from (5,6) :

$$\sigma(T) = \theta(T) \int_0^\infty \varphi(\varepsilon) \cdot (-\partial f^0/\partial \varepsilon) \, d\varepsilon. \qquad (15)$$

$$S(T) = \frac{1}{e.T} \frac{\int_0^\infty \varphi(\varepsilon)(\varepsilon - \mu)(-\partial f^0/\partial \varepsilon) d\varepsilon}{\int_0^\infty \varphi(\varepsilon)(-\partial f^0/\partial \varepsilon) d\varepsilon} \qquad (16)$$

Now we can consider (14, 15) as a coupled integral equations for the function φ(ε). Solving these equations one can obtain from experimental data on the temperature dependences of the electrical conductivity σ(T) and thermopower $S(T)$ an electron energy-dependent function φ(ε) which can be compared with the electronic structure characteristics of a metal. One of the possible ways of solving eqs.(15,16) consists in constructing an appropriate function φ(ε) with a certain (not too large) number of fitting parameter to chosen such that eqs.(14,15) provide the best fit to the experimental temperature dependences of thermopower and electrical resistivity. One has naturally to use some physical considerations in order to choose the most suitable function out of the infinite multiplicity of possible functions. One can, for instance, employ the considerations underlying the well Known Mott's model of s-d scattering (Mott[6]) . In accordance with this model, the relaxation time (in our model, the function φ(ε)) is inversely proportional to the density of d- states in the vicinity of Fermi energy. Thus this approach permits one to use the results of realistic calculations of the electronic structure of metals and alloys for the interpretation of experimental data on the electronic transport properties. On the other hand, by using this procedure one can obtain an useful information on electronic structure of a metal or alloy from the experimental data on the temperature dependences of the thermopower and resistivity. Basing on the function φ(ε) for pure metals , one can , at least in framework of simple models of the electronic structure of alloys, such as the rigid band approximation[7], or more sophisticated coherent potential approximation[64], interpret experimental data for alloys.

6.5 CONCLUSION

In the course of performed by us study, electrical resistivity and thermopower of metallic binary continuous solid solution were investigated intensively. It is possible to find in the literature references on the existence of 29 BCSS systems. The present paper includes sufficiently detailed data on temperature and composition dependences of the thermopower and resistivity for 23 of them. For the first time such representative analysis of the electronic transport properties of the simplest class of disordered systems was carried out. It has shown a restricted use of old Kurnakov-Nordheim's rule. It was revealed that all the BCSS are divided into two types according to the character of composition dependences of their transport properties. Thus the problem is suggested to theoreticians to develop an general physical interpretation of the whole set of electrical and thermoelectrical properties instead of special calculation of a particular property of a given alloy. We also proposed a simple phenomenological approach which can be useful in the interpretation of the experimental data on the electronic transport properties.

REFERENCES

1. M.V. Vedernikov, V.G. Dvunitkin and A. Zhumagulov, *Fizika tverdogo tela*, **20** (1978) **3302** ; In Russian. *Sov. Journal Sol. State Phys.*
2. A. Zhumagulov and M.V. Vedernikov, *Fizika Metallovi Metallov* , **49** (1980) 892; In Russian *Sov. Journal Metal Phys. and Metal Material Science*.
3. M.V. Vedernikov, V.G. Dvunitkin and A. Zhumagulov In *Therrmoelectricity in metallic conductors*. ed. by F.J. Blatt and P.A. Shroeder (London, Plenum Press, 1978) p. 117-123.
4. M.V. Vedernikov, V.G. Dvunitkin and A.T. Burkov, In Russian (Institute for High Temperatures of the Academy of Sciences, Moscow USSR ,1990) N5,**85** (1990) p.45-92 in series *Reviews on thermophysical properties of substances*.
5. Landolt - Bornstein. *Numerical Data and Functional Relationships in Science and Technology*. New Series. Group III. **15**, Metals. N.Y. Tokio (Springer Verlag, 1985).
6. N.F. Mott and H. Jones, *The thoery of the properties of Metals and Alloys* (Oxford, 1936).
7. J.M. Ziman, *Electrons and Phonons* (Cambridge University Press,1960) .
8. M.V. Vedernikov, *Adv. Phys.*, **18** (1969) 337.
9. A.T. Burkov, M.V. Vedernikov and ETF Zh., **85** (1983)1821.; In Russian *Sov. JETP*
10. A.T. Burkov, M.V. Vedernikov and E. Gratz, *Solid State Comm.*, **67** (1988)1109.
11. D. Barnard, *Thermoelectricty in metals and alloys* (Taylor & Frances, London ,1972).
12. D.A. Papaconstantopoulos, *Handbook of the Band Structure of Elemental Solids* (Plenum Press, N.Y., London ,1986).
13. J. Bass, *Adv Phys.* , **21** (1972) 431.
14. R.B. Roberts, *Phil. Mag.* , **36** (1977) 91.
15. J.J. Lander, *Phys. Rev.*, **74** (1948) 479.
16. N.E. Cusack and P.W. Kendall, *Proc. Phys. Soc.* , **72** (1958) 898.
17. A.E. Vol and I.K. Kagan, *Structure and properties of binary metallic system;* In Russian (Nauka., Moscow, 1979).
18. D.K.C. Mc Donald, and W.B. Pearson, *Proc. Roy. Soc.* , **A219** (1953) 373.
19. D.K.C. Mc Donald and W.B. Pearson, *Proc. Roy. Soc.* , **A221** (1954) 534.
20. A.M. Guenault and D.K.C. Mc Donald, *Proc. Roy. Soc.*, **A274** (1963) 154.

21. V.I. Savin, M.I. Lesnaja and B.N. Ovdei, *Izvestia Academii Nauk*, *Seria Metally*, **6** (1979) 195. In Russian. *Sov. Journal Proc. Acad. Sc. USSR, series Metals.*
22. M.A. Tilkina, A.I. Pekarev and E.M. Savitskii *Zhurnal Neorg. Himii,***4** (1959) 2320; In Russian. *Sov. Journal of Inorganic Chem.*
23. I.P. Druzhinina, In *Alloys of rare metals with special physical properties.* (Nauka, Moscow,1974) p. 166. In Russian.
24. N.N. Sirota, E.A. Ovseichuk and AN. Docklady SSSR, **174** (1967) 570. In Russian. *Sov. Journ. Reports Acad. Sc. USSR.*
25. I.P. Druzhinina, *Izmeritelnaja Technika,* **8** (1969) 36. In Russian. *Sov. Journ. Measur. Techn.*
26. O. Rapp and M. Pokorny, *Physica Scripta*, **6** (1972) 200.
27. MaMiys and Y. Masuda, *J.Phys. Soc. Japan,* **40** (1976) 390.
28. K. Schroder and H. Tomaschke, *Z. Phys.*, **B7** (1968) 318.
29. A.L. Trego and A.R. Mackintosh, *Phys. Rev.*, **166** (1968) 495.
30. S. Arajs, E.E. Anderson and K.V. Rao, *J. Less. Comm. Met.*, **26**(1971)157.
31. R. Kieffer, *Z. Mettalkunde*, **50** (1959) 18.
32. I.F. Beliaev, V.M. Karbolin and N.I. Timofeev, In *Alloys of rare metals with special physical properties* (Nauka, Moscow, 1974) p. 144. In Russian.
33. A.A. Rudnitskii, *Thermoelectric properties of noble metals and their alloys.* (A cad Sc. USSR Pub. House Moscow, 1956) In Russian.
34. W. Koster, W. Gmohling and D. Hagmann, *Z. Metallkunde* **54** (1963) 325.
35. P. Blood and D. Greig, *J. phys.F.* **2** (1972) 79.
36. S. Crisp and J. Rungis, *Phil. Mag.* **22** (1970) 217.
37. S. Tamaki, *Phys. Stat. Sol.* **a18** (1973) 597.
38. T. Ricker, *Z. Metallkunde,* **54** (1963) 718.
39. S. Kashido, *J. Phys.Soc. Japan,* **28** (1970) 261.
40. A. Giannuzi, H. Tomashke and K. Schorder, *Phil. Mag.*, **21** (1970) 279.
41. J. C. H. Chiu, *Phys. Rev.* **B13** (1976) 1507.
42. G. I. Terechov, M. Ja. Altzitcer and O. S. Ivanov, In *Structure and properties of alloys for nuclear energetics* (Nauka, Moscow, 1973) p. 178. In Russian.
43. W. R. Cox, D. J. Hayes and F. R, Brotzen, *Phys. Rev.* **B7** (1973) 3580.
44. A. C. Lin and F. R. Brotzen, *J. Appl. Phys.*, **51** (1980) 669.
45. V. I. Krimer *Zhurnal Neorgan. Himii,* **3** (1958) 895. In Russian *Sov. Journ. Inorg, Chem.*
46. V. S. Micheev and D. M.. Pevtcov *Zhurnal Neorgan. Himii,* **3** (1958) 861. In Russian. *Sov. Journ. Inorg, Chem.*
47. L.J. Torn van, *Phys. Stat. Sol.*, **13** (1966) 345.
48. L. Tomas, *Z. Metallkunde,* **59** (1968) 127.
49. J. Nystrom, In *Landolt-Bornstein. Zahlenwerte and Funktionen,* **Bd..2**, teil 6, (Springer, Berlin, 1959).
50. W.R.G. Kemp., *Proc.Roy.Soc.*, **A233** (1956) 480.
51. T. Ricker and E. Pfluger, *Z. Metallkunde*, **B57** (1966) 39.
52. H. J. Seemann and G. Rennolet, *Z. Phys.* **196** (1966) 486.

53. B. R. Coles and J. C. Taylor, *Proc. Roy. Soc.* **A267** (1962) 139.
54. H. M. Ahmad and D. Greig, *Phys. Rev. Letters*, **32** (1974) 833.
55. S. Arajas, K. V. Rao, Y. D. Yao and W. Teoh, *Phys. Rev.* **B15** (1977) 2429.
56. W. K. Chen and N. E. Nicholson, *Acta Metallurgica*, **12** (1964) 687.
57. J. Taylor and B. R. Coles, *Phys. Rev.* **102** (1956) 27.
58. L. I. Chupina and V. E. Zinoviev, *Fizika Metallov i Metalloved.*, **48** (1979) 476. In Russian. *Sov.Jorn. Metal Phys. and Metal Materials Sc.*
59. W. Koster and D. Hagmann, *Z. Metallkunde*, **52** (1961) 721.
60. R. Fletcher and D. Greig, *Phys. Lett.*, **17** (1965) 6.
61. R. Fletcher and D. Greig, *Phil.Mag.*, **17** (1968) 21.
62. M. J. Kim and W. F. Flanagon, *Acta Metallurgica*, **15** (1967) 735.
63. P. L. Rossiter, *The electical resistivity of metals and alloys* (Cambridge University Press, Cambridge, 1987).
64. H. Ehrenreich and L. M. Schwartz, *Solid St. Phys.*, **31** (1976) 149.

CHAPTER 7

INTERPLAY BETWEEN ELECTRON CORRELATION AND DISORDER

Fabio Siringo
Dipartimento di Fisica, Università di Catania, Italy

7.1 Introduction

In 1958 Anderson[1] presented a theorem stating the possible absence of delocalized electronic states in disordered systems. The context was that of a pure independent particle approximation, and since then a considerable effort has been made in order to extend the theorem to a more realistic correlated ensemble of electrons. A key point in Anderson's approach was the choice of a 'local' set of states compared to the usual plane wave representation in $\vec{k}$-space. The last allows a perturbative description valid in the limit of weak disorder. Conversely the 'local' point of view is more realistic in a regime of strong disorder where the kinetic energy may be regarded as a perturbative correction to the Hamiltonian.

The natural implementation of such local description consists in adopting a tight-binding Hubbard Hamiltonian which contains a simple (but subdolous) correlation energy for electrons occupying the same local site at the same time. While the presence of the Hubbard correlation energy is of basilar importance for predicting realistic magnetic properties, it turns out that generally the localization theorem does not break down in presence of correlation.

What comes out[2] is that the presence of disorder strongly enhances the effect of correlation on the magnetic properties, thus giving rise to a deep interplay between disorder and magnetism.

After briefly reviewing the localization theorem and some of the current tools for calculating the average Density of States (DoS) of disordered independent-electron systems, we will dedicate to the Hubbard Hamiltonian in order to point out its usefulness and physical richness, always having in mind the original 'local' point of view. The final section will be devoted to some attempts to put together the Hubbard Hamiltonian and the presence of strong disorder, and to the comparison of the emerging scenario with some experimental data on ionic liquid alloys.

7.2 Anderson Localization

In order to describe the electronic properties of a monatomic liquid or amorphous solid, it is quite reasonable to idealize the system as a random ensemble of N_o localized Wannier states filled with N_e electrons. The filling fraction $y = N_e/N_o$ may differ from 1 in presence of doping. Transport is allowed by a non zero matrix element of the Hamiltonian between different states. Ignoring any interaction among the electrons we can write the simple tight-binding Hamiltonian as

$$\hat{H} = \sum_{i\sigma} \epsilon_i C^+_{i\sigma} C_{i\sigma} + \sum_{i\neq j,\sigma} V_{ij} C^+_{i\sigma} C_{j\sigma} \qquad (7.1)$$

where $C_{i\sigma}$, $C^+_{i\sigma}$ are annihilation and creation operators for an electron of spin σ on the local site i. V_{ij} are regarded as non-random functions of the relative positions $\vec{r}_{ij}$. The interest here is in the presence of a random diagonal term ϵ_i (diagonal disorder): its range of variation is supposed to be large compared to the hopping off-diagonal term V_{ij} which in turn may be regarded as the perturbative parameter for a simplified Hamiltonian diagonal in the 'local' representation of Wannier states.

The Hamiltonian (7.1) implies the linear equation of motion

$$i\dot{C}^+_j = \epsilon_j C^+_j + \sum_k V_{j,k} C^+_k. \qquad (7.2)$$

The Green function $G^\sigma_{ij}(z)$ is defined as

$$G^\sigma_{ij}(z) = \langle 0| C_{i\sigma} \frac{1}{z - \hat{H}} C^+_{j\sigma} |0\rangle \qquad (7.3)$$

(where $|0\rangle$ is the vacuum) and its equation of motion follows from eq.(7.2)

$$(z - \epsilon_j) G^\sigma_{jk}(z) = \delta_{jk} + \sum_{i\neq j} V_{ji} G^\sigma_{ik}(z). \qquad (7.4)$$

The locator series for G^σ_{ij} can be found by iterating eq.(7.4):

$$G^\sigma_{ij} = \frac{\delta_{ij}}{z_i} + \frac{V_{ij}}{z_i z_j} + \sum_k \frac{V_{ik} V_{kj}}{z_i z_k z_j} + \sum_{kl} \frac{V_{ik} V_{kl} V_{lj}}{z_i z_k z_l z_j} + \ldots \qquad (7.5)$$

(where $z_i = z - \epsilon_i$). The diagonal Green functions G^σ_{ii} may be formally written as

$$G^\sigma_{ii}(z) = [z - \epsilon_i - S^\sigma_i(z)]^{-1} \qquad (7.6)$$

where the self-energy $S^\sigma_i(z)$ is defined by the expansion

$$S_i^\sigma(z) = \left\{ \sum_{k \neq i} \frac{V_{ik} V_{ki}}{z_k} + \sum_{k,l \neq i} \frac{V_{ik} V_{kl} V_{li}}{z_k z_l} + \ldots \right\}. \tag{7.7}$$

The functions G_{ii}^σ and S_i^σ are related to important physical observables: the Density of States (DoS) projected on the local state i is given by

$$D_i^\sigma(E) = -\frac{1}{\pi} Im G_{ii}^\sigma(E + i0^+). \tag{7.8}$$

Making use of eq.(7.6) this reads

$$D_i^\sigma(E) = \frac{1}{\pi} \frac{|Im S_i^\sigma(E)|}{[E - \epsilon_i - Re S_i^\sigma(E)]^2 + [Im S_i^\sigma(E)]^2} \tag{7.9}$$

which looks like a Lorentzian distribution with half-width $|Im S_i^\sigma(E)|$ (at least for $E \approx [\epsilon_i + Re S_i^\sigma(E)]$). The projected DoS gives the probability of having an energy E, for an electron 'prepared' on site i with energy ϵ_i. The resonance at $E = \epsilon_i + Re S_i^\sigma(E)$ is characterized by a rate $1/\tau = |Im S_i^\sigma(E)|$ which gives information about the time spent by the electron on the site i before decaying. For a delocalized electron τ is finite while for a 'localized' state ($\tau \to \infty$) $D_i^\sigma(E) = \delta(E - \epsilon_i - Re S_i^\sigma(E))$ and S_i^σ is a simple perturbative correction to the level ϵ_i.

From the experimental point of view one measures the total averaged DoS (for instance by photoemission) which is defined as

$$D^\sigma(E) = \frac{1}{N_o} \sum_i D_i^\sigma(E). \tag{7.10}$$

Even in presence of localization $D^\sigma(E)$ obviously is a continuous function and does not contain any information about localization. Any such information must be extracted before performing the average. The next section will be dedicated to some computational schemes for the averaged DoS.

Going back to the localization problem we would like to know something more about the self-energy $S_i^\sigma(z)$ before averaging over the disorder. Following Anderson[1], assuming the convergence of the expansion (7.7), let us approximate $S_i^\sigma(z)$ by the first order term

$$S_i^\sigma(z) \approx \sum_{k \neq i} \frac{|V_{ik}|^2}{z - \epsilon_k}. \tag{7.11}$$

The imaginary part is

$$\frac{1}{\tau} = |Im S_i^\sigma(E + i\eta)| = \sum_{k \neq i} |V_{ik}|^2 \frac{\eta}{(E - \epsilon_k)^2 + \eta^2} \quad (\eta \to 0^+) \tag{7.12}$$

and the condition for localization ($1/\tau = 0$) becomes

$$\lim_{\eta \to 0^+} \sum_{k \neq i} \frac{|V_{ik}|^2}{(E - \epsilon_k)^2 + \eta^2} < +\infty. \tag{7.13}$$

If we average this quantity over the disorder we loose any information about localization: the average value is always divergent. However the random quantity $X_i = \sum_{k \neq i} |V_{ik}|^2/(E - \epsilon_k)^2$ follows the probability distribution

$$P(X) = const. \times \frac{1}{X^{\frac{3}{2}}} e^{-\alpha/X} \tag{7.14}$$

provided that the hopping terms V_{ij} fall off as $r_{ij} \to \infty$ faster than $1/r_{ij}^3$. The distribution $P(X)$ has a well defined maximum, even if the average value is divergent, and the most probable value of the rate (7.12) is then $1/\tau = 0$.

The theorem is verified to any order of the perturbative expansion (7.7) and the states are localized whenever the same expansion results to be convergent. In fact the known existence of delocalized states in liquids tells us that most of the time the expansion (7.7) is not convergent at all. It turns out that generally the series converges when the total averaged DoS is small enough: typically we observe long tails of localized states at the edges of the DoS in presence of strong diagonal disorder. Localized states exist up to a critical value of E, called the mobility edge, which is of basic importance for determinating the conductivity properties of the system. Generally speaking, if we are able to change the filling fraction y, then the Fermi level E_F can be pushed up and down across the mobility edge thus giving rise to metal-insulator transitions. However, even a semi-quantitative description of the transition requires a detailed knowledge of the total averaged DoS of the system.

7.3 The Averaged Density Of States

We formally divide the average process of eq.(7.10) in two steps. We define $\bar{G}^\sigma(\epsilon_i, \epsilon_j, \vec{r}_{ij}, z)$ as an ensemble average of the Green function $G_{ij}^\sigma(z)$ with the levels ϵ_i, ϵ_j and the relative position $\vec{r}_{ij}$ constrained to fixed values. In particular for $i = j$, $\epsilon_i = \epsilon_j$ we have the diagonal element $\bar{G}^\sigma(\epsilon_i; z)$. If the system is homogenous then both these averaged functions are translationally invariant. Most of the time the levels $\{\epsilon_i\}$ can be regarded as independent random variables with a given probability distribution $f(\epsilon_i)$, then the fully averaged Green functions follow

$$\bar{G}^\sigma(\vec{r}_{ij}; z) = \int d\epsilon_i \int d\epsilon_j f(\epsilon_i) f(\epsilon_j) \bar{G}^\sigma(\epsilon_i, \epsilon_j, \vec{r}_{ij}, z) \tag{7.15a}$$

$$\bar{G}^\sigma(z) = \int d\epsilon_i f(\epsilon_i) \bar{G}^\sigma(\epsilon_i; z). \qquad (7.15b)$$

This last is related to the averaged DoS by

$$D^\sigma(E) = -\frac{1}{\pi} Im \bar{G}^\sigma(E + i0^+). \qquad (7.16)$$

A systematical derivation of various approximate theories for $\bar{G}^\sigma(z)$ has been given by Logan and Winn[3,4].

From the equation of motion (7.4) we obtain

$$z_i \bar{G}^\sigma(\epsilon_i; z) = 1 + \rho \int d\vec{r}^j \int d\epsilon_j f(\epsilon_j) \bar{G}^\sigma(\epsilon_i, \epsilon_j, (\vec{r^j} - \vec{r}); z) V(\vec{r} - \vec{r^j}) \qquad (7.17)$$

where ρ is the density of sites and $V(\vec{r}_{ij}) = V_{ij}$.

The locator series (7.5) can be used in eq.(7.17) giving rise to an exact perturbative expansion. In the present brief introduction only an approximate treatment will be considered, the so called single-site[3] approximation which allows us to wright

$$\bar{G}^\sigma(\epsilon_i, \epsilon_j, \vec{r}_{ij}; z) = \bar{G}^\sigma(\epsilon_i; z) H(\vec{r}_{ij}, z) \bar{G}^\sigma(\epsilon_j, z) \qquad (7.18)$$

where the function $H(\vec{r}_{ij}, z)$ does not depend on the fixed levels ϵ_i, ϵ_j, and satisfies a Dyson equation (Ornstein-Zernike analogue equation) which sums up the perturbative series. To first order, for a complete random system, $H(\vec{r}_{ij}, z)$ reduces to the matrix element $V(\vec{r}_{ij})$. Inserting eq.(7.18) in eq.(7.17), and making use of eq.(7.15b) we find

$$z_i \bar{G}^\sigma(\epsilon_i, z) = 1 + \bar{G}^\sigma(\epsilon_i, z) \left[\rho \bar{G}^\sigma(z) \int d\vec{r} H(\vec{r}) V(-\vec{r}) \right] \qquad (7.19)$$

which can be formally solved as

$$\bar{G}^\sigma(\epsilon_i, z) = [z_i - S^\sigma(z)]^{-1}. \qquad (7.20)$$

A general result of single-site theories is that the self-energy $S^\sigma(z)$ is independent of the choice of the level ϵ_i, and does not have any explicit dependence on z: $S^\sigma(z) = S^\sigma[\bar{G}^\sigma(z)]$.

To first order, for a complete random system, we simply find

$$S^\sigma(z) = \rho \bar{G}^\sigma(z) \int d\vec{r} H(\vec{r}) V(-\vec{r}) \approx \rho \bar{G}^\sigma(z) \int d\vec{r} |V(\vec{r})|^2 \qquad (7.21)$$

which is the so called Hubbard approximation

$$S^\sigma(z) = J \bar{G}^\sigma(z) \qquad (7.22)$$

with J defined by eq.(7.21).

In general for any single-site theory eq.(7.22) is replaced by a proper closure relation which allows a self-consistent solution for $\bar{G}^\sigma(z)$ through eq.(7.20) and (7.15b):

$$\bar{G}^\sigma(z) = \int \frac{f(\epsilon)d\epsilon}{z - \epsilon - S^\sigma[\bar{G}^\sigma(z)]}. \tag{7.23}$$

The approximation of neglecting the presence of diagonal disorder is equivalent to the choice $f(\epsilon) = \delta(\epsilon - \epsilon_o)$, and the self-consistence equation reduces to

$$\bar{G}^\sigma(z) = \left(z - \epsilon_o - S^\sigma[\bar{G}^\sigma(z)]\right)^{-1}. \tag{7.24}$$

In the Hubbard approximation (7.22) the solution of eq.(7.24) is analytical: the DoS is easily found

$$D^\sigma(E) = \frac{1}{2\pi J}\left[4J - (E - \epsilon_o)^2\right]^{1/2} \tag{7.25}$$

and is a semielliptical band centered around ϵ_o, with bandwidth $W = 4J^{1/2}$.

A simple generalization is the binary alloy for which $f(\epsilon) = n_A\delta(\epsilon - \epsilon_A) + n_B\delta(\epsilon - \epsilon_B)$. In this case, as shown in Fig.7.1, the DoS is basically given by a couple of semiellipses which overlap if the difference $|\epsilon_A - \epsilon_B| \ll W$. Conversely for $|\epsilon_A - \epsilon_B| \gg W$ the two sub-bands split and the alloy behaves like an insulator whenever the Fermi level E_F falls inside the gap. If each atom contributes one valence electron ($y = 1$) then the Fermi level falls inside the gap only at a stoichiometric composition, namely $n_A = n_B$. This is the basic technique employed by some authors[5,6] for describing the occurrence of a metal-insulator transition in the series of Gold-Alkali liquid alloys. Au_x-B_{1-x} (B=Cs, Rb, K, Na, Li) is a metal at any composition $x \neq 1/2$; at stoichiometry ($x = 1/2$) Au-Cs and Au-Rb are insulators, Au-K is a semimetal while Au-Na and Au-Li behave like metals.

All this is readily explained in terms of our previous discussion on the occurrence of a gap in the DoS. The difference $|\epsilon_A - \epsilon_B|$ can be estimated in terms of the electronegativity difference $\Delta\chi$ between Gold and Alkali metals. $\Delta\chi$ continuously increases going from Li to Cs, thus giving rise to a clear gap for Cs and a full overlap of the sub-bands for Li.

The Au-Cs case has been extensively studied both from the theoretical and experimental point of view[2,5-13]. At stoichiometry the liquid is an ionic insulator[7] with a massive charge transfer from Cs to Au. In the DoS the lower sub-band is mainly generated by Au-states while the upper sub-bands originates from the presence of Cs-states: the ionic charge transfer reflects a lack of electrons in the upper sub-band, and a filled lower sub-band. In the Cs rich phase ($x < 1/2$), some exceeding electrons populate the upper band giving rise to normal metallic conduction. However something is missing in this simple

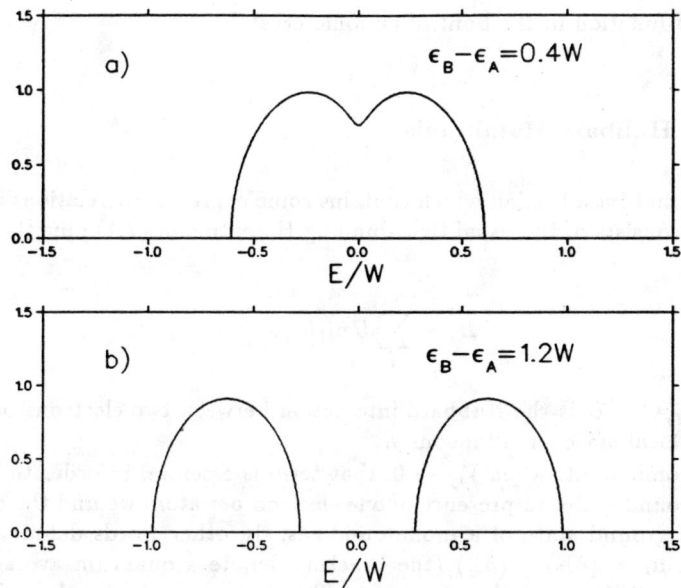

Fig.7.1 - The DoS for a binary alloy at stoichiometry ($n_A = n_B$), in the Hubbard single-site approximation, for $\epsilon_B - \epsilon_A = 0.4W$ (a), and for $\epsilon_B - \epsilon_A = 1.2W$ (b). In (b) the Fermi level falls inside the gap only if $y = 1$ (half-filling).

interpretation of the metal-insulator transition: NMR experiments[11,12] on the Cs rich phase of Au_x-Cs_{1-x} show the occurrence of an intermediate regime for $0.48 < x < 0.50$. This regime is characterized by a very small spin relaxation rate together with a Curie-law paramagnetic susceptibility. Both these are signatures of the presence of localized magnetic moments, i.e. unpaired electrons localized in the 'Anderson sense'. The importance of disorder in this context is obvious; and even if it were not, we know[12] that at the same composition the solid does not show any localization. However we cannot reproduce the presence of unpaired electrons without taking in due account the existence of a Coulomb repulsion among the electrons: we should face the hard task of incorporating some degree of correlation in the localization problem. On the other hand, as it will be shown below, the presence of strong diagonal disorder greatly enhances the role of correlations; and this is indeed the case for an ionic liquid like Au-Cs: the random distribution of charged ions around any Wannier localized state (disordered Madelung potentials) generates a random shift of the level ϵ_i. Under some general assumptions[13] the probability distribution $f(\epsilon)$ is Gaussian, and reduces to a δ-function in the limit of an ionic crystal.

7.4 The Hubbard Hamiltonian

The minimal Hamiltonian which contains some degree of correlations among the electrons consists of the usual tight-binding Hamiltonian (7.1) plus the interaction term

$$\hat{H}_1 = \sum_i U \hat{n}_{i\uparrow} \hat{n}_{i\downarrow} \qquad (7.26)$$

with $\hat{n}_{i\sigma} = C_{i\sigma}^+ C_{i\sigma}$. U is the Hubbard interaction between two electrons occupying the same local state simultaneously.

In the atomic limit, when $V_{ij} \to 0$, that term is essential in order to realize the correct ground state: in presence of one electron per atom we find the correct paramagnetic ground state of a monatomic gas. In other words defining $n_i = \langle \hat{n}_{i\uparrow} \rangle + \langle \hat{n}_{i\downarrow} \rangle$, $\mu_i = \langle \hat{n}_{i\uparrow} \rangle - \langle \hat{n}_{i\downarrow} \rangle$ (the brackets denote a quantum average), in the limit $V_{ij} \to 0$ if $n_i = 1$ then $\mu_i = \pm 1$. This is guaranteed by the existence of a value $U > 0$. The Hamiltonian $\hat{H}_U = \hat{H} + \hat{H}_1$ is the so called Hubbard Hamiltonian which is a useful starting point for the description of some magnetic properties. The most studied case is when $n_i \approx 1$, and for the simple reason that in general $\mu_i \leq min[n_i, (2 - n_i)]$, so that the correlation effects are negligible if $n_i \to 0$ or $n_i \to 2$. For $n_i = 1$ the behaviour of the model is determined by the ratio between the essential energy parameters U and W. When the bandwidth W is large compared to U, the system behaves like an usual metal with $\mu_i = 0$. A perturbative mean-field analysis to first order in U/W gives a semielliptical DoS which is merely shifted in energy with respect to the unperturbated case.

Conversely in the strong U domain the DoS is strongly modified and a gap opens at the Fermi level. The system behaves like a Mott insulator with local moments $\mu_i \approx n_i = 1$ (see for instance Fig.7.2). Any elementary excitation costs an energy U since in the ground state each site is filled with an unpaired electron. In a liquid the unperturbated bandwidth W is an increasing function of the density ρ; this is obvious in the Hubbard approximation since from eq.(7.21)

$$W = 4J^{1/2} = const \times \rho^{1/2}. \qquad (7.27)$$

So we expect, in an expansion experiment, a metal-insulator transition to occur whenever ρ is reduced enough. This is indeed the case as observed[14] for liquid Cs, and discussed by several authors[15,16]. We must emphasize that in the low density domain the magnetic susceptibility shows a Curie-like behaviour which is what we should expect from a Mott insulator.

In order to fully understand the Hubbard model it is quite instructive to take under consideration the single impurity problem[17] (the so called Anderson model). Suppose that the Hubbard interaction energy U is zero everywhere except than for a single site $i = 0$ whose energy level ϵ_0 is fixed. The corresponding occupation number $n_0 = \sum_\sigma \langle \hat{n}_{0\sigma} \rangle$ will be dictated by the relative position between the energy ϵ_0 and the full band. At the Hartree-Fock level the Hamiltonian reads

$$\hat{H} = \sum_{i\sigma} \epsilon_i \hat{n}_{i\sigma} + \sum_{i \neq j, \sigma} V_{ij} C_{i\sigma}^+ C_{j\sigma} + \sum_\sigma \epsilon_0^\sigma \hat{n}_{0\sigma} \qquad (7.28)$$

where the self-consistent σ-spin level $\epsilon_0^\sigma = \epsilon_0 + U \langle \hat{n}_{0,-\sigma} \rangle$. If $(\epsilon_0 + U)$ is under the bottom of the band then $n_0 \approx 2$, while if ϵ_0 is over the top of the band $n_0 \approx 0$. For intermediate values of ϵ_0 the σ-spin levels ϵ_0^σ split: for instance we can have $\langle \hat{n}_{0\uparrow} \rangle \approx 1$, $\langle \hat{n}_{0\downarrow} \rangle \approx 0$ and $\epsilon_0^\uparrow \approx \epsilon_0$, $\epsilon_0^\downarrow \approx \epsilon_0 + U$. These values are self-consistent provided that the band is roughly contained between ϵ_0 and $\epsilon_0 + U$. In this case on the site $i = 0$ (single impurity) we find a magnetic moment $\mu_0 \approx n_0 \approx 1$. Two important messages emerge: (i) In presence of a strong U a local moment is present if $n_0 \approx 1$ (correlations are suppressed if $n_0 \approx 0$ or $n_0 \approx 2$); (ii) The occupation number n_0 is basically determined by the position of the energy level ϵ_0.

These conclusions are relevant for a full understanding of the disordered Hubbard model in the less studied case $y \ll 1$. For a filling fraction $y \approx 1$ the Hubbard model, even in presence of strong disorder, does not show any relevant difference from the ordered case characterized by a probability distribution $f(\epsilon) = \delta(\epsilon)$. A strong U stabilizes the occupation numbers $n_i \approx y$, thus reducing the fluctuations of the charge. On the contrary in the domain $y \ll 1$ the existence of strong charge fluctuations does not allow us to write $n_i \approx y \to 0$ (y is the ensemble average of n_i). At the light of the previous discussion on the single impurity problem it is quite evident that in presence of strong diagonal disorder a fluctuating level ϵ_i will imply a fluctuating charge $n_i \neq y$. Even if most of the sites will have a negligible charge, a few sites with the lowest levels ϵ_i will possess a charge $n_i \approx 1$ and thus a local moment $\mu_i \approx 1$. Important magnetic properties show up in the disordered Hubbard model just in that domain $y \to 0$ where the usual ordered model has negligible correlation effects. From this point of view the disordered Hubbard model can be regarded as a multi-component impurity analogue[18] of the single impurity model.

7.5 Disorder And Correlation

The disordered Hubbard model is the most obvious candidate for a discussion of the interplay between disorder and correlation. However it presents some disadvantages because we are far from an exact treatment of the model, and the approximations adopted can in principle change or hide the rich physical content. An interesting point of view has been considered by Kamimura[19] who choices, as a set of basis functions, the eigenstates of the tight-binding Hamiltonian (7.1) which are Anderson-localized states. However the nature of these states is merely assumed from the beginning and the results are strongly dependent from this initial assumption. Moreover this approach has some validity only in the localization regime, and fails to provide a linkage between the metallic and the insulator phases.

A more flexible treatment of the Hubbard model can be achieved by use of mean-field theories, but as discussed in section 7.2, over averaging can conduct to a useless oversimplified theory. In presence of disorder the formulation of a mean-field theory requires a double averaging process: for instance we can approximate the operator $\hat{n}_i$ with its quantum expectation value $\langle \hat{n}_i \rangle$; then we are tempted to pose $\langle \hat{n}_i \rangle = y$ for any site in the system. The filling fraction y is exactly equal to the average value of $\langle \hat{n}_i \rangle$, but we cannot neglect charge fluctuation at all without loosing any information about the disordered nature of the environment. When these problems arise it is quite better to introduce a proper distribution function, trying to work with distributions instead of averages, as stressed in the localization problem of section 7.2.

At the light of our considerations on the single impurity problem it is quite reasonable to assume that the fluctuations of the level ϵ are the main cause of fluctuation for the charge $\langle \hat{n}_i \rangle$. A mean-field theory could be formulated by requiring the same amount of charge $n(\epsilon)$ to be found on the sites which have the same level ϵ. This position allows for some degree of fluctuation of the charge among the local sites. The single impurity problem tells us that we should expect a strong local moment if $n(\epsilon) \approx 1$, and a negligible role of on-site correlations if $n(\epsilon) \approx 0$ or $n(\epsilon) \approx 2$. In general the local moment should be at least a function of the level ϵ, and we are forced to introduce the corresponding function $\mu(\epsilon)$. The proper definition of these functions is

$$n(\epsilon) = \langle [\langle \hat{n}_{i\uparrow} \rangle + \langle \hat{n}_{i\downarrow} \rangle] \rangle_{\epsilon_i = \epsilon} \qquad (7.29a)$$

$$\mu(\epsilon) = \langle |\langle \hat{n}_{i\uparrow} \rangle - \langle \hat{n}_{i\downarrow} \rangle| \rangle_{\epsilon_i = \epsilon} \qquad (7.29b)$$

where $\langle \ldots \rangle_{\epsilon_i = \epsilon}$ is an ensemble average over all sites with $\epsilon_i = \epsilon$. The full average over ϵ must give

$$y = \int f(\epsilon) n(\epsilon) d\epsilon \qquad (7.30a)$$

$$\bar{\mu} = \int f(\epsilon) \mu(\epsilon) d\epsilon \qquad (7.30b)$$

with $\bar{\mu} \neq 0$ in general since $\mu(\epsilon)$ is positive defined. The single impurity self-consistent σ-spin level of eq.(7.28) generalizes to

$$\epsilon_i^\sigma = \epsilon_i + U\langle \hat{n}_{i,-\sigma}\rangle = \epsilon_i + \frac{1}{2}n_i U - \frac{\sigma}{2}\mu_i U \qquad (7.31)$$

where $\sigma = \pm 1$. Making use of the functions (7.29) we approximate the σ-spin level with its ensemble averaged value

$$\epsilon_i^\sigma = \epsilon_i + \frac{1}{2}n(\epsilon_i)U \mp \frac{\sigma}{2}\mu(\epsilon_i)U \qquad (7.32)$$

where the $\mp$ sign arises from the uncertainty about the orientation of the local moment on the local site. In order to describe a disordered paramagnetic system we regard both the signs with the same probability.

The formalism of single-site theories, exposed in section 7.3, can be easily applied to the case of the random σ-spin level (7.32): from equation (7.20) the ensemble averaged Green function reads

$$\bar{G}_\pm^\sigma(\epsilon, z) = \left[z - \epsilon - \frac{1}{2}U n(\epsilon) \pm \frac{\sigma}{2}U\mu(\epsilon) - S^\sigma(z) \right]^{-1} \qquad (7.33)$$

where $S^\sigma(z)$ is a function of the fully averaged Green function

$$\bar{G}^\sigma(z) = \int d\epsilon f(\epsilon) \left[\frac{1}{2}\bar{G}_+^\sigma(\epsilon, z) + \frac{1}{2}\bar{G}_-^\sigma(\epsilon, z) \right]. \qquad (7.34)$$

The DoS can be extracted from the function $\bar{G}^\sigma(z)$ according to eq.(7.16). Moreover even the partial density of states

$$d_\pm^\sigma(\epsilon, E) = -\frac{1}{\pi} Im \bar{G}_\pm^\sigma(\epsilon, E + i0^+) \qquad (7.35)$$

derives from the knowledge of $\bar{G}^\sigma(z)$ and $S^\sigma(z)$. The functions $d_\pm^\sigma(\epsilon, E)$ give the DoS projected on the local site with level ϵ and moment $\mu = \pm\mu(\epsilon)$. The theory is fully self-contained if

$$n(\epsilon) = \int_{-\infty}^{E_F} \left[d_\pm^\uparrow(\epsilon, E) + d_\pm^\downarrow(\epsilon, E) \right] dE \qquad (7.36a)$$

$$\mu(\epsilon) = \int_{-\infty}^{E_F} |d_\pm^\uparrow(\epsilon, E) - d_\pm^\downarrow(\epsilon, E)| dE \qquad (7.36b)$$

in accordance with the definitions in eq.(7.29). The Fermi level E_F is naturally defined by

$$y = \int_{-\infty}^{E_F} \left[D^\uparrow(E) + D^\downarrow(E) \right] dE. \qquad (7.37)$$

This multi-component mean-field theory only requires a knowledge of the distribution $f(\epsilon)$, and of the closure relation for the self-energy $S^\sigma(z) = S^\sigma\left[\bar{G}^\sigma(z)\right]$.

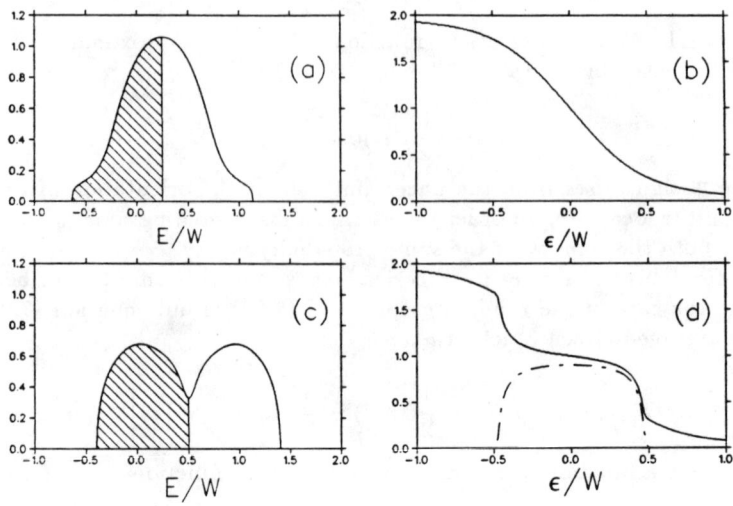

Fig.7.2 - The DoS for $\tilde{U} = 0.5$ (a) and $\tilde{U} = 1.0$ (c); The shaded areas represent the filled portion of the bands. The charge $n(\epsilon)$ (solid line) and moment $\mu(\epsilon)$ (broken line) distributions are also reported for $\tilde{U} = 0.5$ (b) and $\tilde{U} = 1.0$ (d). $\tilde{\lambda}$ is fixed to the value 0.25.

This can be taken to be the Hubbard approximation (7.22) or some more sophisticated condition[3,4]. While a full discussion of the model is reported in Ref.18, we would like to point out the main results emerging from such a mean-field approximation.

In the following discussion we approximate the diagonal disorder distribution $f(\epsilon)$ with a cut Lorentzian

$$f(\epsilon) = \frac{const.}{\lambda^2 + \epsilon^2} \qquad (7.38)$$

defined and normalized in the range $-W \leq \epsilon \leq W$. In the simple Hubbard single-site approximation of eq.(7.22) the problem is characterized by the filling fraction y and by three energies: (i) the bandwidth W of the unperturbed semi-elliptical DoS; (ii) the halfwidth λ of the diagonal disorder distribution $f(\epsilon)$; (iii) the Hubbard on-site Coulomb interaction U. Alternatively we can make use of the three dimensionless variables y, $\tilde{\lambda} = \lambda/W$ and $\tilde{U} = U/W$. For a narrow-band disordered system $\tilde{\lambda} = 0.25$ seems a reasonable value, and U can be comparable

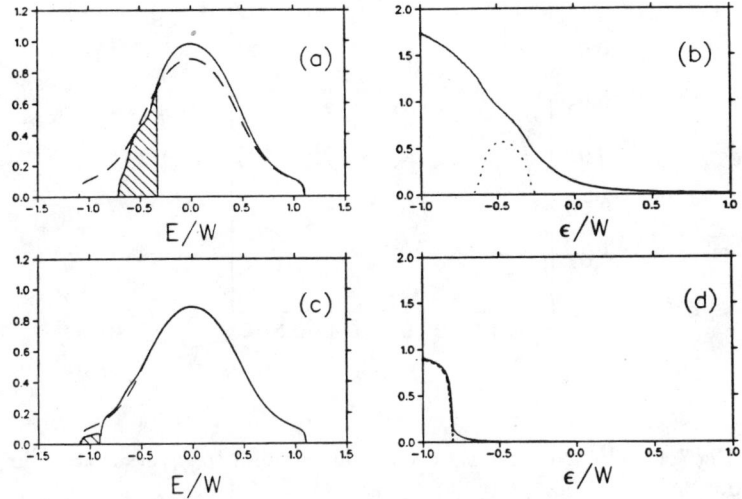

Fig.7.3 - The DoS for $\tilde{U} = 0.5$, $\tilde{\lambda} = 0.25$ $y = 0.3$ (a) and $y = 0.02$ (c); the dashed lines are the corresponding DoS for $U = 0$, while the shaded areas represent the filled portion of the bands. The charge $n(\epsilon)$ (solid line) and moment $\mu(\epsilon)$ (broken line) distributions are also reported for $\tilde{U} = 0.5$ (b) and $\tilde{U} = 1.0$ (d).

with the bandwidth W.

First of all we consider the case $y = 1$ (half-filling). Fig.7.2 shows the DoS and the respective self-consistent charge-moment distributions $n(\epsilon)$, $\mu(\epsilon)$ for $\tilde{U} = 0.5$ and $\tilde{U} = 1.0$. In this case $\tilde{\lambda}$ does not affect too much the result which resembles the usual $\tilde{\lambda} \to 0$ limit of no disorder. The charge is spread over all sites, and as we expected $\tilde{U}$ is the relevant parameter in this limit. When $\tilde{U}$ is small the moments are unstable and the normal restricted Hartree-Fock solution is achieved. $\tilde{U} = 1.0$ is enough for stabilizing the presence of local moments and in this case we predict a paramagnetic ground state with an average local moment $\bar{\mu} \approx 0.7$. A pseudogap opens in the middle of the band and when $\tilde{U}$ is large the system becomes a Mott insulator.

Once we have made sure that the known phenomenology is recovered in the well studied case $y = 1$, we proceed to examine in detail the limit $y \to 0$. In this limit we always predict the existence of local moments and a paramagnetic ground state provided that $\tilde{U}$ is non zero. Let us consider again the case $\tilde{U} = 0.5$, $\tilde{\lambda} = 0.25$

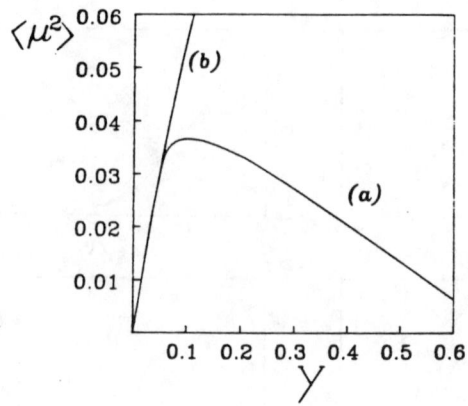

Fig.7.4 - The averaged squared local moment $\langle \mu^2 \rangle$ versus the filling fraction y for $\tilde{\lambda} = 0.25$, $\tilde{U} = 0.5$ (a) and $\tilde{U} = 1.0$ (b).

in which correlations don't play any major role at half-filling (Fig.7.2a,b). As it is shown in Fig(s). 7.3, 7.4, at low filling local moments are stable and they approach the charge value when $y < 0.08$. In the very low filling limit $y \approx 0.02$ we find $n(\epsilon) \approx \mu(\epsilon)$ for most of the filled states thus indicating a strongly correlated ground state. Charge fluctuations play here a major role since the charge is not evenly spread over all local sites but only those sites with a very low energy level ϵ are significantly filled. On the other hand these sites are only a few in the system and it is unlikely that the corresponding local states could overlap each other. In other words the small amount of charge present in this low-filling limit is trapped in a few deep localized states which are singly occupied if the correlation energy U is non zero.

As $y \to 0$ the $n(\epsilon)$ distribution approaches a step-function thus indicating that the atomic limit is correctly achieved. In this limit the $n(\epsilon)$ distribution can be regarded as a quasi-particle distribution which becomes a step-function when the local states are exact eigenstates of the Hamiltonian, or in other words when the filled eigenstates of the Hamiltonian are exactly the local atomic states.

Fig.7.4 shows a remarkable turnover for the average value $\langle \mu^2 \rangle$ vs. y: when y is small enough the model is not sensitive to the choice of the parameter $\tilde{U}$, but provided that $U \neq 0$ local moments stabilize in the system. It is the analogue of an atomic limit since some local deep states are filled with one electron which cannot jump over the surrounding sites because these have quite higher energies. Even a small value for U prevents from any double occupancy thus giving rise to a Curie-law paramagnetic ground state. For large values of the filling fraction y, Fig.7.4 shows a different behaviour for different values of $\tilde{U}$: if $\tilde{U}$ is large the intensity of local moments increases with y and persists even at half-filling where the usual Mott transition is recovered; if $\tilde{U}$ is small any local moments disappear when the band is filled and the usual metallic restricted Hartree-Fock solution is reached around half-filling.

We would like to map the model on the known experimental results for liquids and amorphous solids, by a proper choice of the parameters. We noticed in section 7.3 that the presence of strong diagonal disorder dominates the intermediate regime occurring at the metal-insulator transition of the liquid alloy Au-Cs. As previously discussed in the Cs-rich phase $(Au-Cs)_{1-x}Cs_x$ with $0 < x < 0.07$, some experimental evidence has been reported[11] for the existence of localized unpaired electrons. The absence of localization in the solid[12] allows us to talk about a 'melting-induced' localization, or more properly 'disorder-induced'. We focus here to the upper sub-band of the alloy, mainly generated by Cs atomic states, and filled with a filling fraction $y = x$ varying with composition. If we ignore the presence of the lower sub-band (for $x > 0$), the relevant physical aspects can be described by the one-band model discussed in this section[2]. Moreover the presence of a random distribution of Au^-, Cs^+ ions gives rise to disordered Madelung potentials[13] that contribute to enlarge greatly the half-width λ of the probability distribution $f(\epsilon)$: we expect to deal with a strong diagonal disorder induced by the random ionic field. The choice $\tilde{\lambda} = 0.25$, $\tilde{U} = 0.5$ of Fig(s). 7.2-7.4 is quite reasonable for a binary liquid alloy like Au-Cs at ordinary density. The case $y = 1$ (Fig.7.2) thus reproduces the correct absence of local moments in pure Cs; conversely in the range $0 < y < 0.07$ the observed localization of unpaired electrons is a consequence of the diagonal disorder in the Cs-rich ionic alloy. The presence of disorder greatly enhances the correlative effect of the Coulomb repulsion and the observed Curie-law paramagnetic susceptibility is an expected consequence. As appearing from Fig.7.4 a larger filling fraction y (i.e. a larger amount x of excess Cs) produces a smooth transition towards a different regime which is metallic and poorly correlated in agreement with the experimental findings.

Given the general nature of the approximations adopted for developing such multi-component mean-field theory, the model by itself could be a useful starting point for understanding the physics of other liquid metals or even very different materials like amorphous semiconductors. For instance Si:P;B which behaves like a quenched solid state gas of Phosphorus and Boron impurities embedded in a Silicon background. Even larger magnetic effects are expected in this material given the narrow impurity band W which yields larger values for $\tilde{U}$ and $\tilde{\lambda}$.

Last of all it is remarkable that the multi-component mean-field theory developed in this section clearly shows the importance of correlations in the limit of an almost empty band for a disordered system. When charge fluctuations are negligible, as it is often the case in crystals for symmetry reasons, we may assume $n_i = y$ and the largest effect of correlations is expected to appear near the half-filling of the band ($y \approx 1$). It is then commonly believed that in a one-band problem correlation effects are really important only when y is close to 1 and U is large compared with the bandwidth W. We want to emphasize here that this is not a general statement at all, and that correlation effects can become really important in the limit $y \to 0$ in a disordered system. First of all when the band is almost empty a disordered system behaves like an insulator or a dirty metal so that any screening is deeply reduced and the effective value of U is enhanced with respect to the half-filled band case. But even assuming U constant, in a disordered system we cannot neglect charge fluctuations any longer when y is very small. While in a perfect crystal the charge is spread over Bloch delocalized states at any small value of the filling y, we find it unreasonable to think the same for a disordered system where the charge localizes somewhere in atomic-like states. In this atomic limit n_i is zero almost everywhere, but it is $n_i \approx 1$ on a few atomic sites where correlation effects become extremely important.

References

1) Anderson,P.W. *Phys. Rev.* **109**, 1492 (1958).
2) Siringo,F. and Logan,D.E., *J.Phys.: Condens.Matter* **3**, 4747 (1991).
3) Logan,D.E. and Winn,M.D., *J.Phys.C: Solid State Phys.* **21**, 5773 (1988).
4) Winn,M.D. and Logan,D.E., *J.Phys.: Condens. Matter* **1**, 1753 (1989).
5) Franz,J.R., Brouers,F., and Holzhey,C., *J.Phys. F: Metal Phys.* **10**, 235 (1980).
6) Holzhey,C., Brouers,F., Franz,J.R., and Schirmacher,W., *J.Phys. F: Metal Phys.* **12**, 2601 (1982).
7) Evans,R., and Telo Da Gama,M.M., *Phil. Magazine B* **41**, 351 (1980).
8) Freyland,W., and Steinleitner,G., *Ber. Bunsenges. Phys. Chem.* **80**, 810 (1976).
9) Ten Bosch,A., Moran-Lopez,J.L., and Benmann,K.H., *J.Phys.C* **11**, 2959 (1978).
10) Kittler,R.C., and Falicov,L.M., *J.Phys.C: Solid State Phys.* **9**, 4259 (1976).
11) Dupree,R., Kirby,D.J., Freyland,W. and Warren,W.W.,Jr, *Phys. Rev. Lett.* **45**, 130 (1980).
12) Dupree,R., Kirby,D.J., and Warren,W.W.,Jr, *Phys.Rev. B* **31**, 5597 (1985).
13) Logan,D.E., and Siringo,F., *J.Phys.: Condens. Matter* **4**, 3695 (1992).
14) Freyland,W., *Phys.Rev.B* **20**, 5104 (1979).
15) Chapman,R.G., and March,N.H., *Phys.Rev.B* **38**, 792 (1988).
16) Logan,D.E., *J.Chem.Phys.* **94**, 628 (1991).
17) Anderson,P.W., *Phys.Rev.* **124**, 41 (1961).
18) Logan,D.E., and Siringo,F., *J.Phys.: Condens. Matter* **5**, 1841 (1993).
19) Kamimura,H., *Phil. Magazine B* **42**, 763 (1980).

CHAPTER 8

HIGH RESOLUTION SCANNING TUNNELLING MICROSCOPY OF DEFECT STRUCTURES AND DISTORTION OF THE CARBON CAGE C_{60} FORMING FULLERENE LATTICE

A.V.Narlikar, S.B.Samanta and P.K.Duttta
National Physical Laboratory, Dr. K.S.Krishnan Road,
New Delhi-110012, INDIA

ABSTRACT

Scanning tunnelling microscopic studies of C_{60} films carried out under ambient conditions, have revealed the characteristic FCC lattice of the C_{60} clusters with intermolecular distances in close accord with the reported data. The crystal lattice is found to contain defects, crystallographically consistent with the FCC structure. The buckyball diameter matches with the size given by the positions of the carbon atoms than by the pi-electron cloud. The submolecuar features of the carbon cage in the finest details, as inferred from the diffraction data, in the form of hexagons and pentagons, have been directly observed. Since the images obtained are distinct, the orientational motion of the molecules at the ambient temperature is believed to be frozen out. The three possible reasons advanced to explain the retardation or freezing are (i) the change in the electronic structure of the fullerene films due to close proximity of metallic species of ions, (ii) the presence of higher fullerene derivatives C_h in the sample such that C_{60} -C_h interaction overpowers the rotational energy of 50 meV, and (iii) the existence of large electric field gradients between the STM tip and the sample. A closer examination of the observed buckyballs has revealed them to be distorted. The effect is briefly discussed.

INTRODUCTION

The achievements of many new materials such as quasi-crystals, oxide-superconductors and fullerenes etc. have presented a new interests in the field of condensed matter sciences and technology. The crystallographic geometry of 2-d and 3-d particles packings configurations and the dynamics of lowering of periodic structures may be helpful in describing such disordered structures.

The recent discovery of fullerenes [1] as a third form of carbon and the exciting observation of superconductivity at appreciable temperatures in C_{60} solids doped with alkali metals [2] have spurred a wide spread interest in the chemical, electronic and physical properties of these novel materials. Various spectroscopic and diffraction studies [3-6] have predicted and corroborated the truncated icosahedral structure (soccer ball) of the C_{60} carbon cluster (CC) cage of the buckyball which has also been beautifully reproduced by computer simulation [7]. More recently [8,9], in a broad way, some indications of the submolecular features of the carbon cage of C_{60} may be found in the STM (scanning tunnelling microscopy) images of fullerene samples while the structure in the finest details revealing the well resolve hexagons and pentagons has been reported in the STM studies of deposited films by the present group [10].

Host of X-ray and neutron diffraction data [11,12] of crystalline C_{60} have shown that at ambient temperature the molecules are orientationally disordered and the crystal structure may be described as a FCC configuration of C_{60} spheres. When the system is undoped, below 249K, the molecules become orientationally ordered with a change in crystallographic symmetry from FCC to premitive (P) cubic. The low temperature neutron diffraction data show [12] the ensuing ordering scheme is one involving a rotational alignment of the bukyballs such that electron rich short inter-pentagon bonds face the electron deficient pentagon centres of the neighbouring C cages. The recent NMR experiments of ^{13}C [13] relate the ambient temperature disorder to the observed rapid reorientational motion of individual molecules and, at a lower temperature corresponding to the orientational order, the rotational diffusion gets singnifcantly slowed down. The molecules execute jumps between symmetry equivalent orientations which presumably show up as rotational alignment, as canonical snapshots, in the diffraction data [11].

The STM technique is now extensively being used as a versatile tool to probe the surface morphology on an atomic scale. However, there are two obvious drawbacks in the use of STM for investigating pure C_{60} samples. Firstly, their poor conductivity, with a band gap of about 1.5 eV, precludes setting up of any measurable tunnelling currents needed for imaging. Secondly, the aforesaid rotational diffusion of the individual molecules is believed to smear-out the microscopic details of the carbon cage morphology, although this need not seriously impair the imaging of the lattice framework. Interestingly, however, both these problems are circumvented in the case of samples synthesized by the variable pressure graphite arc method. The presence of higher fullerene derivatives C_h in these samples can lower the effective energy gap due to hybradization of the electronic bands of the constituent clusters. This would make charge transfer possible, resulting in the formation of conducting islands where the STM observations become feasible [14,15]. Also, the presence of C_h in the vicinity of C_{60}, as pointed out earlier [10,16], can significantly retard or freeze the rotational diffusion, making the individual carbon cage structures visible on nanometric scale in STM observations.

In this paper we report a consolidated account of the results of STM studies [10,14,15,17] of C_{60} films prepared by the graphite arc deposition technique. The resolved images of the molecular lattice have been related to the prominent crystallographic planes of the C_{60} struture and the presence of defect structures has been identified. The carbon cage of the individual buckyballs has been resolved and attempt is made to observe the assembleage of such resolved cages forming the C_{60} lattice.

EXPERIMENTAL

The samples were prepared by variable pressure graphite arc deposition technique [1] under the inert atmosphere of helium or argon, using graphite of spectroscopic purity. The details of the preparation are given elsewhere [18]. The main difference between our technique and that reported in ref.[1] was that we intentionally carried out deposition under different inert gas pressures for successive gas discharges whose time duration and numbers were also varied. It has been observed that by varying the growth conditions [18,19], the composition of clusters, viz., C_{60}/C_h, can be suitably controlled and in our case, the pertinent growth conditions were the gas pressure, duration and the number of discharges. The carbon soot of several micron thickness was deposited over silver electrodes on a glass substrate. The characterization of these samples with XRD, SEM, electron diffraction [18] and Raman spectroscopy [20] has been described elsewhere.

RESULTS AND DISCUSSION

Owing to their high over-all resistivity and inhomogeneous nature, the operation of engaging the tip to the samples and to establish and sustain the required constant tunnelling current for imaging was a paramount problem in the STM study of the deposited films. However, with repeated attempts regions were found where the local resistivity was presumably low enough to achieve the optimum tunnelling conditions for atomic level (in the present case molecular level) resolution.

Fig.1a shows a real time gray scale STM image in 3D-perspective at 60° pitch over a scan area of 15 nm X 15 nm showing the CC lattice. As may be seen, the central portion of the image appears distorted while the outer periphery exhibits a nearly square lattice pattern. Interestingly, the pertinent regions when zoomed down to a lower scan size of 7 nm X 7 nm and examined using the top view illumination mode software, revealed the former showing up as a triangular (or hexagonal) pattern while the latter as a square one (Fig.1b). A closer study of the square lattice (of Fig.1a as well of figures to follow), in the FCC configuration,

445

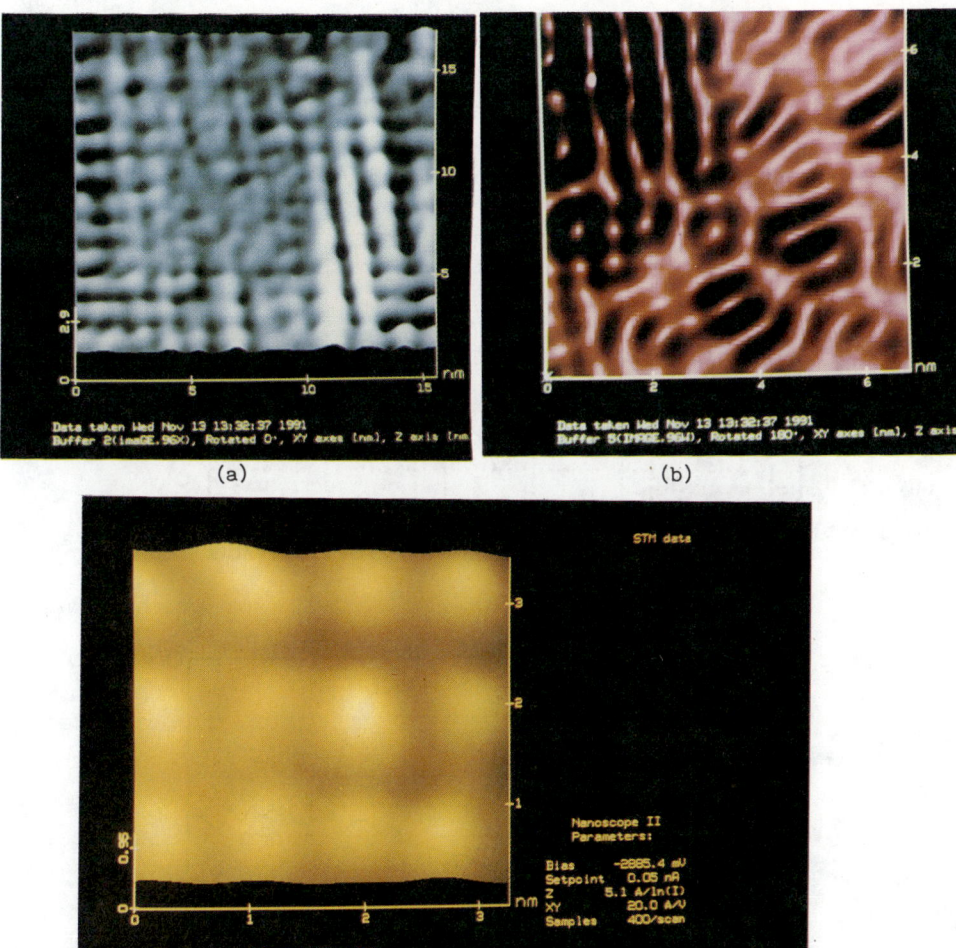

Fig.1

(a) The lattice image of CCs in 3D perspective at 60° pitch over a scan area of 15 nm X 15 nm (bias voltage:-2400 mV and current: 50 pA,unfiltered), showing a square lattice arrangement at the outer periphery and unresolved pattern at the centre; (b) Zoomed image (scan size: 7 nm X 7 nm) of Fig.1a covering parts the square lattice and unresolved patterns of Fig.1a, in the top view illumination mode revealing the unresolved pattern to be hexagonal (H) along with the squre (S) lattice pattern; (c) Zoomed image (scan size: 3.2 nm X 3.2 nm; bias voltage:-2885 mV and current:50 pA; moderately filtered) showing both smaller and larger size CCs on face corners and face centres of the FCC lattice.

corroborates the lattice parameter of about 1.4 nm of the C_{60} lattice, with the diameter of each carbon cluster agreeing with the value of 7 nm [21]. Higher CCs, for example of C_{70}, on the other hand, are known to form a hexagonal lattice [21], and our above observation gives credence to the simultaneous presence of both C_{60} and C_{70} clusters. This apart, we could resolve local regions on the

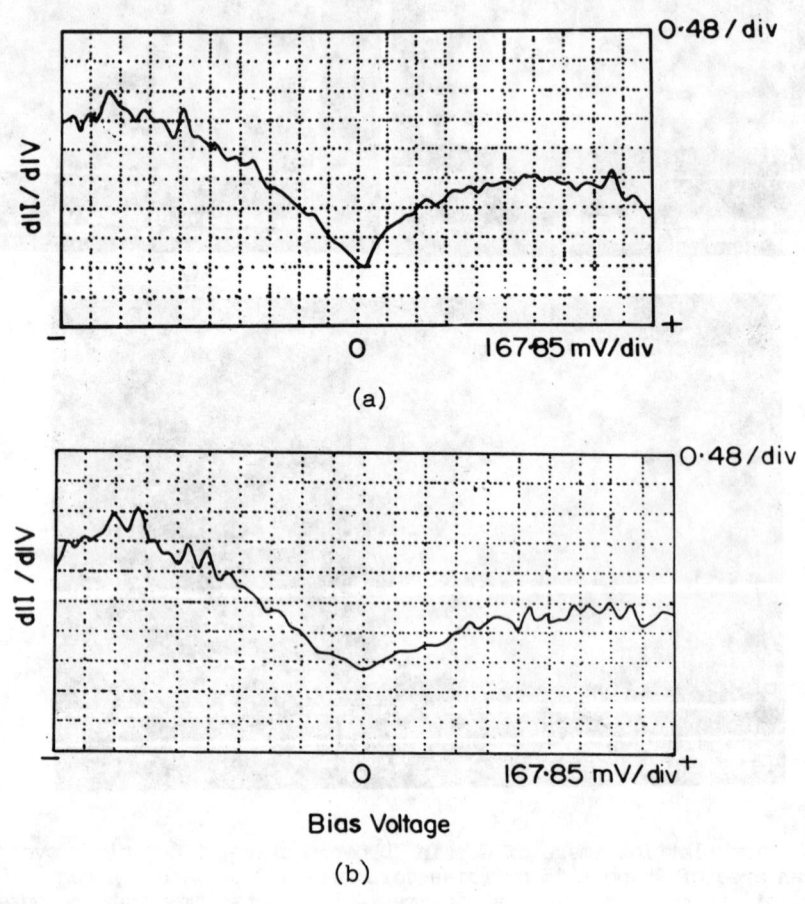

Fig.2

(a) Normalised conductance spectra (dlI/dlV) versus scanning bias voltage for the image of Fig.1a showing the presence of a small energy gap; (b) Normalised conductance spectra of a different region of Fig.1a showing a quasi metallic energy band structure with no visible energy gap.

447

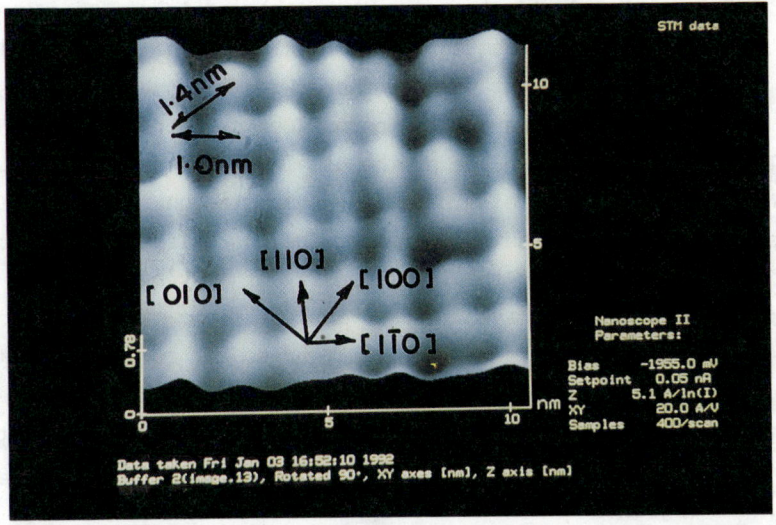

Fig.3

Typical square lattice arrangement of CCs with no obvious lattice defects on a scan size: 10 nm X 10 nm; 3D perspective image in 60° pitch (bias voltage:-1955 mV, current:50 pA, moderate filtering).

nanometric scale where one set of lattice sites, viz., the corner lattice points, comprised smaller size clusters (i.e., of C_{60}) while the face centring ones were distinctly larger (i.e., of C_{70}), and vice-versa (Fig.1c), indicating formation of an ordered structure of C_{60} and C_{70}. This is depicted in the image of Fig.1c of the scan size of 3.2 nm X 3.2 nm in 3D perspective at 60° pitch. The fact that the STM images are obtained for these samples is itself indicative of the imaged areas being electrically conducting. This is corroborated by scanning tunnelling spectroscopic (STS) studies. The STS was performed simultaneously by first resolving the images as obtained above and subsequently by breaking up the feed-back loop system at a particular instance and then switching over to the STS mode through the software. The normalised conductance spectra of dlI/dlV were recorded in place of the usual dI/dV spectra, against the bias voltage, as it eliminates any transmission factor involved between the tip and the sample surface. The normalised conductance is directly related to the local density of states, LDOS, and the spectra so obtained can reveal some insight into the electronic structure of the surface layer. Figures 2a an 2b show the conductance spectra with drastically reduced (0.3 eV or even zero) energy gap between valence and conduction bands. We attribute this to the presence of higher carbon cluster derivatives, C_h , revealed in the STM studies.

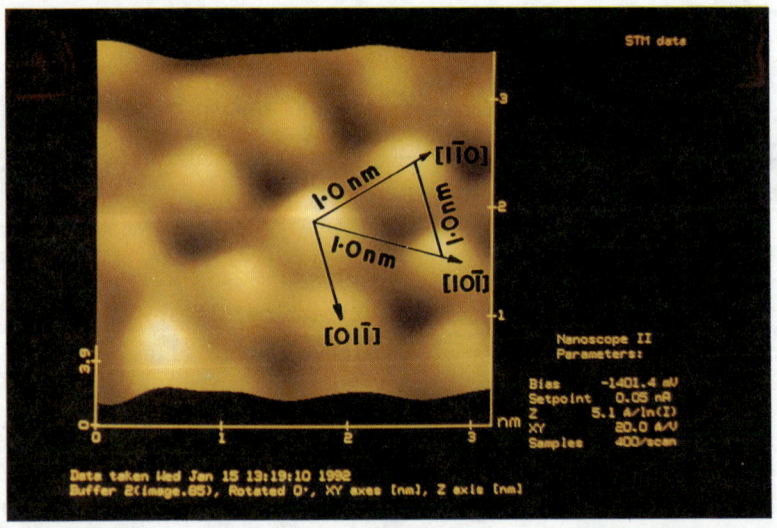

Fig.4a

(a) Typical triangular lattice arrangement of densely packed (111) planes of FCC structure in the zoomed image over a scan area of about 3.2 nm X 3.2 nm, in 3D perspective at 60° pitch (bias voltage: -1400 mV, current:50 pA).

449

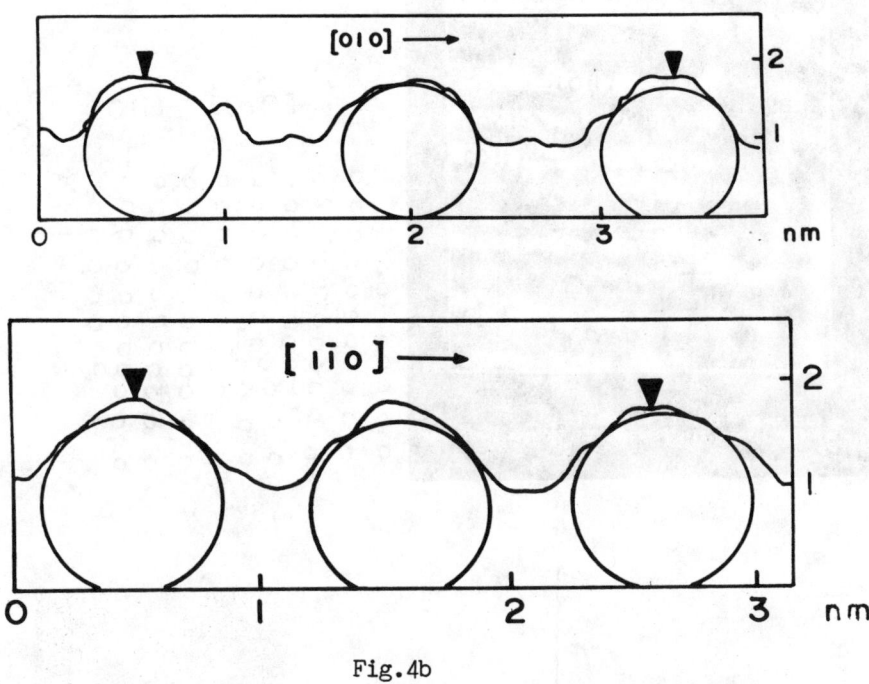

Fig.4b

Section mode plots, in two directions showing the buckyball diameter to be close to 0.7 nm.

In contrast to Fig.1a and Fig.1c, the image of Fig.3 shows a zoomed portion of a more regular square lattice formed over a scan area of about 10 nm X 10 nm, presented in 3D perspective at 60° pitch. As may be seen, the distances between lattice points along prominent directions can be ideally fitted with respect to $\{001\}$

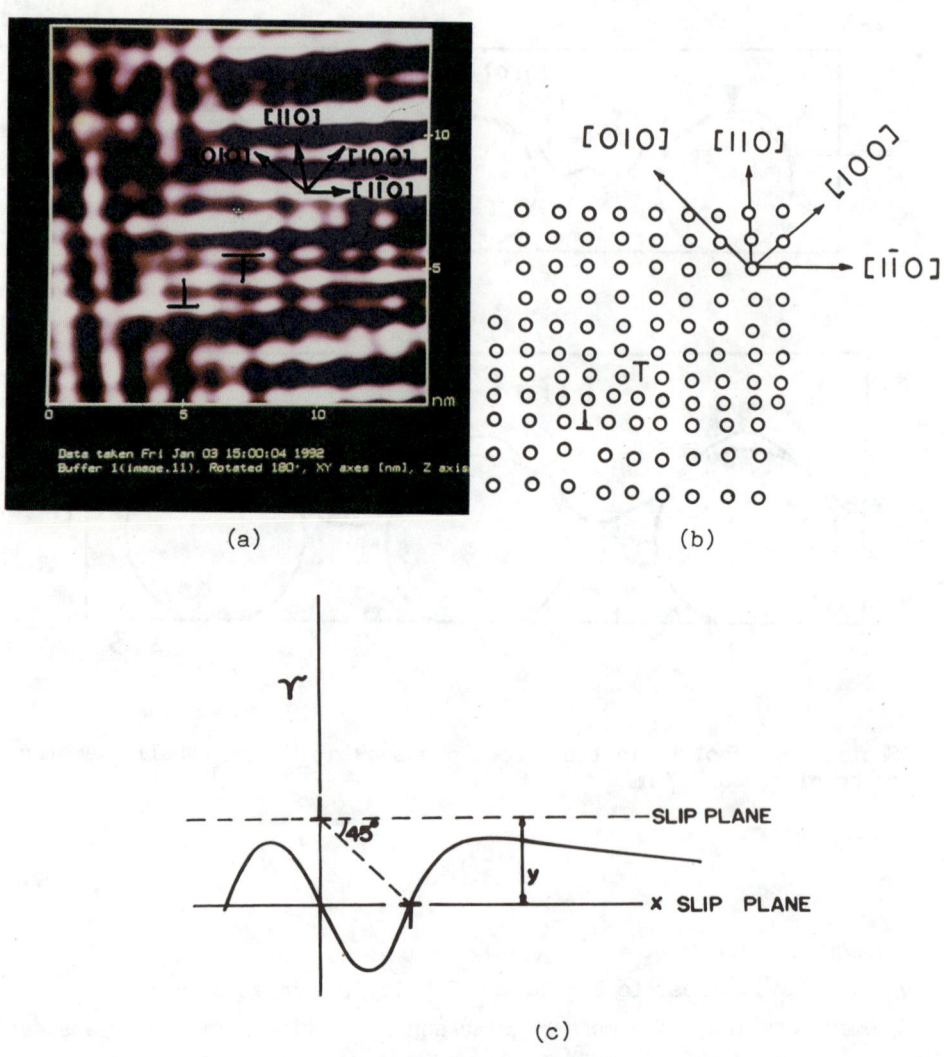

Fig.5

(a) Top view image of (001) plane over a scan area of 15 nm X 15 nm showing two unlike edge dislocations on the adjacent slip planes; their strain-fields tend to cancel each other as seen from the straightening-up of the lattice planes on farther sides (bias voltage:-1955 mV, curren:50 pA, moderate filtering); .(b) Sketch of Fig.5a; the Burgers vector of the dislocations of Fig.5a is b= ± ½[110] ≈ a/√2 =1.0 nm; (c) Interaction between two unlike edge dislocations on neighbouring slip plane. The force is attractive and the stable equilibrium is attained for the 45° position shown which is corroborated with the image of Fig.5a (for details, see ref.[24]).

planes of the C_{60} FCC lattice [21] although some larger clusters are also seen in the lattice. In accordance with the C_{60} lattice, the clusters are spaced 1 nm apart along two orthogonal $\langle 110 \rangle$ directions while their separation along two orthogonal $\langle 100 \rangle$ directions is close to 1.4 nm. Similarly, in a different region of the sample, a typical lattice pattern, characteristic of $\{111\}$ planes in FCC structure, observed over a small area zoomed scan (size about 3 nm X 3 nm) is shown in 3D perspective at 60° pitch in Fig.4a. The figure shows almost spherical 0.7 nm diameter clusters placed 1 nm apart along three 110 directions, forming equilateral triangles in (111) plane. The above value of the buckyball diameter can be readily seen in the section mode plot of Fig.4b.

It is worth noting that our value of the buckyball diameter is close to the one reported in the STM studies of Lang et al.[9] and is smaller than 1.0 nm diameter of the C cage observed by Wragg et al.[22]. The larger value is ascribed to the protruding pi-electron lobes of the carbon atoms forming the C_{60} cluster and may well be ascribed to a rather large tunnelling current of 1 nA used by these authors, which make the submolecular details more obscure in the STM observations. Since the present observations show that the carbon cage diameter is closer to 0.71 nm, the value which is given by the positions of the carbon atoms [21] it seems that our system is perhaps better described in terms of the pi- molecular orbitals rather than pi-valence orbitals [23]. Furthermore, it is known that the lobes tend to get flattened when the system turns more conducting or with the application of the electric field which should make the underlying submolecular features more transparent to tunnelling. In this cntext, it is worth stressing that for the reasons mentioned below, the present samples were conducting and the images were obtained at a relatively higher bias voltage and extremely low tunnelling current of 50 pA. These conditions, we believe, were conducive to a better molecular imaging free from instabilty in the present study.

Well resoved images of Figs.3 and 4 reveal a high degree of lattice regularity free from any obvious lattice imperfections. However, the sample did reveal regions where the lattice contained defects. An example of this is shown in the top view image of Fig.5a obtained over a scan size of 15 nm X 15 nm. As may be noticed from its illustrated sketch of Fig.5b, the regular arrangement of (001) plane is disturbed by the presence of lattice defects in the sample. Two extra half planes, characteristic of edge dislocations of opposite sign are clearly visible. As their strain fields tend to cancel each other, the bent lattice planes in their immediate vicinity are gradually seen to get straightened as one moves away from them. From the crystallographic directions indicated in Fig.5b, it is apparent that their Burgers vector $b = \pm \frac{1}{2}[1\bar{1}0] = a/\sqrt{2}$ which is conformity with the characteristic Burgers vector for a dislocation in the FCC lattice having $\langle 110 \rangle$ as the close packed direction for slip to occur. Since for a C_{60} solid, a = 1.4 nm, the magnitude of b = 1.0 nm. Incidently, this value is approximately three to four times larger than for the conventional metallic materials. Since the strain energy per unit length of the dislocation is proportional to μb^2, where μ is the shear modulus, a dislocaion would possess a greater line tension and as such we expect the dislocation lines in the C_{60} solid to be straight. It therefore seems unlikely that the two dislocations seen in Fig.5 would be the entry and exit parts of one and the same dislocation loop of about 2 nm diameter. Such a loop would be unstable as under its own large line tension it would collapse. Large strain energy should make the dislocations in C_{60} solid to be more interactive. It is well established [24] that the two unlike edge dislocations attract each other while the like ones repel, and the stable equilibrium position of the former on the parallel slip planes is at a 45° position, corresponding to $x = \pm y$ (Fig.5c). Interestingly, the two dislocations of Fig.5, observed on the adjacent parallel slip planes, are placed close to the 45° position, in accord with the above contention. A mutual annihilation of these two edge dislocation can occur only by climb which is a thermally activated process. Interestingly, the two dislocations of Fig.5a correspond to a large dislocation density of $10^{12}/cm^2$ whose origin may be ascribed to the lattice mismatch between the fullerene film and the silver layer.

Since in FCC crystals, the most densely packed $\{111\}$ planes undergo slip, we expect these planes to be most susceptable to defect formation. In contrast to Fig.4, showing a near ideal $\{111\}$ configuration over a small area scan, the larger area scans obtained have revealed extensive defect structures in $\{111\}$ planes, as depicted in Fig.6. The figure is a top view scan over a size of 13 nm X 13 nm. A preliminary study of the relative heights of individual clusters located on equilateral triangles has suggested the possibility of a disorder in the neighbouring stacks of 111 planes in the z-direction resulting in the formation of surface defects such as twins and stacking faults, which need to be further studied. Incidently, the recent electron diffraction studies [25] have indicated the presence of planar faults in the fullerene samples.

In the images presented above although the lattice structure of the C_{60} buckyballs is clearly observed, the microscopic details of the buckyballs themselves are not resolved. With our earlier efforts [10], forgoing the former, we had succeeded in imaging the characteristic arrangement of hexagons and pentagons of an individual carbon cage. It is exciting to notice that the characteristic submolecular features, comprising hexagons and pentagons are better resolved in the zoomed image of the individual buckyball, shown in Fig.7a. As may be seen, the pentagons and hexagons are found to follow the characteristic soccer ball pattern over the sphere of about 0.7 nm diameter. A typical individual pentagon has been zoomed in Fig.7b. The bond length is found close to the reported value of 0.14 nm [21], though a small difference in respect of its one or two other sides is not unreasonable, considering that the image is a projection of a spherical surface

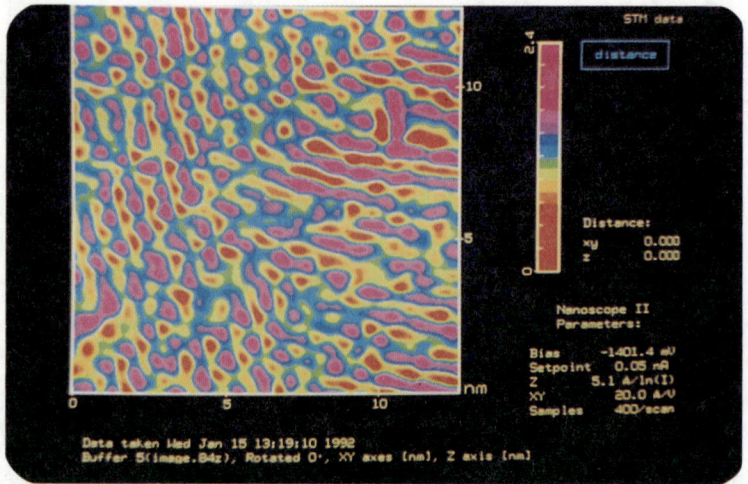

Fig.6

Top view image of (111) planes showing triangular arrangement of clusters with stacking disorders, zoomed over a scan area of 13 nm X 13 nm (bias voltage: -1401 mV and current : 50 pA; moderate filtering).

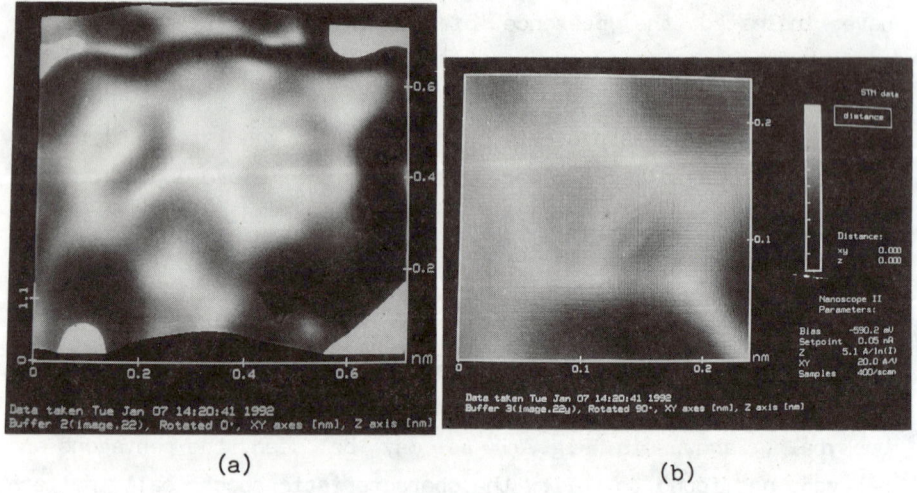

(a) (b)

Fig.7

(a) Zoomed image (scan size: 0.6 nm X 0.6 nm) of individual buckyball in 3D perspective at 60° pitch showing submoleculer features (-550 mV, 50 pA, low pass filtering); (b) Zoomed image of a typical pentagon on the buckyball sphere, in top view mode; the bond length is close to 0.14 nm [21] (imaging conditions same as in Fig.(a).

onto a planar surface.

Following the imaging procedure essentially as described earlier [10] we have attempted to see the relative orientations of the assemblage buckyballs located at corner and face-centring sites of the C_{60} lattice. The image thus obtained, showing 12 buckyballs in a scan area of about 3 nm X 3 nm, in 3D perspective at 60° pitch is presented in Fig.8. Unlike the lattice images of the previous figures, the buckyballs forming the lattice of Fig.8 seem to carry features of the truncated icosahedral structure containing pentagons and hexagons. Four such balls have been zoomed in Fig.9a having the scan size of 1.6 nm X 1.6 nm and observed in 3D perspective at 60° pitch. One of the pairs of the diagonally opposite ones may be

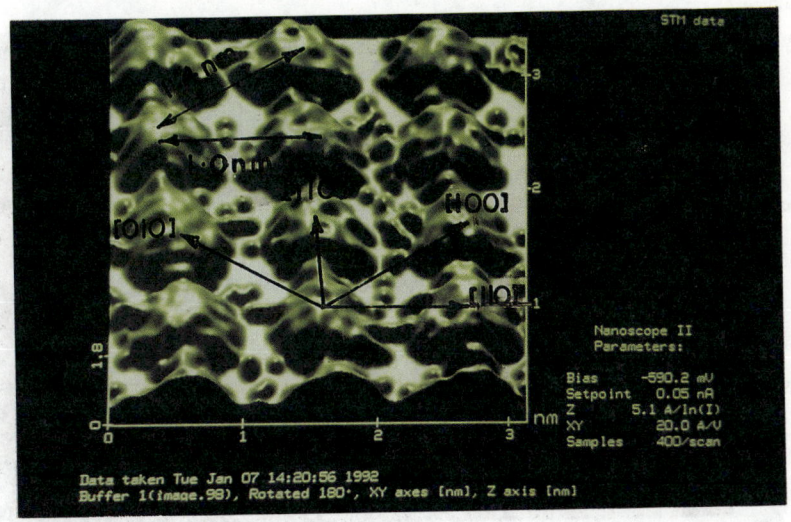

Fig.8

3D perspective image in 60° pitch showing 12 bucky balls of 0.7 nm diameter with truncated icosahedral structure, over a scan area of about 3 nm X 3 nm (-550 mV and current: 50 pA, moderate filtering).

considered as the corner molecules of the FCC unitcell while the other pair is formed by face centring ones of the adjoining unitcells. Figs.9b-e show the four bucky balls A, B, C and D of Fig.8a individually zoomed over a scan area of about 0.7 nm X 0.7nm, in 3D perspective at 60° pitch. There seems no orientational relationship between the resolved features seen in the four carbon cage images, indicationg absence of any correlation between them. This is consistent with each molecule of the C_{60} solid, at ambient temprature, undergoing rotational diffusion which is uncorrelated with the motion of its neighbours and such an orientational disorder has presumably been retarded or frozen out in the present samples.

One of the reasons for retardation or freezing of the rotational component has been ascribed in the beginning to the presence of higher fullerene derivatives in the samples. It was

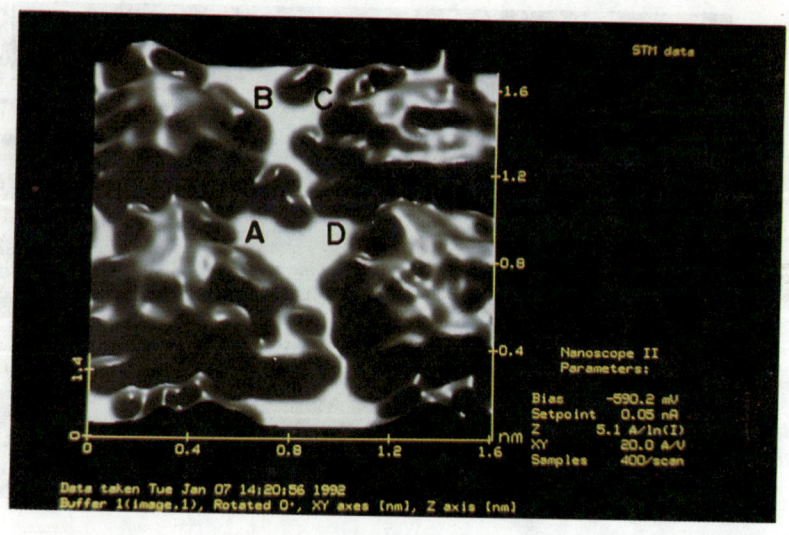

Fig.9a

Zoomed image of 4 bucky balls from Fig.8 (scan size: 1.6 nm X 1.6 nm), the two diagonally opposite balls A and C may be considered the corner sites while the the other two, B and D as face centring ones of the neighbouring unit cells(or vice versa).

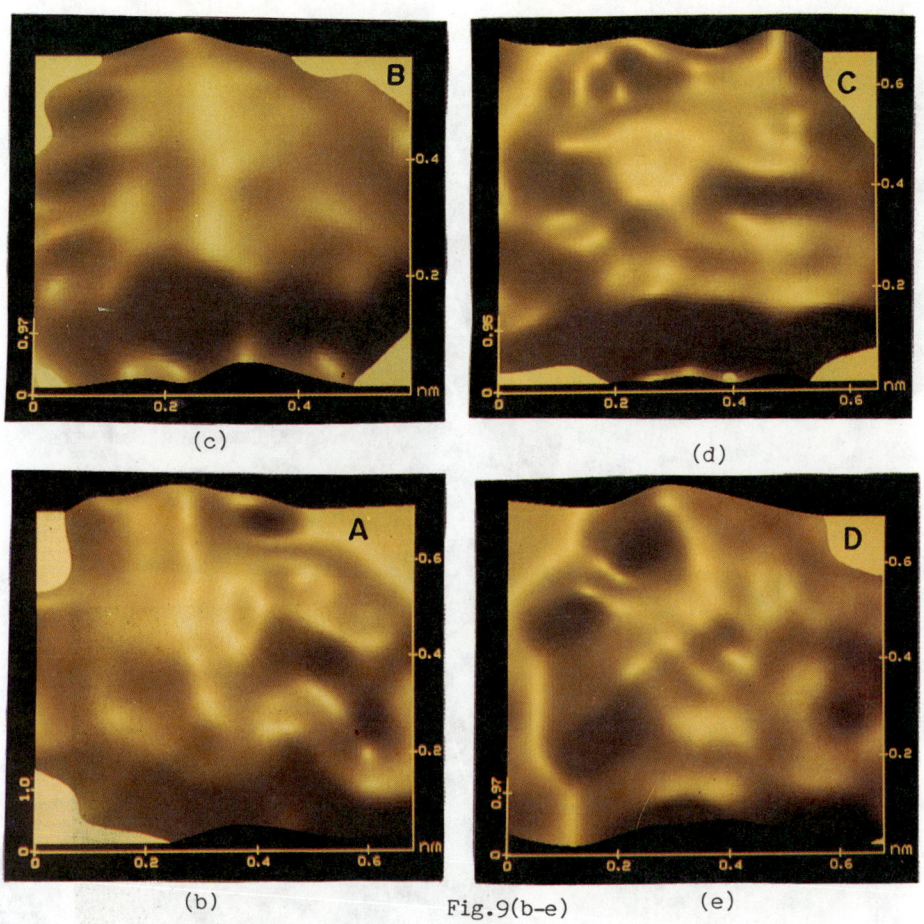

Fig.9(b-e)

The four buckyballs A, B, C and D of Fig.9a separately zoomed

pointed out in our earlier communication [8] that in such samples exhibiting a reasonable conductivity the C_{60}-C_h interaction, may be dominating over the rotational interaction, which is of the order of 50 meV/C molecule [26], and be of the order of the band width of 0.1 eV. This apart, as mentioned earlier, the present films had been deposited on silver electrodes over a glass surface, and as such the C_{60} molecules had silver atoms in their neighbourhood. Their

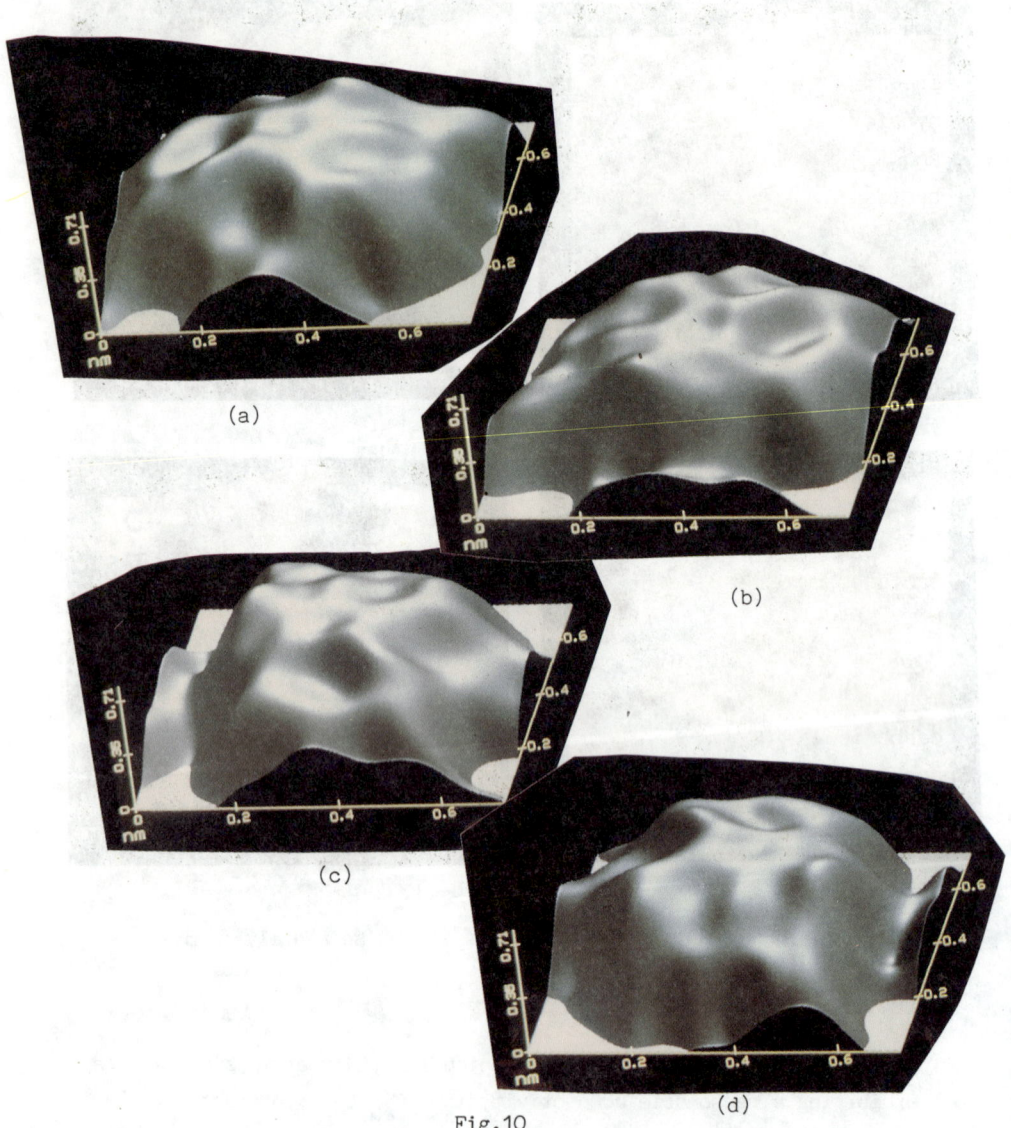

Fig.10

(a-d) 3D perspective image in 30° pitch of a single buckyball seen with successive 90° rotation showing distortions (scan size: 0.7 nm X 0.7 nm; bias voltage:-590 mV, current 50 pA, moderate filtering).

interaction, as pointed out by Wilson et al.[27] in case of C_{60} films on gold electrodes, can change the electronic structure of the films to make them conducting and simultaneously cause some retardation or freezing of the free rotation. This essentially gives rise to hybridization amongst the highest occupied molecular orbitals (HOMO) and the lowest unoccupied molecular orbitals (LUMO),leading to the broadening of the band.

Yet another possible cause of freezing may be related to the STM imaging itself. The electric field gradients between the STM tip and the sample surface are known influence the molecular motion [28,29] and, in the present case, if they are sufficiently large they may polarise the bucky balls and thereby freeze or significantly retard their rotation. The issue is currently being examined by imaging the sample by using an atomic force microscope (AFM) where, between the vibrating canty lever and the sample surface, no such electric fields exist.

Apart from the defects in the C_{60} lattice, it is worth mentioning that, in the present investigation, a closer study of the carbon cage made after zooming has revealed that the images obtained are invariably distorted. The effect is seen when the buckyball is viewed in 3D perspective at 30° pitch from different directions. Figs.10a-d depict four images of the same bucky ball over a scan size of about 0.7 nm X 0.7 nm, in 3D perspective at 30° pitch, as seen on successive 90° rotation. Clearly, the observed images are asymmetric, revealing the presence of distortion. The imperfect cage structures have recently been theoretically examined by Raghavachari and Rohlfing [30], and currently an attempt is being made to relate the observed distortion to their contentions. Its origin may either be intrinsic to the experimental conditions used in the synthesis of the fullerene films or alternatively it may be related to Kekule' type distortion [28] analogous to that previously observed in the STM images of the benzene molecules, resulting from the electric field gradients between the tip and the sample surface. This aspect invites a further study using the atomic force microscope which is free from the constraints of the electric field gradient.

SUMMARY

To sum-up, the described STM studies of C_{60} films containing conducting islands have revealed the FCC lattice of the C_{60} clusters with intermolecular distances in various planes, along different directions, in close accord with the reported data based on XRD and neutron diffraction analysis. The molecular lattice is found to contain extended defects, crystallographically consistent with the FCC structure. The characteristic cage morphology of the buckyballs in the form of hexagons and pentagons has been resolved individually as well as in the lattice. The high resolution zoomed images of the latter indicate that the orientational disorder has been frozen out from a state in which each C_{60} molecule of the FCC lattice is undergoing a rotational motion which is uncorrelated with the motion of its neighbours. It is suggested that the retardation or freezing of the rotational diffusion is perhaps due to the presence of higher fullerene derivatives C_h in the sample such that $C_{60}-C_h$ interaction overpowers the rotational energy of 50 meV. An alternate explanation of reorientational freezing is attributed to the presence of large electric field gradients between the STM tip and the sample. It is further suggested that the latter effect is particularly important in producing some distortions observed in the buckyball images.

Acknowledgement: The authors thank their colleague Dr Ratan Lal of the Theory Group for many stimulating discussions and they are indebted to Professor S.K. Joshi, Director General, CSIR for his continued interest. Some useful suggestions made by Professor M.V. Sadovskii of the Russian Academy of Sciences (Ural Branch) are acknowledged. The STM studies at NPL were carried out under the Indo-French cooperation funded by the EEC contract No. CII.0339 IND(H).

REFERENCES

[1] W. Kratschner, L.D. Lamb, K. Fostiropoulos, D.R. Huffman, Nature, 347 (1990) 354

[2] A.F. Hebard, M.J. Rosseinsky, R.C. Haddon, D.W. Murphy, S.H. Glarum,T.T.M. Palstra, A.P. ramirez and A.R. Kortan, Nature, 350 (1991) 600.

[3] R. Taylor, J.P. Hare, A.K. Abdul-Sada and H.W. Kroto, J. Chem.Soc.Chem.Commun.,20 (1990), 1423.

[4] D.S. Bethune, G. Meijer, W.C. Tong and H. Rosen, Chem. Phys.Lett.,174 (1990) 219

[5] R. Tycko, R.C. Haddon, G. Dabbagh, S.H. Glarum, D.C. Douglass and A.M. Mujsce, J.Phys.Chem. 95 (1991) 518

[6] J.M. Hawkins, A. Meyer, T.A. Lewes,S. Loren and F.J. Hollander, Science 252 (1991) 312

[7] T. Suzuki, Q. Li, K.C. Khemani, F. Wudl and O. Almarsson, Science, 254 (1991) 1186.

[8] Y. Zhang, X. Gao and M.J. Weaver, J. Phys.Chem.,96 (1992) 510

[9] H.P. Lang, V. Thommmen-Geiser, J. Frommer, A. Zahab, P. Bernier and H.J. Guntherodt, Europhys.Letters, 18 (1992) 29.

[10] A.V. Narlikar, S.B. Samanta, P.K. Dutta, L.G. Grigoryan, and A.K. Mazumdar, Phil Mag Letts.,65 (August 1992)-in press.

[11] P.A. Heiney, J.E. Fischer, A.R. McGhie, W.J. Romanow, A.M. Denenstein, J.P. McCauley, Jr., A.B. Smith and D.E. Cox, Phys.Rev.Lett., 66 (1991) 2911.

[12] W.I.F. David, R.M. Ibberson, J.C. matthewman, K. Prassides, T.J.S. Denis, J.P. Hare, H.W. Kroto, R. Taylor and D.R.M.Walton, Nature 353 (1991) 147.

[13] R. Tyco, G. Dabbagh, R.M. Fleming, R.C. Haddon, A.V. Makhija and S.M. Zahurk, Phys.Rev.Lett., 67 (1991) 1886

[14] N. Sudhakar, S.V. Sharma, L.G. Grigoryan, Prem Chand,

A.K. Majumdar, P.K. Dutta, S.B. Samanta and A.V. Narlikar, to be published.

[15] G.M. Vaughan, P.A. Heiney, J.E. Fischer, D.E. Luzzi, D.A. Ricketts-Foot, A.R. McGhie, Y.W. Hui, A.L. Smith, D.E. Cox, W.E. Romanow, B.H. Allen, N. Coustel, J.P. McCauley, Jr. and A.B. Smith, Science, 254 (1991), 1350.

[16] A.V. Narlikar, P.K. Dutta, S.B. Samanta, L.G. Grigoryan and A.K. Mazumdar, Cond.Matter and Mat.Communs., accepted (1992)

[17] A.V. Narlikar, P.K. Dutta, S.B. Samanta, L.G. Grigoryan and A.K. Mazumdar, to be published.

[18] L.S. Grigoryan, Prem Chand, S.V. Sharma and A.K. Majumdar, Sol.State Commun., 81 (1992) 853.

[19] Y. Maniwa, K. Mizoguchi, K. Kume, K. Kikuchi, I. Ikemoto, S. Suzuki and Y. Achiba, Sol State Commu.,80 (1991), 609

[20] L.S. Grigoryan, H.D. Bist, S. Sathaiah, S.V. Sharma, H. Clara and A.K.Majumdar, J. Raman Spectroscopy, 23 (1992) 127.

[21] K. Raghavachari and C.M. Rohlfing, J. Phys.Chem., 95 (1991) 5768

[22] J.L. Wragg, J.E. Chamberlain, H.W. White, W. Kratischmer and D.R. Huffman, Nature, 348 (1990) 623.

[23] B.E. Douglas and D.H. McDaniel, Concept and Model of Inorganic Chemistry, Blaisdell Pub.Co., London (1965) p. 76.

[24] A.H. Cottrell, Dislocations and Plastic Flow in Crystals, The Clarendon Press, Oxford, (1953).

[25] D.E. Luzzi, J.E. Fischer, X.Q. Wang, D.A. Ricketts-Foot, A.R. McGhie and W.J. Ramanow, J. Mat.Res., 7 (1992) 335.

[26] R. Tyeko, G. Dabbagh, R.M. Fleming, R.C. Huddon, A.V. Makhija and S.M. Zahurak, Phys.Rev.Lett, 67 (1991), 1886

[27] R.J. Wilson, G. Meijer, D.S. Bethune, R.D. Johnson,

D.D. Chambliss, M.S. de Vries, H.E. Hunziker and H.R. Wendt, Nature, 348, (1990), 621.
[28] H.Ohtani, R.J. Wilson, S. Chiang and C.M. Mate, Phys. Rev.Letts, 60 (1988) 2398.
[29] P.H. Lippel, R.J. Wilson, M.D. Muller, Ch.Woll and S. Chiang, Phys.Rev.Letts., 62 (1989) 171.
[30] K. Raghavachari and C.M. Rohlfing, J. Phys.Chem., 96 (1992) 2463.